Ecology and Evolution of Communities

Ecology and Evolution of Communities

*Martin L. Cody
and Jared M. Diamond,
Editors*

The Belknap Press of Harvard University Press

Cambridge, Massachusetts, and London, England
1975

Dedicated
to
Robert MacArthur

In November 1972 a brief but remarkable era in the development of ecology came to a tragic, premature close with the death of Robert MacArthur at the age of 42. When this era began in the 1950s, ecology was still mainly a descriptive science. It consisted of qualitative, situation-bound statements that had low predictive value, plus empirical facts and numbers that often seemed to defy generalization. Within two decades new paradigms had transformed large areas of ecology into a structured, predictive science that combined powerful quantitative theories with the recognition of widespread patterns in nature. This revolution in ecology had been due largely to the work of Robert MacArthur.

Robert MacArthur's study of ecology began formally at Yale University under the guidance of G. Evelyn Hutchinson, founder of modern niche theory and the leading ecology teacher of his generation. But Robert's ecological experience began informally much earlier, through the stimulation of parents devoted to biology and through observations of birds, mammals, and trees in Ontario and rural Vermont. He was also fascinated as a child by practical problems of devising strategies for reconciling competing interests— e.g., how to divide a cake to the satisfaction of several consumers of differing capacities. This fascination developed into the trenchant and delightful formulations of competing strategies (generalists vs. specialists, pursuers vs. searchers, users of coarse-grained vs. fine-grained resources, r-selection vs. K-selection, etc.) that came to pervade his ecological writings.

After changing from mathematics to biology at the end of his second year of graduate school, Robert selected as his thesis problem the relations of five coexisting warbler species that abounded in New England forests. These warblers presented a paradox to the emerging niche theory of the 1950s, for they seemed very similar to each other ecologically and were suspected of violating the competitive exclusion principle. While spending his two years of military service at mathematical tasks for the army, Robert guessed how the paradox must be resolved, and upon his return to New England speedily made the observations that confirmed his guess. The five warbler species, which have become known affectionately in the ecological literature as "MacArthur's warblers," were indeed remarkable, but only in the unexpectedly subtle differences in foraging strategies that they had evolved for dividing the insect resources of New England forests. The result of the study, Robert's doctoral dissertation, was the classical article (listed as 1958c in his bibliography on p. 13 of this book) that earned him the Mercer Award (best ecology paper, 1957–1958) and established new standards of sophistication for studying niche relations.

During the next 15 years, first at the University of Pennsylvania and then at Princeton, Robert MacArthur realigned our perspective in almost every aspect of ecological endeavor. Core concepts that he revitalized or created de novo include species diversity, relative abundances of species, species packing, competition and niche theory, life histories and strategies of single species, and island biogeography. Some of his main contributions are summarized briefly in the introduction to this volume, which also provides a complete bibliography of his writings. His outlook and much of his work are also portrayed in his book *Geographical Ecology,* written within the space of a few months immediately following the diagnosis of his fatal illness.

The value and scope of Robert Mac-Arthur's contributions to ecology were made possible by his unique combination of several attributes. One was a keen ability as a field naturalist and observer with a deep appreciation for nature, as exemplified by his recognition of the subtly different foraging patterns of "his" warblers. A second, contrasting ability was as a mathematician, exemplified by his models describing growth of colonist populations, competition along a resource axis, species-abundance relations, or immigration and extinction rates as a function of island area and distance. The models he constructed are realistic and detailed enough to be able to describe complex natural phenomena. Yet they are charac-

teristically simple and elegant, and are based on concepts sufficiently universal to be of general applicability. Another attribute was his ability to keep in mind the costs and benefits of competing strategies, and to capture the essence of these problems by formulating vivid dichotomous choices. Finally, he could abstract and define, from the initially bewildering complexity of nature, important parameters susceptible to quantitative measurement. For example, many ornithologists had recognized that structurally complex habitats like forests harbor more bird species than structurally simple habitats like grassland. However, only when Robert MacArthur devised a simple index of vegetational complexity (called foliage height diversity) did it become possible to translate ornithological lore into an equation by which habitat structures are compared and their bird species diversities predicted.

Robert MacArthur stimulated ecology not only through his own work, and through inspiring the research of others, but also through his collaborative efforts. He was gifted at exchanging and jointly developing ideas with a wide range of people—from pure theoreticians to empirical naturalists, from beginning students to established senior scientists, and from ornithologists, mammalogists, and herpetologists, to entomologists, botanists, and limnologists. Many of these joint studies, such as the book on island biogeography written with E. O. Wilson, opened new areas of ecology.

A year after Robert's death, in November 1973, a group of us—his friends, colleagues, and relatives—returned to

Princeton in order to remember him and his impact on each of us and on ecology. In papers presented at this symposium we shared with each other the progress of the studies that he had inspired us to pursue. Coming together made us realize again not only our sadness at his loss but also the strengths of the man we had come to honor. We were missing the person uniquely capable of catalyzing, focusing, and leading such a diverse gathering.

The present volume grew out of this memorial symposium. All contributors to the volume were at one time closely associated scientifically with Robert Mac-Arthur. Most of us had coauthored papers with him, and about half of us had studied under him. Those who participated in the symposium, and either presented papers, helped reformulate other papers, or shared in other ways in the preparation of this tribute to Robert MacArthur, are: William Bennett, James Brown, Robert Colwell, Joseph Connell, Kenneth Crowell, Stephen Fretwell, Henry Hespenheide, Henry Horn, G. Evelyn Hutchinson, James Karr, Peter Klopfer, Egbert Leigh, Jr., Richard Levins, Richard Lewontin, Gerald Lieberman, Betsy Mac-Arthur, John MacArthur, Duncan Mac-Arthur, Alan MacArthur, Lizzie Mac-Arthur, Donnie MacArthur, Edward Maly, Robert May, Gordon Orians, Ruth Patrick, Eric Pianka, Robert Ricklefs, Michael Rosenzweig, William Schaffer, John Terborgh, Edwin Willis, and Edward Wilson.

The Editors

Contents

Ecology and Evolution of Communities

Introduction

This book developed from papers presented and discussed at the symposium in memory of Robert MacArthur. The range of topics covered in the papers is broad, from single-species strategies of energy allocation on the one hand to the evolution of communities under both mendelian and group selection on the other. Thus, the volume depicts or exemplifies the status of many key aspects of evolutionary ecology and many of the problems that engaged Robert MacArthur.

The volume considers three broad classes of ecological problems, and is divided into sections that correspond to these classes. The first section discusses the diversity and relative abundance of species, how these measures fluctuate over time and how they evolved, how species are associated into sets or communities, under what conditions the communities are stable, and what courses the evolution of communities can take. Next is a section on the strategies that coexisting species evolve, under the pressures of competition and natural selection, for the division of resources. The following section, on community structure, considers the parallel ways in which species communities are assembled in response to similar selection pressures, and the roles of competition and predation in structuring communities. This section is divided into two parts, reflecting differences in organisms studied,

in methods, and in the main determinants of community structure. The last two chapters discuss the development and application of ecological principles from a broader perspective, and look to some future extensions of Robert MacArthur's ideas.

In this introduction to the volume we provide a brief background to the subject matter of each section and its overall significance to ecology. We then mention some specific contributions of Robert MacArthur to the topics of the section and show how his work is extended in the chapters that follow. Finally, we provide brief summaries of each chapter. A few of the chapters are primarily reviews; others focus on new methods of analysis, new facts, or new theories. While each chapter is self-contained and can be read by itself, an effort has been made to make clear the relations among the chapters as well as to express the content of evolutionary ecology in the organization and chapter sequence of the volume.

The Evolution of Species Abundance and Diversity

The most basic ecological data consist of counts of individuals and of the species to which these individuals belong, of the trophic relations between these species, and of the way that these counts and rela-

1

tions vary with time. It is obvious that the numbers of locally coexistent species vary from place to place, that the local abundances of species fluctuate with time, and that the distribution of individuals among species likewise varies in different places, communities, and species groups. The patterns and details of these variations, their explanations, and the evolutionary factors producing them occupy a central place in ecology. It is against a background of these facts of species diversities and abundances that all ecological endeavors are conducted.

Much ecological thought on species numbers and abundances has attempted to relate their stability to the properties of environments which themselves fluctuate. In his first published paper, an early attempt to derive stability criteria for species sets, Robert MacArthur (1955—see MacArthur bibliography at end of this introduction) sought these criteria in the connectivity of species associated into food webs. This notion is dramatically developed in the first chapter of the present volume by Richard Levins' elegant qualitative analysis of food webs, trophic levels, and their stability, using loop theory. The stability and survival probabilities of fluctuating populations stem from birth and death schedules, which therefore play a key role in the equilibrium theory of island species diversity that MacArthur and E. O. Wilson developed jointly and that ranks as one of the triumphs of modern ecology. In *The Theory of Island Biogeography* MacArthur and Wilson (1967b) used birth and death schedules to analyze the fluctuations of small populations and

to calculate their probability of survival. Later, with Robert May, MacArthur related the degree of niche overlap between competing species to the magnitude of environmental fluctuations to which the species are exposed, and thus cast species coexistence as a function of a variable environment (May and MacArthur, 1972d). In Chapter 2 of the volume Egbert Leigh, Jr., extends the theoretical analysis of population fluctuations and of species coexistence begun by the MacArthur–Wilson and May–MacArthur treatments, while Robert's brother John MacArthur uncovers in Chapter 3 a simple empirical relation between species diversity and environmental stability.

In his last two field studies (MacArthur, Diamond, and Karr, 1972a; MacArthur, MacArthur, MacArthur, and MacArthur, 1973) Robert MacArthur returned to the aspect of species abundance called density compensation, by which the species of an impoverished community may compensate in abundance for the absent species found only in richer communities. However, his interest in abundances had already been evidenced in his second published paper, which together with its sequel developed the now classical "broken-stick" model of species-abundance relations (MacArthur, 1957, 1960a). A thorough analysis of this and other species-abundance models is given by Robert May in Chapter 4 of this volume, and their relationship to the species-area curves that contributed so much to the island biogeography theory is derived. The last chapter in this section of the volume (Chapter 5, by Michael Rosenzweig) extends the equilibrium theory of species diversity from islands to continents.

Let us now briefly summarize the contents of each chapter of this section in turn. In Chapter 1 Richard Levins develops a powerful new technique, loop analysis, for representing evolution in communities near equilibrium. Communities are affected by natural selection for particular traits in their component species, and by changes in variables such as nutrient levels or predation pressure. Loop analysis is concerned with how such changes affect the stability of the community and the abundance of each component species. Selection in a species affects the whole community in ways that depend on the interactions of that species with other species. Thus, it is often not possible to anticipate even the direction of evolutionary change if one considers species in isolation. Loop analysis permits conclusions about incompletely specified communities near equilibrium and requires knowledge only of the sign of the interaction between variables (positive, negative, or zero). One can assess whether a community is stable or what type of instability it exhibits; how mendelian selection in one species affects the abundance of that species and of other species; and how mendelian selection and group selection are related. This chapter will surely be an important paper for evolutionary biology in the next decade, especially for understanding such controversial problems as group selection or the regulation of population numbers.

Following in Chapter 2, Egbert Leigh, Jr., presents a theoretical analysis of how population fluctuations or steadiness (as opposed to the ability of a species to recover from a major catastrophe) affect a species' probability of survival, and how the fluctuations are related to community structure and to fluctuations in the environment. The population's coefficient of variation is identified as overwhelmingly the most important variable affecting a population's chances of extinction. Population fluctuations due to environmental variability result in a reduced capacity for population increase. Empirical evidence, though weak, suggests no obvious difference in steadiness between tropical populations and temperature populations. Leigh predicts that increased specialization rather than increased niche overlap will characterize species in more constant environments. Extending these results to communities with several trophic levels, he finds that stable environments will favor more complex food webs, but that only under restricted circumstances will these complicated webs stabilize the steadiness of their component populations. Thus, stable environments do not necessarily mean stable populations.

Complementing Leigh's theoretical treatment, John MacArthur in Chapter 3 examines the empirical relation between environmental variability and species diversity. Taking winter–summer differences in mean temperature as his measure of variability, he shows that a simple equation gives remarkably good fits to the latitudinal diversity gradients of birds, mammals, and gastropods in western North America. This equation has the form of the equation describing information transmission through noisy channels in Shannon–Wiener information theory, and environmental variability plays a biological role analogous to noise. The equation is alternatively suggested by a logseries distribution of species abundances. Whatever the interpretation of the equation's parameters, its beguilingly simple form and good fit to ten gradients of species diversity call for further testing of its meaning and range of validity.

Species-abundance relations are subjected by Robert May (Chapter 4) to a comprehensive reexamination that supersedes previous scattered and incomplete derivations. In any community there are some rare species and some common species, but different detailed distributions of species abundances are observed in different natural communities and are predicted on different theoretical grounds. May discusses the three most significant such distributions—the lognormal distribution, the broken-stick distribution, and the geometric and logseries distributions—in order to disentangle features of biological significance from features that are simply consequences of statistics. In particular, statistical explanations are provided for two long-standing puzzles, Hutchinson's observation of relatively constant width of lognormal distributions and the validity of Preston's so-called canonical hypothesis. For each of the three distributions May also contrasts the merits of some alternative measures of species diversity and discusses sampling problems. He shows how each species-abundance relation gives rise to a particular predicted species-area relation, and he

derives the somewhat mystical exponents of these relations.

Since the analysis of species-area-distance relations in *The Theory of Island Biogeography,* it has been generally accepted that species numbers on islands approach equilibria between rates of immigration and of extinction. Whether similar equilibria between rates of speciation and of extinction are approached on continents, however, has been unclear. This subject is examined in the concluding chapter (Chapter 5) of the first section of the volume, by Michael Rosenzweig, who considers the sources of feedback from species diversity into speciation and extinction, the two continental processes that determine species diversity. He finds that both rates will increase with diversity, but that the former increases at a decreasing rate, the latter at an increasing rate. Thus, there must be an intersection between the two curves, representing a steady state of species diversity. This treatment can be applied, for example, to analyzing whether continental faunas have actually reached steady states, and why the faunas of the tropics or the abyss are so rich.

Competitive Strategies of Resource Allocation

It is manifestly impossible for a single species to be superior to other species in all respects under all circumstances. The metabolic energy and genetic information that go into one set of adaptations ultimately come at the expense of energy and information that could go into another set of adaptations. Hence the benefits derived from any trait must be weighed against the costs of maintaining that trait and also against the abandoned benefits of alternative traits. Furthermore, the balance of benefits and costs may vary under different conditions. Thus, species must evolve strategies that make trade-offs and solve optimization problems.

The evolution of optimal strategies comes about through natural selection, via competition for the resources that two or more phenotypes or species attempt to use. If a group of species (or phenotypes) with independently evolved strategies are thrown together, those whose resource utilizations are closest will compete most severely, and the least efficient at using a given resource will be eliminated. Because of the trade-offs involved in strategies, a species that is most efficient at one strategy will inevitably differ in its resource utilization from a species that is most efficient at some other strategy. Thus, it is natural selection, operating through competition, that makes the strategic decisions on how sets of species allocate their time and energy; the outcome of this process is the segregation of species along resource-utilization axes.

Robert MacArthur had a flair for capturing the essence of such strategic decisions by means of trenchant dichotomies. One such example is the contrast between r-selection and K-selection employed in *The Theory of Island Biogeography*. At the one extreme a species can maximize its growth rate (r-selection), at the other extreme its competitive ability (K-selection); the former strategy is favored in an uncrowded or unstable environment, the latter in a crowded or stable environment. Another example is the trade-off between a generalist and a specialist, between being moderately efficient at many tasks or very efficient at a few of them (MacArthur, 1965a). An extension of the generalist–specialist dichotomy concerns the choice of axis along which to specialize: should increased competition cause a species to narrow its diet while continuing to forage in the same space, or to restrict its habitat range while maintaining the same breadth of diet (MacArthur and Pianka, 1966c)? In the same paper MacArthur and Pianka consider the trade-offs involved in allocating foraging time to search time or to pursuit time. Beginning with his doctoral studies on the subtly distinct strategies of "MacArthur's warblers," MacArthur (1958c) saw the central role that such strategic decisions played in segregation of species along resource axes. This interest in wedding strategic thinking to competition theory and niche theory continued through the following decade and a half of MacArthur's work [as expressed, for instance, in his papers on species packing (1969e, 1970b)], and culminated in the discussions of competitive strategies of resource allocation that permeate the pages of *Geographical Ecology* (1972g).

The next four chapters of this volume continue this tradition of deriving insight from competitive strategies of resource allocation. In Chapter 6, by William Schaffer and Madhav Gadgil, on selection for optimal life histories in plants, competition is intraspecific for such resources as moisture and nutrients, and strategy success is measured by the seeds, flowers, and fruits produced. The other three chapters are concerned with interspecific competition. The resource spectrum considered by Henry Hespenheide in Chapter 7 is food size and food type, and the strategists are bird consumers. In Chapter 8, by Arthur Shapiro, the spectrum is seasonal time, along which the resources of nectar and good flying weather are distributed, and the consumers are butterflies. Time is again the axis in Henry Horn's analysis of forest succession (Chapter 9), but the changing resource distribution along this axis is now generated by the successive consumers themselves, whose shade determines what subsequent consumers with the right strategies of shade tolerance will be next to convert the resources of light and nutrients into growth.

In the first of these chapters (Chapter 6), reproductive strategies in plants are analyzed in the light of direct, spartan equations by William Schaffer and Madhav Gadgil. The trade-offs between increased adult survival, on the one hand, and increased seed production, on the other, shift with environmental conditions and determine the relative values of an annual, biennial, or perennial strategy. Similar trade-offs apply to the dichotomy of repeated (iteroparous) reproduction versus "big-bang" reproduction (a single intense reproductive episode followed by death). Schaffer and Gadgil finally examine in what ways crowding affects the optimal output of total reproductive effort, the optimal energy expenditure per seed, and the optimal balance between sexual and asexual reproduction.

Henry Hespenheide examines dietary strategies of birds, especially of insectivorous guilds, in Chapter 7. To those of us who have inferred simple differences in prey size from differences in predator size, his results are a revelation of overlooked complexities. Yes, larger birds do choose larger food items, but . . . . The form of the relation varies with the bird guild and with the food item, depending on several factors: the ease with which the item can be captured and processed by means of the foraging tactics employed by the particular bird species; the nutritive benefits of the food item; and the resource spectrum in the bird's preferred habitat. The niche-variation hypothesis of Van Valen has led workers to seek evidence of broader diets reflected in more variable bills, as morphological evidence of broader niches. The results to date have been equivocal, and Hespenheide suggests that such evidence is less likely to be forthcoming for territorial species than for social species.

In Chapter 8, Arthur Shapiro discusses the seasonal distributions of adult butterflies, to identify the selective factors affecting how time is allocated as a resource. Most species do not utilize the full flight seasons that physiology, warm weather, and nectar availability would permit. Instead, they adopt a conservative strategy and stop flying earlier than necessary in most years; this is the price they pay to avoid being wiped out in the years when cold weather arrives earlier than usual. K-selected butterfly species generally have a single annual peak flying time, while r-selected species have several peaks. Shapiro compares the flight patterns at localities with short and long summers, and dissects the factors controlling adult flight times from those related to larval feeding times and overwintering strategy. Interspecific competition for "flying time" or for nectar sources fails to explain why sets of species have evolved coincident flying peaks, and the origin of these peaks remains a major unresolved problem.

The last chapter on strategies considers another and much more familiar situation in which time is allocated as a resource among species. One of the oldest ecological problems is why regrowth of vegetation on cleared land, after proceeding through sequences of successional states, tends to converge on the same climax community independent of the starting condition. Ultimately, succession is related to changes in environments with time as they are occupied by species that differ in the shade they cast or tolerate. In Chapter 9, Henry Horn provides a strikingly simple and illuminating analysis of this classical problem, using the theory of Markov processes. Defining a matrix of transition probabilities that each tree species will be succeeded by each other tree species, he determines from Markov analysis the conditions under which repeated application of this matrix to any starting state (each application representing one generation of trees) converges on the same final state. The analysis considers stability conditions, effects of patch-cutting vs. clearcutting, and species diversity as a function of successional stage. Horn has estimated a transition matrix from the frequencies of saplings of each tree species under established trees of each species in the Insti-

tute Woods at Princeton. The model can be made increasingly sophisticated by incorporating extra assumptions about seed production, but even the most restrictive assumptions result in remarkably good predictions of the species composition of forest tracts of different ages.

Community Structure

The pattern of resource allocation among the species of a community, and the patterns of their spatial and temporal abundance, constitute the community's structure. Interspecific interactions, namely, competition and predation, generally compress the niche of each species below the limits imposed by physiology or morphology in the absence of such interactions. If the observed patterns in community structure are products of natural selection, then similar selection by similar environments should produce similar optimal solutions to community structure. In particular, if species are assembled nonrandomly into communities and if the fine structure of such assemblages is determined by the physical and biological environment, then patterns of community structure should be reproducible, independent of the species pool from which the component species (and the biological components of the selective background) are drawn.

A powerful technique for examining whether there are optimal solutions to community structure is to compare communities in different geographical areas that provide similar physical conditions but independently evolved faunas. Differences between the structures of such replicate communities might arise either through effects of history and different lengths of time available to achieve optimal solutions, or else through effects of chance leading to alternate stable equilibrium structures.

Convergence in community structure between similar habitats or resource sets in different geographical areas implies that resources have been correctly identified for the consumer species and that a general species-packing theory applies to the consumers. Robert MacArthur's work was fundamental to both of these assumptions. His doctoral study on warblers (MacArthur, 1958c) examined the structure of a small guild and set new standards for disentangling complex and multi-dimensional resource sets. In his field work from 1959 to 1966 (e.g., MacArthur and MacArthur, 1961c; MacArthur, Recher, and Cody, 1966a) he pioneered in developing measures of foliage height diversity to express how bird species diversity increases in habitats of increasingly complex vegetational structure. Subsequently he developed species-packing theory (e.g., MacArthur, 1968b, 1969e, 1970b) by investigating the limiting similarity between species and the simultaneous equilibria in consumer and resource populations. These studies culminated in *Geographical Ecology* (1972g), which brilliantly demonstrated how chance or history aided by diffuse competition could lead to alternate, stable, invasion-resistant communities, and which incorporated the results of Robert MacArthur's last two field studies on density compensation.

The seven chapters on community structure in this volume are divided into two groups. The first five chapters (Chapters 10–14) share several characteristics: the species discussed are vertebrates; niches of the species are compressed much more by competition than by predation; community structure is relatively stable over short times; and the methods of study used are primarily observational. These chapters (by Martin Cody, James Karr and Frances James, Eric Pianka, James Brown, and Jared Diamond) examine the detailed convergences in structure of bird, lizard, or rodent communities on different continents, islands, or at different localities on the same land mass. Differences due to history are recognizable in the communities treated in all five chapters; the phenomenon of diffuse competition is also pervasive, most explicitly in the chapters by Pianka and by Diamond; density compensation is treated by Cody and by Brown; and morphological correlates of convergent community structure are subjected to quantitative analysis by Karr and James. In contrast, Chapters 15 and 16 focus on the lower trophic levels, on invertebrates and plants; effects of predation are often conspicuous; short-term changes in structure are sufficiently conspicuous that understanding of succession becomes a central problem; and the methods of study used have often been experimental. These two chapters provide syntheses of stream communities (by Ruth Patrick) and of the relative roles of predation, environmental harshness, and competition in determining community structure (by Joseph Connell).

Although a general framework now exists for describing species diversities on islands, both at equilibrium and in a non-steady state, patterns of species diversity on continents are much more complex. In the first of these chapters on community structure, Chapter 10, Martin Cody assembles elements of a continental theory of species diversity by contrasting bird distributions in three of the widely separated areas of the world with Mediterranean climates. For each continent he used bird distributions over the habitat gradient to construct curves of α-diversity (number of species coexisting in a plot of uniform habitat) and β-diversity (rate of species turnover between habitats) as a function of position on the habitat gradient. Intercontinental differences in these curves can be attributed in large part to past and present differences in the available areas of each habitat type. A striking continental analogue of the recently discovered nonequilibrium "relaxation" effects in island faunas is that California oak woodlands and forests are still supersaturated with bird species, a legacy of the expanded Pleistocene areas of these habitats. Despite intercontinental differences in the α- and β-diversity curves, density compensation and niche-breadth compensation make the curves of total bird abundance quite similar among continents. The diversity curves are used to assess in detail the convergence between ecologically equivalent species and guilds on different continents.

The classical problem of convergent evolution in body form is reexamined in Chapter 11 through new quantitative approaches developed by James Karr and Frances James. They consider to what extent the morphologies and ecological roles of species are correlated, and to what degree these correlations exhibit intercontinental convergence, in the forest bird faunas of three continents, two of them tropical and one temperate. Of the powerful multivariate statistical techniques Karr and James use, one involves plotting species in morphologically determined two-dimensional spaces, with further morphological dimensions specifying shapes of each point; another method utilizes principal components analysis to extract biologically meaningful axes determined by sets of correlated morphological attributes; and a third introduces canonical correlation analysis to identify variates composed of constellations of correlated ecological and morphological attributes. A morphological space is used to assess the extent to which morphological configurations absent in temperate-zone birds appear in the tropics, and to identify the relative contributions of increased species packing and increased community-niche volume to tropical-temperate differences in species diversity.

Chapter 12, by Eric Pianka, and Chapter 13, by James Brown, both contrast desert faunas, the former using intercontinental comparisons of lizard communities (Australia, Africa, North America), the latter using intracontinental comparisons of seed-eating rodent communities in North American desert basins. For both types of desert communities, productivity and its annual variability, as estimated by rainfall, are the most important determinants of species diversity. Among the three continents the lizard faunas differ markedly in species number and in proportional representation of ecological types, probably because of historical factors. For example, Australia's inaccessibility to mammals and snakes and paucity of intercontinental bird migrants may have permitted the lizard fauna there to diversify and exploit resources that go to other vertebrate classes on other continents. History has also affected rodent faunas, by causing species diversity in relatively inaccessible desert basins to be low because of island-like effects, and evidence of the diversion of rodent resources to ants appears in the Sonoran Desert. Coexistence of granivorous rodent species depends mainly on differences in body size: rodents of different sizes eat seeds of different sizes and forage at different distances from bushes. Competitive spacing of species by size has produced striking intracontinental parallels in rodent community structure. Different species of similar body size replace each other geographically to produce similar size sequences in different areas; as resources dwindle, rodents are successively removed from these size sequences in a predictable order. Niche relations among desert lizard species appear to be more complex and to depend on segregation along spatial, temporal, and dietary axes. Mean niche overlap increases with species packing among the rodents but, for the lizards, is lowest on the continent with the richest fauna. The lizard overlap pattern is interpreted in terms of diffuse competition.

New methods of analyzing the assembly of species communities arise from Jared Diamond's work on the bird communities of New Guinea and its satellite islands (Chapter 14). To quantify how nonrandomly the faunas of islands are drawn from the available species pool, he introduces the concept of "incidence functions," which relate for each species the probability of occurrence on an island to the total species number on the island. The remarkably orderly patterns that emerge can be related to each species' habitat preference, dispersal ability, reproductive strategy, area requirement, and level of endemism. Most species can be assigned unambiguously to one of six categories, ranging from "supertramps" (confined to the most impoverished communities) to "high-S species" (confined to the richest communities). Within a guild of ecologically related species, only a few of the theoretically possible combinations of species actually exist on islands, and analysis of these permitted and forbidden combinations leads to the recognition of several types of "assembly rules." These rules are determined partly by selection of colonists, compression of their niches, and adjustment of their abundances so as to match the consumption curve of the colonists to the resource production curve of the island. Other factors underlying assembly rules include hierarchies of overexploitation strategies, by which the species of richer faunas starve out invaders from poorer faunas; dispersal strategies and perhaps also transition probabilities among permitted sequences may play further roles. Although developed for island communities, incidence functions and assembly rules may also apply to assembly of habitat communities and of geographically patchy communities. This chapter begins to unravel the complexities of diffuse competition, one of the most challenging of ecological problems.

In Chapter 15, the first of the two chapters emphasizing experimental studies of community structure, Ruth Patrick reviews the natural history of fresh-water stream communities and discusses the factors affecting their species diversity and structure. Stream organisms respond to the structure of a stream (e.g., its current profile and substrate texture) just as birds respond to the foliage structure of terrestrial habitats. But stream systems, unlike bird communities, are amenable to experimental manipulation and therefore provided early testing grounds of the MacArthur–Wilson island-equilibrium model. Glass slides can be used to create artificial islands of different areas, and colonization can be manipulated by changing stream velocity. In addition, effects of nutrient levels and trace metals on the equilibrium levels of competitors or of predators and prey can be tested directly. Stream organisms prove to exhibit high diversity, short life cycles, high reproductive rates, and marked effects of predation. These stream communities offer many advantages as experimental systems.

Chapter 16 by Joseph Connell, the final chapter on community structure, presents an overview of the relative roles of predation, harsh physical conditions, and competition in determining community structure. This evaluation is derived mainly from the results of controlled field experiments on invertebrates and plants. The hazards of predation or harsh physical conditions usually prevent many species from reaching population densities high enough so that competition for resources could be significant. Juveniles of species with large adults but small, unprotected young may escape these hazards to reach adulthood and become sufficiently abundant to compete only during occasional reductions in predation or in harshness of physical conditions, resulting in widely spaced "dominant year classes." Small species may escape predation and compete as juveniles, but depend as adults on refuges, or on dispersing between patches more rapidly than predators can, to escape these hazards. Since predation seems more intense under benign physical conditions, competition between species susceptible to predation is more likely in harsher regimes. Because of strong parental care many vertebrates are less susceptible to these hazards as juveniles, and regularly compete. Connell summarizes his chapter by graphs depicting how environmental conditions and the body size of a prey species yield sets of contours, determined by mortality from predation and from harsh physical conditions, that delimit the zone in which the species can exist.

Outlook

The last two chapters consider some prospects for future advances in understanding the types of problems to which Robert MacArthur devoted himself.

In Chapter 17, G. Evelyn Hutchinson presents four case histories illustrating how advances in ecology depend on the interaction of theory with field and laboratory observations. His first case is, appropriately, the competitive exclusion principle, where the mutual support of theory and observation has been symmetrical and nearly ideal and has facilitated the rapid progress of both. In contrast, the marriage of theory and fact in understanding predator-prey oscillations has been unhappy; the difficulties in documenting the existence of such oscillations remain substantial, and the existence of simple and even convergently similar theories may have hampered the search for the facts. Further theoretical discoveries may be lurking in the extensive observations relevant to the ecological control of polymorphism. In the last problem area, the topic of upper and lower limits to body size and of rates of evolution itself, whole fields of potentially fertile observation and theory remain to be opened. Detailed familiarity with nature is as important a prerequisite for these advances as is mathematical competence.

The concluding chapter (Chapter 18), by Edward Wilson and Edwin Willis, explores practical applications of the biogeographic revolution arising from Robert MacArthur's work to the problem of managing what is left of our natural heritage. As long as so much of the natural world remains unknown and adequate management techniques cannot yet be devised, a "diversity ethic" requires that we act as stewards, not managers. For instance, it is simply not yet possible to predict the ecological consequences of the proposed sea-level Panama Canal, and so the canal should incorporate a barrier to prevent exchange of the Atlantic and Pacific marine biota. More immediate progress can be made, however, in applying results of equilibrium and non-equilibrium island biogeography to planning the optimal shape and size of natural preserves. The selective loss of species on Barro Colorado Island since its isolation early in this century dramatically illustrates how island phenomena operate in preserves. Further detailed advances in understanding community structure may eventually make it possible to enrich communities selectively in desired ways, and even to assemble new communities.

MacArthur, R. H. 1955. Fluctuations of animal populations, and a measure of community stability. *Ecology* 36:533–536.

MacArthur, R. H. 1957. On the relative abundance of bird species. *Proc. Nat. Acad. Sci. U.S.A.* 43:293–295.

MacArthur, R. H., and P. Klopfer. 1958a. North American birds staying on board ship during Atlantic crossing. *Brit. Birds* 51:358.

MacArthur, R. H. 1958b. A note on stationary age distributions in single species populations and stationary species populations in a community. *Ecology* 39:146–147.

MacArthur, R. H. 1958c. Population ecology of some warblers of northeastern coniferous forests. *Ecology* 39:599–619.

Hutchinson, G. E., and R. H. MacArthur. 1959a. A theoretical ecological model of size distributions among species of animals. *Amer. Natur.* 93:117–125.

Hutchinson, G. E., and R. H. MacArthur. 1959b. On the theoretical significance of aggressive neglect in interspecific competition. *Amer. Natur.* 93:133–134.

MacArthur, R. H. 1959c. On the breeding distribution pattern of North American migrant birds. *Auk* 76:318–325.

MacArthur, R. H. 1960a. On the relative abundance of species. *Amer. Natur.* 94:25–36.

Klopfer, P., and R. H. MacArthur. 1960b. Niche size and faunal diversity. *Amer. Natur.* 94:293–300.

MacArthur, R. H. 1960c. On Dr. Birch's article on population ecology. *Amer. Natur.* 94:313.

MacArthur, R. H. 1960d. On the relation between reproductive value and optimal predation. *Proc. Nat. Acad. Sci. U.S.A.* 46:144–145.

MacArthur, R. H. 1960e. Population studies: Animal ecology and demography. *Quart. Rev. Biol.* 35:82–83. Review of Cold Spring Harbor Symposium, vol. 22.

MacArthur, R. H. 1961a. Population effects of natural selection. *Amer. Natur.* 95:195–199.

Klopfer, P., and R. H. MacArthur. 1961b. On the causes of tropical species diversity: niche overlap. *Amer. Natur.* 95:223–226.

MacArthur, R. H., and J. W. MacArthur. 1961c. On bird species diversity. *Ecology* 42:594–598.

MacArthur, R. H., J. W. MacArthur, and J. Preer. 1962a. On bird species diversity. II. Prediction of bird censuses from habitat measurements. *Amer. Natur.* 96:167–174.

MacArthur, R. H. 1962b. Some generalized theorems of natural selection. *Proc. Nat. Acad. Sci. U.S.A.* 48:1893–1897.

Rosenzweig, M. L., and R. H. MacArthur. 1963a. Graphical representation and stability conditions of predator-prey interactions. *Amer. Natur.* 97:209–223.

MacArthur, R. H., and E. O. Wilson. 1963b. An equilibrium theory of insular zoogeography. *Evolution* 17:373–387.

MacArthur, R. H. 1964a. Environmental factors affecting bird species diversity. *Amer. Natur.* 98:387–397.

Garfinkel, D., R. H. MacArthur, and R. Sack. 1964b. Computer simulation and analysis of simple ecological systems. *Ann. N.Y. Acad. Sci.* 115:943–951.

Dethier, V. G., and R. H. MacArthur. 1964c. A field's capacity to support a butterfly population. *Nature* 201:728–729.

MacArthur, R. H., and R. Levins. 1964d. Competition, habitat selection, and character displacement in a patchy environment. *Proc. Nat. Acad. Sci. U.S.A.* 51:1207–1210.

MacArthur, R. H. 1964e. Ecology. *In* A. L. Thompson, ed., *New Dictionary of Birds*. Nelson, London.

MacArthur, R. H. 1965a. Patterns of species diversity. *Biol. Rev.* 40:510–533.

MacArthur, R. H. 1965b. Ecological consequences of natural selection. *In* T. H. Waterman and H. J. Morowitz, eds., *Theoretical and Mathematical Biology*. Blaisdell, New York.

MacArthur, R. H., H. Recher, and M. Cody. 1966a. On the relation between habitat selection and species diversity. *Amer. Natur.* 100:319–332.

Levins, R., and R. H. MacArthur. 1966b. The maintainence of genetic polymorphism in a spatially heterogeneous environment: Variations on a theme by Howard Levine. *Amer. Natur.* 100:585–589.

MacArthur, R. H., and E. R. Pianka. 1966c. On optimal use of a patchy environment. *Amer. Natur.* 100:603–609.

MacArthur, R. H. 1966d. A review of *The Pattern of Animal Communities* by C. S. Elton. *Amer. Sci.* 54:497A. (Additional book reviews by MacArthur are in *Amer. Sci.* 54:106A, 139A, 228A, 230A, 349A; 55:78A, 102A, 353A, 498A, 513A, 514A; 56:58A, 280A; 57:126A, 162A, 234A, 244A; and 58:447A.)

Vandermeer, J., and R. H. MacArthur. 1966e. A reformulation of alternative *b* of the broken stick model of species abundance. *Ecology* 47:139–140.

MacArthur, R. H. 1966f. Note on Mrs. Pielou's comments. *Ecology* 47:1074.

MacArthur, R. H., and J. H. Connell. 1966g. *The Biology of Populations.* Wiley, New York.

MacArthur, R. H., and R. Levins. 1967a. The limiting similarity, convergence and divergence of coexisting species. *Amer. Natur.* 101:377–385.

MacArthur, R. H., and E. O. Wilson. 1967b. *The Theory of Island Biogeography.* Princeton University Press, Princeton.

MacArthur, R. H. 1968a. Selection for life tables in periodic environments. *Amer. Natur.* 102:381–383.

MacArthur, R. H. 1968b. The theory of the niche. *In* R. C. Lewontin, ed., *Population Biology and Evolution,* pp. 159–176. Syracuse University Press, Syracuse.

MacArthur, R. H. 1969a. Patterns of communities in the tropics. *Biol. J. Linn. Soc.* 1:19–30.

MacArthur, R. H. 1969b. The ecologist's telescope. *Ecology* 50:353.

MacArthur, R. H., and H. S. Horn. 1969c. Foliage profile by vertical measurements. *Ecology* 50:802–804.

Levins, R., and R. H. MacArthur. 1969d. An hypothesis to explain the incidence of monophagy. *Ecology* 50:910–911.

MacArthur, R. H. 1969e. Species packing and what competition minimizes. *Proc. Nat. Acad. Sci. U.S.A.* 64:1369–1371.

MacArthur, R. H. 1970a. Graphical analysis of ecological systems. *In* J. D. Cowan, ed., *Some Mathematical Questions in Biology,* pp. 61–73. Amer. Math. Soc., Providence.

MacArthur, R. H. 1970b. Species packing and competitive equilibrium for many species. *Theoret. Pop. Biol.* 1:1–11.

MacArthur, R. H. 1971. Patterns of terrestrial bird communities. *In* D. S. Farner and J. R. King, eds., *Avian Biology,* vol. 1, pp. 189–221. Academic Press, New York.

MacArthur, R. H., and J. M. Diamond, and J. R. Karr. 1972a. Density compensation in island faunas. *Ecology* 53:330–342.

Horn, H. S., and R. H. MacArthur. 1972b. Competition among fugitive species in a Harlequin environment. *Ecology* 53:749–752.

MacArthur, R. H., and D. MacArthur. 1972c. Efficiency and preference at a bird feeder. *J. Ariz. Acad. Sci.* 7:3–5.

May, R. M., and R. H. MacArthur. 1972d. Niche overlap as a function of environmental variability. *Proc. Nat. Acad. Sci. U.S.A.* 69:1109–1113.

MacArthur, R. H. 1972e. Strong, or weak, interactions. *Trans. Conn. Acad. Arts Sci.* 44:177–188.

MacArthur, R. H. 1972f. Coexistence of species. *In* J. Behnke, ed., *Challenging Biological Problems,* pp. 253–259. Oxford University Press, New York.

MacArthur, R. H. 1972g. *Geographical Ecology.* Harper and Row, New York.

MacArthur, R. H., and J. MacArthur, D. MacArthur, and A. MacArthur. 1973. The effect of island area on population densities. *Ecology* 54:657–658.

MacArthur, R. H., and A. T. MacArthur, 1974. On the use of mist nets for population studies of birds. *Proc. Nat. Acad. Sci. U.S.A.* 71:3230–3233.

I

The Evolution of Species Abundance and Diversity

1 Evolution in Communities Near Equilibrium

Richard Levins

The structure and dynamics of ecological communities depend on the biological properties of the component species, each of which is evolving under selection pressures that depend on the circumstances of the species. These circumstances include the constellation of other species in the community. Therefore, selection on any given species necessarily affects, as well as depends on, the community as a whole, in ways that cannot be understood by looking at species in isolation.

This chapter will present a new method of analysis for the study of partially specified systems. The method, called "loop analysis," proves particularly useful for examining the properties of biological communities in which the interactions between species can be specified in a qualitative but not a quantitative way. I will show that much can be deduced about the structure and behavior of such systems merely by using the sign of an interspecific interaction. The required information is simply whether species X_i is helped, harmed, or not affected by the presence of each species X_j in the community.

Using the method of loop analysis, I shall consider the impact of selection within each species of a community on the stability of the community as a whole, and on the abundances of the other coexisting species. It will be possible to determine whether mendelian selection at the species level has any inevitable consequences at the community level. Insofar as the possibility arises that intraspecific mendelian selection may increase the probability of extinction of local populations or communities, the course of such selection will oppose the group selection[1] that operates through biogeographic dynamics to delay group extinction and promote rapid recolonization.

If we restrict consideration to communities that are at or near equilibrium, we can examine local stability properties. The position of the equilibrium point itself may change under selection. This is relevant to the present analysis as long as the community is at equilibrium or close to it on a limit cycle of small amplitude, which would be dragged along by a shifting equilibrium point in its interior.

The presentation in this chapter is organized as follows. First, I present the methods and formalism of loop analysis and illustrate them by some simple ex-

[1] There has been much discussion about group selection, its definition, its reality, and the need for acknowledging it even if it is real. For the purposes of this chapter, group selection will refer to the existence, for some local population or group, of a property that reflects gene frequency of the group, and provides feedback into the environment of the group in such a way that the group's size or probability of survival is enhanced. A formal manipulation of symbols can translate group selection into mendelian selection, but such a manipulation obscures the significance of the processes involved. Parsimony is no virtue where it masks understanding.

16

amples. Next, these methods are used to derive general conditions that must be met if a system of interacting species and variables is to be stable. The application of these stability conditions is then illustrated by a series of biological examples. The remainder of the chapter is devoted to some general conclusions that follow from loop analysis. It is shown that the course of mendelian selection in a single species may stabilize or destabilize the community as a whole, may introduce oscillations, and may affect community stability in ways that are not intuitively obvious. Similarly, it is by no means inevitable that mendelian selection serves to increase the abundance of the species undergoing selection: the abundance of the species may decrease, and it is even possible for a species to select itself to extinction. The reality of group selection, which may oppose the direction of mendelian selection, is shown to follow from the analysis. It is found that the major cause of mortality need not be among the factors that determine the abundance of a species. Finally, the conclusions of biological interest are summarized.

The analysis involves the use of matrices. Readers who do not wish to go into the formalism of the analysis may skip to the results discussed in later sections on mendelian selection and group selection, or to the summary in the conclusions section. However, an effort is made to explain derivations in detail, and to represent matrices alternatively as diagrams, in such a way that a reader not familiar with matrices can follow the analysis with a little effort.

Methods of Loop Analysis

The mathematical methods used below depend on the equivalence between differential equations near equilibrium, on the one hand, and matrices and their diagrams, on the other hand. An elementary work such as Searles's *Matrix Algebra for Biologists* (1966) provides the technical background for most of what follows.

We begin by considering the dynamics of a community of n variables as represented by a set of n differential equations. The variables are usually species abundances, the levels of resources used by the species, or some factor produced by a species such as a toxic by-product or a predation pressure. The differential equation that describes the rate of change of the abundance or level X_i of the ith variable is

$$\frac{dX_i}{dt} = f_i(X_1, X_2, X_3, \ldots, X_n; C_1, C_2, \ldots) \quad (1)$$

That is, the growth rate of X_i is some function f_i of the levels of X_i, of all the other variables in the system $X_1,$ $X_2, \ldots, X_n$, and of a set of parameters C. The Cs characterize such potentially relevant aspects of the environment as temperature or rainfall, and such biological properties of the component species as growth rate constants or feeding efficiencies. There may be one or more equilibrium points of the system at which all the dX_i/dt are zero. In the neighborhood of any equilibrium point the behavior of the system depends on the properties of the so-called community matrix,

$$A = \begin{bmatrix} a_{11} & a_{12} & a_{13} & \cdots & a_{1n} \\ a_{21} & a_{22} & a_{23} & \cdots & \\ a_{31} & & & & \\ \vdots & \vdots & \vdots & & \vdots \\ a_{n1} & & & & a_{nn} \end{bmatrix} \quad (2)$$

where the elements a_{ij} are given by

$$a_{ij} = \frac{\partial f_i}{\partial X_j} \quad (3)$$

and are the coefficients of the X_j in the eqs. 1 for dX_i/dt, evaluated at the equilibrium point.

The qualitative properties of this system are represented by a "diagram," i.e., a picture that illustrates just where each variable fits into the system. Each variable is represented by a point or *vertex* in the diagram. The relations amongst variables appear as oriented *links* connecting the vertices, or variables, so that the line connecting X_j to X_i represents the interaction or effect of X_j on X_i and corresponds to the matrix element a_{ij}. A link or series of links that leaves and eventually reenters the same vertex is called a *loop*. Corresponding to the diagonal matrix elements a_{ii} there are loops that connect each X_i to itself, termed *self-loops*. These self-loops will be considered to be of unit *length*, and by analogy other loops will be of length 2, 3, or more depending on whether they are composed of two, three, or more of the elements a_{ij}. Examples of some simple systems represented in this way are shown in Figures 1 and 2, and these will be further discussed below. For the present, note that in Figure 1 one can leave and eventually reenter a vertex along the following paths or loops: three self-loops (a_{11}, a_{22}, a_{33}), each of whose lengths is by definition 1; six loops ($a_{12}a_{21}$, $a_{21}a_{12}$, $a_{13}a_{31}$, $a_{31}a_{13}$, $a_{23}a_{32}$, and $a_{32}a_{23}$), each of whose lengths is 2 because the loop consists of two coefficients a_{ij}; and two loops ($a_{21}a_{32}a_{13}$ and $a_{31}a_{23}a_{12}$, actually the same loop traversed in opposite directions), each of whose lengths is 3 because the loop consists of three coefficients a_{ij}.

The elements a_{ij} are readily interpreted as the effect of X_j on the level of X_i. The values of the a_{ij} may be difficult to obtain, and may be different at different equilibrium points. But it is often easier to determine the sign of each of a_{ij} than its numerical value, and further analysis of the systems will depend on this information only. Thus, if X_j preys on X_i, a_{ij} is negative and a_{ji} is positive. As another example, if X_i and X_j compete with each other, a_{ij} and a_{ji} are both negative. In the diagrams we distinguish between positive and negative a_{ij} by an arrow and a circle, respectively (cf. Figures 2a and 2b).

The interpretation of the loops of length 1 corresponding to diagonal elements a_{ii} is less obvious. If the function $f_i(X_i, X_2, X_3, \ldots, X_n; C_1, C_2, \ldots)$ is of the form

$$\frac{dX_i}{dt} = X_i[g_i(X_{\text{all } j \neq i}); C_1, C_2, \ldots)], \quad (4)$$

in which X_i does not appear in g_i, then an equilibrium (i.e., $dX_i/dt = 0$) at which $X_i \neq 0$ must have $g_i = 0$. Hence the coefficient of X_i, which is a_{ii}, is zero at equilibrium. Qualitatively, this means that the growth rate of X_i is independent of X_i, i.e., is neither self-damped nor self-accelerating. But if a species is self-damped owing to crowding or other effects, a_{ii} will be negative, while if the growth rate is enhanced by the level of X_i, there is positive feedback from X_i to itself and a_{ii} is positive. However, the meaning of the a_{ii} terms is sensitive to the way the model is set up. If a species uses a depletable resource which is not explicitly represented as a variable in the system, then the species is necessarily self-damped and a_{ii} is negative. But if we introduce the resource itself as a variable in the system, the self-damping vanishes and is replaced by the loop connecting the species and its resource.

In every ecosystem there are resources that are not self-reproducing in simple ways, such as mineral nutrients, organic matter, or detritus. Therefore, at the bottom of any trophic structure there are either self-damped variables that stand for resources, or else species that use these resources and incorporate the damping effect of the resources in the form of a self-damping effect on the species.

The determinant of a matrix is a number that is a sum of products of the elements of the matrix. For instance, a 3×3 matrix, which could represent a system of three species, is a sum of $3! = 6$ such products. In each product three matrix elements are multiplied:

$$\begin{vmatrix} a_{11} & a_{12} & a_{13} \\ a_{21} & a_{22} & a_{23} \\ a_{31} & a_{32} & a_{33} \end{vmatrix}$$

$$= a_{11}a_{22}a_{33} - a_{11}a_{23}a_{32} - a_{12}a_{21}a_{33}$$
$$+ a_{12}a_{23}a_{31} + a_{13}a_{21}a_{32} - a_{13}a_{22}a_{31} \quad (5)$$

In each element a_{ij} the first subscript i is a row index and the second subscript j is a column index. Thus, in each product of three elements and six subscripts each row and each column is represented just once. But each first subscript i stands for an input to X_i in a system diagram, and each second subscript j stands for an output from X_j. Therefore, each product in the expansion of the determinant is a product of loops in the diagram which go into each variable once and leave it once.

Let us examine Figure 1, the diagram of the three-variable system represented

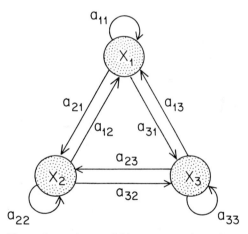

Figure 1 A three-variable system, such as three competing species X_1, X_2, and X_3, is represented in diagrammatic form. The interactions of the species with themselves are represented by self-loops (a_{11}, a_{22}, and a_{33}), and of the species with each other are represented as connecting lines (a_{12}, a_{21}, a_{32}, a_{13}, a_{31}). The magnitudes of these interactions are the coefficients a_{ij} of the X_j in the growth equation of X_j (eq. 4). See text for the explanation of how various interconnecting links form loops in the system, which can be derived from the community matrix **A** of coefficients a_{ij}, and for the explanation of what each loop means in terms of the stability of and evolution within the system.

in matrix terminology by the above matrix, in order to illustrate the relation between loops in the diagram (Figure 1) and the six terms in the matrix expansion (eq. 5). The six expansion terms are of three types. Two of the terms ($a_{12}a_{31}a_{23}$ and $a_{13}a_{21}a_{32}$) each represent a product of three elements a_{ij} around a single loop of length 3 (because the loop consists of three coefficients a_{ij}); the term $a_{12}a_{31}a_{23}$ represents a clockwise traverse of the loop, and $a_{13}a_{21}a_{32}$ is a counter-clockwise traverse. Three of the terms ($a_{11}a_{23}a_{32}$, $a_{22}a_{13}a_{31}$, and $a_{33}a_{12}a_{21}$) do not represent a single loop because each term does not correspond to a continuous path leaving and reentering the same variable. Instead, these terms correspond to the product of a self-loop of length 1 leaving and reentering one vertex (e.g., a_{11}), and another loop of length 2 leaving and reentering another two vertices (e.g., $a_{23}a_{32}$). When two loops share no vertices in common, we refer to the loops as *disjunct*. Two loops that share a vertex, such as a_{11} and $a_{12}a_{21}$, are called *conjunct*. Thus, each of the three terms $a_{11}a_{23}a_{32}$, $a_{22}a_{13}a_{31}$, and $a_{33}a_{12}a_{21}$, represents a product of two disjunct loops. The last term, $a_{11}a_{22}a_{33}$, represents a product around three disjunct loops, each a self-loop of length 1. Thus, each of the six terms in the matrix expansion is the product around one or more disjunct loops of the diagram.

The sign of a term in the expansion of the determinant is determined by the number of permutations of second subscripts as follows. Take the product of the principal diagonal terms as having a positive sign, $+a_{11}a_{22}a_{33}$. Then each term in the expansion is generated by permuting second subscripts, and each single permutation (at most two permutations will be necessary in our 3×3 example) will reverse the sign of the product. Thus, permuting the first pair of second subscripts gives us $-a_{12}a_{21}a_{33}$, which has a negative sign because there has been only one permutation. If one then permutes the second pair of second subscripts, one obtains $+a_{12}a_{23}a_{31}$, where the sign has been restored to positive because two permutations introduce two minus signs, which cancel. (The reader can check that similar permutations show the term $a_{13}a_{21}a_{32}$ to be positive, and $a_{11}a_{23}a_{32}$ and $a_{13}a_{22}a_{31}$ to be negative.) Note that the two disjunct loops a_{11}, a_{22} were fused into a single loop $a_{12}a_{21}$ of length 2 by the first permutation, and that the second permutation created a single loop of length 3 ($a_{12}a_{23}a_{31}$) from two disjunct loops of lengths 2 and 1 ($a_{12}a_{21}$ and a_{33}, respectively). Each permutation in fact either fuses two loops into one or splits one loop into two; that is, the number of loops is changed by one. Each permutation also changes the sign of the term. Therefore, we can express the whole determinant as

$$D_n = \sum (-1)^{n-m} L(m,n) \qquad (6)$$

where $L(m,n)$ is the product of n links (i.e., coefficients a_{ij}) which form m disjunct loops. Since the product may be a product of $1, 2, \ldots, m, \ldots$, up to n loops, and a product of any given number of loops m may be composed in several different ways, the summation is taken over all m

and all possible products involving m loops.

This interpretation of a determinant as a sum of products of loops in a diagram was used by Mason (1954) to compute electrical circuits, and much earlier by Sewall Wright (1921; review volume 1968) as a method for the path analysis for heritable factors in inbreeding systems. However, the specialized purposes for which this method has previously been used have prevented biologists from gaining a general appreciation of the method and its significance for qualitative analysis (see Levins, 1974).

It is convenient to introduce next an expression F_k, which describes the "feedback" at level k in a system, and is defined by

$$F_k = \sum (-1)^{m+1} L(m,k) \qquad (7)$$

That is, F_k is the sum of all possible products that involve just k vertices or variables and consist of disjunct loops. The sign is adjusted in such a way that if all loops are negative, F_k is also negative. A product that is a single positive loop contributes a positive term to F_k; a product of $m = 2$ positive disjunct loops makes F_k negative; a product of three positive disjunct loops makes F_k positive again; and so on. By these definitions eq. 7 yields

$$F_1 = \sum_i a_{ii} \qquad (8)$$

That is, the feedback at level $k = 1$ is the sum of all self-loops, since these are the sole loops of length 1. Since m, the number of disjunct loops in each term of the

summation of eq. 8, is 1, $(-1)^{m+1} = +1$, and the sign of each term is positive. At level $k = 2$, eq. 7 becomes

$$F_2 = \sum a_{ij} a_{ji} - \sum a_{ii} a_{jj} \qquad (9)$$

That is, there are two types of terms that represent products of disjunct loops involving two vertices. One type comprises the terms $a_{ij} a_{ji}$, each of which is a single loop ($m = 1$) of length 2. Since $(-1)^{m+1} = +1$, these terms have positive signs. The other type comprises the terms $a_{ii} a_{jj}$, each of which is a product of two ($m = 2$) disjunct self-loops. Since $(-1)^{m+1} = -1$, these terms have negative signs. At level $k = 3$, eq. 7 becomes

$$F_3 = \sum a_{ii} a_{jj} a_{kk} - \sum a_{ii} a_{jk} a_{kj}$$
$$+ \sum a_{ij} a_{jk} a_{ki} \qquad (10)$$

That is, products of disjunct loops involving three vertices are of three types: products of three self-loops [$m = 3$, $(-1)^{m+1} = 1$], the terms $a_{ii} a_{jj} a_{kk}$; products of self-loops times disjunct loops of length 2 [i.e., products of two disjunct loops, $m = 2$, $(-1)^{m+1} = -1$], the terms $-a_{ii} a_{jk} a_{kj}$; and single loops [$m = 1$, $(-1)^{m+1} = 1$] through three vertices, the terms $a_{ij} a_{jk} a_{ki}$. Feedback at higher levels is given by analogous general formulas. For consistency we define

$$F_0 = -1 \qquad (11)$$

As an illustration, let us compute the feedback at all possible levels in Figure 1. At level 1, there are three loops each involving just one vertex, the self-loops

a_{11}, a_{22}, and a_{33}; eq. 8 yields

$$F_1 = a_{11} + a_{22} + a_{33} \qquad (12)$$

At level 2, all possible products involving two vertices are $a_{12}a_{21}$, $a_{23}a_{32}$, $a_{13}a_{31}$, $a_{11}a_{22}$, $a_{11}a_{33}$, and $a_{22}a_{33}$; the first three terms have a positive sign, since they involve a single loop; hence $m = 1$, and $(-1)^{m+1}$ is positive; the second three terms are negative, since each involves two disjunct loops (e.g., a_{11} and a_{22}), $m = 2$, and $(-1)^{m+1}$ is negative; hence we have

$$F_2 = a_{12}a_{21} + a_{23}a_{32} + a_{13}a_{31}$$
$$- a_{11}a_{22} - a_{11}a_{33} - a_{22}a_{33} \qquad (13)$$

Equation 13 also follows directly from eq. 9. Finally, at level 3, F_3 is just the full expansion of the 3×3 determinant as given above by eq. 5, which can also be obtained from eq. 10.

We can now look at some simple systems of biological relevance, as illustrated in Figure 2, in order to make clearer the computation and significance of feedback. In each example our main purpose for the moment will be to determine whether feedback is positive or negative at each level. Later we shall show that a system is stable only if all feedbacks F_k, the net feedbacks at each level k, are negative.

Figure 2a represents a simple predator-prey system, in which the growth of neither species is density-dependent. Since the prey X_1 promotes the growth of the predator X_2, the coefficient a_{21} (effect of X_1 on X_2) is positive and is symbolized by an arrow from X_1 to X_2. Since the predator X_2 depresses the growth of the prey X_1, the coefficient a_{12} (effect of X_2 on X_1)

is negative and is symbolized by a circle from X_2 to X_1. Since neither species is self-damped, the coefficients a_{11} and a_{22} are zero, and there are no self-loops. The left side of Figure 2a is the diagram of this system, and the right side is the matrix of the system with the signs of each term specified (i.e., the coefficients themselves are all positive numbers; we follow this convention throughout). Since Figure 2a has no nonzero products involving just one vertex ($a_{11} = a_{22} = 0$), the feedback at level 1, F_1, is zero. At level 2, of the two possible products of loops involving two vertices, one ($a_{11}a_{22}$) is zero; the other is ($-a_{12}a_{21}$); this constitutes a single loop, hence $m = 1$, and $(-1)^{m+1}$ is positive; and $F_2 = (-a_{12}a_{21})$, i.e., is negative, as also follows from eq. 9. Since there are only two vertices, feedback does not exist above level 2.

Figure 2b is similar to Figure 2a except that the growth of the prey X_1 is self-damped, as expressed by the self-loop with a circle on X_1 and by the negative sign of a_{11}. At level 1, the only non-zero self-loop is $-a_{11}$; since this is a single loop, $(-1)^{m+1}$ is positive; and $F_1 = -a_{11}$, which is negative. F_2 is the same as in Figure 2a, negative.

Figure 2c represents two competing species, so that each depresses the growth of the other, the links in both directions are circles, and a_{12} and a_{21} are both negative. Both species are assumed to be self-damped, symbolized by the self-loops with circles and by the negative signs of a_{11} and a_{22}. At level 1, $F_1 = -a_{11} - a_{22}$ and is negative. At level 2 there are two loops involving two vertices: $(-a_{12})(-a_{21})$,

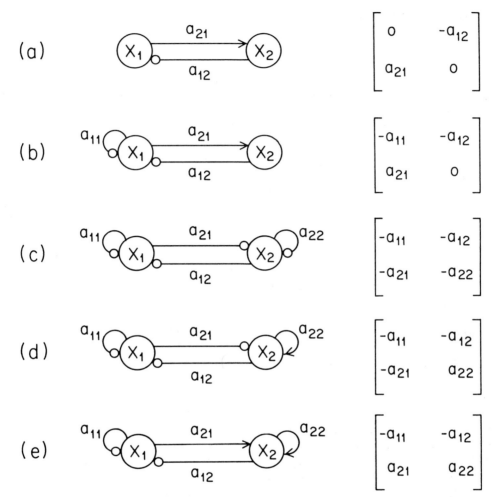

Figure 2 Some examples of two-variable systems. An arrow from variable X_1 to variable X_2 means that the sign of a_{21} is positive. Where circles replace arrowheads, as in the effect of X_2 on X_1 in Figure 2a, the sign of a_{12} is negative. The matrix to the right of each figure gives the interactions and their signs. Figure 2a is a predator-prey system without density dependence. Figure 2b is the same system except that the prey is self-damped, as perhaps with a nutrient and its user. In Figure 2c we show a system with a pair of self-damped competitors. In Figure 2d one of these (X_2) is autocatalytic rather than self-damped (i.e., has positive feedback in its growth). Figure 2e shows the same self-loops as 2d, but applied to a predator-prey system. See text for further explanation.

which is a single loop, hence $(-1)^{m+1}$ is positive; and $(-a_{11})(-a_{22})$, which consists of two disjunct loops, hence $(-1)^{m+1} = (-1)^{2+1}$ is negative. Therefore (as may be confirmed by eq. 9) $F_2 = (-a_{12})(-a_{21}) - (-a_{11})(-a_{22}) = a_{12}a_{21} - a_{11}a_{22}$, and the sign of F_2 is ambiguous: i.e., it cannot be determined from qualitative considerations alone but requires knowledge of the numerical values of the four coefficients.

Figure 2d also depicts two competitors and differs from Figure 2c only in that growth of species X_2 is now self-accelerating (self-loop with arrow, a_{22} positive) instead of self-damped. Feedback at level 1, $F_1 = -a_{11} + a_{22}$, has an ambiguous sign that cannot be determined without knowing the values of a_{11} and a_{22} (i.e., whether X_1 is more self-damped than X_2 is self-accelerating, or vice versa). F_2 is unambiguous and positive, since

$$F_2 = (-1)^{1+1}(-a_{12})(-a_{21})$$
$$+ (-1)^{2+1}(-a_{11})(a_{22})$$
$$= a_{12}a_{21} + a_{11}a_{22}$$

Since F_2 is positive, the system is unstable. One might not have suspected intuitively that a system consisting of a self-damped species and a self-accelerating competitor is in general unstable, regardless of the magnitudes of the competition coefficients and self-feedback coefficients.

Finally, Figure 2e is a predator-prey system differing from Figure 2b only in that the predator's growth is self-accelerating (self-loop with arrow, a_{22} positive). Figure 2e differs from Figure 2d only in that a_{21} is positive rather than negative.

At level 1, $F_1 = -a_{11} + a_{22}$ just as in Figure 2d, and is ambiguous. Unlike Figure 2d, however, F_2 is now also ambiguous:

$$F_2 = (-1)^{1+1}(-a_{12})(a_{21})$$
$$+ (-1)^{2+1}(-a_{11})(a_{22})$$
$$= -a_{12}a_{21} + a_{11}a_{22}$$

Note that the self-damping of species X_1, represented by $-a_{11}$, contributes negative feedback at level 1 but positive feedback at level 2 (the term in which it appears in F_2, $a_{11}a_{22}$, is positive). In order for both F_1 and F_2 in Figure 2e to be negative, we require two conditions to be met simultaneously:

$$F_1 = -a_{11} + a_{22} < 0,$$

or

$$a_{22} < a_{11};$$
$$F_2 = -a_{12}a_{21} + a_{11}a_{22} < 0,$$

or

$$a_{11}a_{22} < a_{12}a_{21}$$

If $a_{11} < \sqrt{a_{12}a_{21}}$, these two conditions are incompatible.

Conditions for Stability

The local behavior of a system of differential equations near an equilibrium point can be understood by initially considering two simpler systems. First I will describe a discrete time process in one variable, and then a differential equation, or a continuous time process, also in just one variable.

The discrete time process shows how a

variable X changes over time when these changes are effected at discrete time intervals Δt. The value of X at time $(t + \Delta t)$ depends on its value at time t according to:

$$X(t + \Delta t) = X(t) + r\,\Delta t \cdot X(t) \quad (14)$$

This means that the magnitude of the change in X depends upon the value of X at time t, on a rate constant or response rate r, and on the duration of its action Δt. It is clear that $X = 0$ is an equilibrium, since then there can be no further change. If r is positive, then X increases when X is positive and decreases when X is negative. In either case X moves away from zero so that the equilibrium point is unstable.

If r is negative, the behavior of the system is more complex. First let $-r\,\Delta t$ be less than one, say $\frac{1}{2}$. Then if we start at $X = 1$, in successive time intervals we have $X = 1, \frac{1}{2}, \frac{1}{4}, \frac{1}{8}, \frac{1}{16}, \dots$ Thus, X approaches its equilibrium value asymptotically from above, and similarly approaches equilibrium asymptotically from below if we start with negative X. This system is stable. If $1 < -r\,\Delta t < 2$, a new behavior appears. Suppose that $-r\,\Delta t = 1\frac{1}{2}$ and we start with $X = 1$. The sequence of values generated by the system is now $1, -\frac{1}{2}, +\frac{1}{4}, -\frac{1}{8}, +\frac{1}{16}, \dots X$ still returns to an equilibrium point, and that point is therefore a stable one. But X approaches the equilibrium in an oscillatory manner. If $-r\,\Delta t$ increases to 2, we are at the boundary of two behaviors, and X oscillates back and forth between its initial value $X(0)$ and the negative of this value $-X(0)$. Finally, for $-r\,\Delta t > 2$ we

get oscillations of increasing amplitude. Thus, at $-r\,\Delta t = 3$ the sequence of states of X, still beginning at $X = 1$, is $1, -2, +4, -8, +16, \dots$ Thus, stability requires negative feedback ($r\,\Delta t$ negative), but a system may be unstable despite negative feedback. Instability may be attributed, then, to two circumstances: positive feedback, or negative feedback in which the product of the feedback and its time lag is too large.

As Δt shrinks, $r\,\Delta t$ eventually becomes smaller than 1, and the system approaches the continuous time system represented by the differential equation

$$\frac{dX}{dt} = rX \quad (15)$$

Now the time lag has disappeared, and the behavior of the system depends only on whether r is positive or negative. The solution of eq. 15 is

$$X(t) = X(0)e^{rt} \quad (16)$$

so that if $r < 0$, X returns to its equilibrium level, which from eq. 15 is at zero. If $r > 0$, X moves away from equilibrium; again there is a boundary between these two behaviors at $r = 0$, in which case X remains at zero. There is no possibility of oscillatory behavior.

Now we can proceed to look at a system of differential equations that describe the simultaneous behavior of a number of variables X_i. Instead of the local behavior depending on a single exponential, as in eq. 11, the solution is now a sum of exponential terms $e^{\lambda_i t}$ with coefficients c_i:

$$X_i = \sum_i c_i e^{\lambda_i t} \quad (17)$$

The λ_i are the roots of the so-called characteristic equation $|\mathbf{A} - \lambda\mathbf{I}| = 0$, where as before $\mathbf{A}$ is the community matrix of coefficients a_{ij} for the X_j in the differential equation $dX_i/dt = 0$, and $\mathbf{I}$ is the so-called identity matrix (in which all diagonal elements are 1 and all other elements are zero). The λ_i are therefore found from

$$\begin{bmatrix} a_{11}-\lambda & a_{12} & a_{13} & \cdots & a_{1n} \\ a_{21} & a_{22}-\lambda & a_{23} & \cdots & a_{2n} \\ \vdots & & & & \vdots \\ a_{n1} & \cdots & & \cdots & a_{nn}-\lambda \end{bmatrix} = 0 \quad (18)$$

Leigh (Chapter 2) also discusses the solution of this equation and its bearing on community stability.

As before, the system is stable if $\lambda < 0$, but now all the λs must be negative. And a new possibility has been introduced: λ may be a complex number. Since

$$e^{(r+i\theta)t} = e^{rt}[i \sin\theta t + \cos\theta t] \quad (19)$$

and since complex roots occur in conjugate pairs $r \pm i\theta$ (where i now stands for $\sqrt{-1}$), a complex root introduces the term $e^{rt}\cos\theta t$ into the solution of the characteristic equation. Thus, a complex root means oscillatory behavior, whereas the real part of the root, r, determines whether the amplitudes of the oscillations increase, decrease, or remain unchanged. The oscillatory behavior that disappeared when we went from a discrete-time process to a continuous process has now returned. Time lags have been reintroduced into the system by virtue of a variable X_i affecting itself indirectly, by way of inter-

actions through several intervening variables X_j.

These results can now be conveniently summarized in terms of the nature of the real and imaginary parts of the roots $\lambda = r \pm i\theta$, as shown in the following:

r	θ	System behavior
< 0	0	Stable; returns to equilibrium asymptotically.
< 0	$\neq 0$	Stable; returns to equilibrium through damped oscillations.
0	0	Neutral stability; any displacement from equilibrium persists.
0	$\neq 0$	Neutral stability; the system oscillates with a persistent amplitude that depends on the initial displacement.
> 0	0	Unstable; system moves away from equilibrium in the direction of the initial displacement.
> 0	$\neq 0$	Unstable; oscillates about equilibrium with increasing amplitude.

We can now determine the conditions of a system under which these alternative stability outcomes may arise.

The matrix equation (eq. 18) can be multiplied out to give a polynomial in λ, $P(\lambda)$, of order equal to the number of variables X in the system. For a system with only two variables we have

$$\begin{vmatrix} a_{11} - \lambda & a_{12} \\ a_{21} & a_{22} - \lambda \end{vmatrix}$$
$$= (a_{11} - \lambda)(a_{22} - \lambda) - a_{12}a_{21} \quad (20)$$

or

$$P(\lambda) = \lambda^2 - (a_{11} + a_{22})\lambda + a_{11}a_{22} - a_{12}a_{21} \quad (21)$$

Similarly for the three-variable system the characteristic equation becomes

$$P(\lambda) = \begin{vmatrix} a_{11} - \lambda & a_{12} & a_{13} \\ a_{21} & a_{22} - \lambda & a_{23} \\ a_{31} & a_{32} & a_{33} - \lambda \end{vmatrix}$$

and on expansion this becomes

$$\begin{aligned} \lambda^3 &- \lambda^2[a_{11} + a_{22} + a_{33}] \\ &+ \lambda[(a_{11}a_{22} - a_{12}a_{21}) \\ &+ (a_{11}a_{33} - a_{13}a_{31}) \\ &+ (a_{22}a_{33} - a_{23}a_{32})] \\ &- [(a_{11})(a_{22}a_{33} - a_{23}a_{32}) \\ &- (a_{12})(a_{23}a_{31} - a_{21}a_{33}) \\ &+ (a_{13})(a_{21}a_{32} - a_{22}a_{31})] \quad (22) \end{aligned}$$

The general expression for the polynomial in λ is

$$P(\lambda) = \lambda^n + \sum_k (-1)^k D_k \lambda^{n-k} \quad (23)$$

where by D_k we mean the sum of all principal kth order determinants, or the sum of those determinants corresponding to kth order subsystems. D_k can be directly translated into our diagram terminology by means of eq. 6, whence the polynomial becomes

$$P(\lambda) = \lambda^n + \sum_k (-1)^k(-1)^{k-m} L(m,k)\lambda^{n-k} \quad (24)$$

Since $(-1)^{2k-m} = (-1)^{-m} = (-1)^{m+2}$ if k and m are integers, eq. 24 can be rewritten as

$$P(\lambda) = \lambda^n + \sum (-1)^{m+1}(-1)L(m,k)\lambda^{n-k} \quad (25)$$

And because the right-hand side of eq. 25 contains the expression for feedback as defined by eq. 7, it can be rewritten as

$$P(\lambda) = \lambda^n - \sum_k F_k \lambda^{n-k} \quad (26)$$

Thus, the polynomial $P(\lambda)$ involves λ, which is related to the stability of the system as indicated above, as well as feedback, which enters as a coefficient of λ. The Routh-Hurwitz theorem provides rules for relating the coefficients of an equation to its roots. There are two kinds of conditions for all roots having negative real parts, which is a prerequisite for a stable system that returns to its equilibrium value after a perturbation. The first condition is that all coefficients be positive. This is equivalent to requiring that all feedbacks F_k, the net feedback at each level k, must be negative. The second requirement involves a sequence of inequalities that relate various coefficients in the equation. In a qualitative sense, these inequalities demand that negative feedback coming from long loops cannot be too strong compared to the negative feedback from shorter loops (see the following section for examples and further discussion). The first of these inequalities is sufficient for our purposes:

$$F_1 F_2 + F_3 > 0 \quad (27)$$

For the three-variable system whose feedbacks F_1, F_2, and F_3 we derived earlier as eqs. 12, 13, and 5,

$$\{(a_{11} + a_{22} + a_{33})(a_{12}a_{21} + a_{13}a_{31}$$
$$+ a_{23}a_{32} - a_{11}a_{22} - a_{11}a_{33} - a_{22}a_{33})\}$$
$$+ \{(a_{11}a_{22}a_{33}) + (a_{12}a_{23}a_{31} + a_{13}a_{21}a_{32})$$
$$- [a_{11}(a_{23}a_{32}) + a_{22}(a_{13}a_{31})$$
$$+ a_{33}(a_{12}a_{21})]\} > 0 \qquad (28)$$

This can be written in shorthand notation, using L_i for self-loops involving just vertex i, L_{ij} for loops of length 2 involving just vertices i and j, and L_{ijk} for loops of length 3 involving all three vertices. The expression eq. 28 then becomes

$$\left\{ \left(\sum L_i \right) \left(\sum L_{ij} - \sum L_i L_j \right) \right\}$$
$$+ \left\{ \sum L_i L_j L_k + \sum L_{ijk} - \left[\sum L_i L_{jk} \right] \right\}$$
$$> 0 \qquad (29)$$

where the terms that are in braces or square brackets in eqs. 28 and 29 correspond to each other. We note in eq. 28 that all the terms in F_3 consisting of products of disjunct loops also occur in F_1F_2, but with the opposite sign. Thus, in multiplying out eq. 28 the first product of disjunct loops is $a_{11}a_{23}a_{32}$ in the product F_1F_2, and this occurs later in F_3 as $-a_{11}a_{23}a_{32}$; the second product of disjunct loops is $-3a_{11}a_{22}a_{33}$, which likewise shows up in F_3 as $a_{11}a_{22}a_{33}$. Cancelling these terms, we are left with only the products of conjunct loops (i.e., such terms as $a_{11}a_{12}a_{21}$, $a_{11}a_{13}a_{31}$, etc.) from the F_1F_2 product, and with the loops of length 3 from F_3. The inequality eq. 27 or eq. 29 becomes

$$[-a_{11}^2(a_{22} + a_{33}) - a_{22}^2(a_{11} + a_{33})$$
$$- a_{33}^2(a_{11} + a_{22})]$$
$$+ [a_{11}(a_{12}a_{21} + a_{13}a_{31})$$
$$+ a_{22}(a_{12}a_{21} + a_{23}a_{32})$$
$$+ a_{33}(a_{13}a_{31} + a_{23}a_{32})] - [2a_{11}a_{22}a_{33}]$$
$$+ [a_{12}a_{23}a_{31} + a_{13}a_{21}a_{32}] > 0 \qquad (30)$$

or in the loop subscript shorthand,

$$\left[-\sum L_i^2 \sum L_j \right] + \left[\sum L_i \sum L_{ij} \right]$$
$$- \left[\sum 2L_i L_j L_k \right] + \left[\sum L_{ijk} \right] > 0 \qquad (31)$$

Examples of Stability Conditions Applied to Biological Systems

We shall now examine the implications of these results in some simple biological systems that involve three or four variables. Several examples are shown in Figure 3. In all three examples we shall deduce instabilities or constraints on stability that might not have been guessed at or understood without analysis.

The system of Figure 3a consists of a predator X_2, which feeds on X_1, which in turn feeds on X_4 (i.e., a_{21}, a_{14} positive; a_{12}, a_{41} negative). In addition, X_2 boosts the level of X_3, which in turn boosts X_4, which in turn boosts X_2—a one-way chain of interactions (i.e., a_{32}, a_{43}, a_{24} positive; a_{23}, a_{34}, a_{42} zero). The only species with density-dependent growth is X_1, which is self-damped (a_{11} negative; a_{22}, a_{33}, a_{44} zero). The existence and sign of the coefficients a_{ij} are summarized in Figure 3a in the matrix to the right of the diagram.

Feedback at level 1 in the system of Figure 3a, given by the sum of the self-loops (cf. eq. 8), comes only from $-a_{11}$

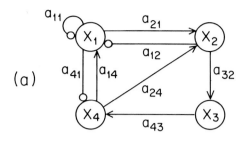

(a)

$$\begin{bmatrix} -a_{11} & -a_{12} & 0 & a_{14} \\ a_{21} & 0 & 0 & a_{24} \\ 0 & a_{32} & 0 & 0 \\ -a_{41} & 0 & a_{43} & 0 \end{bmatrix}$$

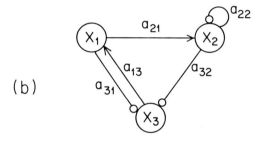

(b)

$$\begin{bmatrix} 0 & 0 & a_{13} \\ a_{21} & -a_{22} & 0 \\ -a_{31} & -a_{32} & 0 \end{bmatrix}$$

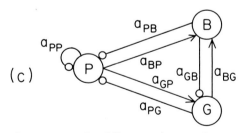

(c)

$$\begin{bmatrix} -a_{PP} & -a_{PB} & -a_{PG} \\ a_{BP} & 0 & a_{BG} \\ a_{GP} & -a_{GB} & 0 \end{bmatrix}$$

Figure 3 More examples of diagrams of systems (on the left), with the corresponding matrices on the right. These systems show various types of instability. Above, Figure 3a (consisting of a three-level predator-prey system $X_2 - X_1 - X_4$ plus another variable or species X_3) is unstable because it violates the condition that feedback in the system at all levels must be negative. While F_1 and F_2 are negative, F_3 and F_4 are positive. In the center, the system of Figure 3b (a predator X_1, prey X_3, and self-damped variable X_2) violates the condition that $F_1 F_2 + F_3 > 0$, so that this system produces an oscillatory instability. Figure 3c (below) shows blue-green algae B and green algae G both exploiting the self-damped nutrient phosphate P. The green algae supply the blue-greens with vitamins but receive a toxin in return. This and the other systems of the figure are discussed in detail in the text.

and is therefore negative. Feedback at level 2, given by eq. 9, is negative:

$$F_2 = (-a_{12})(a_{21}) + (a_{14})(-a_{41})$$
$$= -a_{12}a_{21} - a_{14}a_{41} \quad (32)$$

or, indicating loops by the vertices they traverse in parentheses,

$$F_2 = (X_1 X_2) + (X_1 X_4) \quad (33)$$

At level 3 the only way that we can form loops or products of disjunct loops involving three vertices (cf. eq. 10) is by

$$F_3 = (X_1 X_4 X_2) + (X_2 X_3 X_4)$$
$$= (-a_{41})(a_{24})(-a_{12})$$
$$+ (a_{32})(a_{43})(a_{24})$$
$$= a_{41}a_{24}a_{12} + a_{32}a_{43}a_{24} \quad (34)$$

Since both terms are positive, F_3 is positive. (Each of these terms corresponds to a single loop. Had a_{42} been nonzero, there would also have been a term $(a_{21})(a_{42})(a_{14})$ representing another single loop. Had a_{22}, a_{33}, or a_{44} been nonzero, there would have been terms such as $(a_{22})(a_{14})(-a_{41})$, $(a_{33})(a_{21})(-a_{12})$ and $(a_{33})(a_{14})(-a_{41})$, or $(a_{44})(a_{21})(-a_{12})$ representing products of two disjunct loops, a self-loop and a loop of length 2.) At level 4 the only way that we can form loops or products of loops involving four vertices is by

$$F_4 = (X_1 X_2 X_3 X_4) + (X_1)(X_2 X_3 X_4)$$
$$= (a_{21})(a_{32})(a_{43})(a_{14})$$
$$- (-a_{11})(a_{32})(a_{43})(a_{24})$$
$$= a_{21}a_{32}a_{43}a_{14} + a_{11}a_{32}a_{43}a_{24} \quad (35)$$

Since both terms are positive, F_4 is positive. (The first of these terms corresponds to a single loop of length 4; the second

term, to the product of two disjunct loops, a self-loop and a loop of length 3. Had a_{34} been nonzero, there would also have been a term $(a_{34})(a_{43})(-a_{12})(a_{21})$ corresponding to the product of two disjunct loops. Had a_{34} and a_{22} been nonzero, there would have been a term $(a_{11})(a_{22})$ $(a_{34})(a_{43})$ corresponding to the product of three disjunct loops.)

The conclusion to be drawn from this analysis is that, despite the negative feedback at levels 1 and 2, the system is unstable because of positive feedback at both levels 3 and 4.

The system of Figure 3b consists of a predator X_1 which feeds on X_3 (i.e., a_{13} positive, a_{31} negative), plus a self-damped variable X_2 (a_{22} negative) that is boosted by X_1 and inhibits X_3 (a_{21} positive, a_{32} negative, a_{12} and a_{23} zero). Since the only self-loop is $(-a_{22})$, feedback at level 1 is $F_1 = -a_{22}$. From eq. 9, feedback at level 2 is $F_2 = (a_{13})(-a_{31})$, the other terms of eq. 9 being zero. To obtain F_3, we note (or see eq. 10) that loops or products of disjunct loops involving three vertices can be formed in two ways:

$$F_3 = (X_1 X_2 X_3) - (X_2)(X_1 X_3)$$
$$= (a_{21})(-a_{32})(a_{13})$$
$$- (-a_{22})(a_{13})(-a_{31})$$
$$= -a_{21}a_{32}a_{13} - a_{22}a_{13}a_{31}$$

Thus, the stability requirement that feedback at all three levels be negative is satisfied. But the requirement of eq. 27 that $F_1 F_2 + F_3$ be positive is violated, since

$$F_1 F_2 + F_3 = (-a_{22})(a_{13})(-a_{31})$$
$$- a_{21}a_{32}a_{13} - a_{22}a_{13}a_{31} = -a_{21}a_{32}a_{13}$$

Alternatively, one can reach this conclusion as follows from eq. 31, which states the same condition in a different form. Starting with the first term in eq. 31, there are no combinations $L_i^2 L_j$, since there is only one self-loop L_i or L_j, $-a_{22}$. The second term, $L_i L_{ij}$, is also represented by no combination, since the sole self-loop L_i, at vertex X_2, is disjunct from the sole loop of length 2, connecting vertices X_1 and X_3. Thus, the sole nonzero combination in eq. 31 arises from the last term, L_{ijk}, represented by the loop L_{123}, and this is negative $[(a_{13})(a_{21})(-a_{32})]$, violating the inequality condition eq. 31. Therefore, the system of Figure 3b shows oscillatory instability.

In the system of Figure 3c, phosphate (P) is utilized as a nutrient both by green algae (G) and by blue-green algae (B). Thus, a_{BP} and a_{GP} are positive, while a_{PB} and a_{PG} are negative. Neither alga is self-damped ($a_{BB} = 0 = a_{GG}$), but the phosphate level is self-damped by negative feedback from its own concentration (i.e., a_{PP} is negative). In addition, the green algae provide vitamins to the blue-green algae, which in turn produce a substance with a toxic effect on the green algae (i.e., a_{BG} positive, a_{GB} negative). F_1, given by the sole self-loop ($F_1 = -a_{PP}$), is negative. Three loops of length 2, all negative, contribute to F_2, which is thus also negative $[F_2 = (a_{BP})(-a_{PB}) + (a_{BG})(-a_{GB}) + (a_{GP})(-a_{PG})]$. F_3, consisting of loops or products of disjunct loops involving three vertices, comprises three terms: the loop PBG clockwise, the same loop counter-clockwise, and the self-loop at P times the

disjunct loop of length 2 at BG. That is,

$$
\begin{aligned}
F_3 &= (PBG) + (PGB) - (P)(BG) \\
&= (a_{BP})(-a_{GB})(-a_{PG}) \\
&\quad + (a_{GP})(a_{BG})(-a_{PB}) \\
&\quad - (-a_{PP})(a_{BG})(-a_{GB}) \\
&= a_{BP} a_{GB} a_{PG} \\
&\quad - a_{GP} a_{BG} a_{PB} - a_{PP} a_{BG} a_{GB} \quad (36)
\end{aligned}
$$

The first term of eq. 36 is positive and the second two are negative, giving an ambiguous result. If the toxic effect of the blue-greens is much weaker than the stimulation of the blue-greens by vitamins from the greens (i.e., $a_{BG} \gg a_{GB}$), then the second two terms of eq. 36 exceed the first term, and F_3 as well as F_1 and F_2 is negative. Expressed alternatively, the net effects of the loops of length 3 ($a_{BP} a_{GB} a_{PG} - a_{GP} a_{BG} a_{PB}$), as well as the effect of the product of the disjunct self-loop and length-2 loop ($-a_{PP} a_{BG} a_{GB}$), on F_3 are negative. But, although $F_3 < 0$ is one condition for stability, $F_1 F_2 + F_3 > 0$ is another condition for stability, so that F_3 must be negative but not too negative. The condition $F_1 F_2 + F_3 > 0$ is more likely to be satisfied if the net effect of the length-3 loops is positive and F_3 is not too negative, hence the condition $F_1 F_2 + F_3 > 0$ may be violated if the stimulation of the blue-greens by the greens is strong enough. Thus, the ambiguity remains, and it indicates what measurements must be made on the system to resolve the question of its stability. Clearly some of the subtle effects of nutrient and trace-metal levels on algal competition described by Patrick in Chapter 15 would be amenable to this type of analysis.

Mendelian Selection

When a species is at equilibrium population size, eq. 1 is zero:

$$\frac{dX_i}{dt} = f_i(X_1, X_2, \ldots, X_n; C_1, C_2, \ldots)$$
$$= 0$$

The parameters C_h are of several kinds. Some are environmental parameters beyond the influence of species $X_1, \ldots, X_n$, such as daily maximum temperature or vegetation density. Some depend on the genotype of species i, on such heritable traits as its fecundity or heat tolerance. Still others are dependent on more than one species. For instance, the rate of predation of X_j on X_i depends on the biology of both. Finally, some of the C_h may depend on species in whose equations they do not appear, as they might represent an environmental effect on one species that is mediated by another.

If a genetic variant arises which alters a parameter C_h, it can of course affect many of the species in the system, both directly and indirectly. But it will be selected only if

$$\frac{\partial f_i}{\partial C_h} > 0 \qquad (37)$$

Therefore the action and direction of mendelian selection is determined by the function f_i for the species in question. But the consequences of that selection depend on the whole ensemble of species and their interrelations, since f_i for one species may depend on all the X_i. We will examine two aspects of this dependence: the effect of selection on the stability of the system, and on the abundance of the species at equilibrium.

Evolution of Local Dynamics

We can now look at the effects of evolution within the component species on the behavior of the whole system. First, consider the several model systems or communities illustrated in Figure 4.

Figure 4a illustrates a system of blue-green algae and green algae similar to that discussed previously in connection with Figure 3c, but with the addition of a second nutrient (nitrate, abbreviated N) secreted by the blue-green algae (a_{NB} positive, a_{BN} zero) and utilized by the green algae (a_{GN} positive, a_{NG} negative), with a herbivore (H) eating green algae (a_{HG} positive, a_{GH} negative), and without the vitamin provided to the blue-green algae by the green algae. $F_1(= -a_{NN} -a_{PP}$, from eq. 8) is negative. $F_2[= (a_{GN})(-a_{NG}) + (a_{BP})(-a_{PB}) + (a_{GP})(-a_{PG}) + (a_{HG})(-a_{GH}) - (-a_{PP})(-a_{NN})$, from eq. 9] is also negative. F_3 (from eq. 10) has five negative terms from self-loops times disjunct loops of length 2; in vertex notation, these are (N)(BP), (N)(GH), (N)(GP), (P)(NG), and (P)(GH). However, the loop of length 3, PBG, contributes a positive term to F_3:

$$(-1)^{m+1}(-a_{PG})(-a_{GB})(a_{BP})$$

where m is 1. The same positive loop makes it possible for F_4 to be positive, since in F_4 this loop will be multiplied (cf. eq. 7) by the negative self-damping of N (i.e., by $-a_{NN}$) and since the product has $(-1)^{m+1} = (-1)^3 = -1$ in front of it. Therefore, selection within the green algae

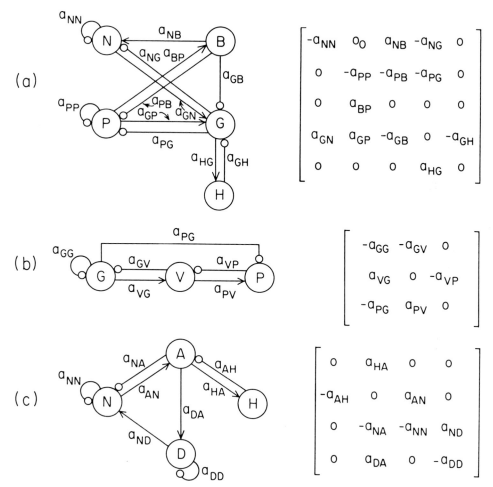

Figure 4 These systems illustrate the various effects that evolution within a component variable of a community may have on community. Figure 4a (above) is an expansion of the blue-green (B) and green (G) algae system of Figure 3c; a second nutrient, nitrate (N), has been added as well as a herbivore (H) on the green algae. This system is stabilized by evolution within the green algae for toxin resistance. Figure 4b (center) shows voles V that eat self-damped grass (G). P is predator intensity. Lush grass growth reduces feeding time and hence predation (a_{PG} negative). Selection for increased predator immunity in the voles increases stability, whereas selection in the predator for increased searching ability destabilizes. Figure 4c (below) is again a modification of the algal system, where algae A die and pass through a detritus stage D, which recycles nutrient N. A herbivore H crops the green algae. Here selection in A for predator resistance destabilizes, whereas selection for greater resistance to nonpredator mortality stabilizes the system.

for resistance to the toxin released by the blue-green algae will stabilize the system by weakening the link BG which contributes a positive term to F_3 and F_4 via the loop PBG. However, selection within the green algae for decreased susceptibility to predation by the herbivore will destabilize the system by weakening the loop GH which contributes a negative term to F_3. Thus, evolution within the green algae may either stabilize or destabilize the community of which they are a part, depending on whether selection is for resistance to toxin or resistance to the herbivore.

Figure 4b depicts a system in which voles, V, eat grass, G (a_{VG} positive, a_{GV} negative). The third variable P is not the level of the predator population but the intensity of predation, which depends not only on the number of predators but also on the amount of time a vole spends feeding. The more grass, the less time the vole must spend feeding and the more time it can spend under cover, safe from the predator. Therefore, the level of G reduces P (a_{PG} negative). The growth of grass is self-damped (a_{GG} negative). From eqs. 8–10,

$$F_1(= -a_{GG}),$$
$$F_2[= (a_{VG})(-a_{GV}) + (a_{PV})(-a_{PV})],$$

and

$$F_3[= -(-a_{GG})(a_{PV})(-a_{VP})$$
$$+ (-a_{VP})(-a_{GV})(-a_{PG})]$$

are all negative. If there is instability, it will be of the oscillatory kind and will

result from the violation of the condition of eq. 27 or 31. Both Hutchinson (Chapter 17) and Leigh (Chapter 2) discuss such cycles in predator-prey organisms. This condition would be violated if, from eq. 31,

$$(-a_{GG})(-a_{GV})(a_{VG}) + \atop (-a_{PG})(-a_{VP})(-a_{GV}) < 0 \qquad (38)$$

since the system of Figure 4b has only one self-loop and thus has no terms corresponding to the first term of eq. 31. Factoring $(-a_{GV})$ from the left side of eq. 38 yields

$$(-a_{GG})(a_{VG}) + (-a_{PG})(-a_{VP}) > 0 \quad (39)$$

as a condition for instability. The first term is negative, the second positive. Selection within the vole population to escape predation will reduce a_{VP} and the second term, hence will make the inequality eq. 39 less likely to hold and will promote stability. Conversely, selection in the predator for increased predation effectiveness will increase a_{VP} and the second term of eq. 39 and thereby destabilize the system. Analysis of this sort could well point out which factors to examine in order to explain, for example, the local absence of oscillation in hare populations in central Norway (Hutchinson, Chapter 17).

The community of Figure 4c includes algae A which use a nutrient N (a_{AN} positive, a_{NA} negative), and die to join the detritus D (a_{DA} positive, a_{AD} zero), which then breaks down to restore nutrient to the system (a_{ND} positive, a_{DN} zero). Detritus and nutrient are self-damped resources

(a_{DD} and a_{NN} negative). From eqs. 8 and 9,

$$F_1 \quad (= -a_{NN} - a_{DD})$$

and

$$F_2 \quad [= (a_{HA})(-a_{AH}) + (a_{AN})(-a_{NA}) - (-a_{NN})(-a_{DD})]$$

are both negative. But, from eq. 10, F_3 includes a positive loop of length 3 $[= (a_{DA})(a_{ND})(a_{AN})]$ as well as three terms of negative feedback from the product of self-loops times disjunct length-2 loops

$$[-(a_{NN})(a_{HA})(-a_{AH}) \\ - (-a_{DD})(a_{HA})(-a_{AH}) \\ - (-a_{DD})(a_{AN})(-a_{NA})]$$

An increased resistance to predation on the part of the algae, by reducing a_{HA} and a_{AH}, reduces the first two of the negative feedback terms and may destabilize the system. In contrast, greater resistance in the algae to mortality from causes other than the herbivore reduces a_{DA}, representing the conversion of algae to detritus, and thereby reduces positive feedback in F_3 and stabilizes the system.

Thus, as shown by Figures 4a–4c, selection within a single system may either stabilize or destabilize the community as a whole, and often affects stability in ways that are not intuitively obvious. Chapter 16, by Connell, reviews in detail the importance of the predator-prey interaction, including grazers and marine algae, and the sensitivity of community stability to these links.

A second aspect of local dynamics or community stability is oscillation. From the characteristic equation $|\mathbf{A} - \lambda \mathbf{I}| = 0$, we can find the variance of the characteristic roots λ of a system:

$$\mathrm{var}(\lambda) = \mathrm{var}(a_{ii}) + (n-1)\overline{a_{ij}a_{ji}} \quad (40)$$

where $\overline{a_{ij}a_{ji}}$ is the average value of loops of length 2. If these loops are negative, as they are in predator-prey interactions, then an increase in these loops will eventually produce a negative variance of the roots λ, indicating complex roots and hence oscillation. (The oscillations may be damped or else of increasing amplitude.) Now consider a two-level trophic structure in which predators exploit one or more species. If genetic variants arise in the predator populations which allow them to use additional prey species without decreasing predation on the prey species already exploited, $a_{ij}a_{ji}$ increases and the community will eventually oscillate.

A more general treatment proceeds by examining the rate at which roots change with respect to change in the feedback level, $d\lambda/dF_k$ (see eq. 27). Differentiating the characteristic equation in the form eq. 26 with respect to F_k gives

$$\frac{\partial P(\lambda)}{\partial F_k} = \lambda^{n-k} = \frac{\partial P(\lambda)}{\partial \lambda} \cdot \frac{\partial \lambda}{\partial F_k} \quad (41)$$

which yields

$$\frac{\partial \lambda}{\partial F_k} = \frac{\lambda^{n-k}}{\partial P/\partial \lambda} \quad (42)$$

The denominator $\partial P/\partial \lambda$ will be of different sign at different roots, and is positive at the largest root, negative at the next largest, and so on. If λ is very small, then small values of k give high exponents of λ in the numerator of eq. 42, and therefore the derivative will be small. Thus, for small λ, $\partial\lambda/\partial F_k$ is most sensitive at large k, and as $\lambda \to 0$, only F_n affects the derivative. If a given short loop does not enter into F_n, it has negligible effect on the small λ. On the other hand, for large roots λ, small values of k give large λ^{n-k}. Hence the largest λ are affected mostly by the short loops of the system; this is intuitively reasonable, since these are the loops which, when they are strong (as they would be when species are strongly self-damped or self-accelerating, or when pairs of species are associated in closely interlocked predator-prey or competitive loops), provide the large values of λ.

Evolution of Equilibrium Numbers

In this section we examine the effect of selection on the abundances of species at equilibrium. That is, we are concerned with the following question: is the driving force of selection within a species generally directed so as to make the species more abundant (as is often assumed intuitively), or can selection leave equilibrium numbers unaffected or even operate so as to decrease abundance? We first present the general equations by which loop analysis treats this problem, and we illustrate these equations by the example of Figure 5. We then apply the analysis to three biological examples illustrated in Figure 6.

If C_h is a parameter under genetic control of species i (e.g., its bill length, rate of larval development, ability to escape predators, etc.), then C_h will increase under selection when it enters the growth equation (eq. 1) for species i with a positive coefficient—that is, when $\partial f_i/\partial C_h > 0$. To determine the effect of C_h on the equilibrium levels of all the species in the community, we must differentiate the system of expressions on the right of eq. 1 (one such expression for each species X_i),

$$f_i(X_1, X_2, X_3, \ldots, X_n; \\ C_1, C_2, C_3, \ldots) \quad (43)$$

with respect to C_h, and set the derivatives equal to zero:

$$\sum\left(\frac{\partial f_i}{\partial X_j}\right)\left(\frac{\partial X_j}{\partial C_h}\right) + \left(\frac{\partial f_i}{\partial C_h}\right) = 0 \quad (44)$$

The $\partial X_j/\partial C_h$ are the unknowns in these equations. Since the coefficients of X_j in eq. 43, $\partial f_i/\partial X_j$, are the familiar a_{ij} of the community matrix [A] (eq. 2), the complete set of equations exemplified by eq. 44 is represented in matrix form as

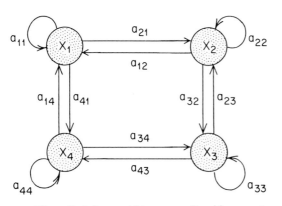

Figure 5 A four-variable community with symmetrical links. See text accompanying eqs. 48–50 for detailed discussion of how to analyse the course of evolution in this community.

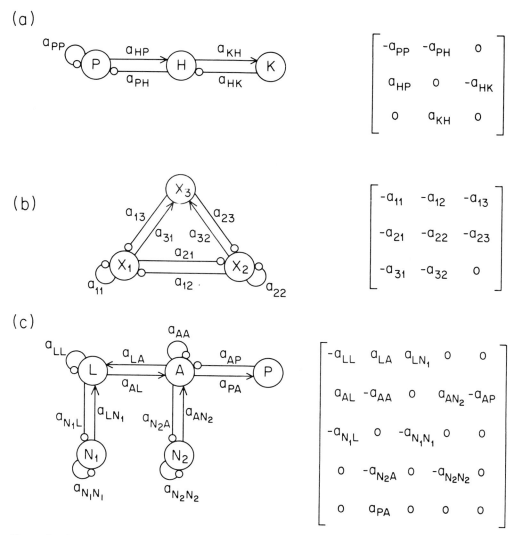

Figure 6 These examples show systems in which evolution in one variable for a particular trait may either increase or decrease the level of the variable on which selection operates, and may either increase or decrease the levels of other variables within the community. Figure 6a (above) exhibits a carnivore K eating a herbivore H eating plants P; Figure 6b (middle), two competitors X_1 and X_2 stabilized by a "keystone predator" X_3; and Figure 6c (below), a species in which the larval (L) and adult (A) stages utilize nutrients N_1 and N_2, respectively, and only the adults are exposed to predators (P). These examples are discussed at length in the text.

$$
\begin{bmatrix} a_{11} & a_{12} & a_{13} & \cdots & a_{1n} \\ a_{21} & a_{22} & \cdot & \cdots & \\ a_{31} & & & & \\ \vdots & \vdots & & & \\ a_{n1} & \cdot & \cdot & \cdots & a_{nn} \end{bmatrix} \begin{bmatrix} \partial X_1/\partial C_h \\ \partial X_2/\partial C_h \\ \vdots \\ \partial X_n/\partial C_h \end{bmatrix}
$$

$$
= \begin{bmatrix} -\partial f_1/\partial C_h \\ -\partial f_2/\partial C_h \\ \vdots \\ -\partial f_n/\partial C_n \end{bmatrix} \quad (45)
$$

The left side of eq. 45 is the product of a matrix and a vector, and the right side is a vector. The solution to eq. 45 for any particular species j gives us $\partial X_j/\partial C_h$, the effect of changing C_h on the growth rate of species j. It may be shown by matrix algebra that the solution is obtained by substituting the vector $[-\partial f_i/\partial C_h]$, which is the right side of eq. 45, for the jth column of the determinant of the community matrix $[\mathbf{A}]$, and dividing this modified determinant by the unmodified determinant of the matrix $[\mathbf{A}]$. Thus, for species j one has

$$
\frac{\partial X_j}{\partial C_h}
$$

$$
= \frac{\begin{vmatrix} a_{11} & a_{12} & \cdots & -\partial f_1/\partial C_h & \cdots & a_{1n} \\ a_{21} & a_{22} & \cdots & -\partial f_2/\partial C_h & \cdots & a_{2n} \\ \vdots & \vdots & & \vdots & & \vdots \\ a_{n1} & a_{n2} & \cdots & -\partial f_n/\partial C_h & \cdots & a_{nn} \end{vmatrix}}{\begin{vmatrix} a_{11} & a_{12} & \cdots & a_{1j} & \cdots & a_{1n} \\ a_{21} & a_{22} & \cdots & a_{2j} & \cdots & a_{2n} \\ \vdots & & & & & \vdots \\ a_{n1} & a_{n2} & \cdots & a_{nj} & \cdots & a_{nn} \end{vmatrix}} \quad (46)
$$

The denominator of eq. 46 can be immediately expressed in terms of diagrams exemplified by Figures 1–4, since from eq. 6

$$
\begin{vmatrix} a_{11} & a_{12} & \cdots & a_{1n} \\ a_{21} & a_{22} & \cdots & a_{2n} \\ \vdots & \vdots & & \\ a_{n1} & a_{n2} & \cdots & a_{nn} \end{vmatrix}
$$

$$
= \sum (-1)^{n-m} L(m,n) \quad (47)
$$

The numerator of eq. 46 is most easily interpreted in terms of feedback by the use of another example. The substitution of the terms $-\partial f_i/\partial C_h$ for the jth column a_{ij} in the determinant of $|\mathbf{A}|$ has the effect of breaking all the closed loops that had a_{ij} as a link. In the four-variable symmetrical system represented in Figure 5, for instance, the effect of C_h on X_2 is given by

$$
\frac{\partial X_2}{\partial C_h} = \frac{\begin{vmatrix} a_{11} & -\partial f_1/\partial C_h & a_{13} & a_{14} \\ a_{21} & -\partial f_2/\partial C_h & a_{23} & a_{24} \\ a_{31} & -\partial f_3/\partial C_h & a_{33} & a_{34} \\ a_{41} & -\partial f_4/\partial C_h & a_{43} & a_{44} \end{vmatrix}}{\begin{vmatrix} a_{11} & a_{12} & a_{13} & a_{14} \\ a_{21} & a_{22} & a_{23} & a_{24} \\ a_{31} & a_{32} & a_{33} & a_{34} \\ a_{41} & a_{42} & a_{43} & a_{44} \end{vmatrix}} \quad (48)
$$

The numerator of eq. 48 expands to

$$
\begin{aligned}
&- (-\partial f_1/\partial C_h)[a_{21}(a_{33}a_{44} - a_{43}a_{34}) \\
&- a_{23}(a_{31}a_{44} - a_{34}a_{41}) \\
&+ a_{24}(a_{31}a_{43} - a_{41}a_{33})] \\
&+ (-\partial f_2/\partial C_h)[a_{11}(a_{33}a_{44} - a_{34}a_{43}) \\
&- a_{13}(a_{31}a_{44} - a_{34}a_{41}) \\
&+ a_{14}(a_{31}a_{43} - a_{33}a_{41})] \\
&- (-\partial f_3/\partial C_h)[a_{11}(a_{23}a_{44} - a_{24}a_{43}) \\
&- a_{13}(a_{21}a_{44} - a_{24}a_{41}) \\
&+ a_{14}(a_{21}a_{43} - a_{23}a_{41})]
\end{aligned}
$$

$$+ (-\partial f_4/\partial C_h)[a_{11}(a_{23}a_{34} - a_{24}a_{33})$$
$$- a_{13}(a_{21}a_{34} - a_{24}a_{31})$$
$$+ a_{14}(a_{21}a_{33} - a_{23}a_{31})] \qquad (49)$$

Notice that the coefficient of the first term of eq. 49, $\partial f_1/\partial C_h$, has three components. The first of these components is $a_{21}(a_{33}a_{44} - a_{43}a_{34})$. This is the product of a_{21}, the remaining link on the now-open direct loop between X_1 and X_2 that was broken by replacement of a_{12}, times the sum of products of disjunct loops that involve all the variables except X_1 and X_2, that is, $(a_{33})(a_{44})$ and $(a_{34})(a_{43})$ involving X_3 and X_4. Now take the second component, which is $a_{23}(a_{31}a_{44} - a_{34}a_{41})$. The first term of this second component is $a_{31}a_{23}a_{44}$, which is the product of $a_{31}a_{23}$, the product of links on a different path via X_3 along the now-open loop from X_1 to X_2 (i.e., $a_{31}a_{23}$ is analogous to a_{21} in the first component), times a_{44}, which is the only loop involving X_4, the sole variable not on this path from X_1 to X_2 (i.e., a_{44} is analogous to $a_{33}a_{44} - a_{34}a_{43}$ in the first component). The second term of the second component is $a_{23}a_{41}a_{34}$, which is the product of three links via X_4 and X_3 along the now-open loop from X_1 to X_2 (analogous to a_{21} or $a_{31}a_{23}$). Since this path involves all four variables, there are no loops from variables not on the path, analogous to $a_{33}a_{44} - a_{43}a_{34}$ or a_{44}. Similarly, the third component, $a_{24}(a_{31}a_{43} - a_{41}a_{33})$, follows two more paths along the now-open loop from X_1 to X_2: via X_3 and X_4, yielding $a_{31}a_{43}a_{24}$ (there are no variables not on this path, hence no disjunct loops); and via X_4, yielding $a_{41}a_{24}$, multiplied by a_{33}, the only

loop involving X_3, the sole variable not on this path.

Thus, we see that the coefficient of $\partial f_1/\partial C_h$ in eq. 49 is the sum of all possible products of open paths between X_1 and X_2, times all the closed loops sharing no variables with the open path considered. We symbolize the open paths from variable i to j as P_{ji}, and in particular, the open paths composed of k links as $P_{ji}^{(k)}$. There is just one path of one link $P_{21}^{(1)}$, two paths of two links $P_{21}^{(2)}$ (via X_3 or X_4), and two paths of three links $P_{21}^{(3)}$, via X_3 then X_4 or via X_4 then X_3. In the coefficient of $\partial f_1/\partial C_h$ in eq. 49, $P_{21}^{(1)}$ is multiplied by the sum of two disjunct loops, each of the $P_{21}^{(2)}$ is multiplied by one disjunct loop, and each of the $P_{21}^{(3)}$ is multiplied by unity since there is no disjunct loop. Thus, the coefficient of $\partial f_1/\partial C_h$ can be represented as

$$\sum_{k,m} P_{21}^{(k)} L(m - 1, 3 - k) \qquad (50)$$

where $L(m - 1, 3 - k)$ means the products of disjunct loops that can involve up to $(3 - k)$ variables (2 at most) and $(m - 1)$ links (2 at most). Similarly, the reader can verify that the coefficients of $\partial f_3/C_h$ or $\partial f_4/C_h$ in eq. 49 represent sums of products of open paths from X_3 or X_4, respectively, to X_2, times the disjunct loops involving variables not on the path. The reader can also verify that the coefficient of $\partial f_2/\partial C_h$ comprises all the possible products of disjunct loops involving X_1, X_3, and X_4, but no open paths, since there can be no open path to X_2 from itself. Thus, by defining $P_{22}^{(k)}$ and in general $P_{jj}^{(1)}$ as unity, we can compactly express eq. 46, the expansion of whose numerator was exemplified by eq. 49, as

$$-\frac{\partial X_j}{\partial C_h} = \qquad (51)$$

$$\frac{\sum (-1)^{n-m}\left(\dfrac{\partial f_i}{\partial C_h}\right)P_{ji}^{(k)}L(m-1, n-1-k)}{\sum_m (-1)^{n-m}L(m, n)}$$

The summation in the numerator of eq. 51 is taken over all m, k, and i for a particular n ($n = 4$ in the case worked out above), and the summation in the denominator is taken over all m for a particular n. We take $L(m - 1, 0)$ to be unity, as will be the case when $P_{ji}^{(k)}$ is the path from i to j involving all variables in the system (i.e., when $n = k$). Thus, the numerator of eq. 51 contains: $-\partial f_i/\partial C_h$, whose substitution for a_{ij} broke the loops between i and j; $P_{ji}^{(k)}$, the open paths from i to j remaining after the originally closed loops were thus broken; and $L(m - 1, n - k)$, the closed loops disjunct from (containing no vertices in common with) these open paths and not broken by this substitution.

Equation 51 can now be reexpressed in terms of feedback. Note first that by eq. 6 the denominator of eq. 51 is simply the feedback of the community at level n, since F_n equals D_n [with the sign adjusted by $(-1)^{n-1}$]. To interpret the numerator, we define the *complementary subsystem* to the open path $P_{ji}^{(k)}$ as the subsystem including all vertices not on the path $P_{ji}^{(k)}$. The complementary subsystem must therefore contain $(n - k)$ vertices, and we refer to its feedback as $F_{n-k}(\text{comp})$. The numerator of eq. 51 then becomes

$\sum_{i,k} (\partial f_i/\partial C_h)P_{ji}^{(k)}F_{n-k}(\text{comp})$, and eq. 51 becomes

$$\frac{\partial X_j}{\partial C_h} = \frac{\displaystyle\sum_{i,k} (\partial f_i/\partial C_h) P_{ji}^{(k)}F_{n-k}(\text{comp})}{F_n} \qquad (52)$$

Suppose that the parameter C_h refers to a property of just one of the species or variables (e.g., its variability or fecundity), and not to a property of an interaction between species, such as predation or competition. Then C_h appears only in the equation dX_h/dt for that species X_h, and not for other species. (We postpone until later a discussion of parameters that depend on the biology of, and enter into the equations of, more than one species.) Let C_h be such a parameter for X_h. Then, since $P_{hh}^{(1)} = 1$ and since $\partial f_h/\partial C_h$ is by definition positive (C_h is selected for, not against), $\partial X_h/\partial C_h$ must have the same sign as $F_{n-1}(\text{comp})/F_n$ (from eq. 52). But if the system as a whole is stable, F_n must be negative. Therefore, if the complementary subsystem to X_h has negative feedback $F_{n-1}(\text{comp})$, selection increasing C_h will also increase the level of X_h. If the complement has zero feedback, then selection for increasing C_h will not affect the equilibrium level of X_h. If the complement has positive feedback $F_{n-1}(\text{comp})$, we obtain the surprising result that selection for the parameter C_h, some aspect of the biology of X_h, will result in a decrease in its abundance. (An instance of this last outcome is discussed in detail by Diamond in Chapter 14: high-S bird species evolve overexploitation strategies to starve out competing supertramps, but in so doing reduce their own abundance.) Figure 16 shows some simple systems that illustrate these points.

In Figure 6a we have a three-species hierarchy of plants P, a herbivore H, and a carnivore K. Evolution for increased overall viability or fecundity can occur in any of the three species. Since $n = 3$, we must examine $F_{n-1}(\text{comp}) = F_2(\text{comp})$ to determine what effect selection in any species will have on its abundance and that of other species. The complementary subsystem for the plant is (H, K), for which $F_2 = (a_{KH})(-a_{HK})$ by eq. 9. Similarly, the complementary subsystem for the carnivore is (P, H), for which $F_2 = (a_{HP})(-a_{PH})$ by eq. 9. Since these complementary feedbacks are negative, evolution in the plant (e.g., for increased photosynthetic efficiency) will increase its abundance, and evolution in the carnivore (e.g., for resistance to disease) will increase its abundance, by the reasoning of the preceding paragraph. But the complementary subsystem for the herbivore, (P, K), has $F_2 = 0$ from eq. 9, since $a_{PK} = a_{KP} = a_{KK} = 0$. Therefore evolution in the herbivore (e.g., for increased efficiency of digestion) does not increase its own population level, and does not affect the plant population, but instead increases the abundance of the carnivore.

We verify this conclusion as follows. The determinant of the whole community, which equals F_n, is just $-(-a_{PP})(-a_{HK})(a_{KH}) = -a_{PP}a_{HK}a_{KH}$, the sole nonzero term in the expansion of the community matrix at the right of Figure 6a. If selection occurs for some trait C_H that directly affects only the biology

of the herbivore, then $\partial f_H/\partial C_H$ is positive and $\partial f_K/\partial C_H = 0 = \partial f_P/\partial C_H$. Substituting the vector $[-\partial f_i/\partial C_H]$ for the second column (the column pertaining to H) of the community matrix (Figure 6a), and dividing by F_n, give eq. 46 for $\partial H/\partial C_H$:

$$\frac{\partial H}{\partial C_H} = \frac{\begin{vmatrix} -a_{PP} & 0 & 0 \\ a_{HP} & -\partial f_H/\partial C_H & -a_{HK} \\ 0 & 0 & 0 \end{vmatrix}}{-a_{PP}a_{HK}a_{KH}} \quad (53)$$

That is, the abundance of the herbivore is unaffected. Similarly, calculating $\partial P/\partial C_H$, by substituting the vector $[-\partial f_i/\partial C_H]$ into the first column of the community matrix and dividing by F_n, shows the plant to be unaffected. However, the same substitution into column three, the carnivore's column, gives

$$\begin{aligned} \frac{\partial K}{\partial C_H} &= \frac{\begin{vmatrix} -a_{PP} & -a_{PH} & 0 \\ a_{HP} & 0 & -\partial f_H/\partial C_H \\ 0 & a_{KH} & 0 \end{vmatrix}}{-a_{PP}a_{HK}a_{KH}} \\ &= \frac{-a_{PP}a_{KH}(\partial f_H/\partial C_H)}{-a_{PP}a_{HK}a_{KH}} \\ &= \frac{\partial f_H/\partial C_H}{a_{HK}} \quad (54) \end{aligned}$$

Thus, $\partial K/\partial C_H$ is positive, and the carnivore population increases. Evolution in the herbivore goes only to feed the carnivore.

We can similarly verify the consequences of evolution in the plant of the community illustrated in Figure 6a. Evolution in the plant means $\partial f_P/\partial C_P$ positive, $\partial f_H/\partial C_P = 0 = \partial f_K/\partial C_K$. Substituting the new vector, $[-\partial f_i/\partial C_P]$, into the first column (the plant's column) of the matrix of Figure 6a and evaluating eq. 46 for $\partial P/\partial C_P$ yield $(\partial f_P/\partial C_P)/a_{PP} = \partial P/\partial C_P$; since this is positive, the plants increase. Substituting the vector $[-\partial f_i/\partial C_P]$ into the third column of the matrix of Figure 6a and evaluating the effect on the carnivore's abundance from eq. 46 yield

$$\frac{\partial K}{\partial C_P} = \frac{\begin{vmatrix} -a_{PP} & -a_{PH} & -\partial f_P/\partial C_P \\ a_{HP} & 0 & 0 \\ 0 & a_{KH} & 0 \end{vmatrix}}{-a_{PP}a_{HK}a_{KH}}$$

$$= \frac{(-\partial f_P/\partial C_P)a_{HP}a_{KH}}{-a_{PP}a_{HK}a_{KH}}$$

$$= \frac{(\partial f_P/\partial C_P)a_{HP}}{a_{PP}a_{HK}} \tag{55}$$

Since this is positive, the carnivore increases. A similar substitution into column 2 of the matrix shows $\partial H/\partial C_P$ to be zero. Thus, evolution in the plant increases the abundance of the plant and leaves the abundance of the herbivore unchanged; the herbivore has more food but suffers heavier predation and merely has a higher turnover rate, increasing the abundance of the carnivore. The reader can verify by a similar analysis that evolution in the carnivore $(\partial f_K/\partial C_K > 0, \partial f_H/\partial C_K = 0 = \partial f_P/\partial C_K)$ affects all three levels: the carnivores increase

$(\partial K/\partial C_K$
$= (a_{PH}a_{HP}(\partial f_K/\partial C_K)/a_{PP}a_{HK}a_{KH}),$

the herbivores decrease $(\partial H/\partial C_K = -(\partial f_K/\partial C_K)/a_{KH})$, and the plants increase $(\partial P/\partial C_K = a_{PH}(\partial f_K/\partial C_K)/a_{PP}a_{KH})$.

Figure 6b shows a pair of competing prey species X_1 and X_2, plus a predator X_3. As discussed in connection with Figure 2c, in the absence of X_3, F_2 for the system (X_1, X_2) is $a_{12}a_{21} - a_{11}a_{22}$ and may be positive unless self-damping (a_{11} and a_{22}) is very strong in both species. Thus, the competing prey species are unlikely to coexist in the absence of the predator. The presence of the predator adds two more negative terms to F_2: $(a_3a_1)(-a_1a_3) + (a_3a_2)(-a_2a_3)$. Since the predator may thereby stabilize the community, it is termed a "keystone predator." Chapter 16, by Connell, discusses numerous examples of keystone predators that permit the coexistence of several competing species and in whose absence the community is unstable and becomes dominated by one of the competitors. If F_2 for the system (X_1, X_2) is in fact positive, then X_3 would have a complementary subsystem (X_1, X_2) with positive feedback. The reasoning given earlier therefore means that evolution within X_3 leads to a decrease in its own abundance: it can select itself to extinction. In more detail, suppose that selection for some trait C_3 of the predator increases its viability or fecundity. Then $\partial f_3/\partial C_3 > 0$, $\partial f_2/\partial C_3 = 0 = \partial f_1/\partial C_1$. When one expands the community matrix on the right of Figure 6b as the denominator of eq. 46, and replaces the third column by the vector $-\partial f_i/\partial C_3$ to obtain the numerator of eq. 46, eq. 46 then becomes for the predator:

$$\frac{\partial X_3}{\partial C_3} = \frac{\begin{vmatrix} -a_{11} & -a_{12} & 0 \\ -a_{21} & -a_{22} & 0 \\ a_{31} & a_{32} & -\partial f_3/\partial C_3 \end{vmatrix}}{(-a_{11}a_{23}a_{32} + a_{12}a_{23}a_{31} + a_{13}a_{21}a_{32} - a_{13}a_{22}a_{31})}$$

$$= \frac{(\partial f_3/\partial C_3)(a_{12}a_{21} - a_{11}a_{22})}{(-a_{11}a_{23}a_{32} + a_{12}a_{23}a_{31} + a_{13}a_{21}a_{32} - a_{13}a_{22}a_{31})} \quad (56)$$

For the community to be stable at all, the denominator, which is F_3, must be negative. X_3 is by definition a keystone predator if the coefficient of $\partial f_3/\partial C_3$ in the numerator, which is F_2 for the system (X_1, X_2), is positive. Thus, $\partial X_3/\partial C_3$ is negative for a keystone predator: selection on it reduces its abundance. In the case of the prey species, however, F_2 of the complement is negative, and selection increases the abundance of the selected species and reduces the abundance of its competitor. For instance, the reader may use eq. 46 and the community matrix of Figure 6b to verify that selection within $X_1(\partial f_1/\partial C_1 > 0, \ \partial f_2/\partial C_1 = 0 = \partial f_3/\partial C_1)$ yields $\partial X_1/\partial C_1 = -a_{23}a_{32}(\partial f_1/\partial C_1)/F_3$, $\partial X_2/\partial C_1 = a_{23}a_{31}(\partial f_1/\partial C_1)/F_3, \partial X_3/\partial C_1 = (a_{21}a_{32} - a_{22}a_{31})(\partial f_1/\partial C_1)/F_3$, abbreviating the community matrix as F_3. Since F_3 must be negative, $\partial X_1/\partial C_1$ must be positive and $\partial X_2/\partial C_1$ negative for selection in X_1, whereas X_3 can either decrease or increase, depending on the relative magnitudes of the coefficients a_{21}, a_{32}, a_{22}, and a_{31}. For example, if X_3 preys more on X_1 than on X_2 (i.e., $a_{31} > a_{32}$), then $\partial X_3/\partial C_1$ is likely to be positive, and selection for

increased viability of X_1 will tend to make X_3 more abundant.

In Figure 6c we subdivide a population into two age classes, labeled larvae (L) and adults (A). Both stages require nutrients which here are treated as distinct; only the adult stage suffers from predation (P). Genetic changes that affect only the adult do not change the equilibrium levels of any variable except the predator (only the AP path has nonzero level-3 feedback in its complement), which is increased by improved viability in the adult prey. But an increase in the viability of the larvae will increase the larval population and the predator, and result in a greater turnover in the adult population. That is, the adult population will be younger. Since L and A have symmetrical interactions, an analogous set of conclusions would apply if it were the larval stage that is vulnerable to the predator. As discussed by Connell (Chapter 16), a species may be exposed to a sequence of different predators and different intensities of predation at different stages in its life history. In general, selection that increases the viability of the preyed-on stage will not increase the level of that stage, whereas selection in the predator-free stage increases the level of that stage, increases the predator population, and decreases the life expectancy but not the numbers of the other life history stages.

For the last example in this sequence, refer again to the nutrient-algae-herbivore system pictured in Figure 4a. In this system both green and blue-green algae require phosphate; the blue-greens secrete nitrate, which serves as a nutrient for the green algae, and they also secrete a toxin that inhibits the growth of the greens. We can contrast the effects of evolution with and without the presence of a herbivore that preys on green algae. Without the herbivore H, evolution within the blue-green algae increases the level of their population (because of negative feedback in the complementary subsystem). This evolution also results in an increase in the nitrate level, but in ambiguous effects on the green algal and phosphate levels of the system. Evolution in the green algae produces an increase in the level of the greens [because of net negative feedback in the complement (N)(P, B)], a decrease in blue-greens and nitrate, and no change in the phosphate level (which has zero feedback at level 3 in its complement due to the isolation of the blue-greens in the system). When the herbivore is present, the blue-greens still increase when selection increases their reproductive rate [F_4(comp) is negative, from (P)(N, G, H)]. Nitrate still increases in this case, but now the phosphate is reduced, there is no effect on the green algae, and the herbivore may either increase or decrease. Evolution in the green algae, which in the absence of herbivores affected all variables in the system except phosphate, now affects only the herbivores, which increase in abundance. Finally, the herbivore itself may evolve. This decreases the greens, increases the blue-greens and nitrate, has no effect on phosphate, and may either increase or decrease the abundance of the herbivore itself, because its complement has both positive and negative compo-

nents. The two negative components are (N)(P, B, G) and (B, N, G, P), and the single positive component is (B, P)(N, G). If the toxic effect of the blue-greens on the greens is strong enough, the (P, G, B) loop which is involved in a net negative loop may dominate the complementary subsystem and promote a herbivore increase. Examples of the effects of changes of nutrient input level on the equilibrium levels of variables in aquatic systems are given and discussed in Lane and Levins (in preparation).

So far, we have examined the consequences of changing parameters that depend directly on the biology of only one of the species in the system and enter the system only through the growth equation for that species. But some of the parameters of the system depend directly on the biology of two or more species, and appear in the equations for these two or more species. Let us consider examples of such parameters in Figures 6a, 6c, 4b, and 4c. A further example is derived in Chapter 14 (see Figure 50), by Diamond.

A predation rate depends on the genotypes of both predator and prey. An increase in predation rate provides a positive input to the predator's growth rate and a negative input to the prey's growth rate. Suppose that, in the system of three trophic levels illustrated in Figure 6a, a genotype arises that makes the herbivore H more resistant to predation. This genotype provides a positive input to the growth of H and a negative input to the growth of the carnivore K. The former input has no effect on the abundance of H, because the feedback of its complement is zero; the latter input does increase the abundance of H (refer to an earlier part of this section for the derivation of these conclusions). Therefore, even though selection for increased litter size or viability in H and selection for increased predator avoidance in H both act similarly in the growth equation for H, only the latter selection would increase the equilibrium population level of H, given the system in which H lives. For example, in many species the whole litter, whatever its size, normally succumbs to predation or the effects of a harsh environment, and only occasionally does an age cohort escape predators or environmental effects to survive as adults (Connell, Chapter 16).

In the model of discrete age classes of Figure 6c, let a gene arise that speeds up the rate of larval development. This increases the rate of removal of larvae from the population of L, and appears in its equation as a negative input. But the gene increases the rate of adult formation, and therefore is a positive input to A. The complement of L has negative feedback $(-1)^4 (N_1)(N_2)(A, P)$, so that the gene is favored and L is reduced. But the complement to A has zero feedback, and so the adult population is not increased. Instead, the predator population increases, and there is an increased turnover rate of adults but no increase in adult numbers. Therefore, evolution for faster development rate may reduce the net abundance of the species. Intuitively, this occurs because acceleration of development shifts the population from the predation-free larval stage to the predation-sensitive adult stage. Note that if the predator were

self-damped, A's complement would be negative, and an accelerated development rate would cause A to increase along with L. If predation by predators that are not self-limiting were in the larval stage, increased development rate would also increase the adult population.

Now suppose that adults and larvae compete for the same nutrient ($N_1 = N_2$). Here an increase in the rate of larval development still increases larval abundance and has no effect on the adult population. But now the effect of larval development on the predator is ambiguous. The open path (L, A, P) with its complement (N) is positive, while the open path (L, N, A, P) is negative. Therefore, although larvae turn into adults more rapidly, the adults will die faster because of diminished nutrient levels (some food has been preempted by larvae), and perhaps also because of increased predation. The (L, A, N) loops of length 3 make it possible for the system to show unstable oscillations.

For another example of a parameter that appears in two growth equations, consider again the system of grass G, voles V, and predation rate P in Figure 4b. Here P is equivalent to the size of the predator population times the intensity of predation. Grass reduces the predation rate by reducing the time during which the voles have to forage, and by making them more difficult for the predators to see. Voles are vulnerable to predation only while they are foraging, and only when the cover where they seek food is sparse. The (G, P) link is the only difference between this system and that of Figure 6a. This link allows an increase in grass growth to increase the vole population by way of the path (G, P, V); recall that in the system of Figure 6a the herbivore population was unchanged following evolution in the plant community.

Figure 4c is a modification of Figure 4a (minus the blue-green algae and nitrate), but differs in that dead algae are recycled to nutrients through detritus stages. This addition introduces the positive loop (N, A, D), which is one component of the complement of H, and may therefore make this complement positive. This is likely to happen if the death rate of the algae is high, in which case selection within the herbivore population will reduce the level of the herbivores.

So far, we have treated the parameters of the growth equations as if they were independent of each other. This need not be the case. For instance, in Figure 6a a genotype that results in increased seed set and therefore a positive input to the plant trophic level may involve a diversion of energy or other resources from leaf structures or substances that serve to repel herbivores. This increases predation on the plants by the herbivore and appears as a negative input to the plant population and a positive input to the herbivores. But the path from H to P has zero complement in Figure 6a and does not affect the equilibrium level of P. In this case, the genotype that is selected within P for increased seed set results in a positive input to P, and hence mendelian selection increases P. But if the carnivore K were absent, the reverse would be true, as input to P would have zero complement: if mendelian selection favors seed set over herbivore resistance, the plant population will decline since this effect would enter as a positive input to H. Evolution within the herbivore produces equally contrasting results. Selection for increased reproductive rate or increased feeding efficiency within H will not alter the level of H in the system, whereas selection in H for increased predator avoidance will increase the abundance of H. Thus, if feeding effectiveness prevails over predator avoidance in selection for energy allocation within H, H will decline.

Thus, we see that selection within a species for such biologically sensitive traits as effective resource utilization, demographic parameters, and predator avoidance has different effects on the abundance of that species, depending on the position of the species in the trophic structure. Although all of these traits may respond to mendelian selection, and evolution produces species that are more efficient, fecund, or crafty, species populations just below the top of the trophic structure will increase only by selecting for predator avoidance. There is also the peculiar result that although the herbivore may be the main cause of death of plants, the plant population level does not respond to changes in the herbivore. On the other hand, although a carnivore has no direct contact with the plants, changes in the carnivore do change the plant populations. This underlines the difference between those factors that directly affect survival of individuals in a population and those that determine the level of that population.

Group Selection

Local populations of a species do not survive indefinitely even if the system of differential equations modeling the community is ostensibly stable. Recent studies in dynamic biogeography emphasize the surprisingly high turnover rates of species on islands (Diamond, 1969, and Chapter 14; Heatwole and Levins, 1973; Levins, Pressick, and Heatwole, 1973; Simberloff and Wilson, 1970; etc.). Since most species occur in a patchy distribution even when the boundaries of the patches are not obvious, it is likely that the island biogeographical patterns are a fair representation of the distributions of organisms on continuous land areas as well. (A number of other chapters in this volume contribute extensively to this reasoning; see Diamond, Chapter 14, Figures 33–38, for examples of patchy distributions in birds; and Cody, Chapter 10, Rosenzweig, Chapter 5, and Brown, Chapter 13, for extrapolation of island patterns and processes to continental situations.)

The turnover rates of species depend on their whole environment, including the structure of the communities in which they live and on the genetic make-up of the local populations. Therefore, in principle, group selection could operate at a fairly high intensity, to perpetuate groups whose component species have attributes that complement each other so as to promote the persistence of the group through time (cf. the discussion of "permissible combinations" of species by Diamond in Chapter 14). Levins (1970) estimated that if the life expectancy of a population of a species is less than ten generations for a given locality, then the strength of group selection is commensurate with that of mendelian selection. And such life expectancies, for instance of birds on small islands, are not uncommonly encountered, according to recent studies. However, it seemed previously that the same genotypes favored by mendelian selection, such as those that increase viability, fecundity, development rate, or predator avoidance, would also be favored by group selection. Therefore it seemed unlikely that group selection, no matter how ubiquitous, would oppose mendelian selection.

The arguments of the previous section prove that no special or exotic kind of biology is necessary for mendelian selection within a species to result in population decrease within that same species. But a smaller local population is more likely to go extinct because of random events. MacArthur and Wilson (1967) suggested that time to extinction is an exponentially increasing function of population ceiling or carrying capacity (see also Chapter 2, by Leigh). Furthermore, smaller populations send out fewer migrants and colonists. Therefore, it can be plausibly (but not rigorously) claimed that group selection favors increased population size, and may frequently oppose mendelian selection. Furthermore, if mendelian selection reduces population size and increases the likelihood of extinction, this will intensify group selection. (We are ignoring, of course, the presumably more frequent but less interesting cases where mendelian selection increases population sizes.)

The keystone predator is a simple case in which mendelian selection within the species leads it to local extinction. This extinction leads in turn to the destabilization of the remaining community of incompatible prey species. Colonization will take place anew from populations and communities in which selection has not gone so far. Therefore we might expect a dynamic equilibrium between mendelian and group selection.

An evolutionary equilibrium under mendelian, group, or both kinds of selection may be of several kinds. If the trait undergoing mendelian selection has an optimal value, such as that of a bird bill size, which maximizes the weight of insects caught, then the selection stops when

$$\partial f_i / \partial C_h = 0$$

If C_h appears only in f_i (as it would in this example if insect populations were only trivially affected by bird predation), then the expression

$$\frac{\partial X_i}{\partial C_h} = \left(\frac{\partial f_i}{\partial C_h} \right) \left(\frac{F_{n-i}(\text{comp})}{F_n} \right)$$

is also zero, so that the stable equilibrium for mendelian selection may be the stable or unstable equilibrium for group selection, depending on whether $F_{n-i}(\text{comp})$ is negative or positive.

Group selection, however, will also have an equilibrium at

$$F_{n-1}(\text{comp}) = 0$$

In systems in which the a_{ij} are not constants but rather are functions of the levels of the component populations, the evolution of the species in the community changes the feedback of the system and its subsystems. The process of group selection (that is, the dynamics of local extinction and recolonization), acting directly on parameters that enter into the equations of only one species, will reach an equilibrium at which the complementary subsystem has zero feedback. This does not necessarily make the community unstable. However, if the same processes operate in all the species of the system, then the whole F_{n-1} approaches zero. In that case the community is destabilized. Therefore mendelian selection and group selection may each act to destabilize the community. But they act in different ways and have different endpoints. Hence the interaction between the two modes of selection can result in the stabilization of communities which, under the influence of either mode alone, would be unstable.

Conclusions

1. The effects of mendelian selection within species, on the community to which these species belong, depend on the structure of the community. This structure may be expressed by a diagram of the community variables and the way in which they are connected. In particular, natural selection may stabilize or destabilize a community, and may increase or decrease the equilibrium levels of the species undergoing the selection. Despite assertions that communities evolve to maximize stability or efficiency or information or complexity or anything else, there is no necessary relation between evolution within the

component species and evolution of macroscopic community properties. Yet such
claims are frequently made, and seem to
be attractive to biologists. Perhaps the
reason for this is a frequent reference by
biologists to a philosophical framework
that seeks harmony in nature. Or it may
be the transfer to ecology of the equally
invalid Adam-Smithian assertion in the
economics of capitalism, that some hidden
hand converts the profit-maximizing activities of individual companies into some
social good.

The actual consequences of selection
depend on the detailed structure of the
diagram of the community. Nevertheless
a few generalizations can be made.

(a) A keystone predator can select itself
to extinction.

(b) In a predator-prey system, selection
within the prey species for increased viability or fecundity will not affect its
abundance, whereas selection for more
effective predator avoidance, even if accompanied by lower fecundity or reduced
resistance to physical factors, will increase
its population size.

(c) In a plant-herbivore-carnivore system, the herbivore will not increase
through an increase in feeding effectiveness or in reproductive rate but only by
avoiding its own predator.

(d) If predation affects only one stage
in the life cycle, selection for increased
viability in any stage will result in population increases only in the stage that is not
subject to predation. If viability in the
predation-susceptible stage is increased at
the expense of lowered viability in the
predator-free stage, overall population

size in the prey organism decreases. When
adults are preyed upon, selection for increased developmental rate in larvae of
the same species reduces the larval population without increasing the adult population.

2. No special exotic biology has to be
invented for group selection to oppose
mendelian selection. Rather, this opposi-
tion may be a frequent consequence of
community structure. Further, it looks as
if group selection for traits that affect directly only the species undergoing selection leads toward destabilization of the
community. We note therefore that group
selection does not necessarily look out for
the best interests of each component species in the group.

3. Although selection for tolerance of
physical conditions, number of offspring,
development rate, feeding efficiency, and
predator avoidance are interchangeable
under mendelian selection, as they can all
be similarly bought in the common coin-
age of mendelian fitness or selective value,
each may act differently at the level of the
community. This allows us to expect
group selection to act differently on these
traits in different trophic levels. For instance, at the highest level, predation efficiency is selected, but at the next lower
level, predator avoidance takes priority
over predation or feeding efficiency.

4. The perennial ecological discussion of
what limits a population contains much
ambiguity. For instance, green algae may
die of poisoning by blue-green algae or
be eaten by a herbivore. But even if the
toxin from the blue-green algae is the
main cause of death, increased toxicity or

increased reproduction by the blue-green algae may not alter the population level of the green algae. The herbivore toll may constitute an almost negligible part of the death rate of green algae, but changes in the predation rate or in the herbivore population will immediately alter the size of the green algal population. Thus, the major cause of mortality does not necessarily regulate population numbers.

References

Diamond, J. 1969. Avifaunal equilibria and species turn-over rates on the channel islands of California. *Proc. Nat. Acad. Sci. U.S.A.* 64:57–63.

Heatwole, H., and R. Levins. 1973. Biogeography of the Puerto Rico bank: species turn-over on a small sandy cay. *Ecology* 54:1042–1055.

Levins, R. 1970. Extinction. In M. L. Gerstenhaber, ed., *Some Mathematical Questions in Biology*. American Mathematical Society, Providence.

Levins, R. 1974. Qualitative analysis of partially specified systems. *Ann. N.Y. Acad. Sci.* 231:123–138.

Levins, R., M. Pressick, and H. Heatwole. 1973. Coexistence patterns in insular ants. *Amer. Sci.* 61(4):463–472.

MacArthur, R. H., and E. O. Wilson, 1967. *The Theory of Island Biogeography*. Princeton University Press, Princeton.

Mason, S. J. 1954. Some properties of signal flow graphs. *Proc. I.R.E.* 41:1144–1156.

Searle, S. R. 1966. *Matrix Algebra for Biological Sciences*. Wiley and Sons, New York.

Simberloff, D. C., and E. O. Wilson. 1970. Experimental zoogeography of islands: a two-year record of colonization. *Ecology* 51:934–937.

Wright, S. 1921. Correlation and causation. *J. Agric. Res.* 20:557–585.

Wright, S. 1968. *Evolution and the Genetics of Populations*. Vol. 1. *Genetic and Biometric Foundations*. University of Chicago Press, Chicago.

2 Population Fluctuations, Community Stability, and Environmental Variability

Egbert G. Leigh, Jr.

Volterra (1931) and Lotka (1925) focused our attention on changes in the numbers of individuals in the different species of a community, simply by making these the subject of suggestive theory. One of Robert MacArthur's achievements was to find new uses for this sort of theory, to ask new questions of its equations. As a result, many workers are now enquiring into those factors—particularly competition—affecting population change, hoping thereby to understand the structure of guilds or communities.

What does interest us about the numbers of the various kinds of organisms in a community? Presumably we want to know what controls these populations, what permits these species to coexist, and why other species cannot establish themselves in the community. Interest in population regulation stimulated Lack's (1954) classic book, and also underlies the experimental manipulations with which Connell (1961 and Chapter 16), Paine (1966), and Dayton (1971) seek to identify the normal modes of population regulation in the marine intertidal. Motivated by an interest in species diversity, Robert MacArthur and his school have concentrated on the coexistence of species and the factors excluding invaders from communities and habitats.

However, MacArthur is also known for his theory of island biogeography.

Although he began studying islands to demonstrate the importance of competition, this work is remembered primarily for the proposition that diversity represents a balance between immigration (and/or speciation) and extinction. It reminds us that populations fluctuate, varying in numbers from year to year, and that small populations can fluctuate out of existence quite rapidly. Thus, to understand species packing and the like, we must ask about population fluctuations as well as about average numbers: we must ask not only whether a species can maintain itself under average conditions but also how likely it is to fluctuate out of existence. This perhaps explains Robert MacArthur's persistent interest in population fluctuation, first signalled in 1955 by his community stability paper, reappearing in his 1967 calculation with Wilson of the average lifetime of a population and again (1972a) in a graceful essay to Hutchinson on the likely strength of species interactions, later elaborated in the better-known paper of May and MacArthur (1972).

This chapter is concerned with various aspects of population fluctuation. First I make some necessary definitions, and discuss how a population's pattern of fluctuation may relate to its chances of extinction. I then present arguments and data to show that tropical populations normally fluctuate no less violently than their appropriate counterparts of higher latitude. Finally, I will discuss how a community's structure might affect the response of its populations to environmental variation.

Definitions

We should distinguish two aspects of population change: the year-to-year fluctuations of a community's populations under normal, "undisturbed" circumstances—in other words, the *steadiness* of these populations—and the resistance or *resilience* of the community's members to major disturbance or real catastrophe. This distinction blurs for the wildly and irregularly fluctuating populations so beloved of Holling (1973) or Andrewartha and Birch (1954). In spite of the ambiguity, many authors have striven for a similar dichotomy: ours is related to Margalef's (1969) distinction between persistence and adjustment stability, and to Holling's (1973) distinction between stability and resilience. Bretsky (1973) thinks that the distinction is relevant to his finding that marine communities of unstable environments such as beach zones last longer in the fossil record than communities of stabler environments such as offshore bottoms. In this chapter we shall be concerned entirely with population steadiness.

Steadiness is intended to capture the meaning of stability as given in MacArthur (1955). Stability has since been used in a variety of senses, and Holling (1973) has, I think, settled the issue by defining stability in accord with

mathematical usage: he measures a community's stability by the smallest in absolute value of the real parts of the eigenvalues of the "community matrix" (sensu May, 1973a, p. 22)[1], provided these real parts are all negative. This measure theoretically determines the speed with which a community's populations would return to their equilibrium levels after a small disturbance in an otherwise constant environment. Holling argues that this measures steadiness, since the stabler the community, the less its populations fluctuate. However, this definition of stability requires knowledge of the community matrix, which we can as yet rarely measure, and which may turn out quite useless in some ecological contexts. It also interposes the stability of the environment be-

[1] The community matrix is calculated under the assumption that the community has an equilibrium state at which, were the environment constant, its populations would not change. Suppose now that these population levels are perturbed ever so slightly, so that the present population of species j is $1 + x_j$ times its equilibrium value. If the equilibrium population is 1000, whilst the actual population is 1012, then the "excess" x is 0.012. The excess x_j in the population of species j then contributes an amount $a_{ij}x_j$ to the rate of change of the excess x_i in the population of species i: if the x are all small, the a_{ij} are constant. In symbols,

$$\frac{dx_i}{dt} = \sum_{j=1} a_{ij}x_j \qquad (1)$$

If populations are measured relative to their means, the community matrix constructed by May (1973a) and by Levins in chapter 1, equations 1–3 is identical to that defined by my eq. 1. In general, they differ only by "scale factors" which do not affect predictions about community stability. Levins's (1968) community matrix, by contrast, is a matrix of overlaps: it differs in sign, and in more essential characteristics, from May's.

tween the theoretical community stability
and the more observable fluctuations of
its populations. We need a more direct
measure of population fluctuation.

We wish to measure those aspects of a
population's steadiness that most affect its
chances of extinction. Of two populations
with equal average numbers, presumably
the one that fluctuates more widely will
go extinct sooner; if they fluctuate over an
equal range of numbers, the one fluctuating more rapidly will die first. This suggests that we should measure both the
amplitude and the frequency of a population's fluctuations.

Can we be more precise? To find out
what does affect the chances of extinction,
we calculate in Appendix A the average
time to extinction of a population obeying
the law of growth

$$\frac{dN}{dt} = \gamma_1(t)\sqrt{N}$$

$$+ [r + \gamma_2(t)]N - aN^2 \quad (2)$$

where time is measured in generations,
and where $N(t)$ is the absolute number of
reproductives in the population at generation t. $\gamma_1(t)$ is a white noise (see May and
MacArthur, 1972) with mean zero and
variance unity, representing the accidents
of who happens to reproduce and who
not, accidents that must happen even in
an utterly constant environment: $\gamma_1(t)$ is
analogous in every way to the sampling
error of the population geneticist (Kimura, 1964, p. 192). It represents the sort
of fluctuation considered by MacArthur
and Wilson (1967) and Richter-Dyn and
Goel (1973) in their discussions of popu-

lation lifetimes. Its provenance is discussed in more detail in Appendix A. $\gamma_2(t)$
is a white noise with mean zero and variance σ^2 (we call σ^2 the "environmental
variance") representing the effects of a
varying environment: using a white noise
for this purpose assumes that the state of
the environment in one generation is uncorrelated with the state of the environment one or more generations earlier. This
sort of noise is discussed in Leigh (1969,
p. 39 ff), May and MacArthur (1972), and
May (1973b, p. 628 ff). I assume that it
is less important than sampling error for
populations of only a few individuals, but
overwhelmingly more important for populations near equilibrium. The other
terms, $rN - aN^2$, are those of an ordinary
logistic equation of population growth.

This model is quite crude, suitable only
for qualitative conclusions. Even if correct, the average lifetime it predicts would
tell little of the fate of any given population, for the lifetimes of such populations
obey the same exponential law[2] as Lack's
(1954) adult songbirds: there is a five percent chance that such a population would
last only a twentieth the expected span,
and an equal chance of lasting three times
the expected span.

The expected lifetime of a population,
now near equilibrium, growing according
to eq. 1, is roughly

---

[2] It is very likely that if a population initially of the
equilibrium size M still survives, its numbers will
increase again to over M before dying out: thus the
expected further lifetime of a population, given that
it still survives, does not change with age. Its chances
of death are accordingly the same whatever its age,
which implies an exponential life table.

$$\frac{1}{r} \Gamma\left(\frac{1}{CV}\right)\left[\frac{2rM(CV)^2}{1 + CV(1 + 2r)}\right]^{1/CV} \quad (3)$$

M is the average number of individuals in the population while it is still fluctuating "normally" about its equilibrium [we represent "normal fluctuations" by setting $\gamma_1(t) = 0$: for most N, $\gamma_1(t)$ is much the least important influence on population change, but the model population cannot go extinct without it]. CV is the population's coefficient of variation: the mean square of the deviation of its numbers from the average level M over the course of its "normal" fluctuations, divided by M^2. Notice that CV is a dimensionless, or purely relative, measure of the extent of the population's fluctuations. r is the population's capacity for increase when rare, which we assume much smaller than 1. Γ is the gamma function, tabulated in the Chemical Rubber Company's *Standard Mathematical Tables* (1959, p. 316): if n is a whole number, then $\Gamma(n) = (n - 1)!$, and $\Gamma(\frac{1}{2}) = \sqrt{\pi}$.

Although our formula for extinction time breaks down for CV much exceeding 1, we may accept its conclusion that CV is overwhelmingly the most critical variable affecting a population's chances of extinction, in the sense that doubling or halving CV will have a far greater effect on extinction time than doubling or halving any other variable. We should therefore measure our population fluctuations relative to the mean, as May (1973b) suggests. If the coefficient of variation exceeds unity, the population will not live overlong: $CV < 1$ when $\sigma^2 < r$, which nicely confirms May's (1973b) criterion for an acceptable level of population fluctuation. The formula gives lifetime in generations; if we shorten generation time, it changes the time unit of the model without altering the size of the fluctuations: it merely "speeds up the film," so that the fluctuations occur more rapidly and the population dies out sooner.

We accordingly measure the frequency aspect of steadiness by the average lifetime of "established" adults (for territorial birds, this would be the average lifetime of territory-holders). We measure amplitude from successive annual censuses of a population, such as Lack's (1966) counts of the great tits breeding in Marley Wood in successive years, starting in 1947: 14, 42, 60, 62, 64, 40, 42, 62, 54, 48, 98, 54, 82, 102, 172, 86, 78. Often we lack a full schedule of censuses and cannot calculate CV, so we measure the relative amplitude of fluctuation of a cycling population by the average ratio of its peaks to the preceding and following lows. For the above protocol (assuming with Lack that 14 is a low), the average amplitude of oscillation is

$$\frac{1}{7}\left[\frac{64}{14} + \frac{64}{40} + \frac{62}{40} + \frac{62}{48} \right.$$
$$\left. + \frac{98}{48} + \frac{98}{54} + \frac{172}{54}\right]$$

or 2.3: the maxima average 2.3 times the neighboring minima. We can use such a measure of amplitude because, in the population records we will be discussing, the peaks are well-marked and unambiguous; for others, we would have to design cleverer measures.

Natural Selection and
Population Fluctuation

The principle that a Jack-of-all-trades is master of none implies, among other things, that a species is liable to replacement by more specialized competitors, provided their specialties are not likely to vanish. In Chapter 14, Diamond discusses how generalist supertramp fruit doves, flycatchers, etc., are excluded from the larger satellite islands of New Guinea by guilds of congeneric or related specialists, but survive nicely on islands too small to assure the specialists a *reliable* future.

The immediate advantages of specialization suggest a tendency for speciation to "split niches," to create specialists that at least partially replace pre-existing generalists: these specialists, however, are rarer or more precarious, unsteady if you will, and thus more liable to extinction. We may see this from Willis's (1967, 1972, 1973) discussion of three species of "ant-following" antbirds of Barro Colorado Island (Canal Zone), which feed primarily on insects flushed by raiding columns of army ants. Large size permits ocellated antbirds to feed at the best columns without fear of competition, but their size and stereotyped feeding (primarily sallies from low perches to the ground) are unsuited for foraging away from the ants, where food is less obvious; moreover, they apparently require moderately thick ground vegetation as cover for predators. Ocellated antbirds are accordingly restricted to forest with frequent and large army ant raids (i.e., forest with leaves falling fairly evenly through the year, thus providing a

reliable supply of decomposers to serve as prey for the ants), and moderately thick ground vegetation (Willis, 1973). The smaller and more adaptable bicolored antbirds can use smaller antswarms and can feed away from swarms, albeit at only one-tenth the rate they would at a good swarm, but they are likely to be excluded from the best feeding zones by the larger ocellateds (Willis, 1967). The yet smaller spotted antbirds spend half their time away from swarms and feed only four times faster at swarms than away, but they are entirely excluded from many good antswarms (Willis, 1972). Since Barro Colorado Island has been cut off from the mainland by the building of the Panama Canal, its ocellated antbirds have nearly died out, and bicolored numbers have varied more than spotted (Willis, 1973 and personal communication): population steadiness has varied inversely with degree of specialization.

We assume with Rosenzweig (Chapter 5) that, at least for birds and mammals, species diversities of continents have attained a steady state where the origin of new species is balanced by the extinction of the over-specialized and the precariously rare. Rosenzweig briefly discusses the evidence for such a balance.

How do such balances work? How are they affected by environmental stability? A new species forms when the two halves of a population first differentiate and then reinvade each other's range, provided that the differences are great enough to cause selection against hybrids, and that one half does not die out in the process. The number of opportunities for new species to form is thus proportional to the number S of species already present. Rosenzweig observes that the numbers of individuals in each species shrink as diversity increases, making it less likely that both halves of a divided species will survive, and that ranges become smaller, and thus less susceptible to division by barriers. I believe these are compensated by the increased likelihood of patchy or "checkerboard" distributions in diverse communities (Diamond, Chapter 14), making it more likely for a species to "split." It thus seems simplest to assume that the number of new species arising each century is ρS, where ρ is a constant and S the number of species already present. The net gain ΔS in a continent's diversity in one century is the number of new species formed less the number of old ones going extinct: in symbols,

$$\Delta S = \rho S - \lambda(S)S \qquad (4)$$

where $\lambda(S)$ is the number of extinctions per species per century, which presumably increases as S grows. Diversity stops changing when $\lambda(S) = \rho$, when diversity is just high enough so that the extinction rate per species balances the speciation rate ρ.

Now suppose the environment is stabilized, lowering the extinction rate for each level of diversity: assume, to be specific, that the new extinction rate, which we call $\lambda'(S)$, is half the old for each S. The new speciation rate is not lower than the old: indeed, stable conditions are usually said to favor speciation. Thus diversity will increase until the new extinction rate $\lambda'(S)$ is equal to the new speciation rate ρ'. The end result is probably an *increase* in extinction rate, a shortening of species lifetimes; for those groups whose diversity is in balance, population lifetimes are no longer, and may indeed be shorter, the stabler the environment. In short, species specialize as much as their environments allow; an environment's stability is reflected by the specialization of its occupants, and perhaps the complexity of their relationships, not by the longevity or steadiness of their populations.

Are populations in fact no steadier in stable environments? The obvious comparison is between tropical and temperate or arctic settings. Tropical forest is fairly clearly a stabler environment, at least for vertebrates, than its temperate or arctic counterparts. Tropical mammals hibernate, or go seasonally torpid, less often than their temperate counterparts, even though some tropical Malagasy tenrecs do go seasonally torpid (Eisenberg and Gould, 1970), and the armadillos of Barro Colorado may also (N. Smythe, personal communication). Some tropical birds, particularly hummingbirds and others dependent on flowers (N. G. Smith, personal communication) do migrate, but seasonal migration is less prevalent, and far less spectacular, among tropical than among temperate-zone birds. [One wonders why butterfly migrations, by contrast, are so much more obvious in the tropics (Williams, 1958).] Diamond (1973) remarks of New Guinea that the birds there need not disperse much, for they are unlikely to be wiped out where they are, and by the same token they will rarely find unoccupied habitats by wandering; so that

many birds there will not trouble themselves to cross a few meters of salt water
to find an unoccupied island (cf. Diamond, Chapter 14). These remarks irritate
many tropical biologists, who feel that
environmental vagaries produce quite as
violent population fluctuations in the
tropics as in the temperate zone, but
MacArthur (1972b, p. 203) has warned us
not to judge environmental stability by
population steadiness.

How do tropical and temperate populations compare in steadiness? Data suitable
even for our rough measures of steadiness
are few and far between. One can only
compare the steadiness of a few conspicuous herbivore populations, which might
be quite unrepresentative of the steadiness
of most populations in their community.
Data on lifetimes are more representative,
but they bear only on the less important
side of steadiness.

Table 1 compares the amplitudes of
fluctuation of some common herbivores.
If we discount the initial irruptions of
newly introduced populations, these animals exhibit a three to tenfold amplitude

of oscillation (that is to say, their peaks
are three to ten times the following lows).
These populations oscillated for quite
different reasons: the Isle Royale moose
cycled with their food (Lack, 1954) or
perhaps more specifically with sodium
availability (Hutchinson, Chapter 17), the
Dall sheep varied with the weather
(Murie, 1944), the quokkas held steady
except for occasional very dry years
(Main, Shield, and Waring, 1959), the
howling monkeys held steady except for
occasional yellow fever epidemics, and
rinderpest caused the lows in the old
world ungulate populations. However
sketchy, these data support our case: the
tropical populations are no steadier than
the temperate ones.

The obvious exceptions to this rule are
the well-marked arctic oscillations of vole
and lemming, lynx and hare (Elton, 1942;
Keith, 1963). Hutchinson in Chapter 17
summarizes both the geographical and
statistical limitations of these cycles. The
oscillations of Canada lynx and snowshoe
hare, at least, show no latitudinal gradient
in amplitude except at the fragmented

Table 1. Amplitude of population fluctuation in different herbivores

Species	Locality	Amplitude	Authority
Alces americana (moose)	Isle Royale	2–4	Lack, 1954
Ovis dalli (dall sheep)	Mount McKinley National Park	2–4	Lack, 1954
Setonyx brachyurus (quokka)	Rottnest Island, West Australia	3	Main et al., 1959
Alouatta palliata (howling monkey)	Barro Colorado Island, Panama	4–6	Various
*Bos gaurus** (gaur)	Mudumalai sanctuary, India	10?	Park authorities
*Connochaetes taurinus** (wildebeest)	Serengeti, Tanzania	3+	Schaller, 1972
Syncerus caffer (buffalo)	Serengeti, Tanzania	4+	Schaller, 1972

Amplitudes are represented as the ratio of maximum population levels to preceding or following minima. An asterisk denotes
that the data in question are based on part of a cycle. Note that the amplitude is no greater for the north temperate moose
or sheep, or the south temperate quokkas, than for the tropical animals (howling monkey, gaur, wildebeest, buffalo).

very southern end of their range (Keith, 1963), where increased hunting pressure and fractionation of available habitat have damped out the oscillation (Berrie, 1973). These northern oscillations are a grand mystery. They are not predator-prey oscillations: the predators cannot account for observed prey declines (Pearson, 1966; Nellis, Wetmore, and Keith, 1972), although predators can amplify a pre-existing cycle and may even be necessary for the oscillation to occur (Pearson, 1966). They cannot all be Volterra-style oscillations between herbivores and their food: Krebs (1971) records that fenced voles prevented from dispersing but exposed to predators rose to higher levels than animals of identical habitat outside the enclosure, but the enclosed animals, even though they more obviously depleted their forage and crashed sooner, did not fall to as low levels as the animals outside. Finally, the theory (Krebs et al., 1973) that selection for aggression at high numbers and against it at low generates an oscillation, can hold only under conditions that are probably so special that they must have been selected for. I suspect that the susceptibility to oscillation, which subjects predators to alternate feast and famine, is a form of predator escape or defense; I therefore feel that these oscillations are a very special adaptation, which do not support a general proposition that unstable environments beget unsteady populations.

Adult lifetimes of vertebrates also fail to exhibit a consistent latitudinal gradient. Tropical birds are longer-lived than their temperate counterparts (Snow, 1962;

Lack, 1966; Fogden, 1972); the data in Tinkle (1969) suggest that tropical lizards are shorter-lived than their temperate counterparts, a conclusion reinforced by the unpublished data on tropical lizards, insular and continental, of Dr. R. Andrews. Adult mammal lifetimes seem to be governed more by size and way of life than by latitude. An 18-month-old timber wolf can look forward to another four and a half years of life (Mech, 1970), whilst a yearling spotted hyaena expects five (Kruuk, 1972); the lion (Schaller, 1972), like the cougar (Hornocker, 1970), lives far longer. In Table 2 we plot average lifetimes in the wild for adult unhunted (not necessarily predator-free) ungulates and other large herbivores, showing that size has more to do with lifetime than latitude does. These data offer no ground for supposing that vertebrate populations fluctuate more rapidly, or have shorter-lived individuals, in unstable temperate environments than in stabler tropical ones.

Community Matrices, Population Fluctuations, and Environmental Variation

How does the community's structure affect the steadiness of its populations? How does the community's web of predatory and competitive relationships affect the response of its populations to environmental variation?

Theories of competition along a single resource gradient will tell us something about the response of competitive communities to stable environments, and sug-

Table 2. Average lifetimes of large herbivores

Authority	Locality	Species	Weight	Age of maturity	Expected further lifetime
Pfeffer, 1967	Corsica	Corsican sheep	100 lb.	1 year	5 years
Pimlott et al., 1966	Algonquin Park	White-tailed deer	150 lb.	1 year	3 years
Murie, 1944	Mount McKinley	Dall sheep	180 lb.	1 year	8 years
Caughley, 1970	New Zealand	Himalayan thar	200 lb.	1 year	6 years
Schaller, 1972	Serengeti	Wildebeest	270 lb.	2 years	5 years
Hornocker, 1970	Idaho	Elk	400 lb.	1 year	6 years
Spinage, 1969	Uganda	Waterbuck	440 lb.	1 year	6 years
Spencer and Lensink, 1970	Alaska	Muskox	800 lb.	1 year	14 years
Jordan et al., 1971	Isle Royale	Moose	1000 lb.	1 year	8 years
Bourliere, 1964, p. 314	Uganda	Elephant	8000 lb.	—	25 years

The data refer to average lifetimes in the wild, for populations not hunted by human beings. Expected further lifetimes are given for individuals "established" in the population, that is to say, individuals past the stage of high juvenile mortality; the stage when this occurs is called the age of maturity. Notice that size is a better predictor of longevity than latitude.

gest that populations are equally sensitive to their competitors, however stable their environment. Models involving predators will show how disproportionately the importance of predatory relationships should increase with environmental stability. They will also suggest obliquely how the responses of a community's populations to environmental disturbance may be affected by the form of the foodweb.

Imagine a community of n species, where the population N_i of species i, measured in kilograms per hectare, obeys the Lotka-Volterra equation

$$\frac{d}{dt} \log N_i = r_i + k_i(t) + \sum_{j=1}^{n} a_{ij} N_j \quad (5)$$

The r_i and a_{ij} are constants, and $k_i(t)$ represents the effect of environmental change on population growth. For the moment we assume k_i is a white noise contributing a variance $\sigma_i^2 N_i^2$ per unit time

to change in population i, where σ_i^2 is constant. To measure population changes relative to their means, we set

$$x_i = \log(N_i/M_i), \, N_i = M_i e^{x_i}, \quad (6)$$

where M_i is the mean population size of species i. The mean values also happen to be the equilibrium values of eqs. 5: the M_j thus obey the relations[3]

[3]To see why environmental variation reduces effective capacity for increase by $\sigma^2/2$, consider a population whose environment has two states: every Δt time units, a fair coin is tossed to determine the state for the next Δt time units. If the coin comes up heads, then the population's numbers multiply by a factor $1 + r \Delta t + (\sigma/\sqrt{\Delta t}) \Delta t$, whilst if this coin comes up tails the factor is $1 + r \Delta t - (\sigma/\sqrt{\Delta t}) \Delta t$. Over $2n \Delta t$ time units, the coin falls heads an average of n times: the modal change in population number during this time will be by a factor $[(1 + r \Delta t)^2 - \sigma^2 \Delta t]^n$. If Δt is sufficiently small, then the average change per Δt time units will be by the factor $1 + r \Delta t - \sigma^2 \Delta t/2$. The factor $\sqrt{\Delta t}$ is needed to make the "environmental variance" introduced each generation independent of how often we toss the coin. For a more detailed discussion of the subject, see Lewontin and Cohen (1969).

$$r_i - \sigma_i^2/2 + \sum_{j=1}^{n} a_{ij}M_j = 0 \qquad (7)$$

If we assume the x_i so small that $e^{x_i} \approx 1 + x_i$, we obtain the log linear approximation

$$dx_i/dt = k_i(t) + \sum_{j=1}^{n} a_{ij}M_j x_j \qquad (8)$$

The coefficients $a_{ij}M_j$ form the community matrix of May (1973a). For the logistic equation, the coefficient of variation of population number is equal to the variance of the logarithm of population number in the log linear approximation. I will assume this is at least approximately true for more general Lotka-Volterra equations.

One Trophic Level

Consider first the case so admirably treated by May and MacArthur (1972): a guild of species, such as fruit-eating birds, competing along a gradient of different-sized foods, where variation in food production is unconnected with the number of consumers. They show that stabilizing the environment does not permit much increase of diversity through increase in niche overlap. Empirical evidence for the same result comes from the bird communities described by Cody (1974). What does this result mean? How generally is it true?

Imagine a guild of birds eating fruit of different sizes. We suppose for symmetry's sake that the smallest birds take pecks from the largest fruit, so that the two ends of the competitive gradient join to form

a circle, like the spectrum of French politics where the extremes of left and right are so nearly indistinguishable. We assume also that our guild contains many species, and that the average population of each (in kilograms per hectare) is the same. We then rewrite eqs. 8 to obtain

$$dx_i/dt = k_i(t) - Ma \sum_{j=1}^{n} b_{ij}x_j \qquad (9)$$

where $a = a_{ii}$ (assumed the same for all species i), and $b_{ij} = a_{ij}/a$ is the niche overlap between species i and j (note that $b_{ii} = 1$); b_{ij} is assumed to depend only on the difference $|i - j|$ between the positions of i and j on the "competitive circle": we write $b_{ij} = b(i - j)$. Circular symmetry enters in assuming that $b(n - i) = b(i)$, which ensures that $b_{1n} = b_{12}$.

To see how the niche overlaps b_{ij} relate to feeding habits, we suppose that species m derives a proportion $f_m(x)\, dx$ (the "utilization function" of May and MacArthur, 1972) of its energy from, for example, fruit weighing between e^x and e^{x+dx} grams: a logarithmic size-scale insures that differences of equal proportion are represented by equal shifts on the scale. Presumably individuals of any size are equally put off by food half or twice their preferred weight, so we assume the distributions $f_m(x)$ differ only in their mean, not in variance or general form. The overlap b_{mk} is then

$$\int f_m(x) f_k(x)\, dx \qquad (10)$$

The b_{ij} of eqs. 9 form the overlap matrix of Levins (1968).

How much niche overlap is consonant with coexistence? Suppose fruit is equally available all along the food-size gradient, and that species m eats only fruit weighing between e^{mD-L} and e^{mD+L} grams, not distinguishing sizes within these limits (i.e., $f_m(x) = 1/2L$ for $m - L < x < m + L$ and zero otherwise). The mean log weight of fruits that species m eats differs by D from those eaten by its closest competitors $m + 1$ and $m - 1$, and the overlap $b_{m,m+1}$ between m and $m + 1$, which we call b, is $1 - D/2L$ for $D < 2L$ and zero otherwise. When $D = L$, $b = \frac{1}{2}$, and there is no overlap between species that are not closest competitors; the community's equilibrium is neutrally stable (see Appendix B), subject to derangement by the slightest disturbance. Wider niches usually *stabilize* the equilibrium, but not very securely: it is dangerous for L to exceed D. If $L < D/2$, there is no overlap at all and no need to decrease L/D further. Thus $\frac{1}{2} < L/D < 1$. No matter what the nature of $k_i(t)$, this small range of values spans the difference between zero and dangerous overlap.

May and MacArthur assume, far more reasonably, that

$$f_m(x) = \frac{1}{w\sqrt{2\pi}} \exp\left[-(x - mD)^2/2w^2\right] \quad (11)$$

The standard deviation w is their measure of niche-width. The overlap b between closest competitors is $e^{-D^2/4w^2}$, and $b_{ij} = b^{(i-j)^2}$ if $|i - j| < n/2$, and $b^{(n-|i-j|)^2}$ if $|i - j| > n/2$. If $D = 2w$, $b = 1/e = 1/2.718$, and all overlaps between species

other than closest competitors are less than 2%. Here, too, overlaps between closest competitors are usually the only ones that matter. Decrease D/w, and b becomes dangerously large for almost any environment (although $b = \frac{1}{2}$ is no longer an absolute upper limit); if, on the other hand, $D = 4w$, *all* overlaps are negligible. Once again, small change in D/w causes great change in overlap, and thus great change in sensitivity to environmental variation. To see how great the change, we apply May and MacArthur's "security condition" (Appendix B) that the community stability (*sensu* Holling, 1973) $Ma\Lambda$ be more than half the environmental variance σ^2, where Λ is the lowest eigenvalue of the matrix b_{ij}. Remember from eq. 7 that $r - \sigma^2/2 = M\Sigma_j a_{ij} = Ma\Sigma_j b_{ij}$. If $D/w < 2\sqrt{3}$, we may approximate this last by $2Ma\sqrt{\pi}w/D$. The condition for coexistence may thus be expressed as

$$\frac{D}{w} > \frac{\pi}{\sqrt{\ln(4r/\sigma^2 - 2)}} \quad (12)$$

D/w should be $2\sqrt{3}$ when $r/\sigma^2 = 1.11$; it should be $\sqrt{3}$ when $r/\sigma^2 = 7.2$; it should be 1 when $r/\sigma^2 = 4800$.

Increase the number of significant competitors each species has, however, and D/w becomes more sensitive to r/σ^2; moreover, a given degree of environmental variability will permit much more niche overlap.[4] This happens, as May

---

[4] If $f_m(x)$ is

$$\frac{1}{2}\sqrt{\frac{a}{\pi|mD - x|}} \exp -a|mD - x| \quad (13)$$

the standard deviation w is $\sqrt{3}/2a$, the overlap b between closest competitors is $e^{-aD} = e^{-\sqrt{3D/2w}}$, and

(1973a) points out, if species are simultaneously segregating along several "resource axes."

In summary, species competing along a food gradient will gain very little by increasing overlap. In a stable environment species presumably evolve so as to maximize the "gain per unit risk," where gain is measured in terms of increased capacity to resist competitive displacement. Accordingly, a community's members would respond to environmental stability with increased specialization rather than increased overlap.

Moreover, if prey are not difficult or dangerous to catch, it pays the predator to eat everything edible he comes across. The diets of such animals are incompressible (MacArthur, 1972b, p. 64): it does not pay them to respond to stable conditions by restricting their diet (see Hespenheide, Chapter 7).[5] In the West Indies (E. Williams, 1972) one often finds three sizes of anole, but never more, in any one habitat. Diamond (1973) observes that the average ratio of weights of closest competitors segregating by size of food compresses from 4 on islands with 30 to 50 species of birds to 2 on islands with 100 species or more, but compresses no further on New Guinea itself, with its 513 species. Beyond this point, diversity on larger islands increases through differentiation with respect to habitat or foraging method. Further, the diets of coexisting fruit pigeons are unexpectedly similar in breadth, despite the fact that some of the species characteristically occur alone or with few competitors, and others with many competitors (Diamond, Chapter 14, Figure 31).

Presumably, as species specialize to a stable environment, r will decrease and the environmental variance σ^2 increase until r is once again only a small multiple of σ^2. (Notice that σ^2 really measures the sensitivity of the population to its environment, which presumably increases as the population specializes.) If this is so, then over a wide range of environments species will be more or less equally sensitive to variations in competing populations. Dobzhansky's famous remark about the increased importance of biotic interactions in the tropics should then imply the increased importance of predator-prey relationships, which leads us to our next topic.

Several Trophic Levels

How does predation affect a community's stability? In other words, how does it affect the power of that community's members to weather environmental variation?

$b_{ij} = b^{|i-j|}$ for $|i - j| < n/2$; close overlap between nearest competitors now implies much stronger overlap between species further apart. If $CV < 1$, then
$$b^2 < 1 - \sigma^2/r, \quad \text{and} \quad D/w < (1/\sqrt{3}) \ln\left(\frac{r}{r - \sigma^2}\right).$$
Even for $D/w = 1$, r must be only $1.22\sigma^2$; for $r > 10\sigma^2$ the security condition is $D/w < \sigma^2/\sqrt{3}r$.

[5] The reader acquainted with the work of Brooks and Dodson (1965) concerning the preference of planktivorous fish for large cladocerans, when they can live on small, may wonder at the "incompressibility" of bird diets. Dr. Thomas Zaret informs me that zooplankters occur in schools, and that a fish, like a cheetah, must fix on a particular individual in order to catch one. Presumably a bird's diet is so incompressible because the bird, unlike the fish, comes across its prey one by one.

First, consider how a single predator that feeds equally on all species affects the competitive community of eq. 9. If the community is diverse and if there is some overlap among competitors, the community's stability (*sensu* Holling, 1973) declines in that proportion by which the predator reduces the average population levels of its prey (see Appendix C). If the original "competitive community" has only one species, or if its members do not overlap, predation more than halves community stability unless the predator's numbers are self-regulated through territoriality.

Next, consider what happens if our community of n competitors supports a circle of n predators that stand in much the same relation to the "resource circle" of prey species as these latter do to their resource circle of food. Let each prey and predator species maintain an average of M' and P kilograms per hectare, respectively, and suppose that predator j kills, on the average, $f_{ij}M'P$ kilograms of competitor i per hectare per unit time. The reader may think of the "prey spectrum" f_{im} as an analogue for predator m of the "utilization function" $f_m(x)$ of competitor m, in which the continuous variable x is replaced by the discrete variable i, for there are only n distinct types of prey to choose from, rather than a continuous array of food sizes. (Note, however, that the f_{ij} do not represent proportions of the predator's total diet.) We assume that these prey spectra all have the same form, but that the prey spectrum for predator 2 is displaced one species around the prey circle from predator 1, etc. (If we assign the prey circle a circumference n, then $D = 1$.) The log linear approximation for this expanded community is

$$dx_i/dt = k_i(t) - M'a \sum_{j=1}^{n} b_{ij}x_j(t)$$

$$- P \sum_{j=1}^{n} f_{ij}\,y_j(t) \qquad (14)$$

$$dy_i/dt = k_{n+i}(t) - ry_i(t)$$

$$+ cM' \sum_{j=1}^{n} f_{ji}x_j(t)$$

Here, x_i, or log N_i/M', measures the relative excess of the competitor i's population (in the predator's presence) over its average level, M'; y_j similarly measures the excess of predator j's population over its average level P; c represents the fraction of prey weight eaten, transformed into predator weight gain; and r represents the strength of self-damping, or territoriality.[6]

We find the stability (Appendix C) of this community by calculating the minimum eigenvalue of the community matrix. If the predators are not territorial, that eigenvalue will be one of the k eigenvalues

[6] Notice that if crowding from conspecifics reduces the per capita growth rate by r when the population is at equilibrium, the population's growth rate should be r when it is rare enough not to suffer from crowding by conspecifics. Our use of the letter "r" was no accident, and points to an ecologically meaningful way of calculating innate capacity for increase. Note, however, that earlier we used r to denote a population's growth rate when freed of competition from members of its own or other species.

$$\lambda_k = -\frac{1}{2}[M'a\lambda_{1k}$$

$$- \sqrt{M'^2a^2\lambda_{1k}^2 - 4cQ^2\lambda_{2k}(M'/PB)}] \quad (15)$$

Here,

$$\lambda_{1k} = \sum_j b_{1j} \exp -2\pi i(k-1)(j-1)/n$$

is the kth eigenvalue of the prey overlap matrix b_{ij};

$$\lambda_{2k} = \sum_j B_{1j} \exp -2\pi i(k-1)(j-1)/n$$

is the kth eigenvalue of the predator overlap matrix

$$B_{ij} = \sum_k f_{ik}f_{jk} \Big/ \sum_k f_{ik}^2;$$

B is the predator niche breadth

$$\left(\sum_j f_{1j}\right)^2 \Big/ \sum_j f_{1j}^2$$

(Levins 1968, p. 43), which is here the same for all species; and Q is the per capita death rate of the prey from predation. The community stability can be no greater than $\frac{1}{2}M'a\lambda_1$, where λ_1 is the minimum eigenvalue of the prey overlap matrix: when some of the square roots are real, the stability may be much less. The square roots are more likely to be real if the overlap matrices differ, so that λ_{1k} and λ_{2k} are minimum for different k, or more particularly, if predator diets overlap greatly, complicating the food web and making some λ_{2k} small. As members of communities everywhere should exhibit roughly equal spectra of steadiness, we

expect the most elaborate webs of predator-prey relationships to develop in the stablest environments (see Connell, 1971, and his remarks in Chapter 16). More generally, we expect that predation restricts the distribution and abundance of animals more strongly in the tropics than in the temperate zone.

If $4cQ^2\lambda_{2k}M'/PB > M'^2a^2\lambda_{1k}^2$, the square root in eq. 15 is imaginary and does not contribute to the diminution of community stability. This condition looks hopelessly complex. Recall, however, that with one predator species for each prey, predation mortality Q should exceed the impact $M'a$ of crowding from conspecifics. $B > 1, 4c \sim \frac{1}{2}$, and in most Eltonian pyramids $M' > P$. The square root is thus unlikely to be real unless $\lambda_{2k} < \lambda_{1k}^2$, or unless B is very large. Predators can overlap a bit more than their prey, perhaps because diversity of diet stabilizes them more. However, the enormous sensitivity of λ_{1k} to niche overlap D/w implies even greater sensitivity in λ_{2k}: D/w should be even less affected by environmental variation in these predators than in their prey. We may, in short, safely apply the results of May and MacArthur (1972) to predators that affect the recruitment of their prey as well as to harvesters of excess. Notice, moreover, that in this example, where the prey are not differentially sensitive to predation, the predators do not greatly alter the D/w ratios suitable for prey coexistence.

Does food web complexity ever contribute to population steadiness? If the populations respond independently to environmental change (i.e., if the $k_i(t)$ for different species are uncorrelated), it would appear not, since a new link in a food web is merely a new channel for the transmission of environmental disturbance. On such grounds May (1973a) overthrew the conventional wisdom that complexity begets stability. However, if the $k_i(t)$ for all populations are the same, one can imagine food webs in which the variations in a predator's prey interfere destructively with the variations in the predator's response to his inorganic environment, "cancelling them out." When this happens the community's food web truly stabilizes its populations.

Imagine a community of n species, each of which eats all the others that do not eat it. Numbering the species from the bottom of the food web upward, species 1 is eaten by all the others; species 2 eats species 1 and is eaten in turn by species 3 through n; species 3 eats species 1 and 2 and is eaten by species 4 through n, etc. Species n is the terminal carnivore, which eats all the others. For simplicity, assume that every species feeds equally on all its prey, each kilogram of predator eating a kilograms per unit time per kilogram standing crop from each of its prey, and assume also that all species maintain the same average weight per unit area, which we call M. The log linear approximation for this community's dynamics is then:

$$dx_1/dt = k(t) - rx_1$$
$$- aM(x_2 + x_3 + \cdots + x_n)$$
$$dx_2/dt = k(t) - rx_2$$
$$+ caMx_1 - aM(x_3 + \cdots + x_n) \quad (16)$$
$$\cdots\cdots\cdots\cdots\cdots\cdots\cdots\cdots\cdots\cdots\cdots\cdots$$
$$dx_n/dt = k(t) - rx_n$$
$$+ caM(x_1 + x_2 + \cdots + x_{n-1})$$

Here $k(t)$ measures the effect of a varying environment (assumed the same for each species), r measures the strength of self-damping for these populations, and c, which is significantly less than 1, usually nearer $\frac{1}{10}$, represents the proportion of prey weight killed, transformed into predator weight gain. Since each kilogram of species j gains $caM(j - 1)$ kilograms per unit time from its prey, while it loses $aM(n - j)$ kilograms per unit time to its predators, most of these species must photosynthesize to balance income against outgo. It turns out that this community is stable only if r exceeds $a/2$: the community's members must be territorial if they are to coexist.

Finally, we suppose that $k(t)$ represents the sum of an indefinite number of factors varying sinusoidally with different frequencies and amplitudes, no one frequency dominating the rest. In symbols,

$$k(t) = \sum_{j=1} b_j \cos(w_j t + \theta_j) \quad (17)$$

Then the variances of x_i (very nearly the coefficient of variation of species i, which we call $CV(i)$), is roughly[7]

$$\sum_{j=1} \frac{b_j^2 A_j^{2(n-i)}}{2[(r + ac)^2 + w_j^2]} \quad (18)$$

where

$$A_j^2 = \frac{(r - a)^2 + w_j^2}{(r + ac)^2 + w_j^2} \quad (19)$$

If $r > a/2$, $A_j^2 \leqslant 1$: it is 1 only if $a = 0$. The coefficient of variation $CV(n)$ for the terminal carnivore (species n) is not much

[7]The derivation is lengthy and is omitted. Details may be obtained from the author on request.

less than it would be were *a* zero: in other words, the steadiness of the terminal carnivore is little less than that of a population with equal damping properties and sensitivity to the inorganic environment, which had no prey or predators. Populations further down the web are progressively steadier than comparable isolated populations: they are stabilized (the bottom primary producer most of all) by their food web relationships. This would not be true were the environmental effects on different species not correlated, nor would it be true for a community neatly stratified into trophic levels. Our very artificial result does suggest, however, that food web links which blur trophic levels may stabilize a community's populations against environmental variations which affect them all the same way.

With this exception, we can only agree with May (1973a) that the idea "food web complexity begets population steadiness," which MacArthur so clearly stated, finally making it susceptible to test, is a misreading of the fact that environmental stability permits food web complexity, and with it, that amazingly intricate delicacy of predatory specializations and anti-predator defenses that make the stable tropics so fascinating to the naturalist.

Acknowledgments

I would like to thank many seminar audiences for their comments, particularly my colleagues at Smithsonian Tropical Research Institute who were very helpful and patient with my first chaotic attempt to discuss the subject.

Dr. N. S. Goel kindly invited me to the University of Rochester, introduced me to a splendid swamp, and discussed with me, to my great profit and advantage, the niceties of extinction times.

It is a pleasure to acknowledge my debt to the writings of R. M. May and C. S. Holling, and a subtler one to those of E. O. Willis, which will all be evident to the discerning reader.

Robert May and Deborah Rabinowitz both read the manuscript with some care, suggesting a number of clarifications: such errors as remain are, of course, my own.

Finally I wish to thank the trees and monkeys of this Island, Barro Colorado, which perpetually remind me what a pale shadow of nature these theories are.

Appendix A.   The Time to Extinction of a Colonizing Population

The Imaginary World of Conveyor-belts

Imagine an infinite conveyor-belt bearing a series of identical islands, all empty, toward the observer, who maintains a pool of potential colonists of an appropriate species. Once during each generation of colonists, an island passes under the observer and receives a single adult hermaphrodite whose progeny multiply according to eq. 2 of the text. All populations thus founded eventually die out: at steady state, when extinctions balance new foundations, the conveyor bears only a finite number of simultaneously surviving populations. We will calculate this number, because it is equal to the expected lifetime, in generations, of a single population so founded.

Let $f(N)$ be the average number of populations on the conveyor with exactly N individuals. We suppose that, for a population of size

N, the probability $P(N, N + 1, \Delta t)$ of increasing by 1 in time Δt is $N b_N \Delta t$, whilst the probability $P(N, N - 1, \Delta t)$ of decreasing by 1 during this time is $N d_N \Delta t$; b_N and d_N are the population's per capita birth and death rates. We assume that the chance of changing by more than 1 during this small time interval is negligible.

At steady state, the number of populations decreasing from N to $N - 1$ during time Δt balances the number then increasing from $N - 1$ to N: in symbols,

$$(N - 1)b_{N-1} \Delta t f(N - 1) = N d_N \Delta t f(N)$$

Then

$$Nf(N) = \frac{b_{N-1}}{d_N} (N - 1) f(N - 1)$$

$$= \frac{b_{N-1} b_{N-2}}{d_N d_{N-1}} (N - 2) f(N - 2)$$

$$= \frac{b_{N-1} b_{N-2} \cdots b_1}{d_N d_{N-1} \cdots d_2} f(1)$$

One new population of a single individual is founded each generation, so there is chance Δt that such a population will be founded during the time interval Δt. As extinctions are assumed to balance foundations, $\Delta t = f(1)d_1 \Delta t$; $f(1) = 1/d_1$, and we may write

$$f(N) = (1/N) \frac{b_{N-1} b_{N-2} \cdots b_1}{d_N d_{N-1} \cdots d_2 d_1} \quad \text{(A.1)}$$

Remembering that the average lifetime L_1 of a population founded by a single individual is the average number of simultaneously settled islands, or $\Sigma_{N=1}^{\infty} f(N)$, we obtain

$$L_1 = \frac{1}{d_1} + \sum_{N=2}^{\infty} \frac{1}{N d_1} \prod_{i=1}^{N-1} (b_i/d_{i+1}) \quad \text{(A.2)}$$

where Π is the multiplicative analogue of Σ. Richter-Dyn and Goel (1972) were the first to calculate such an expression for L_1.

The Algebraic World of MacArthur and Wilson

Following MacArthur and Wilson, we remark first that the average lifetime L_N of a population presently of size N is the average time before the population changes (at which point it either increases or decreases by 1), plus the probability that the change will be an increase, multiplied by the average lifetime L_{N+1} of the increased population, plus the probability that the change will be a decrease, multiplied by the average lifetime L_{N-1} of the decreased population; in symbols,

$$L_N = \frac{1}{N(b_N + d_N)}$$
$$+ \frac{1}{b_N + d_N} (b_N L_{N+1} + d_N L_{N-1}) \quad \text{(A.3)}$$

(cf. MacArthur and Wilson, 1967, p. 70). It takes rather a lot of algebra to calculate L_1 by manipulating this relation, but some readers might gladly accept this chore in order to be free of the imagery of infinite conveyors.

Suppose that $L_{N+1} = F(N + 1) + L_N$, where F is some function of N, at present unknown. Substituting for L_{N+1} in eq. A.3 and rearranging, we find that $L_N = F(N) + L_{N-1}$, where

$$F(N) = \frac{1}{N d_N} + \frac{b_N}{d_N} F(N + 1) \quad \text{(A.4)}$$

Notice that $L_1 = F(1) + L_0$; since $L_0 = 0$ (an extinct population has no life left), $L_1 = F(1)$. Thus, if we can find a suitable expression for $F(1)$, we know L_1.

Let z be the smallest population size for which the birth rate b is zero (following MacArthur and Wilson, we assume such a number exists). Thus z is the maximum number of individuals a population could possibly attain. Since $b_z = 0$, eq. A.4 implies $F(z) = 1/z d_z$. Further applying eq. A.4, we find

$$F(z - 1) = \frac{1}{(z - 1)d_{z-1}} + \frac{b_{z-1}}{zd_z d_{z-1}}$$

$$F(z - 2) = \frac{1}{(z - 2)d_{z-2}}$$

$$+ \frac{b_{z-2}}{(z - 1)d_{z-1}d_{z-2}} + \frac{b_{z-1}b_{z-2}}{zd_z d_{z-1}d_{z-2}}$$

$$F(N) = \frac{1}{Nd_N}\left[1 + \sum_{i=N+1}^{z} \frac{N}{i} \prod_{j=N+1}^{i} (b_{j-1}/d_j)\right]$$

We thereby find

$$L_1 = F(1)$$

$$= 1/d_1 + \sum_{N=2}^{z} \frac{1}{Nd_1} \prod_{j=2}^{N} (b_{j-1}/d_j)$$

$$= 1/d_1 + b_1/2d_1 d_2 + b_1 b_2/3d_1 d_2 d_3 + \cdots$$

This is the expression of eq. A.2: the two approaches yield the same answer.

Putting Algebra to Useful Work

To extract meaning from eq. A.2, we first observe that the expected change $M_N \Delta t$ over time Δt in a population now of size N is

$$M_N \Delta t = P(N, N + 1, \Delta t)$$
$$- P(N, N - 1, \Delta t) = N(b_N - d_N)\, \Delta t \quad (A.5)$$

The contribution $V_N \Delta t$ to population variance from the chances of death and reproduction during this time interval is

$$P(N, N + 1, \Delta t)(1 - M_N \Delta t)^2$$
$$+ P(N, N, \Delta t)(M_N \Delta t)^2$$
$$+ P(N, N - 1, \Delta t)(1 + M_N \Delta t)^2$$

Since we are dealing with infinitesimal time intervals, $M_N \Delta t$ can be neglected in comparison with 1, and we may set

$$V_N \Delta t = P(N, N + 1, \Delta t)$$
$$+ P(N, N - 1, \Delta t)$$
$$= N(b_N + d_N)\, \Delta t \quad (A.6)$$

Here V_N is the "instantaneous rate of increase of population variance." We can solve eqs. A.5 and A.6 for b_N and d_N to obtain

$Nb_N = (V_N + M_N)/2$, $Nd_N = (V_N - M_N)/2$. If V_N greatly exceeds M_N, then

$$b_N/d_N = (V_N + M_N)/(V_N - M_N)$$
$$\sim 1 + 2M_N/V_N \approx \exp{(2M_N/V_N)}$$
$$\log{(b_N/d_N)} \approx 2M_N/V_N$$

From eq. A.1 we may conclude that

$$\log Nd_N f(N) = \sum_{i=1}^{N-1} \log{(b_i/d_i)}$$

$$\approx \sum_{i=1}^{N} 2M_i/V_i \approx \int_1^N \frac{2M(n)}{V(n)}\, dn$$

Setting

$$Nd_N = (V_N - M_N)/2 \approx V_N/2$$

and clearing the logarithm, we obtain

$$f(N) = \frac{2}{V_N} \exp \int_1^N \frac{2M(n)}{V(n)}\, dn \quad (A.7)$$

Equations of this sort have had a long history in population genetics; cf. Wright (1939, p. 12, eq. 8) and Kimura (1964, eq. 9.1).

For a constant environment, where $\gamma_2(t) = 0$, a choice of b_N and d_N for which, at that equilibrium where births balance deaths, a generation is a single time unit, is

$$b_N = \tfrac{1}{2} + r/2 - aN/2$$
$$d_N = \tfrac{1}{2} - r/2 + aN/2$$

To verify that this choice is indeed consistent with the equation in the text, notice that $M_N = N(b_N - d_N) = rN - aN^2$, while $V_N = N(b_N + d_N) = N$. In a variable environment, shifts in r inflate the variance by an amount proportional to N^2; V_N then becomes $N + \sigma^2 N^2$, where σ^2 is the "environmental variance." Plugging $M_N = rN - aN^2$, $V_N = N + \sigma^2 N^2$ into eq. A.7, we find

$$f(N) = \frac{2}{(1 + \sigma^2)N}\left(\frac{1 + \sigma^2 N}{1 + \sigma^2}\right)^{2r/\sigma^2 + 2a/\sigma^4 - 1}$$
$$\exp{[-2a(N - 1)/\sigma^2]}$$

Assuming $2r > \sigma^2 \gg a$, we obtain

$$f(N) = \frac{2}{(1 + \sigma^2)N}\left(\frac{1 + \sigma^2 N}{1 + \sigma^2}\right)^{2r/\sigma^2 - 1}$$

$$\exp\left[-2aN/\sigma^2\right]$$

We do not go far wrong if we set L_1 equal to

$$\frac{2}{(1 + \sigma^2)N}\left[\frac{\sigma^2}{1 + \sigma^2}\right]^{2r/\sigma^2 - 1}\int_1^\infty N^{2r/\sigma^2 - 2}$$

$$\exp\left(-2aN/\sigma^2\right)dN$$

The integrand is, up to a constant, the gamma distribution (Feller, 1971, p. 47); this distribution also arises in Kerner's (1957) theory of population fluctuation. If we take this spectrum of population sizes as representative of the variation of a single population over time (or, speaking more strictly, as representative of its "normal" fluctuations before plunging to extinction), then the average population size M will be $(2r - \sigma^2)/2a$; its variance S^2 will be $\sigma^2(2r - \sigma^2)/4a^2$, and its coefficient of variation CV, or S^2/M^2, will be $\sigma^2/(2r - \sigma^2)$. Making the suitable substitutions, we find

$$L_1 = \frac{2}{(1 + \sigma^2)N}\Gamma\left(\frac{1}{CV}\right)\left[\frac{2rM(CV)^2}{1 + CV(1 + 2r)}\right]^{1/CV}$$

The average lifetime of an established population is L_1/p, where p is the probability that the descendants of a colonizing individual will establish themselves and increase to exploit their island to capacity. According to MacArthur (1972b, pp. 121–123), $p = (b_1 - d_1)/b_1 = 2r/(1 + r)$.

Appendix B. The Stability of Some Purely Competitive Communities

To solve eqs. 9 of the text, we find eigenvectors c_{kj} and eigenvalues λ_k such that

$$\lambda_k c_{kj} = \sum_{m=1}^n c_{km} b_{mj} \quad \text{(B.1)}$$

Then we may obtain from eq. 9

$$\frac{d}{dt}\sum_{j=1}^n c_{kj}x_j = \sum_j c_{kj}\,dx_j/dt$$

$$= \sum_j c_{kj}k_j(t) - Ma\sum_{j,m} c_{kj}b_{jm}x_m(t)$$

Substituting in $\lambda_k c_{kj}$ for $\sum_m c_{km}b_{mj}$ from eq. B.1, the above becomes

$$\frac{d}{dt}\sum_{j=1}^n c_{kj}x_j = -Ma\lambda_k\sum_{j=1}^n c_{kj}x_j$$

$$+ \sum_{j=1}^n c_{kj}k_j(t) \quad \text{(B.2)}$$

Notice that if the k_j suddenly vanish, $Ma\lambda_k$ is the logarithmic rate at which $\sum_j c_{kj}x_j$ decays toward equilibrium. Recalling Liebig's "law of the minimum" (Lotka, 1925), we assume that the rate at which each x_j returns toward equilibrium is governed by the narrowest bottleneck in the system, i.e., by the smallest of the λ_k, which, following May (1973a), we call Λ. In this spirit, May and MacArthur (1972) approximated, or rather caricatured, eqs. 9 of the text by

$$dx_j/dt = -Ma\Lambda x_j + k_j(t) \quad \text{(B.3)}$$

(Like any good cartoon, this caricature threw into sharp—and authentic!—relief the features of overlap they wished to understand). To solve, we rewrite eq. B.3 as

$$x_j(t + dt) = x_j(t)(1 - Ma\Lambda\,dt) + k_j(t) \quad \text{(B.4)}$$

There is no reason for x_j to be more often or more strongly positive than negative: it is equally drawn toward the equilibrium from either side. The average of x_j is thus zero. At steady state the variance, here the mean square, of $x_j(t)$ is the same as that of $x_j(t + dt)$: we call them both $\overline{x_j^2}$. Taking the mean square of both sides of eq. B.4 yields the result

$$\overline{x_j^2} = \overline{x_j^2}(1 - 2Ma\Lambda \; dt) + \overline{k_j^2(t)} \,(dt)^2 \quad \text{(B.5)}$$

Since k_j introduces a variance σ_j^2 per unit time in the logarithmic rate of population change, it introduces variance $\sigma_j^2 \, dt$ during time dt: in other words, $\overline{k_j^2(t)} \,(dt)^2 = \sigma_j^2 \, dt$. Thus eq. B.5 implies

$$\overline{x_j^2} = \sigma_j^2/2Ma$$

If $\Lambda = 0$, there is no steady state; populations vary indefinitely until extinction occurs. Otherwise, if we assume that the variance in logarithm of population size approximates the population's CV, the condition $CV < 1$ implies $2Ma\Lambda > \sigma_j^2$.

Suppose now the overlap matrix b_{ij} is symmetric, so that $b_{ij} = b_{ji}$, and cyclic, so that $b_{ij} = b_{1,j-i+1}$ if $j \geqslant i$ and $b_{1,n+j-i+1}$ if $j < i$. Then b_{ij} depends only on the difference $i - j$, and we may express it as $b(i - j)$. Let $c_{kj} = \exp 2\pi i(k - 1)(j - 1)/n$, where $i = \sqrt{-1}$. Then

$$\sum_m c_{km} b_{mj}$$

$$= c_{kj} \sum_{m=1}^{n} e^{2\pi i(k-1)(m-j)/n} b(m - j) = \lambda_k c_{kj}$$

where

$$\lambda_k = \sum_{m=1}^{n} b_{1m} \exp 2\pi i(k - 1)(1 - m)/n$$

Consider now the specific matrix of p. 60, where $b_{11} = 1$, $b_{12} = b_{1n} = \frac{1}{2}$, and all other $b_{1j} = 0$. Then

$$\lambda_k = 1 + \frac{1}{2} \left[e^{-2\pi i(k-1)/n} + e^{2\pi i(k-1)/n} \right]$$

$$= 1 + \cos 2\pi(k - 1)/n$$

When $(k - 1)/n = \frac{1}{2}$ (which can be exactly true when n is even, and approximately true for any diverse assemblage), $\lambda_k = 0$.

However, if $1 > D/L > \frac{2}{3}$,

$$\lambda_k = 1 + 2\left(1 - \frac{D}{2L}\right) \cos 2\pi(k - 1)/n$$
$$+ 2\left(1 - \frac{D}{L}\right) \cos 4\pi(k - 1)/n$$

Λ rises to about $\frac{1}{6}$ for D/L between 1 and $\frac{2}{3}$ before declining to 0 at $D/L = \frac{2}{3}$. For smaller D/L, Λ exhibits progressively smaller humps.

Appendix C. The Stability of Some Two-Level Communities

Consider first a competitive community with a common predator feeding equally on all the prey, killing Pf grams per unit time for each gram of prey standing crop. The community matrix, which we call A, is then

$$\begin{array}{ccccc} -M'ab_{11} & -M'ab_{12} \cdots & -M'ab_{1n} & -Pf \\ -M'ab_{1n} & -M'ab_{11} \cdots & -M'ab_{1,n-1} & -Pf \\ \cdots\cdots\cdots\cdots\cdots\cdots\cdots\cdots\cdots\cdots\cdots \\ M'cf & M'cf & M'cf & -r \end{array}$$

where b_{ij} is the overlap matrix for the prey, M' the average population of each prey in the predator's presence, P the average predator population, r the strength of the predators' "self-damping" or territoriality, and c the proportion of prey weight killed converted into predator weight gain. Let C be the matrix

$$c_{jm} = \exp 2\pi i(j - 1)(m - 1)/n,$$
$$j, m \leqslant n \quad \text{(C.1)}$$
$$c_{n+1,j} = c_{j,n+1} = \delta_{n+1,j} \quad \text{(C.2)}$$

where $\delta_{jm} = 0$ if $j \neq m$ and 1 otherwise. The eigenvalues of CAC^{-1} are just those of A. Recalling that

$$c_{jk}^{-1} = (1/n)\exp 2\pi i(1 - j)(k - 1)/n, \text{if} j, k \leqslant n$$
$$c_{jk}^{-1} = c_{jk} \text{ if } j \text{ or } k = n + 1$$

we can show by straightforward calculations that the eigenvalues of CAC^{-1} are

$$-M'a\lambda_n, -M'a\lambda_{n-1} \cdots -M'a\lambda_2$$

and

$$-\frac{1}{2}[r + M'a\lambda_1$$

$$\pm \sqrt{(r - M'a\lambda_1)^2 - 4ncPM'f^2}]$$

where $\lambda_1 = \Sigma_j b_{1j}$ is the largest eigenvalue of the overlap matrix b_{ij} and the λ_i, $i > 1$, are the others. In general, I would expect the minimum eigenvalue of the competitive matrix to be lower than one of these last two, in which case the predator affects community stability only through its effect on M'.

Now consider the community of eqs. 14. If the predators are not territorial, the community matrix will be

$$\begin{bmatrix} -M'aB & -PF \\ M'cF & 0 \end{bmatrix}$$

where B is the matrix b_{ij} of competitive overlap among the prey and F the matrix f_{ij} of predator feeding rates: both these matrices are assumed to be symmetric and cyclic. Consider now the vector C_k, qC_k, whose first n components are $c_{k1}, \ldots, c_{kn}$ and whose second n are $qc_{k1}, \ldots, qc_{kn}$: C_k is the kth eigenvector for a cyclic overlap matrix (eq. C.1). Letting

$$\lambda_{1k} = \sum_j (b_{1j}/c_{kj}), \quad \Gamma_k = \sum_j (f_{1j}/c_{kj})$$

we find

$$(C_k, qC_k)\begin{bmatrix} -M'aB & -Pf \\ M'cF & 0 \end{bmatrix}$$

$$= (-M'a\lambda_{1k}C_k + qM'c\Gamma_kC_k, -P\Gamma_kC_k)$$

Our problem thus reduces to finding the eigenvalues of the 2×2 matrix

$$\begin{bmatrix} -M'a\lambda_{1k} & M'c\Gamma_k \\ -P\Gamma_k & 0 \end{bmatrix}$$

which are

$$-\frac{1}{2}[M'a\lambda_{1k} \pm \sqrt{M'^2a^2\lambda_{1k}^2 - 4cM'P\Gamma_k^2}]$$

The predator overlap matrix $B_{ij} = F_{ij}^2/v$, where v is the constant $\Sigma_j f_{1j}^2$, so the eigenvalues λ_{2k} of B_{ij} are Γ_k^2/v. Notice that $v = Q^2/BP^2$, where B is predator niche breadth and Q is predator kill rate per unit standing crop of prey: $Q/P = \Sigma_j f_{1j}$.

References

Andrewartha, H. G., and L. C. Birch. 1954. *The Distribution and Abundance of Animals.* University of Chicago Press, Chicago.

Berrie, P. M. 1973. Ecology and status of the lynx in interior Alaska. *In* R. L. Eaton, ed., *The World's Cats.* Vol. 1, pp. 4–41. World Wildlife Safari, Winston, Ore.

Bourliere, F. 1964. *The Natural History of Mammals.* Alfred A. Knopf, New York.

Bretsky, P., S. Bretsky, J. Levinton, and D. M. Lorenz. 1973. Fragile ecosystems. *Science* 179:1147.

Brooks, J. L., and S. I. Dodson. 1965. Predation, body size, and composition of plankton. *Science* 150:28–35.

Caughley, G. 1970. Eruption of ungulate populations, with special emphasis on the Himalayan thar of New Zealand. *Ecology* 51:53–73.

Cody, M. L. 1974. *Competition and the Structure of Bird Communities.* Princeton University Press, Princeton.

Connell, J. H. 1961. The influence of interspecific competition and other factors on the distribution of the barnacle *Chthalamus stellatus. Ecology* 42:710–733.

Connell, J. H. 1971. On the role of natural enemies in preventing competitive exclusion in some marine animals and rain forest trees. *In* P. J. den Boer and G. Gradwell, eds., *Dynamics of Populations* (Proc. Advan. Study Inst., "Dynamics of Numbers in Populations," Oosterbeek, 1970), pp. 298–312. Centre for Agricultural Publishing and Documentation, Wageningen, The Netherlands.

Dayton, P. K. 1971. Competition, disturbance and community organization: the provision and subsequent utilization of space in a rocky intertidal community. *Ecol. Monogr.* 41:351–389.

Diamond, J. M. 1973. Distributional ecology of New Guinea birds. *Science* 179:759–769.

Eisenberg, J. F., and E. Gould. 1970. The tenrecs: a study of mammalian behavior and evolution. *Smithsonian Contributions to Zoology* 27:1–138.

Elton, C. S. 1942. *Voles, Mice, and Lemmings.* Clarendon Press, Oxford.

Feller, W. 1971. *An Introduction to Probability Theory and Its Applications,* vol. 2, 2nd ed. Wiley, New York.

Fogden, M. P. L. 1972. The seasonality and population dynamics of equatorial forest birds in Sarawak. *Ibis* 114:307–344.

Holling, C. S. 1973. Resilience and stability of ecological systems. Ann. Rev. Ecol. Systematics, 4:1–23.

Hornocker, M. 1970. An analysis of mountain lion predation upon mule deer and elk in the Idaho Primitive Area. *Wildlife Monogr.* 21, The Wildlife Society.

Jordan, P., D. Botkin, and M. Wolfe. 1971. Biomass dynamics in a moose population. *Ecology* 52:147–152.

Keith, L. B. 1963. *Wildlife's Ten-year Cycle.* Wisconsin University Press, Madison.

Kerner, E. H. 1957. A statistical mechanics of interacting biological species. *Bull. Math. Biophysics* 19:121–145.

Kimura, M. 1964. Diffusion models in population genetics. *J. Appl. Prob.* 1:177–232.

Krebs, C. J. 1971. Genetic and behavioral studies in fluctuating vole populations. *In* P. J. den Boer and G. Gradwell, eds., *Dynamics of Populations* (Proc. Advan.

Study Inst., "Dynamics of Numbers in Populations," Oosterbeek, 1970), pp. 243–254. Centre for Agricultural Publishing and Documentation, Wageningen, The Netherlands.

Krebs, C. J., M. S. Gaines, B. L. Keller, J. H. Myers, and R. H. Tamarin. 1973. Population cycles in small rodents. *Science* 179:35–41.

Kruuk, H. 1972. *The Spotted Hyena.* University of Chicago Press, Chicago.

Lack, D. 1954. *The Natural Regulation of Animal Numbers.* Clarendon Press, Oxford.

Lack, D. 1966. *Population Studies of Birds.* Clarendon Press, Oxford.

Leigh, E. G. 1969. The ecological role of Volterra's equations. *In* M. Gerstenhaber, ed., *Some Mathematical Problems in Biology,* pp. 1–61. American Mathematical Society, Providence.

Levins, R. 1968. *Evolution in Changing Environments.* Princeton University Press, Princeton.

Lewontin, R. C., and D. Cohen. 1969. On population growth in a randomly varying environment. *Proc. Nat. Acad. Sci. U.S.A.* 62:1056–1060.

Lotka, A. J. 1925. *Elements of Physical Biology.* Williams and Wilkins, Baltimore.

MacArthur, R. H. 1955. Fluctuations in animal populations and a measure of community stability. *Ecology* 36:533–536.

MacArthur, R. H. 1972a. Strong, or weak, interactions? *In* E. S. Deevey, ed., *Growth by Intussusception,* pp. 179–188. Archon Books, Hamden.

MacArthur, R. H. 1972b. *Geographical Ecology.* Harper and Row, New York.

MacArthur, R. H., and E. O. Wilson. 1967. *The Theory of Island Biogeography.* Princeton University Press, Princeton.

Main, A. R., J. W. Shield, and H. Waring. 1959. Recent studies in marsupial ecology.

In A. Keast, R. Crocker, and C. Christian, eds., *Biogeography and Ecology in Australia,* pp. 315–331. W. Junk, den Haag.

Margalef, R. 1969. Diversity and stability: a practical proposal and a model of interdependence. *In* G. Woodwell and H. H. Smith, eds., *Diversity and Stability in Ecological Systems,* Symposium in Biology No. 22, pp. 25–37. U.S. Department of Commerce, Springfield, Va.

May, R. M. 1973a. *Stability and Complexity in Model Ecosystems.* Princeton University Press, Princeton.

May, R. M. 1973b. Stability in randomly fluctuating versus deterministic environments. *Amer. Natur.* 107:621–650.

May, R. M., and R. H. MacArthur. 1972. Niche overlap as a function of environmental variability. *Proc. Nat. Acad. Sci. U.S.A.* 69:1109–1113.

Mech, L. D. 1970. *The Wolf.* Natural History Press, Garden City.

Murie, A. 1944. The wolves of Mt. McKinley. *Fauna Nat. Parks, U.S. Fauna Series* 5:1–238.

Nellis, C., S. Wetmore, and L. B. Keith. 1972. Lynx-prey interactions in Central Alberta. *J. Wildlife Mgt.* 36:320–328.

Paine, R. T. 1966. Food web complexity and species diversity. *Amer. Natur.* 100:65–75.

Pearson, O. P. 1966. The prey of carnivores during one cycle of mouse abundance. *J. Animal Ecol.* 35:217–233.

Pfeffer, P. 1967. Le mouflon de Corse (*Ovis ammon musimon,* Schreber 1782). Position systematique, écologie et éthologie comparée. *Mammalia* 31, supplement.

Pimlott, D., J. Shannon, and G. Kolenosky. 1966. *The Ecology of the Timber Wolf in Algonquin Park.* Department of Lands and Forests, Ontario.

Richter-Dyn, N., and N. S. Goel. 1972. On the extinction of a colonizing species. *Theor. Pop. Biol.* 3:406–433.

Schaller, G. 1972. *The Serengeti Lion.* University of Chicago Press, Chicago.

Snow, D. W. 1962. A field study of the black and white manakin, *Manacus manacus,* in Trinidad. *Zoologica* 47:65–104.

Spencer, D., and S. Lensink. 1970. The muskox of Nunivak Island, Alaska. *J. Wildlife Mgt.* 34:1–14.

Spinage, C. A. 1969. Population dynamics of the Uganda defassa waterbuck (*Kobus defassa ugandae* Neumann) in the Queen Elizabeth Park, Uganda. *J. Animal Ecol.* 38:51–78.

Tinkle, D. W. 1969. The concept of reproductive effort and its relation to the evolution of life histories of lizards. *Amer. Natur.* 103:501–516.

Volterra, V. 1931. *Leçons sur la Théorie Mathematique de la Lutte pour la Vie.* Gauthier-Villars, Paris.

Williams, C. B. 1958. *Insect Migration.* Collins, London.

Williams, E. E. 1972. The origin of faunas. Evolution of lizard congeners in a complex island fauna: a trial analysis. *Evolutionary Biol.* 6:47–89.

Willis, E. O. 1967. *The Behavior of Bicolored Antbirds.* University of California Publications in Zoology, vol. 79. University of California Press, Berkeley.

Willis, E. O. 1972. *The Behavior of Spotted Antbirds.* Ornithological Monographs No. 10. American Ornithologists' Union, Lawrence.

Willis, E. O. 1973. The behavior of ocellated antbirds. *Smithsonian Contributions to Zoology* 144:1–57.

Wright, S. 1939. *Statistical Genetics in Relation to Evolution.* Exposés de Biométrie et de Statistique Biologique 802. Hermann et Cie, Paris.

3 Environmental Fluctuations and Species Diversity

John W. MacArthur

The problem of what determines the number of species of any given group of organisms that occurs at any given place on the earth has intrigued naturalists for centuries. In particular, anyone who visits the tropics is impressed by the great variety of living things there, expecially in contrast to the temperate or arctic zones. It has long been clear that there is some correlation between harshness of climate and a reduced species diversity, but there has been little success in finding quantitative relationships. In this chapter, I will present a formula that seems to have quite good luck in connecting number of species with climate.

The hope is, of course, that this method can be applied to a wide variety of organisms and climates, from gastropods to birds and from the tropics to the arctic regions. I have tried these examples, and more, and I will show the results of some of them here. The formula that appears to fit well is:

$$S_T = A \ln (1 + B/\eta) \qquad (1)$$

where S_T is the total number of species to be found in a small region, η is some measure of environmental fluctuation or "noise" for that region, and A and B are unknown parameters to be fitted by least squares analysis for each group of organisms. I will give the rationale for this function later.

Data

In each case, I have chosen a transect with several requirements in mind: the habitat must be as uniform as possible in all respects except for the variation of climate, the data for climate must be available, and the observations on numbers of species must be available. In most cases, the measure of climatic variability used is ΔT, the extreme range of monthly mean temperatures (usually mean July temperature less mean January temperature), taken from tables in a meteorology textbook (Miller and Thompson, 1970) or from World Weather Records (Clayton, 1944). This climate index is roughly equivalent to seasonality in the usual sense. The data are mostly in Fahrenheit degrees, so I have retained this scale in the interest of simplicity. The species data come from various sources, as noted.

The transects for the curves in Figures 1 through 9, as explained in the legends, are all in the western half of North America, although they extend from tropical to arctic climates. The motive is simply to maintain uniform physiography—mountains, mountains and adjacent plains, coastal plains, or non-coral coastal waters—while having a wide range of climates. The presence of the Gulf of Mexico plainly makes all of these impossible in the eastern half of the continent. Further-

74

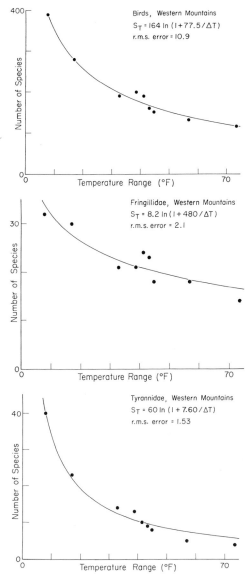

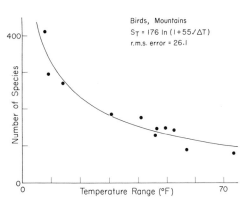

Figure 4 This transect corresponds to the mountainous belt from Coban, Guatemala, to Fairbanks, Alaska, as in Figures 1, 2, 3, but it passes somewhat farther east, through Utah, Montana, and the Canadian Rockies. Species data are from *Geographical Ecology* (MacArthur, 1972, p. 212), squares 300 miles on a side, and climate data are averaged over all available points in each square. Eleven data points and a fitted curve are given.

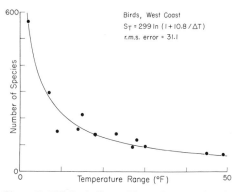

Figures 1, 2, 3 Plots of S_T, the total number of species to be found in a small region, against ΔT, the extreme range of monthly mean temperatures. Figure 1 is based on all bird species; Figures 2 and 3, on the bird families Fringillidae + Emberizidae and Tyrannidae, respectively. The transect is in or near the western mountains of North America, 100 to 300 miles from the Pacific coast; the points chosen are approximately every 500 miles from Coban, Guatemala, through central Mexico, Arizona, Nevada, eastern Washington, central British Columbia, to Fairbanks, Alaska. Species data are from contour maps given by Cook (1969), representing sampling squares 150 miles on a side. Nine data points and a least-squares-fitted curve of the type discussed in the text are shown.

Figure 5 This is similar to Figure 4, except that the squares are along the west coast, not extending far inland. The point at the upper left is Panama, whose species number is taken from *Geographical Ecology* (MacArthur, 1972, p. 135), and the lowest one is Nome, Alaska. Twelve data points, about 500 miles apart as before, and the fitted curve, are shown.

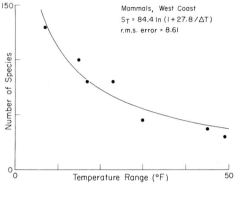

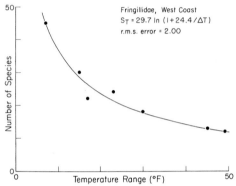

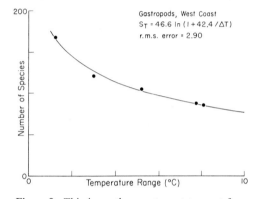

Figure 9 This is another west-coast transect from San Diego to Alaska, where both the temperature and species data have been read from a graph given by Fischer (1960). Only five points are identified on that graph.

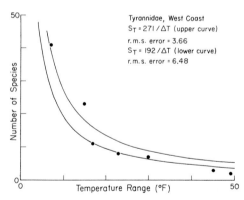

Figures 6, 7, 8 These represent a transect on the west coast, from southern Mexico to Nome, Alaska, with points about every 800 miles. Mammal data are from Simpson (1964); numbers of Fringillidae + Emberizidae and Tyrannidae are from Cook (1969). In Figure 8, the upper curve is a simple least-squares fit, and the lower one is a logarithmic least-squares fit (see the text for discussion of this and of the hyperbolic equation).

76

more, I could not readily obtain the requisite species data for other continents, although I should expect that, say, the Andes or the adjacent coastal plain of western South America would exhibit equally good results.

The root-mean-square (r.m.s.) differences between least-squares fitted curves and data points vary up to about 9% of the maximum species numbers, but more generally are 5% or less. This is good enough to prompt the question: is ΔT just an exceptionally lucky choice for the measure of environmental fluctuation, or would some other measure do just as well? I have tried, and eliminated, several other possible measures: standard deviation of mean temperature, $\sigma(T)$, during the breeding season, an index of weather predictability (Figure 11); reciprocal of actual evapotranspiration (AE) calculated from Turc's (1955) formula, roughly proportional to inverse of productivity (Figure 12); frost-free days can be ruled out since they become constant (365) south of some point well above the tropical end of each curve and, furthermore, they are meaningless for the gastropod case. I also ruled out rainfall, since the curve of rainfall against latitude is strongly bimodal,

Figures 10, 11, 12 These are all along a transect in the western mountains, as for Figures 1, 2, 3, with data points wherever the necessary climate data were available. The species data are from distributions of all bird species in Cook (1969), and the abscissas are ΔT (calculated as for all the other figures), $\sigma(T)$ (the standard deviation of mean temperature in the breeding month), and the reciprocal of actual evapotranspiration, calculated by Turc's (1955) formula. See the text for discussion of the equation for the curve in Figure 11 and the rationale for Figure 12. The curve using ΔT is much the best fit.

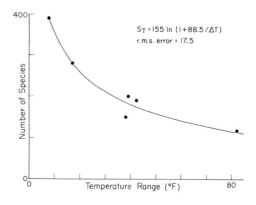

$$S_T = 155 \ln (1 + 88.5/\Delta T)$$
r.m.s. error = 17.5

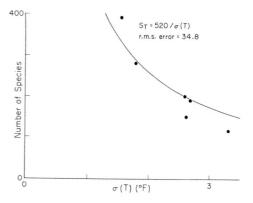

$$S_T = 520/\sigma(T)$$
r.m.s. error = 34.8

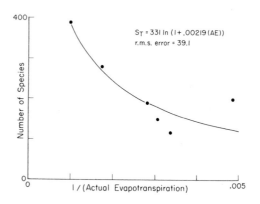

$$S_T = 331 \ln (1 + .00219\,(AE))$$
r.m.s. error = 39.1

whereas of course the species curve is not, and again it is hard to see how rainfall could affect gastropods. In order to exhibit the comparison, I have drawn up Figures 10, 11, and 12 along a single transect in the western mountains, and have plotted data points wherever I could find the necessary detailed climate data, using ΔT, $\sigma(T)$, and AE as the abscissas.

Theory

The formula used in the graphs derives from an intuitive analogy between information theory and evolution. Information must be calculated by a logarithmic measure (often used to find species diversity) because the total content of two messages must be additive, whereas the total probability of the combined message is multiplicative, and a logarithmic function is the only one having this property [see any text on information theory, or Shannon's (1948) original work]. Analogously, the quantity of evolution involved in two successive advancements should be additive, but the overall probability of the evolutionary advance is multiplicative, suggesting a logarithmic function to measure evolution. The fundamental theorem in information theory gives the rate at which information can be transmitted through a noisy channel of communication, and the evolutionary counterpart suggests that

$$R = A \ln (1 + B/\eta) \qquad (2)$$

where R is the rate at which evolution occurs (probably proportional to the number of species present, S_T, for an equilibrium situation between immigration or speciation and extinction), η is some measure of environmental noise, and A (the counterpart of bandwidth) and B (that of signal strength) are to be determined empirically because their biological meanings are not clear. Using the species as a unit for the measure of evolution, we see that eq. 1 follows directly.

It is now evident that η is some measure of whatever it is that opposes the establishment of new species, just as the electrical noise in a channel reduces the rate at which information can be transmitted. In a noisy channel, the signal can not be subdivided as finely as in the absence of noise, reducing the number of distinct levels available to carry information; in a "noisy" environment, the resources can not be subdivided as finely, reducing the number of species that can occupy that environment. Parameter A, the analog of bandwidth, is related to the rate at which potential new species appear (either by speciation or introduction) and to the proportionality constant between rate of establishment of species and their equilibrium number (this proportionality constant also equals the extinction rate per species). B is a measure of the range of resources to be subdivided, taken for the entire group of organisms under consideration. Both, presumably, are held reasonably constant if the habitat is uniform except for climate.

The kinds of events that might exterminate a newly introduced species before it becomes thoroughly established include sudden cold snaps in the breeding season, prolonged frosts, abnormally heavy rain or snow, violent thunderstorms or hurricanes or other such natural phenomena, as well as chance predation or disease. Casual observation in Vermont and elsewhere suggests that it is the first of these that is dominant, and I have assumed that a cold spell in May or June is a minor and temporary reversion toward winter so that the probability of a serious cold snap is proportional to how cold winter is compared to summer. Hence I arrive at ΔT as the easiest measure of environmental noise to obtain. The parameter B has now absorbed another proportionality constant, further obscuring its biological meaning.

The transects used, then, must maintain A and B as close to constant as possible. For example, the Pacific coast has very few violent storms, fairly predictable (although not uniform) rainfall, a narrow coastal plain leading up into the mountains, and the moderating influence of the ocean to windward—all characteristics possessed quite uniformly over its entire length.

Robert May (personal communication) has suggested another approach to the origin of the formula, eq. 1. If η is that which opposes the establishment of a new species, then $1/\eta$ must be related to that which makes the introduction of new species possible—for example, productivity, which is proportional to the total number of individuals present, N_T. Thus eq. 1 becomes equivalent to the logseries distribution of species abundances

$$S_T = A \ln (1 + BN_T)$$

(May, Chapter 4). Assuming that productivity can be estimated by actual evapo-transpiration (*fide* Rosenzweig, 1968), and that the latter can be calculated by Turc's (1955) formula, we are led to the idea tested in Figure 12. The fit is satisfactory but not as good as that obtained by using ΔT for noise. Some of this difference may be due to the fact that ΔT is relatively constant from plains to adjacent mountains, but evapotranspiration depends on mean annual temperature and rainfall, which vary sharply, and the meteorological data are nearly all from the plains.

Three remarks are in order, regarding the fitting of a curve of the family of eq. 1 to a set of data points. First, of course, it is a two-parameter family and thus a perfect fit is possible through any two reasonable points. It is when more than two data points are available that one begins really to test the goodness of fit of the smooth curve to the discrete points. Second, at the limit as $A \to \infty$ and $B \to 0$ simultaneously, the curve goes over into a simple rectangular hyperbola because the first term in the series expansion of $\ln (1 + x)$ is just x, so that $A \ln (1 + B/x) \to (AB)/x$. In two cases (Figures 8 and 11), the process of finding the best-fit curve leads to this limiting form, and it is the hyperbolic formula that I have given. Lastly, it might be argued that it is not the *difference* between data point and curve that should go into the least-squares process but their *ratio* (i.e., difference of logarithms). In most cases, the result is only trivially different, but in Figure 8 I have given both curves, as they differ noticeably.

Conclusions

It is now obvious that, whether or not one accepts a parallelism between the process of evolution of species and the transmission of information in a noisy channel, the formula that results from such a line of thought fits the facts surprisingly well, at least in a limited set of circumstances. For further work, three avenues of approach suggest themselves. First, the range of organisms covered should be greatly extended to include the plant kingdom as well as many other animals; there is no telling what pattern might develop from this extension. Second, as I have suggested, the geographical range should be increased to include other continents, at least wherever the criterion of uniform habitat can be met. Lastly, meanings should be sought for the parameters *A* and *B*, because they do contain biological significance, although it is obscured by the analogy process and by combining these parameters with several proportionality constants.

Acknowledgments

My primary debt of gratitude, of course, is to Robert MacArthur for a lifetime of discussions on this and a virtually infinite range of other subjects. Particular thanks are also due to Richard Lewontin for many helpful suggestions and discussions, to Robert May for suggesting the possible equivalence to the logseries distribution, and to the Sloan Foundation for the mini-computer that made the curve-fitting possible.

References

Clayton, H. H., ed. 1944. *World Weather Records*. Smithsonian Institution, Washington.

Cook, R. 1969. Species density of North American birds. *Syst. Zool.* 18:63–84.

Fischer, A. G. 1960. Latitudinal variation in organic diversity. *Evolution* 14:64–81.

MacArthur, R. H. 1972. *Geographical Ecology*. Harper and Row, New York.

Miller, A., and J. C. Thompson. 1970. *Elements of Meteorology*. Merrill Publishing Company, Columbus.

Rosenzweig, M. R. 1968. Net primary productivity of terrestrial communities: Prediction from climatological data. *Amer. Natur.* 102:67–74.

Shannon, C. E. 1948. A mathematical theory of communication. *Bell Syst. Tech. Jour.* 27:379–423.

Simpson, G. G. 1964. Species density of North American recent mammals. *Syst. Zool.* 13:57–73.

Turc, L. 1955. Le bilan d'eau des sols. *Ann. Agron.* 6:5–131.

4 Patterns of Species Abundance and Diversity

Robert M. May

Contents

1. Introduction

If the relative abundances of the species in a particular plant or animal group in a given community are somehow measured, there will be found some common species, some rare species, and many species of varying intermediate degrees of rareness. These species abundance relations (the relations between abundance and the number of species possessing that abundance) are clearly of fundamental interest in the study of any ecological community. Different types of such species abundance relations have been proposed on theoretical grounds, and are observed in real situations. What these relations mean, and how they are best characterized, has been the subject of considerable discussion, much of it focused on one or another particular aspect of a specific species-abundance relation.

This chapter seeks to give an analytic review of the subject, mainly with the aim of disentangling those features that reflect the biology of the community from those features that reflect little more than the statistical law of large numbers.

This discussion of species-abundance relations also provides the basis for a consideration of species-area relations (the relation between the area of real or virtual islands, and the number of species on the island), and for some rough suggestions

81

relating to practical problems of sampling all the species in a community.

One single number that goes a long way toward characterizing a biological community is simply the total number of species present, S_T. Another interesting single number is the total number of individuals, N_T; alternatively, if the number of individuals in the least abundant species be m, the total population can be expressed as the dimensionless ratio

$$J = N_T/m \qquad (1.1)$$

Although there may be monumental difficulties in determining S_T and J in practice (e.g., Matthew, 10:29–31), such a census is possible in principle, and the bulk of this article deals with properties of the actual species-abundance distribution. Some sketchy remarks on sampling problems are deferred to Section 6.

Going beyond this gross overview of the community in terms of S_T and J, one may ask how the individuals are distributed among the species. That is, what is N_i, the number of individuals in the ith species? Such information may be expressed as a probability distribution function, $S(N)$, where

$$S(N)\, dN = \begin{cases} \text{number of species} \\ \text{each of which} \\ \text{contains} \\ \text{between } N \text{ and} \\ N + dN \\ \text{individuals} \end{cases} \qquad (1.2)$$

That is, roughly speaking, $S(N)$ is the number of species with population N. The quantities S_T and N_T follow immediately from this distribution function:

$$S_T = \int_0^\infty S(N)\, dN \qquad (1.3)$$

$$N_T = \int_0^\infty NS(N)\, dN \qquad (1.4)$$

[$S(N)$ has been expressed here as a continuous distribution. For a discrete distribution $S(N)$, sums replace the integrals in eqs. 1.3 and 1.4, and elsewhere. For relatively large populations this distinction between continuous and discrete distributions is generally unimportant, and it is usual for field data to be plotted as a histogram or as discrete points, and compared with the continuous curves generated by theoretical distributions. For further discussion, see Bliss (1965).]

There are a variety of currently conventional ways in which the properties of a particular species-abundance distribution, $S(N)$, may be displayed. These are surveyed in Section 2. This is largely review material, but it may be illuminating to gather together, and explicitly to relate, these superficially different ways of exhibiting the same information.

In Section 3 the most significant distributions $S(N)$ are singled out. For each distribution, we briefly review underlying theoretical ideas that lead to it, and such evidence as may be culled from appropriate field situations.

First, and most important, is the lognormal distribution (Section 3 A). Theory and observation point to its ubiquity once $S_T \gg 1$, when relative abundances must be governed by the conjunction of a vari-

ety of independent factors. In general, *two*
parameters[1] are needed to characterize a
specific lognormal distribution; these may
be S_T and J. A further assumption as to
details of the shape of the lognormal distribution reduces it to a special *one*-
parameter family of "canonical" lognormal distributions; in this event S_T
alone, or J alone, is sufficient to specify
the distribution uniquely. This empirical
assumption (Preston, 1962) fits a lot of
data, but no explanation has previously
been advanced as to why it may be so.
Another empirical general rule relates to
the width parameter in the conventional
expression for the general lognormal. This
rough rule, $a \sim 0.2$, has invited much
speculation since first enunciated by
Hutchinson in 1953. In Section 3 A it is
argued in some detail that both rules derive from *mathematical* properties of the
distribution, being roughly fulfilled by a
wide range of general lognormal distributions, for a wide range of the values S_T
and J found in nature.

MacArthur's (1957, 1960) "broken-
stick" distribution (Section 3 B) may be
derived in various ways; it is specified by
one parameter, namely S_T. As discussed
most fully by Webb (1973), who calls it
the "proportionality space model," this
distribution of relative abundance is to be

expected whenever an ecologically homogeneous group of species apportion
randomly among themselves a fixed
amount of some governing resource. For
appropriately small and homogeneous
taxa, field observations seem to fit this
distribution.

Two other interesting distributions
(Section 3 C) are the simple geometric
series and the logseries. If the community
ecology is dominated by some single factor, and if division of this niche volume
proceeds in strongly hierarchical fashion
with the most successful species tending
to preempt a fraction k, and the next a
fraction k of the remainder, and so on, we
arrive at a geometric series distribution (as
the ideal case), or a logseries distribution
of relative abundance (as the statistically
realistic expression of this underlying picture). A few natural communities, particularly simple plant communities in harsh
environments, conform to these patterns.
(The logseries also can often arise as a
sampling distribution, as mentioned in
Section 6.) Both geometric and logseries
are *two*-parameter distributions: for the
geometric series the parameters are usually S_T and k (but are alternatively S_T and
J); and for the logseries they are conventionally S_T and "α", or "α" and "x" (but
alternatively S_T and J).

In brief, if the pattern of relative abundance arises from the interplay of many
independent factors, as it must once S_T is
large, a lognormal distribution is both
predicted by theory and usually found in
nature. In relatively small and homogeneous sets of species, where a single factor
can predominate, one limiting case (which

---

[1] A third parameter, namely N_T or m or N_0, would
be needed if one were interested in absolute values
of numbers of individuals, i.e., in the absolute position along the R axis of the peak in the distribution.
By working with the ratio J, eq. 1.1, this need is
circumvented, and for essentially all ecological purposes (rank-abundance curves, diversity and dominance indices, rough estimates of sampling properties, species-area curves) two parameters suffice.
Similar remarks pertain to the other distributions.

may be idealized as a perfectly uniform distribution) leads to MacArthur's broken-stick distribution, whereas the opposite limit (which may be idealized as a geometric series) leads to a logseries distribution. These two extremes correspond to patterns of relative abundance which are, respectively, significantly more even, and significantly less even, than the lognormal pattern. In other words, *the lognormal distribution reflects the statistical Central Limit Theorem; conversely, in those special circumstances where broken-stick, geometric series, or logseries distributions are observed, they reflect features of the community biology.*

Many people have sought to go beyond the simple characterization of a community by the two numbers S_T and N_T (or J), yet stop short of describing the full distribution $S(N)$, by adding one further single number which will describe the "evenness" or "diversity" or "dominance" within the community. In Section 4, the theoretical relationships between such quantities and S_T are exhibited for each of the major species-abundance distributions $S(N)$ mentioned above, and their features are compared. Some of these relations have been explored for particular distributions by previous authors, but a comparative anatomy is lacking, and the lognormal distribution (which I regard as the most important) has received essentially no attention of this kind. The statistical variance to be expected in the usual Shannon-Weaver diversity index, H, is also discussed for the various distributions. We see that for relatively small S_T the various distributions $S(N)$ lead to

H-versus-S_T curves, the differences between which are of the same order as the statistical noise to be expected in any one such relation. Conversely, for $S_T \gg 1$ it is difficult to see how the distribution could be other than lognormal.

The canonical lognormal and broken-stick are one-parameter distributions. Therefore if S_T is given, J may be calculated, leading to unique relations of S_T versus J for these two species-abundance distributions. The general lognormal, geometric series, and logseries are two-parameter distributions, leading in each case to a one-dimensional family of curves of S_T versus J. Unique curves may be specified by assigning a value to the remaining parameter (i.e., by specifying γ for the lognormal, k for the geometric series, α for the logseries). Such S_T-versus-J relations may be converted to species-area relations by adding the independent biological assumption that the number of individuals is roughly proportional to the area, A:

$$J = \rho A \qquad (1.5)$$

where ρ is some constant. This assumption is of doubtful validity, but may serve as a reasonable estimate in island biogeographical contexts, with "island" interpreted in a broad sense (MacArthur and Wilson, 1967). Preston (1962) and MacArthur and Wilson have shown that the canonical lognormal distribution in conjunction with eq. 1.5 leads to a species-area curve that accords with much field data. This work begs the question of where the *canonical* lognormal came from in the first place. In Section 5 it is shown that for all reasonable lognormal distributions (the canonical lognormal being merely one special case) one gets species-area curves in rough agreement with the data. Although none of these relations are simple linear regressions of $\ln S_T$ on $\ln A$ (they have a steeper dependence of S_T on A for small A than for large A), they point to the approximate rule

$$\ln S_T \simeq x \ln A + \text{(constant)} \quad (1.6)$$

with x in the range around 0.2 to 0.3. This agrees with the data (Table 5), and suggests that the property is a rather general consequence of the lognormal species-abundance distribution (not just of the special canonical distribution). We also note the species-area relations predicted by eq. 1.5 in conjunction with the broken-stick distribution (significantly steeper than the lognormal curves), and with the geometric series or logseries distributions (significantly less steep than the lognormal curves).

Section 6 pays very brief attention to some of the problems that arise with samples that are not big enough for all species typically to be represented. Such considerations have been extensively reviewed elsewhere (e.g., Pielou, 1969; Patil, Pielou, and Waters, 1971). It is noted that sampling distributions often tend to be of logseries form (e.g., Boswell and Patil, 1971). One consequence is that if the species-area curve reflects sampling properties, with A being a simple measure of sample size (in contrast to Section 5 where S_T versus A reflected the actual species-abundance pattern along with the biological assumption eq. 1.5), the relation may

be

$$S_T \simeq \alpha \ln A + \text{(constant)} \quad (1.7)$$

Here α is a parameter of the logseries distribution (see Section 3 C). This point is briefly discussed and applied to some field data in Section 6. Estimates of the fraction of species likely to be present in small samples from the lognormal and the broken-stick distributions are also given.

A series of appendices sets out the mathematical properties of the various distributions treated in the main text. Such properties as have been discussed by earlier authors are simply listed, whereas the new work is usually developed somewhat more fully. The appendices are not exercises in mathematical pedantry, but form the backbone of the paper. They are intended to be useful to those who seek a thorough understanding of the morphology of the various distributions. On the other hand, the main text simply quotes the results as they are required, and is designed to present the main points in a self-contained and generally intelligible way, free from mathematical clutter.

Section 7 provides a summary.

2. Various Ways of Presenting $S(N)$

One way of displaying the information inherent in the species abundance distribution function $S(N)$ is to rank the populations N_i $(i = 1, 2, \ldots, S_T)$ in order of decreasing abundance, with the subscript i denoting the rank in this sequence. Abundance-rank diagrams, with abundance N_i/N_T as the y axis, and rank i as the x axis may thus be drawn, and experimental findings compared with theory. Minor variants follow from the use of log (abundance) rather than abundance for the y axis (e.g., Figure 1), and/or log (rank) rather than rank for the x axis.[2]

The characteristic features of such abundance-rank presentations of lognormal, broken-stick, geometric series, and logseries distributions have been clearly and comprehensively reviewed by Whittaker (1965, 1970, 1972). He points out the distinctive character of these various curves. The choice of logarithmic or linear scale for the axes has tended to depend on the predisposition of the author. Broken-stick people tend to use abundance versus log (rank), whereupon the broken-stick $S(N)$ shows up as nearly linear [see eq. D.3, $i \simeq S_T \exp(-S_T N/N_T)$, whence the abundance-ln (rank) curve is approximately linear with slope $-S_T$]. On a log (abundance) versus rank labelling, geometric series are exactly linear [see eq. E.1, whence the ln (abundance)-rank curve is roughly linear with slope $-(1-k)$], and logseries distributions are approximately linear [see eq. F.10, $i \simeq -\alpha \ln(N/N_T)$, whence the ln (abundance)-rank curve is roughly linear with slope $-\alpha$]. The lognormal abundance versus rank curve has a shape intermediate between these two extremes. Figure 1 illustrates this.

$S(N)$ may be displayed more directly by drawing a histogram of the number of species whose populations lie in specific ranges. Williams (1964), for example,

gives many such "frequency plots" of $S(N)$ against N for various assemblages of data. Alternatively, a logarithmic scale may be employed for the abundance; that is, the x axis is log N. Following Preston's (1948, 1962) work, it has become customary to employ logarithms to base 2 (so that the x axis, the logarithmic abundance scale, is divided into "octaves").

Yet again, Williams (1964) has frequently analysed species-abundance data by plotting the logarithm of the abundance (x axis) against the accumulated fraction of species up to that abundance (y axis is $\int_0^N S(N') dN'$). By using so-called probability paper (a gaussian scale on the y axis), Williams thus arranges that a lognormal distribution will show up as a straight line on such a plot. This device for producing a straight line from a lognormal $S(N)$ may be viewed as the analogue of plotting abundance versus log (rank) to get roughly a straight line from the broken-stick $S(N)$, or log (abundance) versus rank to get roughly a straight line from geometric series or logseries $S(N)$.

The essential thing to appreciate here is that *all* such diagrams, whether one or another form of plot of abundance versus rank, or Williams- or Preston-style frequency plot, are simply interchangeable and equivalent ways of expressing the distribution $S(N)$. Intercomparison of different bodies of data would be facilitated, and a certain amount of confusion removed, if some standard format could be agreed on.

This section ends with an explicit formula relating abundance-rank plots to the distribution function $S(N)$. Define $F(N)$ to

---

[2]Throughout, the usual convention is employed whereby log denotes logarithms to base 10, ln denotes logarithms to base e: ln $x = (2.303)$ log x.

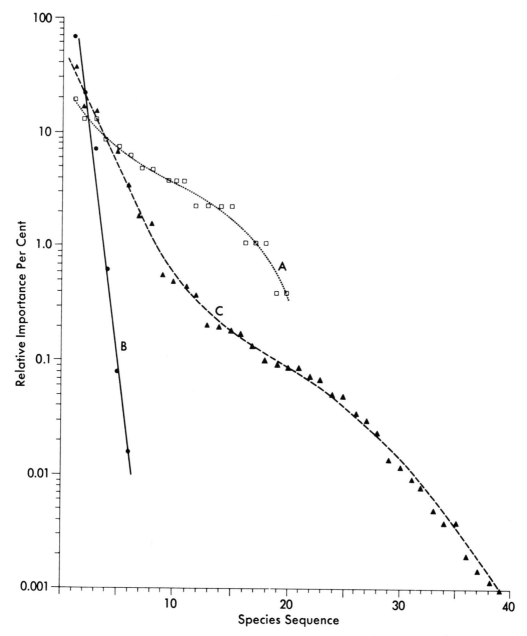

Figure 1 Relationship between relative abundance or importance (expressed as a percentage on a logarithmic scale) and rank of species for three natural communities. This figure, from Whittaker (1970), bears out the remarks in the text. Curve A is for a broken-stick distribution, and is fitted by data from a relatively small community of birds; curve B is for a geometric series distribution, and is fitted by a subalpine plant community; curve C is for a lognormal, and is fitted by the plant species in a deciduous forest.

87

be the total number of species with populations in excess of N:

$$F(N) = \int_N^\infty S(N') \, dN' \qquad (2.1)$$

Although $F(N)$, so defined, is a continuous function, it clearly describes the relation between rank i and abundance N_i; as N decreases from infinity, $F(N)$ attains integer values of 1, 2, etc., until $F(N) \to S_T$ as $N \to 0$. Thus a plot of N (y axis) versus $F(N)$ (x axis, scaled 1 through S_T) *is* the abundance-rank diagram. $F(N)$ may be christened the rank-order function. A geometrical prescription for converting an abundance-rank plot [N versus $F(N)$] into a frequency plot [$S(N)$ versus N] follows from the above remarks: first interchange x and y axes in the abundance-rank figure [to get $F(N)$ versus N], then calculate the slope of this curve at each point [slope $= dF/dN = -S(N)$ from eq. 2.1], and change the sign to arrive at the plot of $S(N)$ versus N. Appendix D applies this procedure explicitly in a discussion of the broken-stick distribution.

3. Specific Distributions

Some of the salient forms proposed for the species abundance distribution $S(N)$ are now reviewed, both with respect to theoretical ideas which lead to the distribution, and to corroborative evidence from field data. Note that while specific biological assumptions imply a unique $S(N)$, the converse is not true; a variety of different circumstances can imply the same $S(N)$. This lack of uniqueness in making ecological deductions from observed $S(N)$ discouraged MacArthur (1966). Even so, worthwhile distinctions can be made between properties which stem from the statistical Central Limit Theorem as opposed to broad biological features.

A: Lognormal Distribution

(i) *Theory.* At the outset, there is a need to distinguish two qualitatively different ecological regimes, commonly referred to as those of "opportunistic" and of "equilibrium" species. In the former limit, the ever-changing hazards of a randomly fluctuating environment can be all-important in determining populations, and thus relative abundances; in the latter limit, a structure of interactions within the community may, at least in principle, control all populations around steady values. One way of expressing this distinction is to define $r_i(t)$ to be the per capita instantaneous growth rate of the ith species:

$$r_i(t) = \frac{1}{N_i(t)} \frac{dN_i(t)}{dt} \qquad (3.1)$$

This growth rate may of course vary systematically or randomly from time to time, and may itself depend on the population of the ith and other species. Formally, however, eq. 3.1 integrates to

$$\ln N_i(t) = \ln N_i(0) + \int_0^t r_i(t') \, dt' \qquad (3.2)$$

MacArthur (1960) discussed the opportunistic limit as being that where on the right hand side in eq. 3.2 the integral is more important than $\ln N_i(0)$, so that the population at time t is essentially unrelated to that at $t = 0$. Conversely, in the equilibrium limit the ln $N_i(0)$ term is more important than the integral; the populations are relatively unvarying, and the relative abundances form a steady pattern.

In the opportunistic regime, environmental vagaries predominate, and the $r_i(t)$ will vary randomly in time. Thus for any one population, labelled i, the accumulated integral of $r_i(t)$ in eq. 3.2 is a sum of random variables. In accord with the Central Limit Theorem (a theorem to the effect that essentially all additive statistical distributions are asymptotically gaussian, or "normal"), this integral will then in general be normally distributed (Cramer, 1948). Hence ln $N_i(t)$ is normally distributed, leading to a lognormal distribution for the population of any one species in time, and consequently a lognormal for the overall community species-abundance distribution at any one time.

The essential point here, and elsewhere throughout Section 3 A, is that populations tend to increase geometrically, rather than arithmetically, so that the natural variable is the *logarithm* of the population density. This central point has been particularly stressed by Williamson (1972, Chapter 1), Williams (1964), Montroll (1972), and others.

In the equilibrium regime, a lognormal distribution of relative abundance among the species is again most likely, once one deals with communities comprising a large number of species fulfilling diverse roles. In this event, Whittaker (1970, 1972) and others have observed that the distribution of relative abundance is liable to be governed by many more-or-less independent factors, compounded multiplicatively rather than additively, and again the Central Limit Theorem applied to such a product of factors suggests the lognormal distribution.

Alternatively, MacArthur (1960) and Williams (1964) have noted that a suggestion of Fisher's (1958) concerning community evolution can imply a lognormal distribution. Assuming roughly that beneficial genes are fixed at a rate proportional to population size, the relative abundances of species in a large community will be lognormally distributed.

In brief, the lognormal distribution is associated with products of random variables, and factors that influence large and heterogeneous assemblies of species indeed tend to do so in this fashion. Such considerations apply quite generally to multiplicative processes where, as it were, the rich grow richer (10% of 10^7 is more exciting than 10% of $10). Thus the distribution of wealth in the United States could be expected to be lognormal, and data in the *Statistical Abstract* (1971) show this to be so. Similarly, McNaughton and Wolf (1973, p. 629) have shown that the international distribution of human populations among the nations of the world, and even the distribution of the gross national products of nations, is lognormal.

(ii) *Mathematical Description of Lognormal Species-Abundance Distribution* Fuller mathematical details are given in Appendix A. Pielou (1969, Chapter 17) discusses the relation between discrete and continuous lognormal distributions.

A lognormal distribution may be written in standard form as

$S(N)$

$$= S_0 \exp\left[-(\ln N - \ln N_0)^2/2\sigma^2\right] \quad (3.3)$$

Here N_0 is the number of individuals in the modal species (the species at the peak of the species-abundance curve in Figure 2), S_0 is the maximum value of $S(N)$ (attained at $N = N_0$), and σ is the gaussian width of the distribution. It has become conventional in much of the ecological literature, following Preston's (1948) work, to plot abundances on a scale, R, of logarithms to the base 2 (so that successive intervals or "octaves" correspond to population doublings):

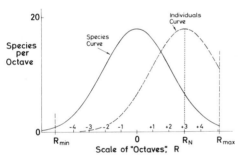

Figure 2 This figure aims to illustrate the features of a lognormal distribution, which are discussed more fully in the text. The solid curve is a lognormal species-abundance distribution, $S(R)$, after eqs. 3.4 and 3.5. The figure is specifically for $a = 0.34$, $S_0 = 18$, and consequently $\gamma = 0.6$ (see eq. 3.8). The dashed curve is the corresponding lognormal distribution in the total number of individuals, $N(R)$, eq. 3.7; the height of this dashed curve is in arbitrary units, leaving N_0 undefined (we are interested only in the shape of the curve). The species' populations are plotted in Preston's "octaves," R, which is to say as logarithms to base 2 (eq. 3.4). The boundary at R_{max} is the approximate position of the last, most abundant species; R_N is the octave in which total numbers peak. Preston's "canonical hypothesis" is that R_{max} and R_N coincide.

$$R = \ln_2 (N/N_0) \quad (3.4)$$

To avoid confusion we shall perpetuate this eccentricity, although it clutters the formulae with factors of $\ln 2$. The lognormal distribution then takes the familiar form

$$S(R)\, dR = S_0 \exp\left(-a^2 R^2\right) dR \quad (3.5)$$

Preston's parameter a is an inverse width of the distribution: $a = (2\sigma^2)^{-1/2}$. Figure 2 aims to illustrate this distribution. Worth mentioning are the values R_{max} and R_{min}, which mark the expected positions of the first (most abundant) and last (least abundant) species; as discussed more fully in Appendix A, $R_{max} = -R_{min}$, and both are related to a and S_0 via $aR_{max} = (\ln S_0)^{1/2}$, eq. A.2. The total number of species in the community, given in principle by eq. 1.3, is now approximately (see eqs. A.7, A.9),

$$S_T \simeq S_0 \pi^{1/2}/a$$
$$\simeq (\pi^{1/2}/a) \exp (a^2 R_{max}^2) \quad (3.6)$$

There is also a lognormal distribution in the total number of individuals in the Rth octave, $N(R)$. Combining eqs. 3.4 and 3.5 gives

$N(R)\, dR$

$$= S_0 N_0 2^R \exp\left(-a^2 R^2\right) dR$$
$$= S_0 N_0 \exp\left[-a^2 R^2 + R \ln 2\right] dR \quad (3.7)$$

This may be seen to be a normal distribution in the variable $(R - \ln 2/2a^2)$, that is, a gaussian with peak displaced a distance $R_N = (\ln 2)/2a^2$ to the right of that of the $S(N)$ distribution. It is now useful to define a quantity γ as the ratio between R_N and R_{max}:

$$\gamma \equiv \frac{R_N}{R_{max}} = \frac{\ln 2}{2a \, (\ln S_0)^{1/2}} \quad (3.8)$$

The relation between the basic lognormal species-abundance distribution and the intervals R_N and R_{max} is perhaps made clear by Figure 2. Finally, noting that the above definitions imply the expected number of individuals in the least abundant species to be

$$m = N_0 2^{R_{min}} = N_0 2^{-R_{max}} \quad (3.9)$$

we may express the quantity $J = N_T/m$ of eqs. 1.1 and 1.4 in a form that involves any two of the interconnected parameters S_0, R_{max}, a, S_T, γ. This is done in Appendix A, to arrive at eqs. A.10 and A.11.

The essential point to grasp is that the general lognormal species abundance distribution requires *two* parameters for a unique specification (see footnote 1). These two parameters may be chosen to be a and γ, in which case other interesting properties of the distribution such as S_0 (from eq. 3.8), R_{max} (from eq. A.2), and the overall S_T and J (from eqs. A.7 and A.10) follow. Alternatively, given the quantities S_T and J, which are of direct biological significance, the distribution is again uniquely described, and all other properties follow. Most commonly in Sections 4 and 5, I shall work with S_T and γ as the parameters.

Preston's (1962) "canonical hypothesis," which will be more fully discussed below, may conveniently be mentioned here, as it is the only reason for the orgy of notation leading up to the definition of the cumbersome parameter γ. This phenomenological hypothesis is that

$$\gamma = 1 \quad (3.10)$$

This is now a *one*-parameter family of canonical lognormal distributions. Given S_T, all else follows: the shape of the distribution is uniquely specified; unique values of J and of various diversity indices may be calculated; there is a unique species-area relation. The hypothesis has a purely empirical basis. In view of its predictive successes, particularly as to species-area relations, it is surprising that no theoretical justification has previously been attempted.

(iii) *Field Data.* As remarked by MacArthur (1960), the lognormal abundance distributions in communities of "opportunistic" creatures reflect nothing about the structure of the community. Dominant species are simply those that recently enjoyed a large r (eq. 3.2), and at different times different species will be most abundant. Such patterns for opportunistic species have been documented by Patrick, Hohn, and Wallace (1954) and Patrick (1968 and Chapter 15).

In steadier ("equilibrium") communities, fits to lognormal species-abundance distributions have been described for a wide variety of circumstances, including geographically diverse communities of birds, intertidal organisms, insects, and plants (Preston, 1948, 1962; Williams, 1953, 1964; Whittaker 1965, 1972; Batzli, 1969). Excellent reviews have been given by Whittaker (1970, 1972), who notes: "When a large sample is taken containing a good number of species, a lognormal distribution is usually observed, whether the sample represents a single community or more than one, whether distributions of the community fractions being combined are of geometric, lognormal or MacArthur form" (Whittaker, 1972, p. 221). Gauch and Chase (1975), who have just produced a useful computer algorithm for fitting normal distributions to ecological data, reexamined the data originally surveyed (and fitted by eye) by Preston (1948), and showed in one typical instance that 96% of the variance in the observed distribution could be accounted for by a lognormal.

In addition to these general fits between theory and data, it has been observed (originally by Preston for γ and by Hutchinson for a) that in most cases the parameters a and γ tend to have special constant values.

Preston (1962) reviewed a considerable body of material and showed that in all cases the shape of the distribution corresponded roughly to the special value $\gamma = 1$. This "canonical hypothesis" has been further discussed by MacArthur and Wilson (1967), and the ensuing unique S_T-versus-J relation applied to explain much species-area data.

Another rough rule, first noted by Hutchinson (1953) and subsequently confirmed by a growing amount of field observation, is that $a \simeq 0.2$. (The rule holds true even for the international distribution of human populations, or of the gross national products, referred to above.) Reviewing the current status of the lognormal distribution, Whittaker (1972, p. 221) observes "the constant a is usually around 0.2." This enigmatic rule has prompted many speculations, from Hutchinson's relatively cautious "it is

likely that something very important is involved here" (Hutchinson, 1953, p. 11), to one recent ecology text that indulges in the thought that "it does seem extraordinary that the constant should have the same value no matter the size or reproductive capacity of the organism, whether we are dealing with diatoms, moths or birds. Perhaps it bears some subtle relationship to the range of variation in the earth's environment, a range set by fundamental properties of the solar system or the elements in the periodic table."

A more prosaic explanation of these two rules will now be offered. They appear to be approximate mathematical properties of the lognormal distribution, once S_T is large.

(iv) *The Canonical Hypothesis and the Rule $a \simeq 0.2$.* (a) First, assume $\gamma = 1$, so that we have a canonical lognormal distribution. There is now a unique relation between a and S_T, the total number of species in the community. This relation (see eq. B.1 with the definition eq. A.6) is depicted in Figure 3. It is obvious that a depends very weakly on the actual value of S_T once there are more than ten or so species. This figure reflects the approximate fact that for $S_T \gg 1$ (see eqs. A.12, A.6)

$$a \sim \frac{\ln 2}{2\sqrt{\ln S_T}} \qquad (3.11)$$

The dependence on S_T as the square root of the logarithm is very weak indeed.

Figure 3 shows that as S_T varies from 20 to 10,000 species, a varies from 0.30 to 0.13. The rule $a \simeq 0.2$ is a mathematical property of the canonical lognormal.

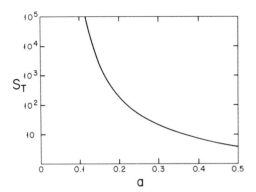

Figure 3 This figure shows the relationship between number of species, S_T, and the parameter a for the canonical lognormal distribution. Note that S_T is plotted on a logarithmic scale, emphasizing the insensitivity of a to changes in S_T.

(b) It remains to consider the canonical hypothesis itself. Without this hypothesis, the general lognormal is characterized by two parameters, conventionally a and S_0 (eq. 3.5), but equivalently a and γ, which can be determined from the total number of species, S_T, and the total number of individuals divided by the population of the rarest species, $J = N_T/m$. These relations between a, γ and S_T, J are given by eqs. A.7 and A.10 along with the definitions eqs. A.6 and A.11 in Appendix A. The form of the relation is illustrated by Figure 4, which shows the range of S_T and J values that are possible if a and γ are allowed to vary independently over the ranges a from 0.1 to 0.4 and γ from 0.5 to 1.8.

It may be observed from the figure that this enormous range of S_T and J values is roughly consistent with the rules $\gamma = 1$ and $a \simeq 0.2$, the agreement becoming more pronounced as S_T becomes larger. As in eq. 3.11, the underlying reason is

that for $S_T \gg 1$, the quantities a and γ depend on S_T and J as $\sqrt{\ln S_T}$ and $\sqrt{\ln J}$ (see eqs. A.6, A.12, A.13, A.14).

The above admittedly constitutes only an imprecise and qualitative explanation of the rules $\gamma \simeq 1$ and $a \simeq 0.2$. However, these empirical rules are themselves only rough ones, and a qualitative theory would seem to represent some advance over the prevailing total absence of any explanation as to such remarkable regularities. I see them as mathematical properties of the lognormal distribution, rather than as reflecting anything biological. This is disappointing.

The emergent moral is, presumably, to characterize lognormal distributions by indices more sensitive than a, γ and their equivalents.

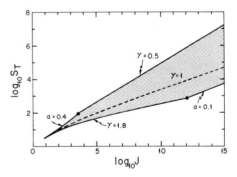

Figure 4 The shaded area illustrates the very wide range of values of S_T and J which can be described by the general lognormal distribution with the parameter a between 0.1 and 0.4 and the parameter γ between 0.5 and 1.8. This shaded area is bounded, as indicated in the figure, by the four line segments along which, respectively, $a = 0.4$ (with γ varying from 1.8 to 0.5), $\gamma = 1.8$ (with a varying from 0.4 to 0.1), $\gamma = 0.5$ (with a varying downward from 0.4), and $a = 0.1$ (with γ varying downward from 1.8). Note that both S_T and J are on logarithmic scales. The dashed line corresponds to Preston's special case, $\gamma = 1$ (and varying a).

B: MacArthur's Broken-Stick
Distribution

(i) *Theory.* If attention is restricted to communities comprising a limited number of taxonomically similar species, in competitive contact with each other in a relatively homogeneous habitat, a more structured pattern of relative abundance may be expected. As pointed out by Mac-Arthur (1957, 1960), and lucidly reviewed by Whittaker (1972), if the underlying picture is one of intrinsically even division of some major environmental resource, the statistical outcome is the well-known "broken-stick" distribution. This is the distribution relevant, for example, to collecting plastic animals out of cornflakes boxes, assuming the various plastic animal species to have a uniform distribution at the factory. The most thorough treatment of the statistical properties of the relative abundances within a group of species, which apportion randomly among themselves a fixed amount of some governing resource, is due to Webb (1974). He not only shows that the familiar broken-stick distribution is the average outcome, but also considers the statistical fluctuations to be expected about this average.

The broken-stick distribution of relative abundance is considerably more even than lognormal ones. It is characterized by a single parameter, S_T. Thus once the number of species in the community is specified, diversity indices and (with eq. 1.5) species-area relations uniquely follow. The mathematical properties of the distribution are catalogued in Appendix D.

(ii) *Field Data.* For appropriately restricted samples, such broken-stick relative-abundance patterns have been found, for example, by MacArthur (1960), King (1964), and Longuet-Higgins (1971) following Tramer (1969), for birds; by Kohn (1959) for some snails; and by Goulden (1969), Deevey (1969), and Tsukada (1972) for microcrustaceans deposited in lake-bed sediments.

As pointed out by Cohen (1968), and reviewed by Pielou (1969) and others, the observation of a broken-stick distribution does not validate the very specific model initially proposed by MacArthur (1957, 1960). It does indicate, however, that some major factor is being roughly evenly apportioned among the community's constituent species (in contrast to the lognormal distribution, which suggests the interplay of many independent factors).

C: Geometric Series and Logseries
Distributions

(i) *Theory.* Let us consider again a relatively small and simple community of species, whose ecology is governed by some dominant factor; the opposite extreme to an intrinsically even (or random) division of resources is one of extreme "niche preemption." In its ideal form, this limit sees the most successful species as preempting a fraction k of the niche, the next a fraction k of the remainder, and so on, to give a geometric series distribution of relative abundance (eq. E.1). An equivalent way of framing this hypothesis is to assume that all species are energetically related to the other species in the community, the magnitude of the relation being proportional to the species abundance (large populations need more energy); the addition of another species then requires the same proportional increase in the abundance of all other species (Odum, Cantlon, and Kornicher, 1960). It does not seem to be commonly appreciated that the ensuing species-abundance distribution discussed by these authors, and in particular their S_T-versus-J relation, is precisely that of the geometric series distribution. Semantically, this identity is not surprising, as their assumptions constitute a form of niche-preemption hypothesis. The formal equivalence between the mathematics of Odum, Cantlon, and Kornicher and that of the conventional geometric-series distribution is established in Appendix E.

If one considers statistically more realistic expressions of the ideas that in their ideal form lead to a geometric series distribution, one is commonly led to a logseries distribution of relative abundance. For example, suppose that the geometric series niche-preemption mechanism stems from the fact that the species arrive at successive uniform time intervals and proceed to preempt a fraction k of the remaining niche before the arrival of the next; randomization of the time intervals leads to a logseries distribution (Boswell and Patil, 1971). Kendall's (1948) remarks as to how an intrinsically geometric series distribution of species per genus is converted into a logseries distribution are also obliquely relevant here (see the discussion in Williams, 1964, Chapter 11, and Boswell and Patil, 1971). Several other ways of arriving at a logseries distribution are comprehensively reviewed by Boswell and Patil (see particularly Sections 5 and 9), and Pielou (1969, Chapter 17). The distribution often arises as a sampling distribution, in which form it was first obtained

by Fisher, Corbet, and Williams (1943). It has the elegant property that samples taken from a population distributed according to a logseries are themselves logseries.

Some mathematical properties of the logseries are listed in Appendix F. Like the geometric series, it is characterized by two parameters, which we usually choose to be S_T and α. The relation between S_T, α and J is very simple:

$$S_T = \alpha \ln (1 + J/\alpha) \qquad (3.12)$$

(ii) *Field Data.* Whittaker (1965, 1970, 1972) has reviewed data from some plant communities, generally with but a few species and either in an early successional stage or in a harsh environment, where the species-abundance distribution approximates a geometric series. The phenomenon of strong dominance may be most expected in such circumstances. McNaughton and Wolf (1973) have also reviewed a series of examples which they interpret as geometric series; however, many of their examples would seem to be fitted better, and certainly at least as well, by lognormal distributions. Some of the systems discussed by Connell (Chapter 16) are likely candidates to fit geometric or logseries distributions, as they would seem to conform to the assumptions. As we would expect, the simple models of Markovian forest succession presented by Horn (Chapter 9) give rise to an explicitly geometric series distribution of relative abundance.

D: Contrast Between these Distributions

In short, the broken-stick and geometric or logseries may be viewed as distributions characteristic of relatively simple communities whose dynamics is dominated by some single factor: the broken-stick is the statistically realistic expression of an intrinsically uniform distribution; and at the opposite extreme the logseries is often the statistical expression of the uneven niche-preemption process, of which the ideal form is the geometric series distribution. Both forms reflect dynamical aspects of the community.

However, if the environment is randomly fluctuating, or alternatively as soon as several factors become significant (as they may in general, and must if $S_T \gg 1$), we expect the statistical Law of Large Numbers to take over and produce the ubiquitous lognormal distribution. This species-abundance distribution is in most respects *intermediate* between the broken-stick and geometric series or logseries extremes, as illustrated clearly by Figure 1 and the tables and figures in Section 4. The empirical rules $a \simeq 0.2$ and $\gamma \simeq 1$ are probably no more than mathematical properties of the lognormal distribution for $S_T \gg 1$.

4. Diversity, Evenness, Dominance

I do not wish to add unnecessarily to the already voluminous literature pertaining to the meaning and relative merit of various diversity indices. Suffice it to say that many people have sought to go beyond S_T and J, yet stop short of the full $S(N)$, by using some single simple number that may characterize the distribution $S(N)$, i.e., some number that will describe whether the J individuals are roughly evenly distributed among the S_T species, or whether they are concentrated into a few dominant species.

Among the many such indices proposed are

(i) The Shannon-Weaver diversity, which we shall refer to as H:

$$H = - \sum_{i=1}^{S_T} p_i \ln p_i \qquad (4.1)$$

Here p_i is the proportion of individuals in the ith species, $p_i = N_i/N_T$. For a given species-abundance distribution $S(N)$, the average value of H will be

$$\langle H \rangle = - \left\langle \sum_i p_i \ln p_i \right\rangle$$
$$= - \sum_i \langle p_i \ln p_i \rangle$$

That is,

$$\langle H \rangle$$
$$= - \int (N/N_T) \ln (N/N_T) S(N) \, dN \qquad (4.2)$$

(ii) Also of considerable interest is the expected magnitude of the statistical fluctuations about this mean value of H. For a given distribution $S(N)$, the expected variance is

$$\sigma_H^2 = \langle (H - \langle H \rangle)^2 \rangle$$
$$= \left\langle \left(\sum_i p_i \ln p_i \right)^2 \right\rangle - \langle H \rangle^2 \qquad (4.3)$$

But we can write

$$\left\langle \left(\sum_i p_i \ln p_i \right)^2 \right\rangle = \sum_i \langle (p_i \ln p_i)^2 \rangle$$
$$+ \sum_i \sum_{j \neq i} \langle (p_i \ln p_i)(p_j \ln p_j) \rangle$$

In the second term on the right-hand side there are no correlations, and consequently the term has the value $[(S_T - 1)/S_T]\langle H\rangle^2$. Therefore

$$\sigma_H^2 = \int [(N/N_T)\ln(N/N_T)]^2 S(N)\,dN - \langle H\rangle^2/S_T \quad (4.4)$$

(iii) Another dominance or diversity index is

$$C = \sum_i p_i^2 \quad (4.5)$$

The reciprocal of C,

$$D = 1/C \quad (4.6)$$

is Simpson's (1949) diversity index, which counts species, weighting them by their abundance, and can vary from $D = 1$ (one dominant species) to $D = S_T$ (completely even distribution). The index $1 - C$ is also widely used, and indeed Hurlbert (1971) has recently decried the popular index H, and has suggested that $1 - C$ may be better. For a specified distribution $S(N)$, the expected value of C is

$$\langle C\rangle = \int (N/N_T)^2 S(N)\,dN \quad (4.7)$$

In analogy to the index (ii) above, the statistical variance in C could also be calculated and displayed for any specified $S(N)$.

(iv) One simple measure of dominance in a community is (Berger and Parker, 1970)

$$d = N_{\max}/N_T \quad (4.8)$$

Here, for given $S(N)$, $N_{\max}$ is the expectation value of the most abundant population.

Various other indices of diversity, dominance, or evenness have been reviewed by Whittaker (1972), Dickman (1968), Pielou (1969), Johnson and Raven (1970), Hurlbert (1971), De Benedictis (1973), and many others.

Our attention here is confined to the four quantities H, σ_H^2, D or C, and d. For each we exhibit and compare the relationships between the index and S_T for the various distributions of Section 3. In this way we aim to clarify the relations between the various species-abundance relations as revealed by any one index, and also the relations between the various diversity indices for any one $S(N)$. Some of these diversity indices have been explored for some of the distributions listed in Section 3 (particularly for the broken-stick), but remarkably little has been done for the lognormal, which is probably the most important one.

A: The Average Value of H

For the broken-stick and the canonical lognormal distributions, unique curves of $\langle H\rangle$ versus S_T can be calculated. The general lognormal, geometric series, or logseries distributions give rise to one-parameter families of curves of $\langle H\rangle$ versus S_T, specified here by the parameters γ, k, and α respectively.

Figure 5 shows such curves for the broken-stick distribution, and for lognormal distributions with $\gamma = 0.7$, 1 (canonical), 1.3. Note that the curves are indistinguishable for relatively small S_T, but that as S_T becomes very large the broken-stick curve leads to significantly larger H values, while the lognormal curves reveal

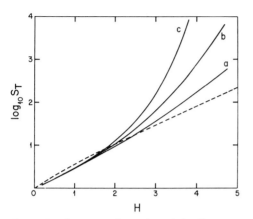

Figure 5 The expectation value of the Shannon-Weaver diversity index, H (see eq. 4.2), as a function of the total number of species, S_T (on a logarithmic scale), for various species-abundance distributions. The solid curves are lognormal distributions with: (a) $\gamma = 0.7$; (b) $\gamma = 1.0$ (canonical); (c) $\gamma = 1.3$; the dashed curve is the relation for the broken-stick distribution.

a very insensitive dependence of H on S_T.

In particular, to achieve H greater than 5 with a canonical lognormal distribution requires $S_T \sim 10^5$ species. As stressed in Section 3, for $S_T \gg 1$ a lognormal distribution is always to be expected, and this

feature of the lognormal distribution may explain the fact, remarked upon at length by Margalef (1972), that $H > 5$ is not observed in data collections.

For $S_T \gg 1$, analytic approximations may be obtained for the average value of H in the various distributions, and these are catalogued in Table 1.

These asymptotic results bear out the general remarks made in Sections 2 and 3, to the effect that uniform or broken-stick distributions are more diverse, which is to say have larger H, than lognormal distributions with $\gamma \simeq 1$, which in turn are more diverse than geometric or logseries distributions. However, these systematic differences are seen from Table 1 to scale only as $\ln S_T$ (broken-stick, uniform) compared with $\sqrt{\ln S_T}$ (lognormal with $\gamma \simeq 1$) compared with a constant (geometric or logseries), even for every large S_T. But as the logarithm of even a quite large number is of the general order of unity, one must conclude that the diversity index H is an insensitive measure of the character of $S(N)$. These comments are true a fortiori when S_T is not large (cf. Figure 5).

Table 1.   Diversity, H, as a function of S_T for $S_T \gg 1$

Species-abundance distribution, $S(N)$	Average value of H, for $S_T \gg 1$	Details in appendix
lognormal (parameter γ)	$(1 - \gamma^2) \ln S_T$; for small γ $2\gamma\pi^{-1/2} \sqrt{\ln S_T}$; for $\gamma \simeq 1$ (constant); for large γ	eqs. A.18, A.19
canonical lognormal	$2\pi^{-1/2} \sqrt{\ln S_T}$	eqs. B.5, B.6
uniform	$\ln S_T$	eq. C.4
broken-stick	$\ln S_T - 0.42$	eq. D.8
geometric series (parameter k)	$\text{constant} = \dfrac{-\ln (1 - k)}{k} + \ln \left(\dfrac{1 - k}{k} \right)$	eq. E.6
logseries (parameter α)	$\text{constant} = \ln \alpha + 0.58$	eqs. F.5, F.6

B: The Variance of H

The variance in H may be calculated from eq. 4.4, and the details are outlined in the various mathematical appendices. Previous work along these lines consists of numerical investigations of the full statistical distribution of H for the broken-stick distribution, carried out by Bowman *et al.* (1971) and by Webb (1974).

Figure 6 shows the standard deviation to be expected in the diversity H for communities whose underlying patterns are broken-stick or canonical lognormal. (As the statistical distribution of H values is skewed, particularly at relatively small S_T values, characterizing its spread by σ_H is not an exact procedure, but should be good enough for all practical purposes; see Webb's computer simulations for the broken-stick case.)

These fluctuations in H are intrinsic properties of the distribution $S(N)$, and it is clear that their magnitude is such as to obscure any differences between the distributions unless S_T is very large.

For completeness, analytic formulae and asymptotic expressions for σ_H^2 and for the root-mean-square relative fluctuations, $\sigma_H/\langle H\rangle$, are given in the appendices. These are messy, and here we note only the asymptotic ($S_T \gg 1$) form of the relative fluctuations in the broken-stick and canonical lognormal distributions:

broken-stick:

$$\frac{\sigma_H}{\langle H\rangle} \sim \frac{1}{\sqrt{S_T}} \qquad (4.9)$$

canonical lognormal:

$$\frac{\sigma_H}{\langle H\rangle} \sim \frac{0.9}{\ln S_T} \qquad (4.10)$$

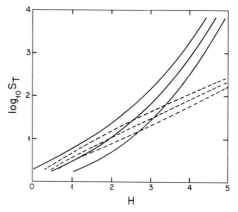

Figure 6 The three solid lines illustrate the average value of H, and the average value of one standard deviation above and below H (i.e., $\langle H\rangle$ and $\langle H\rangle \pm \sigma_H$, from eqs. 4.2 and 4.4), as a function of S_T, for the canonical lognormal distribution of species abundance. Likewise, the three dashed lines show $\langle H\rangle + \sigma_H$, $\langle H\rangle$, and $\langle H\rangle - \sigma_H$ for the broken-stick distribution.

In his computer experiments, De Benedictis (1973) noted a systematic tendency for the variance in the "evenness" (i.e., in $H/\ln S_T$) to decrease as S_T increased, albeit slowly. Figure 6 and the above formulae confirm this tendency.

Figure 7, modified from Webb's (1974) compilation of data, shows the H values for an assortment of communities of corals, copepods, plankton, benthic creatures, trees, and birds, along with the expectation values of H for the broken-stick and the canonical lognormal distributions. From a comparison of Figures 6 and 7, it may be held that the data are more consistent with a lognormal distribution, but certainly no discrimination between the two distributions is possible for small S_T.

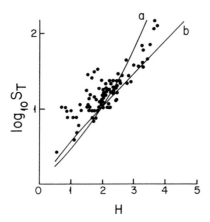

Figure 7 The theoretical curves of diversity versus species number, H versus S_T (on a logarithmic scale), for (a) the canonical lognormal distribution and (b) the broken-stick distribution, are here shown in conjunction with Webb's (1973) compilation of various authors' data (the solid dots) for birds, copepods, corals, plankton, and trees.

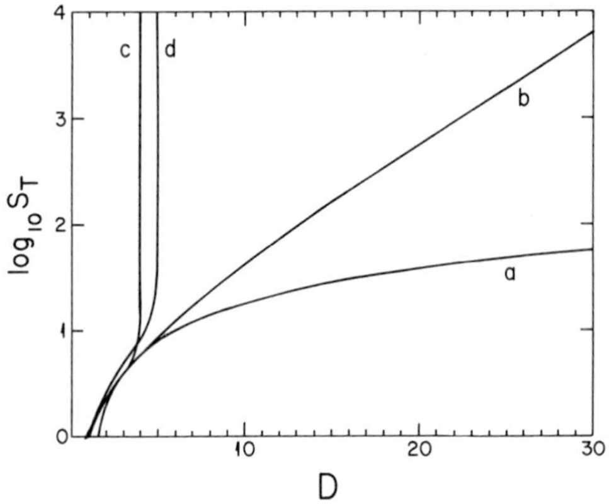

Figure 8 The average value of Simpson's diversity index D (from eqs. 4.6, 4.7) as a function of S_T (on a logarithmic scale) for various distributions of species abundance $S(N)$: (a) broken-stick; (b) canonical lognormal; (c) geometric series with $k = 0.4$; (d) logseries with $\alpha = 5$.

C: The Average Values of D and C

The mathematical appendices contain explicit expressions for the indices C and D of eqs. 4.6 and 4.7, for the various distributions $S(N)$ of Section 3.

Figure 8 illustrates Simpson's diversity index D as a function of S_T for the broken-stick, canonical lognormal, geometric series, and logseries distributions.

(The broken-stick and canonical lognormal curves are unique; the geometric series is for the choice $k = 0.4$, and the logseries for $\alpha = 5$.) Comparison with Figure 5 shows the index D to exhibit a stronger contrast between canonical lognormal and broken-stick distributions than does H.

Table 2 catalogues asymptotic, $S_T \gg 1$,

Table 2.   Simpson's diversity, D, as a function of S_T for $S_T \gg 1$

Species-abundance distribution, $S(N)$	Average value of D, for $S_T \gg 1$	Details in appendix
lognormal (parameter γ)	see Appendix A	eqs. A.22, A.23
canonical lognormal	$\dfrac{\pi \ln S_T}{\ln 2}$	eq. B.10
uniform	S_T	eq. C.5
broken-stick	$\frac{1}{2} S_T$	eq. D.11
geometric series (parameter k)	$(2 - k)/k$	eq. E.8
logseries (parameter α)	α	eq. F.8

expressions for D. Both the figure and the
table again illustrate the properties attributed to the distributions earlier.

D: The Average Value of d

Here Figure 9 contrasts d as a function
of S_T for the canonical lognormal,
broken-stick, geometric series (with
$k = 0.4$), and logseries (with $\alpha = 5$) distributions. The asymptotic formulae relating d to S_T when $S_T \gg 1$ are set out in
Table 3.

Yet again, all the general features are
born out. As S_T becomes large, the geometric and logseries manifest their pattern
of strong dominance, by having d settle
to some characteristic constant value, independent of the increasing number of
species. Conversely, the uniform and
broken-stick distributions have d tending
to zero essentially as $1/S_T$ when S_T becomes large. As ever, the canonical lognormal bestrides these extremes, with d
decreasing, albeit slowly [as $(\ln S_T)^{-1/2}$],
as S_T increases.

We observe that d, which is a pleasingly
simple index from both conceptual and
computational points of view, seems from

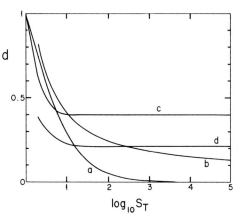

Figure 9 The average value of the simple dominance index d, eq. 4.8, as a function of S_T for various
species-abundance distributions: (a) broken-stick;
(b) canonical lognormal; (c) geometric series with
$k = 0.4$; (d) logseries with $\alpha = 5$.

the foregoing discussion to characterize
the distribution as well as any, and better
than most.

E: Correlations Between H, D, d

It is obvious that, given any particular
species-abundance distribution $S(N)$, one
can for instance take the curves of S_T
versus H and S_T versus d, and eliminate

Table 3. Dominance, d, as a function of S_T for $S_T \gg 1$

Species-abundance distribution, $S(N)$	Average value of d, for $S_T \gg 1$	Details in appendix
lognormal (parameter γ)	see Appendix A	eq. A.24
canonical lognormal	$\dfrac{\ln 2}{\sqrt{\pi \ln S_T}}$	eq. B.11
uniform	$1/S_T$	eq. C.6
broken-stick	$(\ln S_T)/S_T$	eq. D.12
geometric series (parameter k)	k	eq. E.9
logseries (parameter α)	$\dfrac{\ln \alpha}{\alpha}$	eq. F.11, F.12

S_T to provide a unique curve of H versus d. Such relations between the diversity and dominance indices H, D, C, $1 - C$, d, etc., are already implicit in Figures 5–9 and Tables 1–3. That the various indices are correlated is a point that has been elaborated by Johnson and Raven (1970), Berger and Parker (1970), De Benedictis (1973), and others; but an exploration of the precise character of these correlations for the different species-abundance distributions $S(N)$ is lacking.

Table 4 makes this explicit, setting out the relations between H and D, and between D and d, for the various distributions. It may be noticed that (for mathematical rather than biological reasons) broken-stick and logseries distributions coincidentally tend to exhibit similar relations of H versus D and D versus d, whereas those for the lognormal distribution are qualitatively different.

5. Species—Area Relations

A subject of considerable interest, enjoying a growing literature, is the species-versus-area relation for communities of species isolated on real (see, e.g., Diamond, Chapter 14) or virtual (e.g., Cody, Chapter 10; Wilson and Willis, Chapter 18) islands. The above species-abundance distributions $S(N)$ imply predictions as to the relations between S_T and J, and these can be turned into relations between S_T and area A by addition of the biological assumption eq. 1.5, $J = \rho A$, the plausibility of which was discussed briefly in Section 1 (Preston, 1962; MacArthur and Wilson, 1967). As long as we are looking at islands or other isolated biota, this procedure may be justifiable; once A represents areas of different size from a large homogeneous mainland region, the relation of S_T versus A is likely to reflect sam-

Table 4. Asymptotic ($S_T \gg 1$) relations between H, D, d

Species-abundance distribution, $S(N)$	Asymptotic relation between H and D	Asymptotic relation between d and D
lognormal (parameter γ)	depends on γ, see Appendix A	$d^2 = \dfrac{(2\gamma - 1)\ln 2}{\gamma D}$
canonical lognormal	$H = (2/\pi)(D \ln 2)^{1/2}$	$d^2 = \dfrac{\ln 2}{D}$
uniform	$H = \ln D$	$d = 1/D$
broken-stick	$H = \ln D + 0.27$	$d = \dfrac{\ln (2D)}{2D}$
geometric series (parameter k)	$H = \ln D + 0.31 + [\text{order } D^{-2}]$	$d = \dfrac{2}{1 + D}$
logseries (parameter α)	$H = \ln D + 0.58$	$d = \dfrac{\ln D}{D}$

pling properties of the kind discussed in
Section 6.

A unique relation of species versus area
may be obtained from the general lognormal distribution by Preston's canonical
hypothesis, $\gamma = 1$. This curve is illustrated
in Figure 10. The relation has been successfully applied, initially by Preston
(1962) and in more detail by MacArthur
and Wilson (1967), to describe a wide
range of data. These authors approximate
the canonical lognormal relation S_T versus
A by a simple linear regression of $\ln S_T$
on $\ln A$, a procedure that tends to overestimate the slope at large A and underes-
timate it at small A; the actual relation
between S_T and A is not a simple function,
although in the limit $S_T \gg 1$ it takes the
asymptotic form given by eq. 1.6, namely

$$\ln S_T = x \ln A + (\text{constant}) \quad (5.1)$$

with $x = 0.25$. This point is discussed
more fully in Appendix B. We reiterate
that the more exact species-area curve
illustrated in Figure 10, with its steeper
slope at smaller A, tends to give a better
description of real species-area data (see,
e.g., Diamond, Chapter 14, Figure 3); the
approximate eq. 5.1 does not flatter the
theory.

The work of Preston and of MacArthur
and Wilson rests on the canonical hypothesis. What happens for the general lognormal species-abundance distribution?

For each value of γ, the general lognormal gives a particular S_T-versus-A
curve. Two such curves are shown in Fig-
ure 10, for $\gamma = 0.8$ and $\gamma = 1.3$. The shape
is not particularly sensitive to the detailed

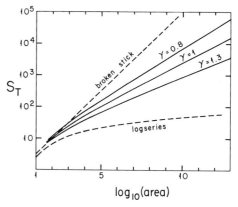

Figure 10 This figure illustrates the species-area (log
S_T versus log A) relations obtained from particular
species-abundance distributions $S(N)$ in conjunction
with the biological assumption eq. 1.5. The solid
curves are for lognormal distributions with $\gamma = 0.8$,
1.0 (Preston and MacArthur-Wilson), eq. 1.3; the
dashed lines are for the broken-stick distribution,
and for a logseries distribution (with $\alpha = 5$), as
indicated.

value of γ, at least for the generous range
of γ values around 0.6 to 1.7, which was
discussed in Section 3 as pertaining to
ecologically reasonable circumstances (see
Figure 4). From eqs. A.15 and A.16, eq.
5.1 provides an asymptotic ($S_T \gg 1$) description of all lognormal species-area
curves, with the quantity x having the
values

$$x = 1/4\gamma \qquad \text{for } \gamma > 1 \quad (5.2a)$$
$$x = 1/(1 + \gamma)^2 \text{ for } \gamma < 1 \quad (5.2b)$$

Thus, roughly speaking, the above range
of γ (0.6 to 1.7) leads to eq. 5.1, with x
values lying between the extremes 0.39 to
0.15.

In brief, the successes of the Preston
and MacArthur-Wilson species-area the-

ories are not pathological consequences of their special canonical lognormal distribution, but rather are robust properties of any reasonable lognormal species-abundance distribution.

Table 5 summarizes this work by showing the x values obtained by various authors from fits of their field data to regressions of $\ln S_T$ versus $\ln A$, along with the asymptotic x values for lognormal distributions of various shapes (as specified by γ). An interesting next step might be to

study these data in greater detail, with the aim of understanding the biological reason why, for example, plants on the California Islands ($x = 0.37$) have a comparatively small value of γ ($\gamma \simeq 0.64$), whereas the Yorkshire nature-reserve plants ($x = 0.21$) have a comparatively large value ($\gamma \simeq 1.2$).

It may be argued that such species-area discussions are likely to involve large numbers of species, and therefore that the lognormal pattern of species abundance is

Table 5. The x values deduced from observations, and from the theoretical lognormal distribution

	Observations		
Source	Flora or fauna	Location	x
Darlington, 1943	beetles	West Indies	0.34
Darlington, 1957	reptiles and amphibians	West Indies	0.30
Hamilton, Barth, and Rubinoff, 1964	birds	West Indies	0.24
Hamilton, Barth, and Rubinoff, 1964	birds	East Indies	0.28
Hamilton, Barth, and Rubinoff, 1964	birds	East-Central Pacific	0.30
MacArthur and Wilson, 1967	ants	Melanesia	0.30
Preston, 1962	land vertebrates	Lake Michigan Islands	0.24
Diamond, Chapter 14, Figure 2	birds	New Guinea Islands	0.22
Diamond, Chapter 14, Figure 3	birds	New Britain Islands	0.18
Cody, Chapter 10	birds	Mediterranean habitat gradients	0.13
Preston, 1962	land plants	Galapagos	0.32
Hamilton, Barth, and Rubinoff, 1963	land plants	Galapagos	0.33
Johnson and Raven, 1973	land plants	Galapagos	0.31
Preston, 1962	land plants	world-wide	0.22
Johnson and Raven, 1970	land plants	British Isles	0.21
Usher, 1973	land plants	Yorkshire nature reserves	0.21
Johnson, Mason, and Raven, 1968	land plants	California Islands	0.37

Theory	
Value of γ (see eq. 3.8)	x
0.6	0.39
0.8	0.31
1.0, *canonical*	0.25
1.2	0.21
1.4	0.18
1.6	0.16

the only one to work with here. Even so, the species-area consequences of the opposite extremes of broken-stick and of geometric series or logseries distributions should be mentioned.

As shown in Appendix D, eq. D.7, the broken-stick distribution in conjunction with eq. 1.5 leads *exactly* to the relation eq. 5.1 with $x = 0.5$. This relation is shown as a dashed line in Figure 10. It is not surprising to find it significantly steeper than any of the observations.

Conversely, the geometric series or logseries distributions tend to give species-area relations of the form set out in eq. 1.7: see eqs. E.5 and F.4 respectively. Such curves lead to a less steep relation than indicated from the data in Table 5. They are discussed in the next section.

6. Sampling Problems and $S(N)$

Up to this point, it has been assumed that we are dealing with situations in which the full values of S_T and N_T or J are known. All species in the community are represented in our samples. This assumption has allowed an unclouded discussion of some issues of principle. The assumption is often unrealistic, however, and in practice there will commonly be a need to work with less complete samples in which not all species are represented. To put it another way, our lognormal distributions are always (in Preston's terminology) fully unveiled; the complications introduced by distributions that are not unveiled would be distracting, and are inessential to the main points of Section 2–5.

The analysis of incomplete samples from distributions $S(N)$ constitutes a large and significant subject, to which the references given in Section 1 constitute an entry. The following discussion is confined to a few brief comments.

As first observed by Fisher, and discussed by Williams (1964), Pielou (1969), and particularly by Boswell and Patil (1971), the logseries distribution can describe sampling distributions under a diversity of circumstances. One consequence is that, if relatively small samples are taken from some large and homogeneous area, the relation between sample area (or volume), A, and the number of species represented in the sample, $S(A)$, is likely to obey the logseries eq. F.13; that is,

$$S(A) = \alpha \ln [1 + \beta A] \qquad (6.1)$$

This is approximately of the form eq. 1.7, $S \sim \alpha \ln A$. If this be the case, a regression of S on $\ln A$ will fit the data better than the $\ln S / \ln A$ regression suggested by eq. 5.1; alternatively, if an attempt is made to fit a relation actually of the form eq. 6.1 with a $\ln S / \ln A$ regression, the coefficient x thus deduced will be small, and the fit poor.

As an example, consider the elegant and much-discussed work of Sanders (1969). He studied marine benthic communities from various parts of the world, and for each particular community he plotted the number of species observed (or expected to be observed) as a function of the number of individuals in the sample. This clearly is a situation where the curves are derived from a sampling distribution. Figure 11 illustrates some of Sanders's

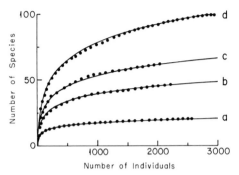

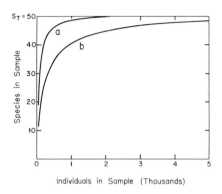

Figure 11 The solid dots are Sanders's (1969) "rarefaction curves" of number of species versus number of individuals in the sample, for data on benthic communities from: (a) boreal shallow water (IRA); (b) deep sea slope off southern New England (DR 33); (c) Walvis bay (190); (d) tropical shallow water (RH 30). The theoretical curves are for a logseries distribution, that is eq. F.13, with the single parameter α having the value: (a) 3.0; (b) 8.2; (c) 12.0 (d) 20.0.

Figure 12 An estimate of the number of species likely to be present in a relatively small sample comprising $\hat{N}$ individuals, when the underlying species-abundance distribution is (a) broken-stick, (b) canonical lognormal. In both cases, the total number of species actually present in the community is 50.

results, and the theoretical fits to them that can be obtained by assuming the sampling distribution to be a logseries, whence eq. F.13 or eq. 6.1. Each curve can be summarized by the single parameter α, which, as we saw in Section 4, is in some rough sense a measure of community diversity.

The appendices contain very crude estimates of the number of species (expressed as a fraction of the total number of species actually present in the community) likely to be observed in small samples drawn from communities whose intrinsic species abundance patterns $S(N)$ are as discussed in Section 3. Figure 12 shows characteristic results of this kind for broken-stick and canonical lognormal distributions. As is to be expected, the broken-stick distribution is more rapidly unveiled with increasing sample size than is the lognormal.

For example, if there are in fact 50 species present, the number of individuals one needs to sample to encounter half the species (25) is of the order of 73 for the broken-stick, 230 for the canonical lognormal distribution.

7. Summary

(1) For large or heterogeneous assemblies of species, a lognormal pattern of relative abundance may be expected. The intriguing rough rule $a \simeq 0.2$, and Preston's canonical hypothesis, are approximate but general mathematical properties of the lognormal distribution (e.g., Figures 3, 4). These rules thus reflect little more than the statistics of the Central Limit Theorem.

(2) For a small, ecologically homogeneous, set of species, which randomly ap-

portion among themselves a fixed amount of some governing resource, one may expect MacArthur's broken-stick distribution. Such a distribution may be thought of as the statistically realistic expression of an ideally uniform ($N_i = N_T/S_T$) distribution; the pattern of relative abundance is significantly more even than the lognormal (e.g., Figures 1, 8, 9).

(3) For a relatively small set of species where "niche-preemption" is likely to have significance, one may expect a logseries distribution. Such a distribution may be thought of as the statistically realistic expression of an ideally geometric series distribution; the pattern of relative abundance is significantly less even than the lognormal (e.g., Figures 1, 8, 9).

(4) The trends summarized in points (1), (2), and (3) are manifested in any measure of diversity or dominance in the community (Tables 1, 2, 3, 4). However, many common measures of species diversity tend not to distinguish these distributions if S_T is relatively small (e.g., Figure 6), while for large S_T we expect a lognormal distribution. Thus as one-parameter characterizations of species-abundance patterns, such indices are often of doubtful value. As a one-parameter description of the distribution, the simple dominance measure, d = (number of individuals in most abundant population)/(total number of individuals), seems as good as any.

(5) Combined with the biological assumption that $J = \rho A$, the general lognormal distribution (of which the canonical distribution of Preston and MacArthur-Wilson is a special case) leads to species-area curves that agree with the bulk of the

pertinent field observations (Figure 10, Table 5). Broken-stick species-area curves are significantly steeper, and logseries or geometric series curves significantly less steep, than lognormal ones (Figure 10).

(6) Sampling problems can becloud species-abundance patterns. In particular, the statistics of sampling processes can often produce species-area curves of the form $S_T \sim \alpha \ln A$ (e.g., Figure 11).

Acknowledgments

I was drawn to this study by Robert MacArthur's seminal 1960 paper. I am grateful to many people, particularly Henry Horn, for conversations on the subject.

Mathematical Appendices

For each of the distributions $S(N)$ under consideration, we (i) make some general comments; then calculate (ii) S_T; (iii) J; (iv) species-area relations; (v) H; (vi) σ_H^2; (vii) D; (viii) d; and finally (ix) make some comments on sampling from the distribution.

A. General Lognormal Distribution

(i) *General.* To facilitate comparison with the conventional ecological literature on the lognormal distribution, we follow Preston's usage of R (see eq. 3.4) rather than N as the basic independent variable. Then the number of species in the interval R to $R + dR$ is, as discussed in the main text (eq. 3.5),

$$S(R) = S_0 \exp(-a^2 R^2) \qquad (A.1)$$

In practice, this distribution will extend not from R equals $-\infty$ to $+\infty$, but rather from the octave $R_{\min}$ wherein lies the least abundant species, to the octave $R_{\max}$ containing the most abundant species. As these symmetrically disposed end points are by definition those where $S(N_{min}) \simeq S(N_{max}) \simeq 1$, we have

$$1 = S_0 \exp[-a^2R_{max}^2]$$
$$= S_0 \exp[-a^2R_{min}^2] \quad (A.2)$$

At this point it is convenient for notational purposes to define a quantity Δ:

$$\Delta \equiv aR_{max} \equiv -aR_{min} \quad (A.3)$$

Consequently eq. A.2 can be rewritten tidily as

$$\Delta = \sqrt{\ln S_0} \quad (A.4)$$

Equation 3.9, which relates the magnitude of the smallest population, m, to that of the modal population, N_0, now takes the form

$$N_0 = m \exp[(\Delta/a) \ln 2] \quad (A.5)$$

and the definition of γ, eq. 3.8, reads

$$2a\,\Delta\gamma = \ln 2 \quad (A.6)$$

With this preliminary festival of notation disposed of, we proceed to catalogue the properties of the general lognormal distribution in terms of the three parameters a, Δ, and γ, *only two of which are independent* (see eq. A.6).

(ii) S_T. Using eq. A.4 for S_0, and taking into account that the range of octaves in practice is from R_{min} to R_{max}, we find that eq. 1.3 for the total number of species becomes

$$S_T = \exp(\Delta^2) \int_{R_{min}}^{R_{max}} \exp(-a^2R^2)\,dR$$
$$= (\pi^{1/2}/a) \exp(\Delta^2) \operatorname{erf}(\Delta) \quad (A.7)$$

Here $\operatorname{erf}(\Delta)$ is the so-called error function (e.g., Abramowitz and Stegun, 1964, Chapter 7), an integral that continually appears when one deals with gaussian probability distributions, defined as

$$\operatorname{erf}(z) = 2\pi^{-1/2} \int_0^z \exp(-t^2)\,dt \quad (A.8)$$

For $\Delta > 1$, as it must be if S_T is not small,

$\operatorname{erf}(\Delta)$ is close to unity, and

$$S_T \simeq (\pi^{1/2}/a) \exp(\Delta^2) \quad (A.9)$$

If the range of integration for R is taken from $-\infty$ to $+\infty$ initially, the result eq. A.9 is exact, and it is widely quoted for the lognormal distribution. [Indeed, as the underlying assumptions leading to the lognormal species-abundance distribution usually require $S_T \gg 1$ (Preston assumes $S_T > 100$ or so), eq. A.9 should always apply; we give the result eq. A.7 only in order to treat modest values of S_T on occasion, even though the lognormal should not then be taken seriously.]

(iii) J. The total number of individuals in the community is given by eq. 1.4, which, after use of eqs. A.3, A.4, A.5, and A.6, becomes

$$N_T = m \exp[\Delta^2(1 + 2\gamma)] \int_{R_{min}}^{R_{max}}$$
$$\exp(-a^2R^2 + R\ln 2)\,dR$$

Changing the variable of integration to $t = (aR - \gamma\Delta)$ reduces this to

$$J = (N_T/m) = (\pi^{1/2}/2a)$$
$$\exp[\Delta^2(1 + \gamma)^2]\,G(\Delta, \gamma) \quad (A.10)$$

Here we have for convenience defined G as

$$G(\Delta, \gamma) = \operatorname{erf}[\Delta(1 - \gamma)]$$
$$+ \operatorname{erf}[\Delta(1 + \gamma)] \quad (A.11)$$

Again, if S_T is not unrealistically small, $\operatorname{erf}[\Delta(1 + \gamma)]$ will be indistinguishable from unity. But as γ may in general be larger or smaller than unity, no such simplifying statement can be made about $\operatorname{erf}[\Delta(1 - \gamma)]$. Referring to Figure 2, this says biologically that the cutoff at the octave R_{max} is important for the $N(R)$ distribution, although the cutoff at R_{min} may just as well be at $-\infty$.

Given any pair of values a and γ (and hence, via eq. A.6, Δ) to characterize the shape of the lognormal distribution, the total numbers of species and of individuals, S_T and J, now follow from eqs. A.7 and A.10. Conversely, given S_T and J, the parameters a, γ, and Δ may be computed from these equations. In this way we arrive at Figure 4. Quite generally, given any two of the parameters S_T, J, a, γ, and Δ, the others may be computed.

In the limit $S_T \gg 1$, a useful analytic approximation is possible. For S_T we have eq. A.9, in which the exponential term must dominate the right-hand side for large S_T, so that roughly

$$\ln S_T$$
$$\simeq \Delta^2[1 + \Delta^{-2} \ln (2\pi^{1/2} \Delta\gamma/\ln 2) + \ldots]$$

That is,

$$\ln S_T \sim \Delta^2 \qquad (A.12)$$

Correspondingly, in eq. A.10 for J, we may distinguish the cases (a) $\gamma < 1$ and (b) $\gamma > 1$. In case (a), $\mathrm{erf}[\Delta(1 - \gamma)]$ lies between 0 and 1, generally being of order unity, and one has the asymptotic approximation

$$\ln J \simeq \Delta^2(1 + \gamma)^2$$
$$[1 + \Delta^{-2}(1 + \gamma)^{-2} \ln (\pi^{1/2}/a) + \ldots]$$

That is, for $\gamma < 1$,

$$\ln J \sim \Delta^2(1 + \gamma)^2 \qquad (A.13)$$

In case (b), $\mathrm{erf}[\Delta(1 - \gamma)]$ is negative, and the result

$$\mathrm{erf}(-z) = -1 + \frac{\exp(-z^2)}{\pi^{1/2}z}\left(1 - \frac{1}{2z^2} + \cdots\right)$$

leads to

$$\ln J \simeq 4\gamma \Delta^2$$
$$[1 - (4\gamma \Delta^2)^{-1} \ln [(\gamma - 1)(\ln 2)/\gamma] + \ldots]$$

That is, for $\gamma > 1$,

$$\ln J \sim 4\gamma \Delta^2 \qquad (A.14)$$

The intermediate case, where $\gamma \simeq 1$ so that $|\Delta(1 - \gamma)|$ is small, is covered under the canonical lognormal in Appendix B. Combining the results eqs. A.12, A.13, and A.14, we have a rough relation between S_T and J which involves only the single parameter γ:

$$S_T \sim J^x \qquad (A.15)$$

where

$$x = (1 + \gamma)^{-2}, \text{ if } \gamma < 1 \qquad (A.16a)$$
$$x = 1/4\gamma, \qquad \text{if } \gamma > 1 \qquad (A.16b)$$

It is strongly to be emphasized that such a linear regression of $\ln S_T$ on $\ln J$ is not an exact result, even for $S_T \gg 1$. For smallish S_T the relation is steeper than indicated by this asymptotic approximation, as is clear from Figures 4 and 10.

(iv) S_T versus A. The species-area curves of Figure 10 are obtained from the additional biological assumption eq. 1.5, which relates J to A. Then, given S_T and γ, A is calculated from eqs. A.7 and A.10 to give Figure 10. The analytic approximation eq. 1.6 presented in the main text follows directly from the discussion in the preceding paragraph, eqs. A.15 and A.16, along with eq. 1.5.

(v) H. To calculate the diversity index H of eq. 4.2 it is first helpful to collect the definitions eqs. 3.4, 1.1, A.4, A.5, and A.6 to write

$$p = N/N_T = J^{-1} \exp[2\gamma \Delta(\Delta + aR)] \qquad (A.17)$$

Then eq. 4.2 for the lognormal distribution takes the form

$$H = \frac{\exp[\Delta^2(1 + \gamma)^2]}{aJ} \int_{-\Delta}^{\Delta} e^{-(s-\Delta\gamma)^2}$$
$$[\ln J - 2\gamma \Delta(s + \Delta)] \, ds$$

We have written $s = aR$. After some manipulation, this reduces to

$$H = \ln\left[\frac{\pi^{1/2}\gamma \Delta}{\ln 2} G(\Delta, \gamma)\right]$$
$$+ \Delta^2(1 - \gamma^2) + \frac{2 \Delta\gamma}{\pi^{1/2}G(\Delta, \gamma)} \{\exp[-\Delta^2$$
$$(1 - \gamma)^2] - \exp[-\Delta^2(1 + \gamma)^2]\} \qquad (A.18)$$

In deriving this, use has been made of eq. A.10 for J, and the definition eq. A.11 for $G(\Delta, \gamma)$.

Once S_T and γ are specified, Δ can be computed from eqs. A.6 and A.7, and thence H from eq. A.18. In this way we obtain for any specified value of γ a unique S_T-versus-H curve, as illustrated in Figure 5.

Limiting approximations can be obtained from eq. A.18 for $S_T \gg 1$, i.e., for Δ significantly in excess of unity. The cases (a) $\gamma \simeq 1$ (where $G \simeq 1$), (b) $\gamma < 1$ (where $G \simeq 2$), and (c) $\gamma > 1$ (where G is small) are to be distinguished, as they were above in obtaining the results of eqs. A.15, A.16 (here, and elsewhere, the symbol $\mathcal{O}$ means "terms of the order of"):

(a) $\gamma \simeq 1$; $H = 2\pi^{-1/2} \Delta \gamma$

$$\left[1 - \frac{2\Delta(1 - \gamma)}{\pi^{1/2}} \left\{ 1 - \frac{\pi(1 + \gamma)}{4\gamma} \right\} \right]$$
$$+ \mathcal{O} \ln (\gamma \Delta) \quad \text{(A.19a)}$$

(b) $\gamma < 1$; $H = \Delta^2(1 - \gamma^2)$
$$+ \mathcal{O} \ln (\gamma \Delta) \quad \text{(A.19b)}$$

(c) $\gamma > 1$; $H = \left[\frac{\gamma}{\gamma - 1} - \ln \left\{ \frac{(\gamma - 1)\ln 2}{\gamma} \right\} \right]$
$$+ \mathcal{O} \Delta^{-2} \quad \text{(A.19c)}$$

(vi) σ_H^2. In a similar way eq. 4.4 can be written down for the general lognormal distribution. The resulting expression is a mess, and we shall not set it out here. However, just as for given γ a unique H-versus-S_T curve can be computed, so here the variance in H can be calculated as a function of S_T. For the special case of the canonical distribution, $\gamma = 1$, the variance has the form given in Appendix B and displayed in Figure 6. The other asymptotic results are worth noting: (a) for $\gamma < 1$ (but still $\gamma > \frac{1}{2}$),

$$\sigma_H^2 = \frac{\Delta^2(1 - \gamma)^4 \ln 2}{4\pi\gamma(2\gamma - 1)}$$
$$\exp \left[-2\Delta^2(1 - \gamma)^2 \right][1 + \mathcal{O}(\ln S_T)^{-1}] \quad \text{(A.20a)}$$

and (b) for $\gamma > 1$,

$$\sigma_H^2 = \frac{(\gamma - 1)^2 \ln 2}{\gamma(2\gamma - 1)}$$
$$\{ \ln [(\ln 2)(\gamma - 1)/\gamma] \}\{ \ln [(\ln 2)(\gamma - 1)/\gamma]$$
$$- (2\gamma)/(2\gamma - 1) \}[1 + \mathcal{O}(\ln S_T)^{-1}] \quad \text{(A.20b)}$$

When we combine these asymptotic results with those for H itself, the root-mean-square relative fluctuations in H are seen to have roughly the limiting forms, for $S_T \gg 1$:

(a) $\gamma \simeq 1$; $\dfrac{\sigma_H}{H} \sim \dfrac{\text{(constant)}}{\ln S_T}$ (A.21a)

(b) $\gamma < 1$; $\dfrac{\sigma_H}{H} \sim (S_T)^{-(1 - \gamma)^2}$ (A.21b)

(c) $\gamma > 1$; $\dfrac{\sigma_H}{H}$

$\sim$ constant (value depends on γ)   (A.21c)

Note that, for all γ, the diversity H displays a significant amount of intrinsic statistical scatter; i.e., the r.m.s. relative variance is of order unity. This tendency is illustrated by Figure 6.

(vii) D. The diversity index D of eqs. 4.6, 4.7 may, with the aid of eqs. A.17 and A.8, be shown to be

$$1/D = \frac{\pi^{1/2} \exp [\Delta^2(1 + 2\gamma)^2]}{2aJ^2}$$
$$\{ \text{erf} [\Delta(2\gamma + 1)] - \text{erf} [\Delta(2\gamma - 1)] \} \quad \text{(A.22)}$$

The case $\gamma = 1$ is discussed further in Appendix B. For any other specified value of γ, eq. A.22 may be used to calculate D as a function of S_T. Again we note the asymptotic results:

(a) $\gamma < 1$; $D \simeq [2\pi \Delta(2\gamma - 1)/a]$
$$\exp [2\Delta^2(1 - \gamma)^2]$$
   i.e., D
$$\sim \text{(constant)}(S_T)^{2(1 - \gamma)^2} \quad \text{(A.23a)}$$

(b) $\gamma > 1$; $D \simeq \dfrac{(2\gamma - 1)\gamma}{(\gamma - 1)^2 \ln 2}$
$$+ \mathcal{O}(\ln S_T)^{-1} \quad \text{(A.23b)}$$

(viii) *d*. The dominance, defined by eq. 4.8, can be expressed immediately from eq. A.17 and the definition that N_{max} occurs in the octave labelled by R_{max} (i.e., by Δ/a):

$$d = J^{-1} \exp(4\gamma \Delta^2)$$

Alternatively, employing eq. A.10,

$$d = 2\pi^{-1/2}a$$
$$\exp[-\Delta^2(1 - \gamma)^2]/G(\Delta, \gamma) \quad (A.24)$$

Again, it is routine to calculate the limiting forms of this expression when $S_T \gg 1$; in this limit the relation

$$D \simeq \frac{(2\gamma - 1) \ln 2}{\gamma \, d^2} \quad (A.25)$$

holds for all γ.

(ix) *Sampling from the lognormal distribution.* A very simplistic approach, which glosses over statistical niceties, may be used to treat incomplete samples from the lognormal (and broken-stick) distributions. We assume that if, in the total sample of $\hat{N}$ individuals, the probable number of individuals from species i exceeds unity we *do* see that species, whereas if the probability corresponds to less than one animal that species is not counted.

In the sample of $\hat{N}$ individuals, let $\hat{m}$ be the average number of individuals from the least abundant species. Then $\hat{N}/\hat{m} = N_T/m$, and recalling the definition of J, eq. (1.1), we have $\hat{m} = \hat{N}/J$. Thus so long as

$$\hat{N} > J, \quad (A.26)$$

we have $\hat{m} > 1$, and all species are represented in our sample, at least in this naive estimate.

For $\hat{N} < J$, write

$$\hat{N} = fJ, \text{ with } f < 1 \quad (A.27)$$

In the sample, the number of individuals per species in the Rth octave, $\hat{N}_R$, is typically a fraction $(\hat{N}/N_T)$ of the actual number. With eqs. 3.4, A.27, 1.1, and 3.9, this comes to

$$\hat{N}_R = f2^{R-R_{min}} \quad (A.28)$$

If only those species that are typically represented by one or more individuals are assumed to be counted, then only those octaves with $R > R_0$ (where $\hat{N}_{R_0} = 1$) are to be counted. From eq. A.28, we have

$$R_0 = -\frac{\ln f}{\ln 2} + R_{min} \quad (A.29)$$

The total number of species, $\hat{S}_T$, represented in this sample is now

$$\hat{S}_T = S_0 \int_{R_0}^{R_{max}} \exp(-a^2R^2) \, dR$$

Alternatively, in the notation used throughout this appendix, the fraction of species represented in the sample is

$$\frac{\hat{S}_T}{S_T} = \frac{\text{erf}(\Delta) + \text{erf}\left[\Delta - \dfrac{\ln(1/f)}{2\gamma \Delta}\right]}{2\,\text{erf}(\Delta)} \quad (A.30)$$

This sampling behavior is illustrated for one particular value of γ in Figure 12. Notice that in general 50% of the total number of species actually present in the community are represented in a sample of fJ individuals, where $f = \exp(-2\gamma \Delta^2)$. Roughly this corresponds (see eqs. A.12, A.13, A.14) to a sample of $\hat{N} \sim (S_T)^{1+\gamma^2}$ individuals if $\gamma < 1$, or $\hat{N} \sim (S_T)^{2\gamma}$ individuals if $\gamma > 1$, which agree at $\hat{N} \sim S_T^2$ if $\gamma = 1$.

B. Canonical Lognormal Distribution: $\gamma = 1$

(i) *General.* In view of Preston's (1962) empirical observations as to the ubiquity of the canonical lognormal distribution, and the present rough explanation (Section 3A) as to why this tendency is likely to be observed, we single out the case $\gamma = 1$ for special review.

(ii) S_T. As Δ and a are now related by eq. A.6 with $\gamma = 1$, eq. A.9 can be put in terms of the *single* parameter Δ:

$$S_T \simeq (2\pi^{1/2}/\ln 2) \Delta \exp(\Delta^2) \quad (B.1)$$

(iii) *J*. Equation A.10 with $\gamma = 1$ becomes

$$J = (N_T/m) = (\pi^{1/2}/\ln 2)$$
$$\Delta \exp(4\,\Delta^2)\,\mathrm{erf}(2\,\Delta) \quad (B.2)$$

Unless S_T has an unreasonably small value, $\mathrm{erf}(2\,\Delta) \simeq 1$ and one has the excellent approximation

$$J \simeq (\pi^{1/2}/\ln 2)\,\Delta \exp(4\,\Delta^2) \quad (B.3)$$

(iv) *S_T versus A*. If we make the biological assumption eq. 1.5, $J = \rho A$, the single parameter Δ may in principle be eliminated between eqs. B.1 and B.3, to give a unique species-area curve. The interesting history of such work is discussed in Section 5, and the curve illustrated in Figure 10.

Note that eqs. B.1 and B.3 do not lead to any simple functional relationship between S_T and A. For $S_T \gg 1$, Δ is significantly greater than unity, and the exponential terms dominate the right-hand side of both equations. This leads for $S_T \gg 1$ to the excellent approximation

$$\ln S_T \simeq (\tfrac{1}{4})\ln A + (\text{constant}) \quad (B.4)$$

However, for smaller values of S_T the relation is steeper than a 0.25 power law (and indeed in the unreasonable limit of $S_T \lesssim 10$ we have roughly from eqs. A.7 and A.10 a linear relation, $S_T \simeq J = \rho A$). It is for this reason that Preston's and MacArthur and Wilson's work, which fitted linear regressions to this theoretical relation $\ln S_T$ versus $\ln A$, lead to regression coefficients slightly larger than the asymptotically exact 0.25 (Preston: 0.262; MacArthur and Wilson: 0.263). It is not a good idea to fit a single such regression to a relation that systematically proceeds from a steeper to a less steep relation as S_T increases.

(v) *H*. For $\gamma = 1$, eq. A.18 reduces to

$$H = \frac{2\,\Delta[1 - \exp(-4\,\Delta^2)]}{\pi^{1/2}\,\mathrm{erf}(2\,\Delta)}$$
$$+ \ln[(\pi^{1/2}\,\Delta/\ln 2)\,\mathrm{erf}(2\,\Delta)] \quad (B.5)$$

Again the approximation $\mathrm{erf}(2\,\Delta) \simeq 1$ will be excellent unless S_T is small, whence

$$H = 2\pi^{-1/2}\,\Delta + \ln(\pi^{1/2}\,\Delta/\ln 2) \quad (B.6)$$

Eliminating Δ by use of eq. B.1 leads to a unique *H*-versus-S_T relationship for the canonical lognormal distribution, as illustrated in Figures 5, 6, and 7. Notice that for $S_T \gg 1$ we have the approximation

$$H \simeq 2\pi^{-1/2}(\ln S_T)^{1/2}$$
$$[1 + \mathcal{O}(\ln S_T)^{-1}] \quad (B.7)$$

As discussed in Section 4, this relation corresponds to a very slow increase in the diversity index *H* as S_T increases.

(vi) σ_H^2. With $\gamma = 1$, eqs. A.1 and A.17 lead to an expression for σ_H^2 for the canonical lognormal distribution: this expression generates the results displayed in Figure 6.

An excellent approximation is again obtained by assuming S_T is not small, so that terms of relative order $\exp(-\Delta^2) \sim 1/S_T$ may be neglected:

$$\sigma_H^2 = \left(\frac{2\ln 2}{\pi\,\Delta}\right)e^{\Delta^2}\int_\Delta^\infty e^{-t^2}$$
$$\left[2\,\Delta^2 - 2\,\Delta t - \ln\left(\frac{\pi^{1/2}\,\Delta}{\ln 2}\right)\right]^2 dt \quad (B.8)$$

To a less accurate approximation, neglecting terms of relative order $1/(\ln S_T)$, this is

$$\sigma_H^2 \simeq \frac{\ln 2}{\pi\,\Delta^2}\left[\ln\left(\frac{\pi^{1/2}\,\Delta}{\ln 2}\right)\right]\left[2 + \ln\left(\frac{\pi^{1/2}\,\Delta}{\ln 2}\right)\right]$$

That is, very roughly, using "$a = 0.2$" in the logarithmic terms, we have

$$\sigma_H^2 \sim \frac{1 \cdot 1}{\ln S_T} \quad (B.9)$$

As noted in Appendix A, eq. 21a, the r.m.s. relative variance is of order $1/(\ln S_T)$, a fact that underlies the statistical dispersion in *H* illustrated and discussed in Figure 6.

(vii) *D*. On putting $\gamma = 1$, and neglecting

terms of relative order $\exp(-4\,\Delta^2) \sim 1/S_T^4$, we have for eq. A.22:

$$D = \pi\,\Delta^2/\ln 2 \qquad \text{(B.10)}$$

This leads to the D-versus-S_T relation illustrated in Figure 8, and has the asymptotic form given in Table 2.

(viii) d. The results displayed in Figure 9 for the canonical lognormal dominance index, d, as a function of S_T, stem from the appropriate form of eq. A.24. This is simply

$$d = \frac{\ln 2}{\pi^{1/2}\,\Delta} \qquad \text{(B.11)}$$

The asymptotic relation to S_T is clearly as given in Table 3.

(ix) *Sampling.* Here the treatment is precisely as in Appendix A, with the simplification that $\gamma = 1$. Recall that for all species to be represented we require $\hat{N} > J$, eq. A.26, which for the canonical lognormal distribution comes down to (see eqs. B.1, B.3) the requirement $\hat{N} \gtrsim S_T^4$.

C. Uniform Distribution

(i) *General.* The ideal uniform distribution,

$$N_i = N_T/S_T \qquad \text{(C.1)}$$

has the following properties.

(ii) S_T. A parameter specifying the distribution.

(iii) J. Clearly $m = N_T/S_T$, so that

$$J = (N_T/m) = S_T \qquad \text{(C.2)}$$

(iv) S_T *versus* A. Assuming eq. 1.5, we have the linear (!) relation

$$S_T = \rho A \qquad \text{(C.3)}$$

(v) H.

$$H = \ln S_T \qquad \text{(C.4)}$$

(vi) σ_H^2. For this ideally even distribution, $\sigma_H^2 = 0$.

(vii) D.

$$D = S_T \qquad \text{(C.5)}$$

(viii) d.

$$d = 1/S_T \qquad \text{(C.6)}$$

D. Broken-Stick Distribution

(i) *General.* Following the initial work of MacArthur (1957, 1960), and the numerical exploration of various aspects of the distribution by Lloyd and Ghelardi (1964), Longuet-Higgins (1971) and Webb (1974) have recently given expositions of the analytic properties of this distribution. There is also a fantastically elaborate formal and numerical exploration of "the distribution of indices of diversity" by Bowman *et al.* (1971) which, despite the generality of its title, is confined to the broken-stick distribution. These results will simply be catalogued under the appropriate headings below. There remain a few aspects of the distribution that do not seem to have been previously discussed (e.g., species-area relations, sampling aspects), and these will be developed more fully.

Unlike the lognormal or the logseries distributions, the broken-stick model is rarely presented in terms of its species-abundance distribution $S(N)$, but rather is conventionally discussed in rank-abundance form. To make concrete the abstract remarks in Section 2, we show how the broken-stick distribution $S(N)$ may be deduced from the more usual description of this distribution.

The routine broken-stick formulation gives the number of individuals in the ith most abundant of S_T species to be

$$N_i = \frac{N_T}{S_T}\sum_{n=i}^{S_T}\frac{1}{n} \qquad \text{(D.1)}$$

This is the sort of information displayed, for example, in Figure 1. For $S_T \gg 1$ this has the approximate form

$$N_i \simeq (N_T/S_T) \ln (S_T/i) \qquad (D.2)$$

Thus the rank-order function of Section 2, $F(N) \equiv i$, is simply

$$F(N) \simeq S_T \exp (-S_T N/N_T) \qquad (D.3)$$

But, as argued in eq. 2.1, $S(N)$ is now the (negative) derivative of $F(N)$:

$$S(N) = -dF/dN = (S_T^2/N_T)$$
$$\exp (-S_T N/N_T) \qquad (D.4)$$

This is exactly the asymptotic distribution function derived by Longuet-Higgins and by Webb; a thorough and elegant derivation is also presented by Cohen (1966). For a full discussion, leading to the exact result

$$S(N) = [S_T(S_T - 1)/N_T]$$
$$(1 - N/N_T)^{S_T-2} \qquad (D.5)$$

see Webb (1973).

(ii) S_T. In the general lognormal distribution, one could choose among several alternative parameters for the two that characterized the distribution; consequently we discussed relations among these parameters. Broken-stick distributions are characterized by *one* parameter, which invariably is just S_T.

(iii) J. Although it does not seem to have been remarked previously, there is a trivial relation between J and S_T for the broken-stick distribution. From eq. D.1, the population of the least abundant species is

$$m = N_T/S_T^2$$

That is,

$$J = S_T^2 \qquad (D.6)$$

(iv) S_T *versus* A. Adding to eq. D.6 the biological assumption of eq. 1.5, we have immediately the broken-stick species-area relation

$$S_T = (\text{constant}) A^{1/2} \qquad (D.7)$$

(v) H. Numerical tables of H as a function of S_T have been given by Lloyd and Ghelardi (1964). The exact analytic result

$$H = \psi(S_T + 1) + (\gamma - 1) \qquad (D.8)$$

has been given by Webb (1974). Here $\psi(z)$ is the logarithmic derivative of the gamma function (Abramowitz and Stegun, 1964, Chapter 6), and $\gamma = 0.577\ldots$ is Euler's constant. Neglecting terms of order $1/S_T$, one has $\psi(S_T + 1) \simeq \ln S_T$, leading to the excellent approximation given in Table 1 (Longuet-Higgins, 1971; Webb, 1974).

(vi) σ_H^2. Substitution of the exact $S(N)$ for the broken-stick distribution, eq. D.5, into eq. 4.4 for the variance of H may be shown to lead to the exact result (I omit the details: these, and any others, will be supplied on request)

$$\sigma_H^2 = \frac{2}{S_T + 1}$$
$$\{[\psi(S_T + 2) - \psi(3)]^2 + [\psi'(3) - \psi'(S_T + 2)]\}$$
$$- \frac{H^2}{S_T} \qquad (D.9)$$

Here $\psi'(z)$ is the derivative of $\psi(z)$. This expression gives the results depicted in Figure 6, which agree with Webb's numerical simulations.

For $S_T \gg 1$ one can obtain [either from eq. D.9, or directly from eq. 4.4, using eq. D.4 for $S(N)$] the asymptotic result

$$\sigma_H^2 \simeq (\ln S_T)^2/S_T \qquad (D.10)$$

The r.m.s. relative fluctuations in H then have the asymptotic form of eq. 4.9.

(vii) D. In a similar fashion, substitution of eq. D.5 for $S(N)$ into eqs. 4.6, 4.7 for D gives the exact and simple result

$$D = \tfrac{1}{2}(S_T + 1) \qquad (D.11)$$

Again the asymptotic form of this result, $D \simeq \tfrac{1}{2}S_T$, may alternatively be obtained by

using the simple asymptotic formula eq. D.4
for $S(N)$.

(viii) *d*. From the basic eq. D.1, the dominance index of eq. 4.8 is

$$d = \frac{1}{S_T} \sum_{n=1}^{S_T} \frac{1}{n}$$

$$= \frac{1}{S_T}[\gamma + \psi(S_T + 1)] \quad \text{(D.12)}$$

Ignoring corrections of relative order $1/S_T$, we
have the excellent approximation given in
Table 3. Notice the neat and exact relation
between H and d for the broken-stick distribution:

$$d = (H + 1)/S_T \quad \text{(D.13)}$$

(ix) *Sampling*. Referring to the discussion
at the end of Appendix A as to rough estimates
of the number of species represented in a
sample of $\hat{N}$ individuals, we first observe that
all S_T species are present once $\hat{N} > J$, eq.
A.26; that is, once $\hat{N} > S_T^2$ (see eq. D.6). This
confirms the features discussed in Section 6
and illustrated in Figure 12: in a sample of
size $\hat{N} \sim S_T^2$, roughly 50% of species are not
represented if the underlying distribution is
the canonical lognormal, whereas essentially
all are represented if the distribution be
broken-stick.

In more detail, we see from eq. D.1 that in
a sample of $\hat{N}$ total individuals, the number
of individuals in the *i*th most abundant species, $\hat{N}_i$, is

$$\hat{N}_i = (\hat{N}/S_T)$$

$$[\psi(S_T + 1) - \psi(i + 1)] \quad \text{(D.14)}$$

That is to say a rough estimate of $\hat{S}_T$ is given
implicitly by the equation

$$\psi(\hat{S}_T + 1)$$

$$= \psi(S_T + 1) - S_T/\hat{N} \quad \text{(D.15)}$$

The relation gives the curve shown in Figure
12. To an excellent approximation, $\psi(z + 1) =$

$\ln z - \gamma$, and so for $\hat{N}$ appreciably less than
S_T^2 we have that the fraction of the actual
species total S_T which is observed in a sample
comprising $\hat{N}$ individuals is

$$\hat{S}_T/S_T = \exp(-S_T/\hat{N}) \quad \text{(D.16)}$$

Roughly 50% of all species actually present in
the community are represented even in a sample as small as $S_T/\ln 2 = (1.4)S_T$ individuals.

E. Geometric Series Distribution

(i) *General*. The "niche-preemption" hypothesis, leading to a geometric series rank-abundance distribution (Motomura, 1932),

$$N_i = N_T C_k k(1 - k)^{i-1} \quad \text{(E.1)}$$

has been reviewed by Whittaker (1970, 1972)
and by McNaughton and Wolf (1970). Here
$k < 1$, and C_k is a normalization constant to
ensure $\Sigma N_i = N_T$:

$$C_k = [1 - (1 - k)^{S_T}]^{-1} \quad \text{(E.2)}$$

Converted to a (continuous) species-abundance distribution $S(N)$, in the manner illustrated at the beginning of Appendix D, this
gives

$$S(N) \, dN = \frac{dN}{N \ln[1/(1 - k)]} \quad \text{(E.3a)}$$

That is, expressed as a relation between
change in the number of species ΔS and
change in the number of individuals ΔN,

$$\frac{\Delta S}{\Delta N} = \frac{K}{N} \quad \text{(E.3b)}$$

This is the basic equation of Odum, Cantlon,
and Kornicher (1960) and we note that their
K and the conventional geometric series k are
related by $K \ln(1 - k) = -1$. [As geometric
series people tend to look at rank-abundance
relations, whereas Odum *et al.* focus on $S(N)$,
it is understandable that their relationship has
not been discussed. An exception is Horn
(1964), who remarked in an empirical way that

the Odum *et al.* S_T-versus-*J* relation is similar to that for the logseries, eq. 3.12.]

In the remainder of this appendix, we emphasize some aspects of the geometric series distribution that do not appear to be widely known.

(ii) S_T. Like the general lognormal and the logseries, the geometric series distribution is characterized by *two parameters:* these are usually taken to be *k*, the "niche-preemption" parameter, and S_T. Note that if the product $kS_T \gg 1$, the normalization constant $C_k \simeq 1$.

(iii) *J.* For the least abundant species, eq. E.1 has $N_i = m$ and $i = S_T$. Consequently

$$\frac{1}{J} = \frac{m}{N_T} = \left(\frac{k}{1-k}\right)\frac{(1-k)^{S_T}}{1-(1-k)^{S_T}}$$

That is,

$$S_T = \frac{\ln[1 + Jk/(1-k)]}{\ln[1/(1-k)]} \qquad (E.4)$$

For $J \gg 1$ and any particular value of *k*, this takes the useful asymptotic form

$$S_T \simeq K \ln J + (\text{constant}) \qquad (E.5)$$

Here the constant *K* is that of eq. E.3.

(iv) S_T *versus A.* When the biological assumption of eq. 1.5 is substituted into the exact eq. E.4, or the excellent approximation eq. E.5, we obtain the species-area relations discussed in Section 5.

(v) *H.* It is routine to calculate the diversity index, eq. 4.1, directly from the distribution eq. E.1. We arrive at a family of curves *H* versus S_T, one for each value of the parameter *k*:

$$H = \ln[(1-\varepsilon)/k]$$
$$- \left(\frac{1-k}{k} - \frac{S_T\varepsilon}{1-\varepsilon}\right)\ln(1-k) \quad (E.6)$$

where

$$\varepsilon \equiv (1-k)^{S_T} \qquad (E.7)$$

For kS_T significantly greater than unity, $\varepsilon \simeq 0$ and eq. E.6 gives the result noted in Table 1.

(vi) σ_H^2. Interpreted strictly, the geometric series distribution of relative abundance, eq. E.1, is rigidly deterministic, and $\sigma_H^2 = 0$. Similarly the exactly even distribution, eq. C.1, has by assumption no variance. As discussed in Section 2, the precise geometric series bears the same sort of relationship to the statistical logseries distribution as the exactly evenly distributed case bears to the statistical broken-stick distribution.

(vii) *D.* Use of eq. E.1 in the definitions eqs. 4.5 and 4.6 leads directly to

$$D = \frac{(2-k)(1-\varepsilon)}{k(1+\varepsilon)} \qquad (E.8)$$

Here ε is as defined in eq. E.7. Again for $S_T \gg 1$ this leads to the asymptotic result quoted in Table 2.

(viii) *d.* From eq. E.1, the dominance measure eq. 4.8 is simply

$$d = k/(1-\varepsilon) \qquad (E.9)$$

This has the value *k* when S_T is large, which essentially constitutes the original definition of *k*.

F. The Logseries Distribution

(i) *General.* This distribution, first discussed in an ecological context by Fisher, Corbet, and Williams (1943), is reviewed in detail by Pielou (1969, Chapter 17) and, complete with numerical tables, by Williams (1964, Appendix A). The distribution is characterized by two parameters, commonly written α and *x*:

$$S(N) = \frac{\alpha x^N}{N} \qquad (F.1)$$

Here *N* runs over the positive integers. A summary of results follows; for details, see the above references.

(ii) S_T. In terms of α and x,

$$S_T = -\alpha \ln (1 - x) \qquad \text{(F.2)}$$

Alternatively, S_T may be chosen as one of the two parameters characterizing the distribution.

(iii) J. The assumptions underlying the logseries distribution imply that it is never fully "unveiled" (see eq. F.1), so that the least abundant species is, or are, represented by a single individual. That is, $m = 1$ and $J = N_T$. For any specific choice of α, it may be seen that eq. 3.12 relates J to S_T.

(iv) S_T *versus A*. Given the biological assumption of eq. 1.5, eq. 3.12 corresponds to a family of species-area curves, one for each value of α:

$$S_T = \alpha \ln [1 + (\rho/\alpha)A] \qquad \text{(F.3)}$$

For $J \gg \alpha$, as it generally will be, we have

$$S_T = \alpha \ln A + \text{(constant)} \qquad \text{(F.4)}$$

This, as remarked in the main text, is of the same form as the geometric series species-area relation eq. E.5.

(v) H. From the definition eq. 4.2 and the logseries distribution eq. F.1 there follows the exact expression

$$H = (\alpha/J) \sum_{i=0}^{\infty} x^i \ln (J/i) \qquad \text{(F.5)}$$

Given S_T and α, the quantities x and J can be computed from eqs. F.2 and 3.12, and a curve H versus S_T computed for each value of α. For the usual limiting case where $J \gg \alpha$, a good approximation is

$$H \simeq (\alpha/J) \int_1^{\infty} x^t \ln (J/t) \, dt$$

That is,

$$H \simeq \ln \alpha + \gamma \qquad \text{(F.6)}$$

Here correction terms of order α/J have been neglected, and $\gamma = 0.577 \ldots$ is Euler's constant.

(vi) σ_H^2. By neglect of terms of relative order $1/\ln (J/\alpha)$, it may be shown that

$$\sigma_H^2 \simeq (\ln \alpha)^2/\alpha \qquad \text{(F.7)}$$

(vii) D. The index defined by eqs. 4.6 and 4.7 is

$$1/D = (\alpha/J^2) \sum_{i=1}^{\infty} ix^i$$

Summing this series, and using eqs. F.2 and F.3, one obtains

$$D = \frac{\alpha}{1 + (\alpha/J)} \simeq \alpha \qquad \text{(F.8)}$$

This makes plain the biological character of the parameter α (as did eq. F.6).

(viii) d. To calculate d, we need an estimate of the average value of the largest population. The most direct way to do this, and one with a clear biological basis, is to calculate the rank-order function $F(N)$ corresponding to eq. F.1 (see eq. 2.1), and then to estimate the maximum population as $F(N_{max}) = 1$. Replacing sums by integrals for $N > N_{max}$, one obtains

$$F(N) = \alpha E_1[N \ln (1 + \alpha/N_T)] \qquad \text{(F.9)}$$

Here E_1 is the standard exponential integral (e.g., Abramowitz and Stegun, 1964, Chapter 5). This expression is useful in itself, as it provides an analytic formula for the logseries rank-abundance computed, for example, by Whittaker (1972); in particular, for $\alpha N/N_T$ small as it will be for all but the most abundant few species, eq. F.9 has the approximate form

$$F(N) \simeq -\alpha \ln (\alpha N/N_T) - \gamma \qquad \text{(F.10)}$$

The dominance index of eq. 4.8 follows from eq. F.9 by putting $F(N_{max}) = 1$, with $N_{max} = dN_T$

In the usual case where N_T/α is large, d is given by

$$1 = \alpha E_1(\alpha d) \qquad \text{(F.11)}$$

That is, d depends only on the diversity index α, and is independent of S_T or N_T. Very roughly, eq. F.11 has the solution

$$d \sim \frac{\ln \alpha}{\alpha} [1 + \mathcal{O} (\ln \alpha)^{-1}] \qquad \text{(F.12)}$$

(ix) *Sampling.* As noted by Fisher, one of the key properties of the logseries distribution is that it is its own sampling distribution. That is, for any specified value of the diversity parameter α, the number of species $\hat{S}_T$ represented in a sample of $\hat{N}$ individuals is (see eq. 3.12)

$$\hat{S}_T = \alpha \ln (1 + \hat{N}/\alpha) \qquad \text{(F.13)}$$

References

Abramowitz, M., and I. A. Stegun. 1964. *Handbook of Mathematical Functions.* Dover, New York.

Batzli, G. O. 1969. Distribution of biomass in rocky intertidal communities on the Pacific Coast of the United States. *J. Anim. Ecol.* 38:531–546.

Berger, W. H., and F. L. Parker. 1970. Diversity of planktonic Foraminifera in deep-sea sediments. *Science* 168:1345–1347.

Bliss, C. I. 1965. An analysis of some insect trap records. *In* G. P. Patil, ed., *Classical and Contagious Discrete Distributions,* pp. 385–397. Statistical Publishing Society, Calcutta.

Boswell, M. T., and G. P. Patil. 1971. Chance mechanisms generating the logarithmic series distribution used in the analysis of number of species and individuals. *In* G. P. Patil, E. C. Pielou, and W. E. Waters, eds., *Statistical Ecology.* Vol. 3, pp. 99–130. Pennsylvania State University Press, University Park, Pa.

Bowman, K. O., K. Hutcheson, E. P. Odum, and L. R. Shenton. 1971. Comments on the distribution of indices of diversity. *In* G. P. Patil, E. C. Pielou, and W. E. Waters, eds., *Statistical Ecology.* Vol. 3., pp. 315–366.

Cohen, J. E. 1966. *A Model of Simple Competition,* pp. 77–80. Harvard University Press, Cambridge, Mass.

Cohen, J. E. 1968. Alternate derivations of a species-abundance relation. *Amer. Natur.* 102:165–172.

Cramer, H. 1948. *Mathematical Methods of Statistics.* Princeton University Press, Princeton.

Darlington, P. J. 1943. Carabidae of mountains and islands: data on the evolution of isolated faunas, and on atrophy of wings. *Ecol. Monogr.* 13:37–61.

Darlington, P. J. 1957. *Zoogeography.* John Wiley and Sons, New York.

De Benedictis, P. A. 1973. On the correlations between certain diversity indices. *Amer. Natur.* 107:295–302.

Deevey, E. S. Jr. 1969. Specific diversity in fossil assemblages. *In* G. M. Woodwell and H. H. Smith, eds., *Diversity and Stability in Ecological Systems,* Brookhaven Symposium in Biology No. 22, pp. 224–241. U.S. Department of Commerce, Springfield, Va.

Dickman, M. 1968. Some indices of diversity. *Ecol.* 49:1191–1193.

Fisher, R. A. 1958. *The Genetical Theory of Natural Selection.* Dover Publ. Inc., New York.

Fisher, R. A., A. S. Corbet, and C. B. Williams. 1943. The relation between the number of species and the number of individuals in a random sample of an animal population. *J. Anim. Ecol.* 12:42–58.

Gauch, H. G., and G. B. Chase. 1975. Fitting

the gaussian curve in ecological applications. *Ecology*. In press.

Goulden, C. E. 1969. Temporal changes in diversity. *In* G. M. Woodwell and H. H. Smith, eds., *Diversity and Stability in Ecological Systems,* Brookhaven Symposium in Biology No. 22, pp. 96–102. U. S. Department of Commerce, Springfield, Va.

Hamilton, T. H., R. H. Barth, and I. Rubinoff. 1964. The environmental control of insular variation in bird species abundance. *Proc. Nat. Acad. Sci. U.S.A.* 52: 132–140.

Hamilton, T. H., I. Rubinoff, R. H. Barth, and G. L. Bush. 1963. Species abundance: natural regulation of insular variation. *Science* 142:1575–1577.

Horn, H. S. 1964. Species diversity indices. Privately circulated manuscript, University of Washington, Seattle.

Hurlbert, S. H. 1971. The nonconcept of species diversity: a critique and alternative parameters. *Ecol.* 52:577–586.

Hutchinson, G. E. 1953. The concept of pattern in Ecology. *Proc. Acad. Nat. Sci. Philadelphia* 105:1–12.

Johnson, M. P., L. G. Mason, and P. H. Raven. 1968. Ecological parameters and plant species diversity. *Amer. Natur.* 102:297–306.

Johnson, M. P. and P. H. Raven. 1970. Natural regulation of plant species diversity. *Evol. Biol.* 4:127–162.

Johnson, M. P. and P. H. Raven. 1973. Species number and endemism: the Galapagos Archipelago revisited. *Science* 179:893–895.

Kendall, D. G. 1948. On some models of population growth leading to R. A. Fisher's logarithmic series distribution. *Biometrica* 35:6–15.

King, C. E. 1964. Relative abundance of species and MacArthur's model. *Ecology* 45:716–727.

Kohn, A. J. 1959. The ecology of Conus in Hawaii. *Ecol. Monogr.* 29:47–90.

Lloyd, M., and R. J. Ghelardi. 1964. A table for calculating the "equitability" component of species diversity. *J. Anim. Ecol.* 33:217–225.

Longuet-Higgins, M. S. 1971. On the Shannon-Weaver index of diversity, in relation to the distribution of species in bird censuses. *Theor. Pop. Biol.* 2:271–289.

MacArthur, R. H. 1957. On the relative abundance of bird species. *Proc. Nat. Acad. Sci. U.S.A.* 43:293–295.

MacArthur, R. H. 1960. On the relative abundance of species. *Amer. Natur.* 94:25–36.

MacArthur, R. H. 1966. A note on Mrs. Pielou's comments. *Ecology* 47:1074.

MacArthur, R. H. and E. O. Wilson. 1967. *The Theory of Island Biogeography*. Princeton University Press, Princeton.

McNaughton, S. J., and L. L. Wolf. 1970. Dominance and the niche in ecological systems. *Science* 167:131–139.

McNaughton, S. J., and L. L. Wolf. 1973. *General Ecology*. Holt, Rinehart, and Winston, New York.

Margalef, R. 1972. Homage to Evelyn Hutchinson, or why there is an upper limit to diversity. *Trans. Conn. Acad. Arts Sci.* 44:211–235.

Montroll, E. W. 1972. On coupled rate equations with quadratic nonlinearities. *Proc. Nat. Acad. Sci. U.S.A.* 69:2532–2536.

Motomura, I. 1932. A statistical treatment of associations [in Japanese]. *Jap. J. Zool.* 44:379–383.

Odum, H. T., J. E. Cantlon, and L. S. Kornicher. 1960. An organizational hierachy postulate for the interpretation of species-individual distributions, species entropy, ecosystem evolution, and the meaning of a species-variety index. *Ecology* 41:395–399.

Patil, G. P., E. C. Pielou, and W. E. Waters (eds.). 1971. *Statistical Ecology*. Vols. 1, 2, 3. Pennsylvania State University Press, University Park, Pa.

Patrick, R. 1968. The structure of diatom communities in similar ecological conditions. *Amer. Natur.* 102:173–183.

Patrick, R., M. Hohn, and J. Wallace. 1954. A new method of determining the pattern of the diatom flora. *Notulae Natura* 259.

Pielou, E. C. 1969. *An Introduction to Mathematical Ecology*. Wiley-Interscience, New York.

Preston, F. W. 1948. The commonness, and rarity, of species. *Ecology* 29:254–283.

Preston, F. W. 1962. The canonical distribution of commonness and rarity. *Ecology* 43:185–215, 410–432.

Sanders, H. L. 1969. Benthic marine diversity and the stability-time hypothesis. *In* G. M. Woodwell and H. H. Smith, eds., *Diversity and Stability in Ecological Systems,* Brookhaven Symposium No. 22, pp. 71–81. U.S. Department of Commerce, Springfield, Va.

Simpson, E. H. 1949. Measurement of diversity. *Nature* 163:688.

Statistical Abstract of the United States. 1971.

P. 327, Tables 522, 523. U.S. Department of Commerce, Springfield, Va.

Tramer, E. J. 1969. Bird species diversity; components of Shannon's formula. *Ecology* 50:927–929.

Tsukada, M. 1972. The history of Lake Nojiri, Japan. *Trans. Conn. Acad. Arts Sci.* 44:337–365.

Usher, M. B. 1973. *Biological Management and Conservation: Ecological Theory, Application and Planning*. Chapman and Hall, London.

Webb, D. J. 1974. The statistics of relative abundance and diversity *J. Theor. Biol.* 43:277–292.

Whittaker, R. H. 1965. Dominance and diversity in land plant communities. *Science* 147:250–260.

Whittaker, R. H. 1970. *Communities and Ecosystems*. Macmillan, New York.

Whittaker, R. H. 1972. Evolution and measurement of species diversity. *Taxon* 21:213–251.

Williams, C. B. 1953. The relative abundance of different species in a wild animal population. *J. Anim. Ecol.* 22:14–31.

Williams, C. B. 1964. *Patterns in the Balance of Nature*. Academic Press, London.

Williamson, M. 1972. *The Analysis of Biological Populations*. Edward Arnold, London.

5 On Continental Steady States of Species Diversity

Michael L. Rosenzweig

Introduction

In 1969, Robert MacArthur published a graph which indicated without comment two striking points: that species diversities on continents should approach steady states, and that this is the case because the total rate of speciation of a community rises at a decreasing rate with diversity, whereas its total extinction rate rises at an increasing rate (see Figure 3). If you are willing to accept the form of these curves without proof, you should immediately proceed to the final section of this chapter, *Synthesis and Discussion*. There you will see that, if MacArthur's graph is correct, it provides a most enlightened understanding of such species-diversity phenomena as adaptive radiation, the diversity-stability hypothesis, and latitudinal diversity gradients. It also leads directly to some tests of the causes of diversity differences.

But if you share with me the feeling that anything as potentially useful as MacArthur's graph warrants intense scrutiny, to discover both if it is true and why it is true, then you will not only want to read the part of the chapter immediately following, but perhaps also two companion papers (Rosenzweig, 1975a, 1975b), which treat the rates of geographical isolate formation and polyploidy more technically. Moreover, what is shown in the first part of the chapter is that, although MacArthur's extinction curve is quite probably correct, his speciation curve could be an oversimplification. If it is, the predictions made from it need to be hedged in some cases. Only careful study of why and how it might be too simple can yield the analytical power an investigator needs to make further progress.

In addition to the questions of whether and why steady states of diversity exist, another question has ultimately to be answered. Have communities already attained their steady-state diversities? Sewall Wright (1941) was probably the first person to suggest the hypothesis that they indeed have. Robert MacArthur made this hypothesis a vital force in evolutionary ecology by collecting data that support it and inspiring others to do the same.

For example, Cody (1966) discovered that grassland birds attain similar diversities in many part of the world. Recher (1969) showed that the patterns of α-diversity (see Cody, Chapter 10, for discussion) obtained for bird species by MacArthur and his colleagues in North America (MacArthur, 1969) also hold for Australian birds. Cody (1973) has recently reviewed the evidence for parallel and convergent evolution in communities.

If one accepts generic diversity as a fair reflection of species diversity—both Valentine (1969) and MacArthur (1969) indicate that this is reasonable—then other compelling evidence exists. Webb (1969) finds that mammalian generic extinction and speciation rates have been similar for the past 10 million years (at least). Despite great turnover in genera, a steady state has been maintained.

If species diversities in two neighboring biogeographical provinces are at a steady state and those provinces become joined by geological processes, one would predict that an increase in extinction rates would ensue as ecological vicars entered each others' ranges. If we allow that familial diversity is a reliable index of species diversity, then there is excellent evidence that, in fact, such waves of extinction *have* taken place. Simpson (1950) points out that a wave of mammalian extinction followed the recent union of North and South America. Valentine and Moore (1972) point out that the catastrophic extinctions of the Permo-Triassic period coincided with the formation of Pangaea. Moreover, they show that the number of families in the shallow marine fauna was at a steady state for perhaps 150 million years before Pangaea began to form. Finally, it appears (from their Figure 7) that about 100 million years after Pangaea disintegrated, a new steady-state of diversity was achieved by this fauna. The data for this fauna are fairly typical of most of the trends shown by the data for all fossil animal families (e.g., Lillegraven, 1972; see Rosenzweig, 1974, for discussion).

Despite the indication that some faunas may have attained their steady-state diversities, the empirical evidence remains inconclusive. This chapter does not try to add to this evidence, nor to review it in detail; in fact, the birds of Californian oak woodlands (Cody, Chapter 10 and Figure 11c) provide an example of a supersaturated fauna above the steady state; Australian desert lizards with few mammal competitors (Pianka, Chapter 12) and granivorous rodents of certain North American desert basins (Brown, Chapter 13) may also not be in a steady state. The function of this chapter is to serve a role analogous to that of an existence theorem in mathematics. Just as an algebraist can prove that a solution to any polynomial equation exists, even though he may not succeed in discovering it, so I hope to show that steady states of diversity exist, even though life may not yet have reached them. Hence, the message of this essay does not depend on whether or not biotas have yet achieved their steady states.

Notes on Methods

How can one hope to understand the steady states of a complex system, each of whose processes is itself intricately complex and whose rates are measured in hundreds of thousands of human generations? The answer is that one need not and, in fact, should not deal with all the variables that affect the production and extinction of species. One should deal only with those variables whose values are affected by diversity; these are the feedback variables through which diversity can affect its own rate of change and whose interrelationships can either produce or prevent steady states of diversity. In mathematical terms, we shall be looking for the signs of the first and second partial derivatives of speciation and extinction rates as functions of diversity.

The four variables that do correlate with diversity are: species abundance, the rate at which barriers produce geographical isolates, the narrowness of specialization, and the probability of competitive exclusion. Of course, since some of these are compounded of others, I have slightly exaggerated the degree to which the problem has herein been simplified.

Moreover, the only origination processes dealt with in this work are geographical speciation and polyploidy. Ultimately there may be no justification for such an emphasis, but speciation following Robertsonian polymorphism (Hall and Selander, 1973) and sympatric speciation by disruptive selection (Smith, 1966) are both too poorly understood to be modeled as functions of diversity.

Even though the problem is now greatly simplified, I shall treat the few remaining variables in only a provisional way. I do hope however to leave with the reader the notion that steady states of species diversity might well be the products of a relatively small number of the variables in the total system. Furthermore, by identifying those variables I hope to open or at least intensify traffic on various promising avenues of research.

Two other methodological points should be mentioned. The first one concerns the taxonomic breadth of the community whose diversity is being modeled. Initially, let us imagine that we are concerned with sets of ecological guilds in one broadly-defined habitat, for example, the grani-

vores of the North American desert. Later in the chapter I shall argue that when one understands diversity feedback and equilibria in such groups of species, one has mastered the rudiments of understanding these processes in whole biotas. Secondly, throughout the chapter, Webb's (1969) rigorous use of the terms "extinction" and "speciation" is followed. Instances of phyletic change—where only nomenclatural extinction and speciation occur—are excluded.

Geographical Speciation

Mayr (1963) has cogently summarized a whole scientific generation of evidence demonstrating the overriding importance of geographical speciation in producing additional species. This process requires the dissection of the range of a species by a geographical barrier and gradual divergence of the isolates. Elsewhere (Rosenzweig, 1975a), I show that, as a first approximation, the rate of geographical speciation equals the rate of isolate formation. In turn, the rate at which each species forms isolates is the rate at which geographical barriers are formed, multiplied by the probability that each barrier succeeds in cutting through the range of the species. These variables are now discussed.

Barrier Formation Rate

Let us define a long-lasting barrier (*llb*) as a line of environments with two properties: the environments are sufficiently inimical to the members of a species that they cannot cross the line, and the line endures for a time sufficient to allow any isolate populations that it might produce to achieve reproductive isolation. Let *B* be the average rate at which *llb*s are formed.

If the species' isolates can diverge rapidly, then, for that species, *B* is larger, because a greater fraction of barriers will be long-lasting. Similarly, if a species recognizes more environments as inimical, *B* for that species is larger. Both influences, divergence rate and barrier effectiveness, will now be examined as functions of diversity.

The rate at which two isolates diverge should depend on the speed at which they can adapt to their different average environments. Sved (1968) and Rosenzweig (1973a) present theoretical evidence that this depends on beneficial mutation rates and not so much on the rate of gene substitution. Since mutation is a random event, the beneficial mutation rate of a whole species must be proportional to population size. Large isolates should therefore speciate faster than small ones. Since increased subdivision of resources between species should accompany increased diversity, populations should be smaller in richer associations and should speciate (once isolated) somewhat more slowly. This effect probably is not very significant until species become quite rare, because most mutations in a large population ought to be duplicates and Fisher (1958) has shown that mutations with even slight advantages do not have to occur very often to become fixed. Hence the rate of divergence of two isolates should be almost independent of diversity until diversity gets large enough to pro-

duce rare species. Then divergence should begin to slow down.

The second influence on *B* is barrier effectiveness. Clearly what constitutes a significant barrier to one species may not to others. Mayr refers to a species' tendency to recognize barriers and to form isolates as its "sedentariness." He hypothesizes that high sedentariness has produced a high species diversity in some taxa such as flightless beetles and grasshoppers (Mayr, 1954, pp. 581–583). Pianka (1969) has evidence that the large diversity of Australian lizards, such as the radiation in the genus *Ctenotus,* is due to the ease with which lizard barriers are formed in Australia. He discusses the resultant ecological segregation in Chapter 12.

But does sedentariness vary predictably with diversity? Because habitat selection should and apparently does become narrower with increasing diversity (Karr and Roth, 1971; Terborgh and Weske, 1969; MacArthur and Levins, 1967; MacArthur, Recher, and Cody, 1966; Cody, Chapter 10), one might speculate that it does.

But there are two components to habitat selection: a discrimination in favor of one or more patches of environment, and an inability or reluctance to traverse unfavorable patches. While we know something about the first component (the relationship of discrimination to diversity), we know little about diversity's relationship to strict habitat taboos; but it is such taboos that are the stuff of effective barriers.

A qualitative model of taboos seems worth an attempt, if only to stimulate

disagreement. Imagine first a truly depauperate community. Since character displacement will not yet have occurred in
such a community, only the most extreme
environments would be effective barriers
to the environmentally promiscuous species that the community contains.

Now imagine the Earth as a hierarchical
set of nested habitats. Whole biomes are
among the largest patches of habitat possible in this hierarchy. As speciation begins to diversify life, sedentariness ought
to be increasing; not needing ever to leave
their native biomes, species can afford to
evolve a loss of ability to tolerate the
unfavorable places. Hence, over low but
increasing diversities, increase in habitat
selection almost certainly is accompanied
by an increased recognition of other habitats as totally unsuitable and even dangerous.

This relationship also yields a clue to
one possible reason for generally greater
diversities in smaller organisms. Such organisms have smaller resource requirements. The minimum habitat patch that
can support an entire species composed of
such organisms is also small. Therefore,
the habitat selection of smaller living
things continues to promote taboos and
thereby continues to promote their speciation rate, even after larger individuals
would no longer be able to afford habitat
intolerance.

For any taxon, however, there should
be some point beyond which increases in
habitat selection can no longer promote
taboos. The reason is that, as specific habitat patches are selected with increasing
precision, the probability is high that indi-

viduals will be required to successfully
traverse inappropriate patches in order to
find the preferred ones. For example, heteromyid rodent species strictly select their
microhabitats, but can often be captured
in an inappropriate one (Rosenzweig,
1973c; Schroder and Rosenzweig, in preparation). At the diversity at which more
restrictive habitat taboos begin to interfere
with an individual's success at finding its
special habitats, sedentariness should stop
increasing. Hence, sedentariness should be
an increasing function of diversity over
lower diversities, but it should reach an
asymptotic value and stop changing after
some threshhold value of diversity is attained.

Levins (personal communication) has
conjectured that as species become rare,
they tend to occur in patches of favorable
habitat. Diamond (Chapter 14, Figures
35–38) shows examples of disjunct ranges
of New Guinea birds, where a species is
absent from the intervening space between the different segments of its range
because of biotic factors, such as diffuse
competition, or because of minor abiotic
factors that could not conceivably lead to
taboos. Thus it would seem possible that
weak habitat preferences also produce
barriers, or that they could be produced
by the very rarity that is an inevitable
result of increased subdivision of the same
abiotic resources among more species. In
either case, diversity would appear to enhance isolate formation rate and contra-
dict the preceding paragraph. However,
this is not likely for the following reasons.

Population geneticists (Moran, 1962;
Maruyama, 1970) have shown that only

a tiny amount of gene flow is necessary to preserve panmixia. Further, Connell's (Chapter 16) and Leigh's work (Chapter 2) suggests that the establishment of a particular community at a place is either a chancy or a cyclic phenomenon. Thus, the particular patches that support a rare species will vary; patches should move about, coalescing and refragmenting so as to prevent long-term isolation. In addition, the studies of Ricklefs and Cox (1972) on taxon cycles indicate that rare species are more likely candidates for extinction than for speciation.

In summary: as diversity begins to increase from low levels, B increases owing to increasing sedentariness, but decreases owing to the decreasing mutation rates of smaller populations. Since the second effect is not likely to be large at low diversities, I will assume that the first dominates it (although assuming the opposite only strengthens the argument in favor of steady states). Thus at low diversities, B is probably an increasing function of diversity. At higher diversities, however, B will be a declining function of diversity, because divergence will be taking longer, whereas sedentariness will have stabilized.

Probability that Barriers Produce Isolated Populations

To produce isolates, a barrier must do more than merely appear. It must appear in an appropriate place. Let D_r be the probability that a random *llb* will appear in such a place as to dissect the geographical range of a species with range size r. Range size (sensu Schmidt, 1950)

r is half the angular distance in radians across the range of a species. A quick method of calculating r, even for irregularly shaped ranges, is to determine the proportion of the Earth's surface covered by a range; this proportion when multiplied by π radians equals r. Now we shall examine the dependence of D on r, and of average r on diversity.

Beginning empirically, we can define the average range size at a point as the mean of the range sizes of all species that occur at that point. We then plot the average range size against the number of species, as I have done for both land turtles and bats in Figure 1.

Apart from other interesting features, Figure 1 shows that average range size declines as diversity increases. Perhaps the decline results from more intense habitat selection in more diverse areas. But even if this is not the explanation, Figure 1 does show that a correlation exists.

Now we investigate D_r, the probability that a geographical barrier of random length and position will dissect a range of size r.

Formally, there are two sorts of barriers, those that are closed curves (like circles) and those that are not. Open curves have to satisfy two conditions to produce isolates: they must find the range and they must cut completely through it. Since cutting through large ranges is harder and finding them is easier, one might suspect that the probability of successful dissection by open-curve barriers would peak over some intermediate range size. Elsewhere (Rosenzweig, 1975a) I show in fact

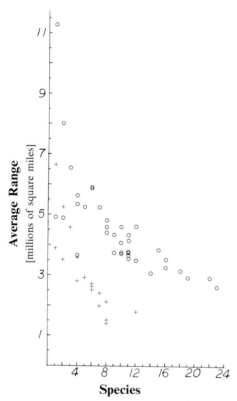

Species

Figure 1 The average geographical range of the species composing a community declines as diversity increases. Crosses represent the land and freshwater turtles listed in Collins (1959). Circles indicate the bats (noninsular) of Canada and the United States as given in Hall and Kelson (1958). Bat ranges include both summer and winter ranges for each species and have been revised to account for changes in nomenclature since 1958. The abscissa gives the number of species coexisting at a point, and the ordinate is the average geographical range of those species.

that this is precisely the case; Figure 2 depicts the actual probabilities obtained from that technical treatment of the problem. The simplifications of the model used

to derive Figure 2 place the peak probability of range dissection over too large a range size. Therefore, I shall assume in this paper that not only the upslope but also the downslope part of the probability distribution occurs over realistically small ranges. As we shall see, this is important to certain tests of the causes of diversity differences, but it does not affect the prediction that steady states of diversity exist.

Closed-curve barriers need only find a range to produce isolates. Thus the chance that a range will be dissected by a closed-curve barrier always grows with range size (Rosenzweig, 1975a).

Without information on the relative frequency of open- and closed-curve barriers and on the actual distribution of barrier lengths, I naturally cannot calculate

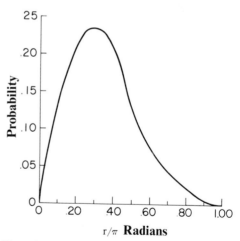

r/π **Radians**

Figure 2 The probability that a random open-curve geographical barrier will successfully dissect the range of a species whose range has angular area $4r$ square radians. The abscissa is r divided by radians. See text for barriers that are closed curves.

the actual total probability that a range of size *r* will be dissected by a random barrier. But one *can* conclude that very small ranges are less likely to be dissected than somewhat larger ones. It turns out that this is all we need in order to predict a steady state of diversity.

The Shape of the Rate of Isolate Production

We must now put what we have learned about *B* and *D* as functions of diversity into a model of total geographical-isolate production (and thus of geographical speciation rate) as functions of diversity.

In a depauperate biota, species will have large ranges and will not be very sedentary. Hence, both *B* and *D* are increasing functions of diversity; further diversification increases the isolate production rate of the average species. Since the rate of speciation for the biota is the number of species multiplied by the isolate production rate per species, the speciation rate for the entire biota must be accelerating (second partial derivative positive). Such acceleration is equivalent to a concave curve (the left-most segment in Figure 3).

As species accumulate, both *B* and *D* become decreasing functions of diversity; maximum *D* is reached and passed (Figure 2); sedentariness stabilizes; and rarity produces species that take longer to speciate. Together, these changes mean that the rate of speciation per species must begin to decline. At the diversity level at which this decline sets in, there will be a point of inflection (second partial derivative zero); beyond this diversity, the total spe-

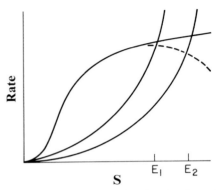

Figure 3 The effect of species diversity *S* on the speciation rates (sigmoid curve) and the extinction rates (concave curves) of a community's species. The dotted segment of the speciation curve is an unlikely modification (see text). Because the curves intersect where the extinction curve has higher slope than does the speciation curve, equilibria E_1 and E_2 (the values for *S* at the intersections) are steady states. E_1 and E_2 differ because of extinction rates. In general, changes in extinction rate should be more effective at altering the steady state than changes in speciation rate (see text).

ciation rate is convex (second partial derivative negative), as MacArthur (1969) indicated.

As higher diversities are achieved, some ambiguity arises. Does the speciation rate per species ever decline fast enough so that additional species actually mean less total speciation? Based on my own acquaintance with range maps, I think the answer is no. Figure 1 indicates that there may be deceleration in the decline of average range size at higher diversities. The reason is that the decline is due mostly to the addition of species with small ranges, not the deletion of those with larger ranges. If this turns out to be generally true—and based on what I know of land mammals and birds, it appears to be—

then even the richest biotas will contain the wide-ranging species that poorer ones do. The total speciation rate of this set of wide-ranging species should be the same in richer as in poorer biotas, but the richer biotas have the additional speciation from those small-ranged species that make them rich. Thus rich biotas should have higher total rates of geographical speciation than poorer ones.

I have indicated my preference in Figure 3 by drawing the declining curve as a dotted line. But those who construct figures like Figure 1 for other taxa will help to resolve the question more securely.

In the plant kingdom, polyploidy is a very common cause of speciation (Stebbins, 1950). Associations of tundra plants are known in which the vast majority of species are polyploid (Morton, 1966). Hence polyploidy cannot be ignored if we are to arrive at an accurate representation of speciation rates in Figure 3. Elsewhere (Rosenzweig, 1975b) I have modeled both autopolyploidy and allopolyploidy. Although these models are very different from those involved in geographical speciation, they also result in a sigmoidal speciation curve. Hence, Figure 3 should be accepted as containing a reasonable depiction of speciation rate as a function of diversity.

Extinction of Species

The second half of the equilibrium model comes from the relation between diversity and species extinctions. A species can become extinct for one of four causes: (1) its fundamental niche might disap-pear; (2) it might suffer a random accident such as a bad storm; (3) it might become involved in overexploitation; (4) it might be outcompeted. Most of these four causes of extinction occur at rates which are a function of species diversity in one way or another.

Niche Disappearance

If character displacement enforces increased specialization on species, then, being narrower, the fundamental niche (sensu Hutchinson, 1957) of a species in a rich biota is more likely to disappear.

Random Accidents

If the resources of a community are subdivided among more species, then each is likely to be less common. Rarity brings an increased likelihood that accidents will occur simultaneously to all individuals of a species. Using the methods of statistical mechanics, Leigh (1971; see also Chapter 2) has explicitly shown that rarity itself leads a population to instability.

Overexploitation

The relationship between species diversity and the likelihood of extinction by overexploitation remains obscure. We need the answers to the following three questions. What is the relationship between diversity and the tendency of the average predator to achieve an overly proficient phenotype in its evolutionary battle with its victims (Rosenzweig, 1973a; Schaffer and Rosenzweig, in preparation)? Do more abundant populations have a better chance of surviving an initially overexploitative encounter with a new species? Is there a relationship between diversity and the tendency of a new species to be involved in an overexploitation?

Recently I have discovered a phenomenon which suggests that the rate of overexploitational extinction may be positively correlated with the rate of extinction from other causes. In a simple 3-species, 3-trophic level system, stability of the top two levels determines the system stability. The plant-herbivore interaction by itself could easily be unstable. Extinction of the carnivore, however, is required to expose the plant and herbivore to their fate (Rosenzweig, 1973b). Thus a certain fraction of extinctions should yield, in turn, further extinctions by overexploitation. Loop analysis tends to produce the same conclusion (Levins, Chapter 1; Diamond, Chapter 14).

Competitive Extinction of Species
on One Trophic Level

May and MacArthur (1972) have recently supplanted previous theories of competitive exclusion. Using a model that allows for environmental uncertainty, they were able to show that only a finite amount of niche overlap is tolerable between competing species. If the resource spectrum is viewed as constant, this implies that probabilities of competitive exclusion are higher in richer biotas. But even if the resource spectrum is allowed to vary in breadth, this conclusion is unaltered (see below).

Evidently the rates of extinction per species are a rising function of diversity for three of the four causes of extinction: accidents, niche disappearance, and competitive exclusion. The fourth, overexploitation, may well have a rate proportional to the total rate of the other three; if so, it too would exhibit a rate positively correlated with diversity.

Since the total rate of extinction is the rate per species multiplied by the number of species, the total extinction rate for any of our ecologically bounded sets of species should be an accelerating function of diversity. A sample of such a curve appears as part of Figure 3. Extinction rates of island biotas have a similar shape (MacArthur and Wilson, 1967; MacArthur, 1969; Wilson, 1969) for similar reasons.

Synthesis and Discussion

In Figure 3 the curve of extinction rates may be seen to intersect that of speciation rates. This intersection is not accidental but is a necessary result of the model presented so far. It occurs because the speciation curve has a negative second derivative (is convex) to the right, whereas the extinction curve has a positive second derivative. This conclusion is somewhat different from that obtained by MacArthur and Wilson (1963) for islands. Island diversities equilibrate because the *first* derivatives of their extinction and colonization rates have opposite signs. However, just as in island equilibria, stability is produced because, at the point of intersection, the slope of the extinction-rate curve exceeds that of the speciation-rate curve.

Does Positive Feedback in the
Production of Diversity Interfere
with the Steady State?

At least since MacArthur's seminal,
inaugural paper of 1955, many ecologists
have been convinced that extinction becomes less probable as diversity increases.
Most have simply stated that as diversity
increases, so does stability. Should this be
the case, the rate of species diversity
would be endowed with positive feedback:
the more species,... the less extinction,... the faster diversity grows.

However, the Brookhaven Symposium
(1969) clearly exhibited the semantic difficulties in the word "stability" itself. Ecologists use it to mean all sorts of related
and unrelated measures (Watt, 1969). Despite this, there has been a veritable stam-
pede of ecologists who are anxious to accept the MacArthur conclusion.

Some ecologists have viewed this conclusion more rationally. Slobodkin (1961),
in a gently agnostic way, questions the
relevance to natural systems of Mac-
Arthur's measure of stability. Laboratory experiments point out that other
features of an ecosystem besides diversity
may have an overriding influence on the
persistence of species within it (Hairston
et al., 1968). May (1971) has reviewed
such evidence and also has shown, using
the admittedly oversimplified Lotka-
Volterra predation equations, that as species are added, exploitational ecosystems
exhibit a weaker capacity to return to their
equilibrium following perturbation. May
(1972, 1973) has recently extended his
model to more realistic situations and has
shown in fact that this measure of stability

does indeed decline with diversity. Diverse systems appear to attain some measure of stability by being organized into
subsystems that are only loosely interconnected ecologically (Gardner and Ashby,
1970).

Clearly the ability of diversity to enhance stability and thus perpetuate itself
is moot. Hence I have felt justified in not
incorporating it into my argument. Subsequent modelers may need to correct this
if it proves to be a deficiency.

Whittaker (1972) takes into account
that new species, being themselves resources, increase the breadth of the resource spectrum for other species. Hence
the resource spectrum is, in an important
sense, infinitely broad and capable of supporting an infinite number of species.
Whittaker makes it clear that he too is
invoking positive feedback: "For terrestrial plants and insects, increase of species
diversity, with elaboration of the niche
hyperspace and division of the habitat
hyperspace, is a self-augmenting evolutionary process without any evident limit"
(p. 214 and similarly on p. 242).

But as long as the world is saddled with
energetic and nutrient limitation and as
long as each species must contain at least
one individual, there must be some limit
to diversity. Let us now construct a model
that combines abiotic limitation with
Whittaker's principle. This model will
show how positive feedback in the evolution of diversity can be incorporated into
the system without producing the coun-
terintuitive result that diversity is limitless.
Let us consider two trophic levels and
assume only that the two variables, num-

ber of niches and number of species, are positively correlated (MacArthur and Levins, 1964; Levin, 1970; Haigh and Smith, 1972). We imagine a state-space in which the number of victim species is the abscissa; that of exploiters the ordinate (Figure 4). Given any fixed number of exploiter species S, the number of victim species will reach a saturation level K_S, according to May and MacArthur (1972) and the arguments of the previous parts of this paper. Above that number, extinction of victim species is faster than speciation. As we add predator species, the number of maintainable victim species K_S increases. This relationship incorporates positive feedback into the system. The collection of all points K_S is an isocline $(dS/dV = 0)$. What is the curvature of this isocline?

The addition of new predators does not add energy, water, or any other abiotic resource to the ecosystem; hence as more victim species are packed into the system, each becomes scarcer and the per-species extinction factors influenced by the population size must increase. Hence the victim's diversity isocline must increase (because of feedback), but with a positive second derivative (because of declining abundance per species). A simple way of summarizing that is: each added predator species allows the coexistence of additional victim species but fewer additional ones than any previous predator addition.

Since a similar argument will apply to any trophic level, the exploiter's isocline must have a negative second derivative. Thus the isoclines must intersect as in Figure 4. This intersection is a steady state, since arrows depicting the qualitative changes of species diversity all point toward a stable equilibrium. The conclusion must be that although increases in diversity in one trophic level do indeed lower the extinction rate in others, the relationship never gets out of hand. If there is a steady-state diversity in each trophic level when diversity in others is imagined to be constant, then one exists when all trophic levels are permitted to vary in diversity.

Niche Space and Speciation Rates

It has become part of conventional evolutionary wisdom to say that speciation has occurred faster when there were many unfilled niches. In fact, Mayr (1963, p. 554) expresses the opinion that *most* geographical isolates fail to found new species because they do not find open niches. If he is correct, Figure 3 is wrong; speciation rate declines with diversity, and a steady-state diversity is produced because extinction rate and speciation rate have first derivatives of opposite sign.

One role of this chapter is to point out that steady states of diversity exist, *even if the conventional wisdom is wrong.* This has been accomplished by deducing that diversity has a steady state without relying on any relationship whatever between speciation rate and emptiness of niche space.

It is important that steady states can and should exist even in the absence of any positive response of speciation rate to niche availability. This is so because it is quite possible that the conventional wisdom is wrong.

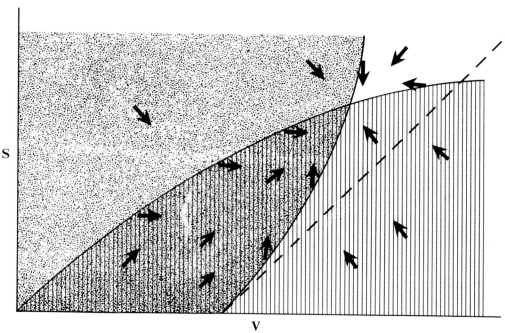

Figure 4 Dynamics of species diversity in two
trophic levels. S is the number of predatory species;
V, the number of victim species, increases in the
stippled region. S increases in the shaded region.
Arrows indicate the approximate direction that the
species diversity will take. The point of intersection
between the two solid lines (isoclines) can thus be
seen to be a steady state. The dashed line shows how
the V isocline would look if each new predator
allowed a constant number of victims to coexist.

Probably most evolutionists readily
agree to the conventional wisdom because
they are aware of the generality of the
phenomenon of adaptive radiation. However, adaptive radiation may be readily
comprehended without recourse to the
proposition that speciation speeds up in
the presence of empty niches.

Adaptive radiation is the observation of
rapid net growth in diversity. But net
growth is speciation minus extinction. So
there are two ways to increase diversity:
increase speciation or decrease extinction.
If Figure 3 is correct, then a decrease from
steady-state diversity produces a decrease
in speciation, but an even larger decrease
in extinction. Therefore adaptive radiation on continents appears likely to be due

largely to decline in extinction. Except in depauperate eras, new species, just like exotic introductions (Elton, 1958), are likely to become extinct quickly when geographical isolation disappears and are unlikely to have diverged sufficiently to be noticed by paleontologists.

The process can be summarized as follows: after geographic isolation, the isolates diverge and form semispecies at a rate independent of the availability of empty niches. After the initially allopatric semispecies are reunited geographically, diversity is enhanced only if one or the other does not go extinct; only here do empty niches affect diversity, by acting through extinction rate.

Adaptive radiations are often observed on islands. In such cases, it appears likely that very high speciation rates should play a role. After all, a species that finds an empty archipelago has also found a ready-made set of geographical barriers. Barrier formation occurs each time a propagule colonizes a new island. Judging from Mayr (1965), Diamond (1969 and Chapter 14), and Heatwole and Levins (1973), that must happen quite a bit more often than orogeny and similar geological and meteorological upheavals. In addition, some forms on islands definitely evolve increased sedentariness (Carlquist, 1966). But there is no indication that high speciation rates on islands are a response to the presence of empty niches.

All this is not to aver that the conventional wisdom is surely wrong. In fact, there is likely to be some truth in it if sympatric speciation (Smith, 1966) occurs (Rosenzweig, in press). But, as Mayr has so often pointed out (e.g., 1963), the evidence for the natural occurrence of sympatric speciation is at best equivocal, whereas that for polyploidy and for geographical speciation is robust.

Diversity Gradients

The model summarized in Figure 3 has two comments to make about the likely causes of gradients in species diversity. The first comment is implicit in the figure 3. Suppose one postulates that the diversity in such and such a place is higher because extinction rates are lower there. Then one actually goes and measures extinction rates. If the hypothesis is right, what will be found? Provided that both places are at a steady state, extinction will be higher in the place with higher diversity! Thus Figure 3 affords us a cheap, a priori lesson: since extinction and speciation rates are both dependent on diversity, simple comparisons of the two in different places should be undertaken only when the diversities of the places are the same.

The second comment is also implicit in Figure 3: diversity's steady state should prove to be much more sensitive to changes in extinction rates than speciation rates. This follows from the slopes of the speciation and extinction curves at the steady state: both are positive, but that of extinction is greater. Thus, any given change in diversity's steady state will require a small alteration in extinction rates, compared to the one required if it were to be produced by a change in speciation rate. The fact that the second derivative of extinction is positive, while that of speciation is negative, accentuates the disparity.

Recognizing the importance of geographic range, Terborgh (1973) attributes the latitudinal diversity gradient (i.e., the greater richness in species of tropical compared to temperate zones) to the fact that the average area of life zones declines as the latitude at which they are normally found increases. Since there is more area per habitat in the tropics, species there should have larger ranges and their speciation should proceed more rapidly. Also, since tropical habitats are more extensive, habitat taboos can evolve to be more restrictive and sedentariness can be higher. This parallels Janzen's (1967) observation that a given altitudinal gradient is biologically more restrictive in the tropics than in the temperate zone.

Terborgh supports his theory by comparing a selection of tropical realms and pointing out that those with larger area do tend to have many more species than their ecological analogues in other continents. For example, the extensive tropical African savannahs, grasslands, and arid lands harbor 94 species of ungulates. In the Neotropics, which is mostly forest (and therefore mostly unsuited to grazing ungulates), there are but 20 species. Cody (Chapter 10) shows similarly that Mediterranean bird diversity is highest in those habitat types that are best represented in the 1000-mile radius of "source habitats."

Another instructive comparison comes from Hessler and Sanders (1967). They discovered that the deep ocean or abyss contains an extraordinarily diverse animal community. The samples from the bottom of the Sargasso Sea have as many species as do comparable samples in the richest tropical, continental-shelf community, and far more species than the high-productivity shelf communities of temperate latitudes. Again, we see the importance of area: the abyss is of great extent; the shelf is narrow; at similar latitudes, diversity is far higher in the larger area. But if area is the only relevant difference, why are the restricted tropical shelves just as rich as the extensive abyss? Not all diversity differences can be explained by noting differences in area.

One hypothesis to explain why even those tropical habitats of limited extent are rich is that these habitats receive more immigrants from their extensive, rich, tropical neighbors than do habitats at higher latitudes. Were this so, their speciation rates would be higher, and differences in extinction rate would still not be required to explain their high diversity.

Yet, if there is merit in the reasoning about extinction rates that is summarized in this chapter, then extinction-rate differences are almost certainly going to be part of a satisfactory explanation of latitudinal gradients. Tropical latitudes have considerably higher productivities than other latitudes (Rosenzweig, 1968). Thus, when temperate and tropical zones have similar diversities, species will be more abundant in the tropics and have extinction rates that are lower because of reduced accidental extinction. Even if the tropics were similar in environmental predictability to other life zones, and they probably are not, their extinction rates should be lower because of extra productivity.

Since the vast abyss is also a vast desert, one might be tempted to conclude that it too should have high extinction rates. However, its extinction rates are probably depressed, owing to its high predictability and constancy (Hessler and Sanders, 1967). Moreover, the whole notion of attributing only high speciation rates to large areas seems one-sided. Who is to say that island effects are negligible on continent-sized areas, that once an area becomes fairly large, further areal increases cease to diminish extinction rates? I submit that such a conclusion lacks support. Thus the larger area of the abyss might well help to depress its extinction rates in addition to and perhaps even much more than it enhances its speciation rates. In fact, it is also possible that Terborgh's (1973) examples of habitats of large area and high diversity will turn out to have been products of low extinction rates.

Provided paleontologists can compare the per-species turnover rates of different life zones by disentangling true extinction from phyletic extinction, the model in this chapter provides a way to assess the relative effects of differences in speciation rate and in extinction rate on differences in diversity. If two places share the same extinction-rate curve, then the one with the higher diversity must show a higher speciation rate and a higher turnover per species if both are at steady state (Figure 5a). Therefore, if turnover is the same or lower in the more diverse area, that area has a lower extinction curve.

Similarly, if two places share the same speciation curve, then the one with higher diversity has a lower extinction rate and, almost surely, a lower turnover (Figure 5b). An exception might occur if the steady state of the poorer place is at a concave part of the speciation-rate curve. Considering the fact that this curve may

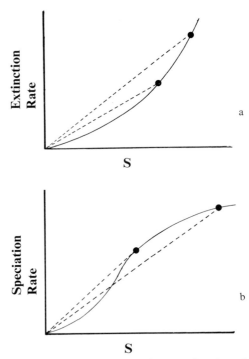

Figure 5 Turnover per species as a function of different steady states of diversity. Coordinates on this figure are like those of Figure 3. In part a, two communities share an extinction rate curve. They have different speciation rate curves, however, and, although these are not indicated in the figure, they intersect at different points on the extinction curve; these points are indicated. The slope of the dashed line connecting an intersection point with the origin is the turnover per species at that steady state. Hence the richer community must have a higher turnover at its steady state. In part b, the communities share a speciation-rate curve and the richer community has the lower species turnover.

not have a concave part and that even if it does, Figure 2 indicates that no modern community is near it, this exception itself is unlikely. Ignoring it, we provisionally conclude that two communities at steady state do not share the same speciation curve if the richer has a higher turnover rate per species.

Since area should affect both speciation and extinction rates, both should be involved in producing latitudinal gradients. Also, since extinction rates should respond to the extra productivity and to the increased predictability of the tropics, both of those phenomena should decrease the rate of accidental extinction. Therefore one must conclude that no single item on Pianka's (1966) well known list of influences on species diversity is going to account entirely for latitudinal gradients. As with most complex systems, several variables are bound to be involved in the complete explanation.

Acknowledgments

This chapter has had the benefit of suggestions and criticism from an unusually large number of colleagues. Since their qualms and comments were so instrumental in guiding its gestation, I list these critics, to all of whom I am deeply indebted: J. H. Brown, M. L. Cody, J. M. Diamond, J. S. Findley, S. D. Fretwell, F. R. Gelbach, E. G. Leigh, Jr., S. Levin, R. Levins, R. H. MacArthur, E. Mayr, E. R. Pianka, W. M. Schaffer, D. Simberloff, N. Slack, F. G. Stehli, J. Terborgh, S. D. Webb, R. H. Whittaker and E. O. Wilson. In view of the reservations of some of these critics, the usual disclaimer must be emphasized: they are not responsible for any of my mistakes. Ryland Loos of SUNY at Albany prepared Figures 2

and 3. The National Science Foundation has continued its support of my work.

References

Brookhaven Symposium in Biology No. 22. 1969. G. M. Woodwell and H. H. Smith, eds. *Diversity and Stability in Ecological Systems.* Brookhaven National Laboratory, Upton, N. Y.

Carlquist, S. 1966. The biota of long-distance dispersal. II. Loss of dispersability in Pacific Compositae. *Evolution* 20:30–48.

Cody, M. 1966. The consistency of intra- and inter-continental grassland bird species counts. *Amer. Natur.* 100:371–376.

Cody, M. 1973. Parallel evolution and bird niches. *In* F. D. Castri and M. Mooney, eds., *Ecological Studies No. 7.* Springer-Verlag, Wien.

Collins, H. H. Jr. 1959. *Complete Field Guide to American Wildlife.* Harper and Row, New York.

Diamond, J. M. 1969. Avifaunal equilibria and species turnover rates on the channel islands of California. *Proc. Nat. Acad. Sci. U.S.A.* 64:57–63.

Elton, C. 1958. *The Ecology of Invasion by Plants and Animals.* Methuen, London.

Fisher, R. A. 1958. *The Genetical Theory of Natural Selection.* 2nd rev. ed. Dover Publications, New York.

Gardner, M. R., and W. R. Ashby. 1970. Connectance of large dynamic (Cybernetic) systems: critical values for stability. *Nature* 228:784.

Haigh, J., and J. M. Smith. 1972. Can there be more predators than prey? *Theor. Pop. Biol.* 3:290–299.

Hairston, N. G., J. D. Allan, R. K. Colwell, D. J. Futuyma, J. Howell, M. D. Lubin, J. Mathias, and J. H. Vandermeer. 1968.

The relationship between species diversity and stability. An experimental approach with protozoa and bacteria. *Ecology* 40:1091–1101.

Hall, E. R., and K. R. Kelson. 1958. *Mammals of North America*. Ronald Press, New York.

Hall, W., and R. Selander. 1973. Hybridization of karyotypically differentiated populations in the *Sceloporus grammicus* complex (Iguanidae). *Evolution* 27:226–242.

Heatwole, H., and R. Levins. 1973. Biogeography of the Puerto Rican Bank: species turnover on a small cay, Cayo Ahogado. *Ecology* 54:1042–1055.

Hessler, R. R., and H. L. Sanders. 1967. Faunal diversity in the deep-sea. *Deep Sea Research* 14:65–79.

Hutchinson, G. E. 1957. Concluding remarks. *Cold Spring Harbor Symp. Quant. Biol.* 22:415–427.

Janzen, D. H. 1967. Why mountain passes are higher in the tropics. *Amer. Natur.* 101:233–249.

Karr, J., and R. Roth. 1971. Vegetation structure and avian diversity in several new world areas. *Amer. Natur.* 105:423–435.

Leigh, E. G. Jr. 1971. *Adaptation and Diversity.* Freeman, Cooper, San Francisco.

Levin, S. 1970. Community equilibria and stability, and an extension of the competitive exclusion principle. *Amer. Natur.* 104:413–423.

Lillegraven, J. A. 1972. Ordinal and familial diversity of cenozoic mammals. *Taxon* 21:261–274.

MacArthur, R. H. 1955. Fluctuation of animal populations and a measure of community stability. *Ecology* 36:533–536.

MacArthur, R. H. 1969. Patterns of communities in the tropics. *Biol. J. Linn. Soc.* 1:19–30.

MacArthur, R. H., and R. Levins. 1964. Competition, habitat selection and character displacement in a patchy environment. *Proc. Nat. Acad. Sci. U.S.A.* 51:1207–1210.

MacArthur, R. H., and R. Levins. 1967. The limiting similarity, convergence and divergence of coexisting species. *Amer. Natur.* 101:377–385.

MacArthur, R. H., H. Recher, and M. Cody. 1966. On the relation between habitat selection and species diversity. *Amer. Natur.* 100:319–332.

MacArthur, R. H., and E. O. Wilson. 1963. An equilibrium theory of insular zoogeography. *Evolution* 17:373–387.

MacArthur, R. H., and E. O. Wilson. 1967. *The Theory of Island Biogeography.* Princeton University Press, Princeton.

Maruyama, T. 1970. On the rate of decrease of heterozygosity in circular stepping stone models of populations. *Theor. Pop. Biol.* 1:101–119.

May, R. M. 1971. Stability in multispecies community models. *Math. Biosci.* 12:59–79.

May, R. M. 1972. Will a large complex system be stable? *Nature* 238:413–414.

May, R. M. 1973. *Stability and Complexity in Model Ecosystems.* Princeton University Press, Princeton.

May, R. M., and R. H. MacArthur. 1972. Niche overlap as a function of environmental variability. *Proc. Nat. Acad. Sci. U.S.A.* 69:1109–1113.

Mayr, E. 1954. Change of genetic environment and evolution. *In* J. Huxley, A. C. Hardy, and E. B. Ford, eds., *Evolution as a Process,* pp. 157–180. Allen and Unwin, London.

Mayr, E. 1963. *Animal Species and Evolution.* Harvard University Press, Cambridge, Mass.

Mayr, E. 1965. Avifauna: turnover on islands. *Science* 150:1587–1588.

Moran, P. A. P. 1962. *The Statistical Processes of Evolutionary Theory.* Clarendon Press, Oxford.

Morton, J. K. 1966. The role of polyploidy in the evolution of a tropical flora. *In* C. D. Darlington and K. R. Lewis, eds., *Chromosomes Today,* vol. 1, pp. 73–76. Oliver and Boyd, Edinburgh.

Pianka, E. R. 1966. Latitudinal gradients in species diversity: a review of concepts. *Amer. Natur.* 100:33–46.

Pianka, E. R. 1969. Habitat specificity, speciation and species diversity in Australian lizards. *Ecology* 50:498–502.

Recher, H. 1969. Bird species diversity and habitat diversity in Australia and North America. *Amer. Natur.* 103:75–80.

Ricklefs, R., and G. Cox, 1972. Taxon cycles in West Indian avifauna. *Amer. Natur.* 106:195–219.

Rosenzweig, M. L. 1968. Net primary productivity of terrestrial communities: prediction from climatological data. *Amer. Natur.* 102:67–74.

Rosenzweig, M. L. 1973a. Evolution of the predator isocline. *Evolution* 27:84–94.

Rosenzweig, M. L. 1973b. Exploitation in three trophic levels. *Amer. Natur.* 107:275–294.

Rosenzweig, M. L. 1973c. Habitat selection experiments with a pair of coexisting heteromyid rodent species. *Ecology* 54:111–117.

Rosenzweig, M. L. 1974. *And Replenish the Earth: The Evolution, Consequences and Prevention of Overpopulation.* Harper and Row, New York.

Rosenzweig, M. L. 1975a. Range size, isolate production and the rate of geographical speciation. In press.

Rosenzweig, M. L. 1975b. How does diversity influence rates of polyploidy? In press.

Schmidt, K. P. 1950. The concept of geographic range, with illustrations from amphibians and reptiles. *Tex. J. Sci.* 2:326–334.

Simpson, G. G. 1950. History of the fauna of Latin America. *Amer. Sci.* 38:361–389.

Slobodkin, L. B. 1961. *Growth and Regulation of Animal Populations.* Holt, Rinehart, and Winston, New York.

Smith, J. M. 1966. Sympatric speciation. *Amer. Natur.* 100:637–650.

Stebbins, G. L. 1950. *Variation and Evolution in Plants.* Columbia University Press, New York.

Sved, J. A. 1968. Possible rates of gene substitution in evolution. *Amer. Natur.* 102:283–293.

Terborgh, J. 1973. On the notion of favorableness in plant ecology. *Amer. Natur.* 107:481–501.

Terborgh, J., and J. S. Weske, 1969. Colonization of secondary habitats by Peruvian birds. *Ecology* 50:765–782.

Valentine, J. W. 1969. Niche diversity and niche size patterns in marine fossils. *J. Paleont.* 43:905–915.

Valentine, J. W., and E. M. Moores, 1972. Global tectonics and the fossil record. *J. Geol.* 80:167–184.

Watt, K. E. F. 1969. A comparative study on the meaning of stability in five biological systems: insect and furbearer populations, influenza, Thai hemorrhagic fever and plague. *In* G. M. Woodwell and H. H. Smith, eds., *Diversity and Stability in Ecological Systems,* Brookhaven Symposium in Biology No. 22, pp. 142–150. U. S. Department of Commerce, Springfield, Va.

Webb, S. D. 1969. Extinction-origination

equilibria in late cenozoic land mammals of North America. *Evolution* 23:688–702.

Whittaker, R. H. 1972. Evolution and measurement of species diversity. *Taxon* 21:213–251.

Wilson, E. O. 1969. The species equilibrium. *In* G. M. Woodwell and H. H. Smith, eds., *Diversity and Stability in Ecological Systems,* Brookhaven Symposium in Biology No. 22, pp. 38–47. U.S. Department of Commerce, Springfield, Va.

Wright, S. 1941. The "age and area" concept extended. *Ecology* 22:345–347.

II

Competitive Strategies of Resource Allocation

6 Selection for Optimal Life Histories in Plants

William M. Schaffer and Madhav D. Gadgil

Some twenty years ago, Lamont Cole (1954) suggested that the comparative study of life tables might well prove as fruitful as had previous comparative endeavors in anatomy and physiology. The term "life table" here refers to the age-specific schedule of births and deaths observable in an undisturbed population, and the great merit of Cole's suggestion is that such schedules are directly relatable to an organism's malthusian fitness. Consider a phenotype, ϕ, such that $l_x(\phi)$ is the probability that a newborn individual of this kind survives to age x, and $b_x(\phi)$ is the annual number of offspring produced by an x-year-old. Then the rate, $\lambda(\phi)$, at which a population of such organisms multiplies yearly is defined by the stable age equation

$$1 = \sum_{x=0}^{\infty} \lambda^{-(x+1)}(\phi) l_x(\phi) b_x(\phi) \qquad (1)$$

(Fisher, 1930; Lotka, 1956). To the demographer, the expectations $l_x(\phi)$ and $b_x(\phi)$ simply reflect current environmental conditions; to the evolutionist, they also suggest the possibility of adaptive strategies. Suppose, for example, we are dealing with a species in which potential fecundity increases with age, say because of year-to-year increases in size. Further suppose that associated with the act of breeding is a risk, perhaps age-specific, involving either increased post-breeding mortality or else reductions in growth and hence in future fecundity. Then the optimal age of first reproduction will depend on the details of this trade-off and also on environmental circumstances. In a harsh environment, i.e., one in which survivorship declines rapidly with age no matter what the organisms' reproductive strategy, we expect selection to favor breeding at the earliest possible age, even though this results in still greater mortality. Conversely, if the environment is benign, making survival to ages of greater potential fecundity more probable, we expect selection to favor a greater age of first reproduction.

The present paper is an attempt to review the evolutionary-theoretical aspects of such phenomena and, where possible, to apply this theory to plants. Such an undertaking involves serious risks, and we acknowledge the resistance in some quarters to the deductive approach which we shall employ. Nonetheless, we feel strongly that in some respects plant systems are ideally suited to the study of life-table evolution and also that life history theory can make a contribution to plant ecology in general. We will be grateful if our efforts are received in this spirit.

Annual Versus Perennial Reproduction

We begin our discussion by considering the relative evolutionary merits of annual and perennial reproduction. The yearly

142

rate, λ_A, at which a population of annuals multiplies is

$$\lambda_A = cB_A \qquad (2)$$

Here c is the probability that a seed successfully germinates and survives to reproduce. B_A is the number of seeds set. For a closely-related perennial, the corresponding rate is

$$\lambda_p = cB_p + p \qquad (3)$$

(we assume identical seed characteristics and hence equal c's) where p is the probability that an adult plant survives from one year to the next. Comparing these expressions, we note that the annual will outcompete the perennial if

$$B_A > B_p + \frac{p}{c} \qquad (4)$$

(Charnov and Schaffer, 1973). In general, we expect survival rates among seeds and seedlings to be lower than those of mature individuals (Harper, 1967; Tamm, 1972). As a result, the annual, if it is to have the greater fitness, must produce a seed crop that exceeds the perennial's by a considerable amount. For example, Sarukhan and Harper (1973) estimate for the perennial buttercup *Ranunculus bulbosus* that $c = 0.05$, $B = 30$, and $p = 0.8$. In this case, an annual mutant would have to increase its seed set by 53% in order to achieve a fitness equal to that of the existing perennial. Since we expect $p \gg c$ to be the usual case, we must therefore not follow previous authors (Cole, 1954; Murphy, 1968) in looking for reasons to explain the great abundance of perennials in the world, but rather must look for the special circumstances (i.e., those in which $p < c$) that favor the annual habit.

One situation in which this is likely to be the case is in deserts. Such habitats favor annual reproduction for two reasons: first, because the extreme aridity probably reduces the survival of a perennating organ such as a tuber or a bulb to a greater extent than that of a seed; and second, because such environments are not only harsh, but also extremely unpredictable. Thus for a 40-year period, Death Valley, California, had a yearly average precipitation of 34 mm, but this was irregularly distributed over the years, ranging from 19 mm to 94 mm (Went and Westergaard, 1949).[1] To a considerable degree, seeds are better able to cope with such variability than mature plants. As Cohen (1966, 1967, 1968) has emphasized, a plant's seed crop can be programmed so that it germinates in response to specific environmental cues, or else so that a fraction of the total crop germinates over successive years, whatever the immediate environmental conditions. Desert plants apparently engage in both these practices (Went, 1949; Went and Westergaard, 1949; Cronin, 1965; Mayer and Poljakoff-Mayber, 1963), which serve to render them relatively immune to environmental vagaries.

These ideas can be made more explicit if we consider the case in which adult survival equals $p(1 + s)$ in good years and $p(1 - s)$ in bad. Then if good and bad years are equally and randomly distributed in time,

---

[1]In the same article, these authors observe that a minimum of 60 mm of winter rainfall is necessary to induce germination of the spring annuals.

$$\bar{\lambda}_p^2 = \{cB_p + p(1 + s)\}\{cB_p(1 - s)\}$$
$$= (cB_p + p)^2 - s^2p^2 \qquad (5)$$

where $\bar{\lambda}_p$ is the mean, yearly rate of multiplication for the perennial. Hence, for the annual to outcompete the perennial, we require that

$$(cB_A)^2 > (cB_p + p)^2 - (sp)^2$$

or

$$B_A > \sqrt{\left(B_p + \frac{p}{c}\right)^2 - \left(\frac{sp}{c}\right)^2} \qquad (6)$$

Clearly, increasing s, which measures the magnitude of temporal fluctuations in adult survivorship, reduces the perennial's fitness vis-à-vis the annual's fitness even though the mean survival rate is unchanged.[2]

Consistent with this analysis are data accumulated by K. T. Harper (unpublished observations) which indicate that the percentage of annuals (as indicated by percent cover) among the herbaceous flora on various undisturbed sites in the southwestern United States increases with increasing aridity and year-to-year variability in rainfall (Figure 1). In the driest and least predictable area (Death Valley, Cali-

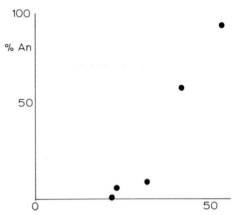

Figure 1 Percentage of herbaceous flora accounted for by annuals (ordinate) plotted against coefficient of variation in total annual rainfall (CV) in five desert habitats in North America. Annuals predominate in less predictable areas. Data provided by K. T. Harper.

fornia), annuals comprise 94% of the flora, whereas in the wettest and most predictable habitat, the figure drops to 1%. Nor is this pattern simply one of biogeography, since the vegetation surrounding the few permanent springs (constant environment) in Death Valley is composed entirely of perennials (Deacon, Bradley, and Moore, 1972).

A second circumstance favoring annuals is environmental disturbance, which reduces competition among seedlings and between seedlings and adults. It seems reasonable to assume (e.g., Harper, 1967) that seedling survival usually decreases more rapidly with increasing population density than does adult survival. Consider a mixed population of annuals and perennials regulated both by density-dependent and density-independent processes. For

---

[2] Conversely, fluctuations in seed (as opposed to adult) survival *increase* the relative fitness of the perennial. In this case,

$$\bar{\lambda}_A^2 = (1 - s^2)c^2B_A^2$$

and

$$\bar{\lambda}_p^2 = (cB_p + p)^2 - s^2c^2B_p^2$$

Then, for λ_A to exceed λ_p, we require that

$$B_A^2 > B_p^2 + \frac{2cpB_p + p^2}{c^2(1 - s^2)}$$

Consequently, increasing s also increases the litter-size differential necessary to favor the annual.

convenience, we assume that seedling survival is the only demographic parameter affected by crowding. Then if $\bar{\lambda}(N)$ is the mean rate of increase of the entire population at density N, we have at equilibrium

$$\bar{\lambda}(N^*) = q\lambda_A + (1 - q)\lambda_p - d = 1 \quad (7)$$

where q is the proportion of annuals; N^*, the equilibrial density; and d, the density-independent death rate. From eq. 7 we calculate $c(N^*)$, the equilibrial rate of seedling survival. Substituting this value into eq. 4, we find that λ_A exceeds λ_p if

$$\frac{B_A}{B_p} > \left(1 - \frac{p}{1 + d}\right)^{-1} \quad (8)$$

Clearly, the greater the density-independent death rate, d, the smaller the ratio B_A/B_p required to favor the annual. It is thus interesting to note that although undisturbed sites in the Great Basin support an herbaceous flora that is almost exclusively perennial, heavily grazed areas are characterized by a considerable abundance of annuals (Holmgren and Hutchings, 1971).

Seeds have an additional advantage over adult plants and perennating organs; they are capable of being dispersed, often over considerable distances. We would therefore expect an annual habit in species occupying transient habitats (Gadgil, 1971). Annual weeds growing along roadsides are an example of such "fugitive" species (cf. discussions of similar fugitive or "supertramp" species among birds, insects, and plankton—Chapters 14, 15, and 17, respectively by Diamond, Patrick, and Hutchinson), and they should be distin-

guished from species whose advantage lies in the resistivity of their seeds. Thus, van der Pijl (1969) has observed that desert annuals sometimes possess mechanisms that ensure that the seeds fall and remain in the vicinity of the parent plant. This further suggests that in such habitats the premium is on the ability to withstand prolonged and unpredictable drought, and not on the ability to colonize unoccupied patches of favorable habitat.

Biennials

Biennials, i.e., plants that store nutrients in a storage organ during the first season and set seed during the second, pose an intriguing problem. Many such plants appear to constitute a rather specialized group of fugitives, as evidenced by their tendency to grow in habitats that are either disturbed or for some other reason unoccupied by potential competitors (e.g., biennial species of the *Isomopsis* complex). The problem then becomes one of determining why such plants are not annuals. One possibility is that starting the growing season as a tuber offers considerable advantages over starting as a seed, advantages which compensate for the increased prereproductive mortality and generation time which necessarily accompany delaying reproduction until the second season. We can exhibit this requirement explicitly by noting that for the annual

$$\lambda_A = cB_1 \quad (9)$$

and for the biennial

$$\lambda_B = \{cp_1B_2\}^{1/2} \quad (10)$$

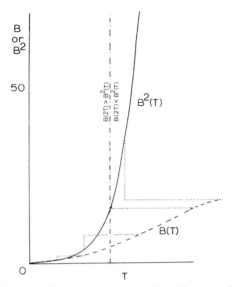

Figure 2 Circumstances favoring biennial or annual reproduction. T (abscissa) is the length of the growing season; $B(T)$, the number of seeds produced; the ordinate is $B(T)$ or $B^2(T)$. Biennial reproduction confers greater fitness when $B(2T) > kB^2(T)$; annual reproduction, when $B(2T) < kB^2(T)$. Dotted lines compare $B(2T)$ and $B^2(T)$; for several values of T; for convenience, we assume $k = 1$. Short T favors biennial habit; long T, annual. The two domains are divided by the vertical line $(- \cdot - \cdot -)$.

Here, p_1 is the probability that the biennial survives from the end of the first season to flower in the second. B_1 and B_2 indicate seed set at ages 1 and 2. Comparing these expressions, we find that the biennial has the greater rate of increase if

$$B_2 > \frac{c}{p_1} B_1^2 \qquad (11)$$

Thus, the second season's growth must roughly square the seed set achievable at the end of the first year, if the biennial is to be favored (Harper, 1967).

Now suppose that we relate seed set to the length or quality of the growing season, which we denote by the symbol T. This permits us to rewrite eq. 11 ($\lambda_B > \lambda_A$) as

$$B(2T - t) > \frac{c}{p_1} B^2(T) \qquad (12)$$

where t represents the cost of manufacturing the storage organ. It is reasonable, we feel, to suppose that $B(T)$ is sigmoidal—a minimum amount of flower and supportive tissue must be manufactured in order to produce the first seed, and there will be mechanical limitations to the number of seeds that can be borne by a single flower stalk. In this case, biennial reproduction will be favored if the growing season is short, and annual breeding if T is long (Figure 2). It is thus interesting to note that in the sweet clover genus *Melilotus,* annuals predominate (13 species vs. 2) at low latitudes and elevations (long growing season), whereas all nine species indigenous to high elevations and latitudes (short growing season) are biennial (Smith, 1927). In the same vein, but in a different kingdom, McLaren (1966) observed that although most arctic forms of the chaetognath worm, *Sagitta elegans,* are biennial, the population inhabiting the warm, partially land-locked fjord, Ogac "Lake," on Baffin Island, is strictly annual.

"Big Bang" Versus Iteroparous Reproduction

Among plants whose generation time exceeds one year, it is convenient to distinguish between so-called iteroparous species that flower repeatedly for several

years and semelparous varieties in which
a single episode of intense reproduction
is typically followed by rapid senescence
and death. To the former category belong
the vast majority of trees and shrubs, as
well as the bulk of the herbs; to the latter,
century plants (genus *Agave*), several genera of palms and bamboos (McClure,
1966), the foxgloves (genus *Digitalis*) and
the fantastic Haleakala silversword, *Argy-
ioxiphium sandwichense,* on the island of
Maui (Carlquist, 1965). Gadgil and
Bossert (1970) were the first to analyze the
conditions under which each of these reproductive strategies might be expected to
evolve. Following Williams (1966a), they
assume that an organism's resources—
time, energy, water, etc.—are finite, and
that, as a consequence, evolutionary-
induced increases in current fecundity are
achievable only at the expense of reductions in survival and reproductive output
(see also Calow, 1973). Intuitively, then,
a mutation resulting in increased reproductive output at a particular age can be
associated with a profit (gains in fertility
at that age) and a cost (reductions in subsequent survival and fecundity). Formally,
this idea can be expressed by noting that
an organism's malthusian fitness (Fisher,
1930) is maximized if at every age the sum

$$b_i + p_i \frac{v_{i+1}}{v_0} \qquad (13)$$

is also maximal (Schaffer, 1974; see also
Williams, 1966b). Here, b_i is the number
of offspring produced at age i; p_i is the
probability of surviving from age i to
$i + 1$; and $\frac{v_{i+1}}{v_0}$ is the reproductive value

of an $i + 1$-year-old. When breeding occurs at discrete (e.g., yearly) intervals, reproductive value is defined by the expression

$$\frac{v_x}{v_0} = \frac{b_x}{\lambda} + \frac{b_{x+1}p_x}{\lambda^2}$$

$$+ \frac{b_{x+2}p_xp_{x+1}}{\lambda^3} + \cdots \quad (14)$$

(Keyfitz, 1968), which is equivalent to
Fisher's (1930) formulation for the continuous case.

Now let us introduce the concept of reproductive effort at age i, signified by E_i.
We define E_i as the percent allocation of
time and materials to breeding. In
general, b_i will be a monotonically increasing function of E_i, and so long as this is
the case, it is possible to show that the

product, $p_i \dfrac{v_{i+1}}{v_0}$—henceforth signified by

pV—is a monotonically decreasing function of E_i (Schaffer, 1974). Thus, we can
define in the following manner the profit
and cost functions associated with an increase in effort at age i from E_i to E_i':

$$\text{Profit} = b_i(E_i') - b_i(E_i) \qquad (15a)$$
$$\text{Cost} = pV(E_i) - pV(E_i') \qquad (15b)$$

Since fitness turns out to be proportional
to the sum of two functions, one of which
increases with effort and the other of
which declines, the optimal effort at age
i, $\hat{E}_i$, can, at least in theory, be determined
by plotting the functions against E_i and
adding them together (Figure 3). Of cru-
cial importance are the shapes of the
functions. If they are concave (Figure 3,
top), the optimal reproductive effort at
each age is either 0 or 100%. Since the

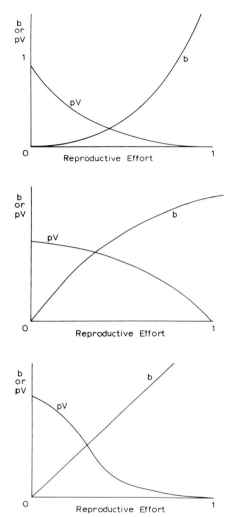

Figure 3 The optimal reproductive effort at age *i* maximizes fecundity, *b*, at that age plus the product-probability of surviving to breed the following year times next year's reproductive value, *pV*. Top: If *b* and *pV* are concave, the optimal effort is 0 or 100%; this selects for semelparity. Center: If *b* and *pV* are convex, the optimal effort is between 0 and 100%, with consequent selection for iteroparity. Bottom: If *b* is linear and *pV* reverse sigmoidal, there will be two values of *E* associated with relative maxima in λ. These values correspond to semelparity, $E = 1$, and iteroparity, $e = E^{\ddagger}$, $0 < E^{\ddagger} < 1$.

chances of post-breeding survival are nil when effort equals 100%, concave curves should favor the evolution of semelparity. On the other hand, if b_i and pV are convex (Figure 3, center), their sum will peak at an effort value intermediate between 0 and 100%, thus favoring the evolution of iteroparity.

Data obtained by Schaffer and Schaffer (unpublished observations) on the reproductive biology of yuccas and agaves are consistent with this analysis. Although similar in basic morphology and often occurring together in the same habitats, these two closely-related genera differ in that most species of *Yucca* are iteroparous, whereas most agaves die after a single flowering. Significantly, within a population, agaves appear to compete with each other for pollinators (mostly bees in the species studied), which prefer the taller flower stalks (Table 1). Presumably, the bees seek out the larger inflorescences in order to minimize the energy expended between visits to successive flowers. Pollinator preference for the taller stalks would, of course, bow the fertility curve upwards for high effort values, and thus tend to select for semelparity. In contrast, the moths which pollinate the iteroparous yuccas do not appear to favor the taller spikes, which in some populations actually receive less attention than flower stalks of intermediate height. In these species, the fertility curve should therefore be either linear or convex, which would tend to favor a less than maximal reproductive effort. Additionally, non-big-bang populations of *Agave parviflora*, which propagates vegetatively for several years after flowering, were discovered and found to resemble the iteroparous yuccas in this regard (the pollinators did not prefer the larger flower stalks), an observation which strengthens the hypothesis that the variation in reproductive strategies is adaptive and not simply a consequence of different ancestry.

Another species in which competition for pollinators may be the selective force responsible for the evolution of semelparity is the monument plant, *Frasera speciosa*. Recently, Beattie, Breedlove, and Ehrlich (1973) have studied its pollination ecology in Colorado. Although numerically a small component of the flora, *Frasera* plants are apparently able to garner the bulk (44–82%) of the available insect pollinators, which are attracted to the large flower stalks and copious amounts of nectar. In addition, colonies of rosettes, the result of caudex branching prior to flowering, were observed to bloom synchronously, an adaptation which may

further improve their ability to attract pollinators away from competing species. In a like vein, one might hazard that the spectacular inflorescences of other big-bang species, such as the Mount Haleakala silversword, serve a similar function. However, this must not always be the case since some semelparous plants, e.g., bamboos, are wind pollinated (Corner, 1964; McClure, 1966).

We should also point out that there often can be more than one set of optimal age-specific effort values (Schaffer, 1974). For example, if $b_i(E_i)$ is linear or concave and $pV(E_i)$ is reverse sigmoidal (see section on Changes in Total Reproductive Effort for discussion), there will be two values of E_i associated with relative maxima in λ (Figure 3, bottom). One of these corresponds to iteroparity, the other to semelparity. In a highly productive environment, fecundity per unit effort will be enhanced and, as a result, the value of λ

Table 1. Intensity of competition for pollinators in populations of Yuccas and Agaves

			Reproductive strategy		
	Big-bang			Non-big-bang and repeated breeding	
Species	N	Competition	Species	N	Competition
Agave schottii	95	0.22	*Agave parviflora*	193	−0.01
A. utahensis	86	.24	*Yucca standleyi*	100	.00
A. deserti	84	.15	*Y. utahensis*	162	.01
A. chrysantha	20	.18	*Y. glauca*	117	− .01
A. palmeri	48	.32	*Y. elata*	84	.01
Yucca whipplei	23	.08			
Mean		.21	Mean		.00
Std. dev.		.09	Std. dev.		.01

Competition was measured as the slope of a regression line of percent flowers fertilized plotted against $(H - \bar{H})/\bar{H}$, where H is stalk height of the plant in question and $\bar{H}$ is mean stalk height in the population. N is the sample size. Note that within populations of big-bang reproducers, competition is significantly more intense ($t = 5.14$; $p(t) < 0.0005$, one-sided) than within populations of repeat breeders.

associated with 100% effort is likely to exceed that associated with the best iteroparous strategy. In a less productive habitat the reverse may be true, and repeated breeding becomes the preferred habit. Robotnov's work (cited by Harper and White, 1971) on red clover, *Trifolium pratense,* may provide an example. A comparison of the populations of flood plain and sub-alpine meadows revealed that individuals growing under favorable conditions in the valley of the Oka River were typically big-bang reproducers, whereas plants growing under less favorable subalpine conditions were generally polycarpic and had a reproductive life span on the order of ten years.

In sum, we conclude that although the formalism developed in this section points to the *mathematical* circumstances that favor the evolution of semelparity, it does not identify the responsible *biological* factors. The elucidation of these, which probably vary from case to case, remains a challenging problem.

Age of First Reproduction

The theory developed in the previous section can also be used to relate the optimal age of first reproduction to other entries in the life table. Suppose that the age of first flowering is α. Then the stable age equation (eq. 1) becomes

$$1 = \sum_{\alpha}^{\infty} \lambda^{-(x+1)} l_x b_x \qquad (16)$$

Charnov (personal communication) has pointed out that this expression can be

usefully rewritten as

$$1 = \frac{l_\alpha}{\lambda^\alpha} \left\{ \frac{b_\alpha}{\lambda} + \frac{b_{\alpha+1} p_\alpha}{\lambda^2} + \cdots \right\}$$

$$= \frac{l_\alpha}{\lambda^\alpha} \frac{v_\alpha}{v_0}(\alpha)$$

or

$$\lambda = \left\{ l_\alpha \frac{v_\alpha}{v_0}(\alpha) \right\}^{1/\alpha} \qquad (17)$$

where $\frac{v_\alpha}{v_0}(\alpha)$ is the reproductive value of

an α-year-old, given that α is the age of first reproduction. This relation suggests a broad correlation between $\hat{\alpha}$, the optimal value of α, and the probability of year-to-year survival and hence mean longevity. Recently, Harper, White, and Sarukhan (White, personal communication) have collated considerable evidence bearing on this point for a large number of species and do, indeed, find such a relationship for trees. For herbs, however, the situation is less clear, being confounded by difficulties in properly defining lifespan in species that propagate vegetatively (see Harberd, 1961).

Response to Density

Much has been written concerning the optimal life-history response to increased crowding and density-dependent mortality. Rather than review this voluminous and often confusing literature, we will here extract what we feel to be the salient points. To do so, we simplify the theory developed in the previous sections by assuming that the fertility and survival functions do not change with age. In this case, the optimal reproductive effort is also age invariant (Schaffer, 1974). Hence, the stable age equation (eq. 1) becomes

$$1 = b(E) \sum_{0}^{\infty} \lambda^{-(x+1)} p^x(E) \qquad (18)$$

which can be rewritten as

$$\lambda = b(E) + p(E) \qquad (19)$$

(Schaffer and Tamarin, 1973). Note that $\hat{E}$, the value of E maximizing λ, satisfies the relation

$$\frac{db(E)}{dE} = -\frac{dp(E)}{dE} \qquad (20)$$

These expressions are more easily handled than eq. 13 because they do not involve reproductive value, which itself contains terms in λ. However, we emphasize that the results we shall derive from eqs. 19 and 20 are qualitatively identical to those deducible from the more complicated expression (Schaffer, 1974; see also Gadgil and Bossert, 1970).

Changes in Total Reproductive Effort

In general it would appear that increased crowding and inter-plant competition can affect the functions, $b(E)$ and $p(E)$, in two ways. To begin with, both effective fecundity and post-breeding survival per unit effort will be reduced. Most often, we expect the effect on $b(E)$ to be more pronounced, because of increased seedling mortality, with the result that the optimal reproductive effort will decline. [If one considers more complicated models

with age structure, the optimal age of first reproduction also increases (Gadgil and Bossert, 1970)]. To show this, suppose that under crowded conditions, $b'(E) = \frac{1}{\Delta} b(E)$ and $p'(E) = \frac{1}{\delta} p(E)$, where Δ and δ are positive constants. Then eq. 20 becomes

$$\frac{1}{\Delta} \frac{db(E)}{dE} = -\frac{1}{\delta} \frac{dp(E)}{dE} \qquad (21)$$

and if $\Delta > \delta$, $\hat{E}$ declines (Figure 4, left). We believe this to be the general case, and this expectation is borne out by the work of Gadgil and his associates (Gadgil and Solbrig, 1972; Abrahamson and Gadgil, 1973; Solbrig, 1972). In the unlikely case that adult survival is more sensitive to increased density than effective fecundity, i.e., $\delta > \Delta$, $\hat{E}$ will increase (Figure 4, right). In either case, $\hat{E}$ varies with increasing density so as to maximize the population's equilibrial size, but at the expense of its capacity to increase when rare. This result accords nicely with MacArthur's (1962; MacArthur and Wilson, 1967) distinction between r- and K-selection (cf. also discussion by Shapiro, Chapter 8), and in light of the continuing commotion over this distinction (Hairston, Tinkle, and Wilbur, 1970; Pianka, 1972) is worth emphasizing.

A second effect of increased crowding is that the shape of the survival function will be altered. In closely packed stands, individuals making a greater reproductive effort than their neighbors are likely to be overtopped by plants producing fewer seeds and allocating a greater percent of

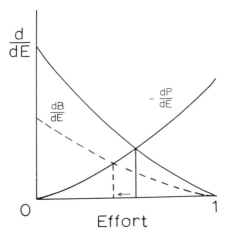 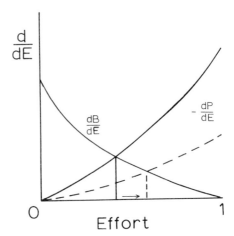

Figure 4 Effect of reductions in effective fecundity, b, and post-breeding survival, p, on optimal reproductive effort, $\hat{E}$, given by the effort (abscissa) value at which the curves db/dE and dp/dE intersect. Left: Reducing effective fecundity per unit effort (dotted line) reduces $\hat{E}$. Right: Reducing post-breeding survival per unit effort (dotted line) increases $\hat{E}$. Arrows indicate change in $\hat{E}$.

their resources to leaf and supportive tissue. As a result, $p(E)$ will come to assume a reverse sigmoidal form. In this case, there can be two values of E associated with relative maxima in λ (Figure 3, bottom). As previously noted, one of these corresponds to semelparity, $E = 1$, the other to iteroparity. As competition becomes more severe, the value of E at which $p(E)$ declines precipitously is lowered and, as a result, the optimal iteroparous effort declines. Additionally, because crowding also reduces effective fecundity, repeated reproduction will generally confer greater fitness than semelparity. In such cases, increasing density will select for reduced reproductive output and, at the same time, increased allocation of re-

sources to leaf and supportive tissue. As Gadgil and Solbrig (1972) have put it, the plants "... enhance their future reproduction by diverting some of their resources to non-reproductive activities ... to combating [the] unfavorable ... biological phenomenon." We should emphasize, however, that sometimes the "phenomenon" is too drastic in its effect to make it worth combating. In this case, $\lambda(100\%)$ is greater than $\lambda(E\ddagger)$, where $E\ddagger$ is the optimal iteroparous effort, and selection will favor maximal reproduction even though this results in a reduction in competitive ability.

An example of what we consider to be the usual case—optimal effort declining with density—emerges from Solbrig's

(1972) study of the dandelion, *Taraxacum officinale*. In this species, isozyme analysis permits the identification of several apomictic strains. One of these, **A,** predominated in disturbed areas where density-independent mortality was rather high and competition for light and resources generally low. Another strain, **D,** was more abundant in less disturbed areas where populations were denser and competition keener. Predictably, **A** was found to mature at an earlier age and set more seed. **D,** on the other hand, produced leaves with longer petioles and generally appears to have traded reproductive output for superior competitive ability. In laboratory experiments, it consistently survived better and produced more biomass than **A** in competitive situations.

This discussion reminds us that, in competitive situations, an individual's optimum phenotype depends in large part on the phenotypes of its competitors (R. Levins, Chapter 1). Thus, as the population's average effort, $\bar{E}$, declines, $\hat{E}(\bar{E})$ will also decline. However, because some reproduction is always advantageous, even when $\bar{E} = 0$, $\hat{E}(\bar{E})$ and $\bar{E}$ will intersect, at which point population evolution comes to a halt. Notice, however, that if at low $\bar{E}$, $p(E)$ falls off sharply enough to cause $\hat{E}(\bar{E})$ to jump to 100%, a polymorphism can result, both high- and low-E individuals being maintained in the population (see Gadgil, 1972).

Optimal Reproductive Expenditure per Seed

In the preceding sections, we have been concerned with total reproductive effort.

This effort can, of course, be allocated in various ways. In particular, it is possible to produce many small seeds or a few large ones. This brings us to the concept of *per seed* caloric expenditure. In general, we expect the rate of seedling survival to increase with seed size. On the other hand, increasing the number of calories allocated to each seed reduces the total number of seeds that can be produced. To determine the optimal per seed expenditure, we suppose that

$$b(E, e) = \frac{c(e)wE}{e} \qquad (22)$$

and thus that

$$\lambda(E, e) = \frac{c(e)wE}{e} + p(E) \qquad (23)$$

Here w is the total number of calories that can be diverted to reproductive functions; e, the number of calories invested in each seed; and $c(e)$, the probability that such a seed successfully germinates and matures. Equation 23 assumes that postbreeding survival is independent of e, and so long as this is the case, the optimal value of e, $\hat{e}$, is independent of E. On the other hand, the optimal total effort does depend on e and, in fact, is maximal when $e = \hat{e}$. To determine $\hat{e}$, we note the value of e at which $c(e)$ is tangent to the line $c = ek$ of greatest k (Figure 5, left). If seedlings compete with each other, or are shaded by mature plants, $c(e)$ will be reduced for all values of e, but especially for low ones. Then $\hat{e}$ will increase (Pianka, 1972). This prediction is consistent with the observation (Salisbury, 1942) that, among the flora of Great Britain, seed size is generally greater in species whose seedlings typically establish themselves in shade than in those that germinate in full sunshine (see also Baker, 1972).

If seedling-seedling, as opposed to seedling-adult, competition is important, $\hat{e}$ will depend on the population average, $\bar{e}$. When $\bar{e}$ is low, $\hat{e}(\bar{e})$ should exceed $\bar{e}$, because even in the absence of competition, there will be a certain minimum seed size necessary to ensure successful germination. As $\bar{e}$ increases, $c(e)$ will be shifted

Table 2.   Comparison of three species of buttercups, *Ranunculus*

Species	Mode of reproduction	C.V. of population growth rates	Effect of density on growth rate
R. repens	Largely vegetative	0.29	Strongly negative
R. acris	About equally by seed and vegetatively	0.45	Weakly negative
R. bulbosus	Exclusively by seed	1.18	Strongly positive

Note that vegetative reproduction is associated with population stability [low coefficient of variation (C.V.) in growth rates] and strong density-dependent regulation of numbers. (After Sarukhan and Gadgil, 1974.)

Figure 5 Evolution of optimal energetic expenditures per seed. Left: $c(e)$ is the survival rate of a seed of caloric value e. $\hat{e}$, the optimal value of e, maximizes the ratio $c(e)/e$. At $\hat{e}$, $c(e)$ is tangent to the line $c = ek$ of largest k. Increasing competition among seedlings [dashed curves $c'(e)$ and $c = ek'$] increases $\hat{e}$ to the value $\hat{e}'$ (dashed line). Right: Evolution of per seed expenditure in a population. $\bar{e}$ is the population average. $\hat{e}(\bar{e})$ and $\bar{e}$ increase until $\hat{e}(\bar{e}) = \bar{e}$.

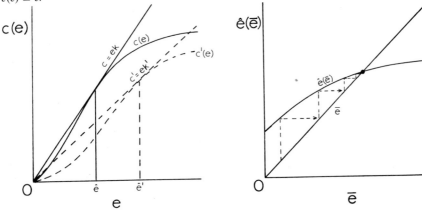

further to the right and $\hat{e}$ will increase.
However, the rate at which $\hat{e}(\bar{e})$ increases
with $\bar{e}$ should ultimately decline—because
competition reduces $c(e)$ at all values, not
just those less than $\bar{e}$. As a result, $\hat{e}(\bar{e})$ and
$\bar{e}$ will intersect (Figure 5, right), at which
point population evolution will cease.

We have suggested that the advantage
of producing large seeds is that it provides
a plant's progeny with an energetic head
start in life. Asexual reproduction, e.g., by
offsets or branching, will have the same
effect but to a far greater degree, and in
very stable situations may be the only
effective means of reproduction (e.g.,
Tamm, 1972). Consequently, we might
expect to find vegetative reproduction
often associated with high and stable densities. This inference is nicely supported
by Sarukhan's studies (Sarukhan and
Gadgil, 1974) of the population dynamics
of buttercups (*Ranunculus* spp.) at Bangor
in North Wales (Table 2). However, we
caution that the optimal balance between
sexual and asexual reproduction can also
be affected by factors other than density—
for example, the abundance of pollinators
and seed predators. Thus, we would not
want to suggest that vegetative propaga-
tion should invariably be associated with
crowding, and, indeed, this is not the case.

Acknowledgments

We are indebted to the following colleagues who read various drafts of this
manuscript and provided valuable comments: H. Bond, J. H. Brown, E. L. Charnov, D. W. Davidson, T. C. Gibson, W. D.
Hamilton, H. S. Horn, L. G. Klikoff, M.
Rosenzweig, J. Sarukhan, and J. White. In
particular, we thank K. T. Harper for allowing us to use unpublished data and
M. V. Schaffer for on-going collaboration
in the yucca-agave study. Parts of this
study were supported by a grant from the
University of Utah Research Committee,
and also by PHS grant No. RR07092.

References

Abrahamson, W. G., and M. Gadgil. 1973.
Growth form and reproductive effort in
goldenrods (*Solidago,* Compositae). *Amer.
Natur.* 107:651–661.

Baker, H. G. 1972. Seed weight in relation to
environmental conditions in California.
Ecology 53:997–1010.

Beattie, A. J., D. E. Breedlove, and P. R.
Ehrlich. 1973. The ecology of the pollina-
tors and predators of *Frasera speciosa.*
Ecology 54:81–91.

Calow, P. 1973. The relationship between fecundity, phenology and longevity: a systems approach. *Amer. Natur.* 107:559–574.

Carlquist, S. 1965. *Island Life.* Natural History
Press, Garden City, N. Y.

Charnov, E. L., and W. M. Schaffer. 1973. Life
history consequences of natural selection:
Cole's result revisited. *Amer. Natur.*
107:791–793.

Cohen, D. 1966. Optimizing reproduction in
a randomly varying environment. *J.
Theor. Biol.* 12:119–129.

Cohen, D. 1967. Optimizing reproduction in
a randomly varying environment when a
correlation may exist between the conditions at the time a choice has to be made
and the subsequent outcome. *J. Theor.
Biol.* 16:1–14.

Cohen, D. 1968. A general model of optimal reproduction in a randomly varying environment. *J. Ecol.* 56:219-228.

Cole, L. C. 1954. The population consequences of life history phenomena. *Quart. Rev. Biol.* 29:103-107.

Corner, E. J. H. 1964. *The Life of Plants.* New American Library, New York.

Cronin, E. H. 1965. Ecological and physiological factors influencing chemical control of *Halogeton glomeratus. USDA Tech. Bull.* 1325:1-65.

Deacon, J. E., W. G. Bradley, and K. S. Moore. 1972. 1971 progress report Saratoga Springs validation site. Desert Biome. *United States International Biological Program.* RM 72-50:1-52.

Fisher, R. A. 1930. *The Genetical Theory of Natural Selection.* Clarendon Press, Oxford.

Gadgil, M. 1971. Dispersal: population consequences and evolution. *Ecology.* 52:253-261.

Gadgil, M. 1972. Male dimorphism as a consequence of sexual selection. *Amer. Natur.* 106:574-580.

Gadgil, M., and W. Bossert. 1970. Life history consequences of natural selection. *Amer. Natur.* 104:1-24.

Gadgil, M., and O. T. Solbrig. 1972. The concept of "*r*" and "*K*" selection: evidence from wildflowers and some theoretical considerations. *Amer. Natur.* 106:14-31.

Hairston, N. G., D. W. Tinkle, and H. M. Wilbur. 1970. Natural selection and the parameters of population growth. *J. Wildl. Mgt.* 34:681-690.

Harberd, D. J. 1961. Observations on population structure and longevity of *Festuca rubra. New Phytol.* 60:184-206.

Harper, J. L. 1967. A Darwinian approach to plant ecology. *J. Ecol.* 55:242-270.

Harper, J. L., and J. White. 1971. The dynamics of plant populations. *In* P. J. den Boer and G. Gradwell, eds., *Dynamics of Populations.* (Proc. Adv. Study Inst. Dynamics of Number in Populations, Oosterbeek), 1970, pp. 41-63. Centre for Agricultural Publishing and Documentation, waginengen, The Netherlands.

Holmgren, R. C., and S. S. Hutchings. 1971. Salt desert shrub response to grazing use, pp. 153-164. *In* McKell, C. M., J. P. Blaisdell, and J. R. Goodin, eds., *Wildland Shrubs—Their Biology and Utilization.* USDA Tech. Rep. Int-1. Ogden, Utah.

Keyfitz, N. 1968. *Introduction to the Mathematics of Population.* Addison-Wesley, Reading, Massachusetts.

Lotka, A. J. 1956. *Elements of Mathematical Biology.* Dover Publ., New York.

MacArthur, R. H. 1962. Some generalized theorems of natural selection. *Proc. Nat. Acad. Sci. U. S. A.* 48:1893-1897.

MacArthur, R. H., and E. O. Wilson. 1967. *The Theory of Island Biogeography.* Princeton University Press, Princeton.

Mayer, A. M., and A. Poljakoff-Mayber. 1963. *The Germination of Seeds.* Pergamon Press, Oxford, New York.

McClure, F. A. 1966. *The Bamboos. A Fresh Perspective.* Harvard University Press, Cambridge, Mass.

McLaren, I. A. 1966. Adaptive significance of large size and long life of the Chaetognath *Sagitta elegans* in the arctic. *Ecology* 47:852-855.

Murphy, G. I. 1968. Pattern in life history and the environment. *Amer. Natur.* 102:390-404.

Pianka, E. R. 1972. *r* and *K* selection or *b* and *d* selection. *Amer. Natur.* 106:581-588.

Pijl, L. van der. 1969. *Principles of Dispersal in Higher Plants.* Springer Verlag, New York.

Salisbury, E. J. 1942. *The Reproductive Capacity of Plants; Studies in Quantitative Biology*. Bell, London.

Sarukhan, J., and M. Gadgil. 1974. Studies on plant demography: *Ranunculus repens* L., *R. bulbosus* L., and *R. acris* L. III: A mathematical model incorporating multiple modes of reproduction. *J. Ecol.* 62: 921–936.

Sarukhan, J., and J. L. Harper. 1973. Studies on plant demography: *Ranunculus repens* L., *R. bulbosus* L., and *R. acris* L. I. Population flux and survivorship. *J. Ecol.* 61: 675–716.

Schaffer, W. M. 1974. The evolution of optimal reproductive strategies: the effects of age structure. *Ecology* 55:291–303.

Schaffer, W. M., and R. H. Tamarin. 1973. Changing reproductive rates and population cycles in lemmings and voles. *Evolution* 27:114–125.

Smith, H. B. 1927. Annual versus biennial growth habit and its inheritance in *Melilotus alba*. *Amer. J. Bot.* 14:129–146.

Solbrig, O. T. 1972. The population biology of dandelions. *Amer. Sci.* 59:686–694.

Tamm, C. O. 1972. Survival and flowering of perennial herbs. III. The behavior of *Primula veris* on permanent plots. *Oikos* 23:159–166.

Went, F. 1949. Ecology of desert plants. II. The effect of rain and temperature on germination and growth. *Ecology* 30:1–13.

Went, F., and M. Westergaard. 1949. Ecology of desert plants. III. Development of plants in Death Valley National Monument, California. *Ecology* 30:26–38.

Williams, G. C. 1966a. *Adaptation and Natural Selection*. Princeton University Press, Princeton.

Williams, G. C. 1966b. Natural selection, the costs of reproduction, and a refinement of Lack's principle. *Amer. Natur.* 100: 687–690.

7 Prey Characteristics and Predator Niche Width

Henry A. Hespenheide

One of the questions to which Robert MacArthur devoted much of his attention was that of how species coexist in a community where resources are limited and competition for them is important. The number of species living together in a community at equilibrium depends on several interrelated factors: the range of resources used by each species (its niche width), the tolerable amount of overlap in the use of resources by two species (the limits to similarity), and the total range of resources available to the community (MacArthur, 1972). For the insectivorous birds that were the object of much of MacArthur's field work, differences in their preferred breeding habitat, in their foraging method and foraging zone within a habitat, and in the size and type of their prey were shown or suggested to be important in permitting coexistence. Of these, there has been considerable detailed study of where birds live (Cody, 1974 and Chapter 10), and how and where they forage for food (Morse, 1971). By contrast, much information on birds' food is qualitative and of limited use in understanding coexistence (e.g., most of the references in Table 2), and recent analyses of bird communities have often had to infer food habits from morphology (Karr and James, Chapter 11; review by Hespenheide, 1973a). Thus, much relevant information on bird diets is lacking, even though one school of thought holds that food is the resource toward which spatial and behavioral means of distinguishing niches are ultimately directed.

The most easily determined characteristics of food items are their size and their taxonomic identity (or type). Regular size differences between closely related, coexisting predators imply corresponding differences in the sizes of their preferred prey ["character difference" of Schoener (1965); the "limiting similarity" of MacArthur and Levins (1967)]. Discussions of the selection of prey by predators that differ in size have usually concentrated on the size characteristics of the prey and ignored differences in the types of prey (Hespenheide, 1971; Schoener, 1969), even though prey type is known empirically to be important (Orians and Horn, 1969). The present chapter will first attempt to assess the relative importance of the size and type of prey items of insectivorous birds for the variability and the overlap of their diets. It will then discuss the relation of prey characteristics and dietary niche-width to foraging behavior. Finally, it will conclude with an evaluation of the idea that the morphological variability of birds is correlated with the variability of their diets, the "niche variation hypothesis" of Van Valen (1965).

Prey Size, Prey Type,
and Predator Niche Width

Prey Size

Study of the prey of insectivorous birds
of a wide variety of foraging types and
taxonomic groups (Hespenheide, 1971,
1974, and unpublished) has shown that
the sizes of prey taken from a single insect
order conform to a frequency distribution
that is normalized when transformed to
logarithms (a log-normal distribution).
Beetle prey (Coleoptera) were used in
these studies because they are common
prey items and are digested slowly, and
can therefore be identified from stomach
contents. For birds that forage in the same
manner, the mean size of prey increases
with both body size and bill size of the
predator, but is more closely correlated
with body size. Figure 1 shows significantly different correlations of prey size
with predator size for birds that forage in
different ways, presumably because
different foraging methods incur different
costs (Schoener, 1971). The positive correlations of prey size with predator size
justify the assumption that differences in
predator sizes confer differences in diets
and thereby allow coexistence in competitive environments (see also Diamond,
Chapter 14, Figures 30 and 31; Brown,
Chapter 13, Figures 6 and 7). However,
Figure 2 demonstrates that the same
difference in diet can be achieved by variable differences in predator sizes. If there
must be a certain minimum difference in
prey size between coexisting species and
there are two types of birds, as repre-

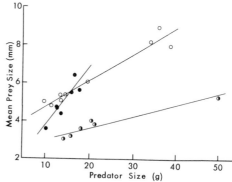

Figure 1 Relation of the mean size of beetle prey
to the weight of bird predators of three different
foraging types: flycatchers (open circles), vireos
(solid circles), and swallows (divided circles; redrawn from Hespenheide, 1971). Mean prey size
increases with predator size within each foraging
type, but along different lines for the different foraging types.

sented by lines *a* and *b* in Figure 2, then
two birds of Type *a* (steep slope) can
achieve the minimum difference with a
small difference in predator sizes, whereas
birds of Type *b* (gentle slope) can achieve
coexistence only with a greater difference
in predator sizes. Conversely, for the same
difference in predator sizes, birds of Type
a show a large difference in prey size,
whereas those of Type *b* show a smaller
difference in prey size.

Figure 2 also suggests that different
slopes relating prey size to predator size
may affect the range of sizes of prey taken,
as well as differences between means: a
small range of predator sizes and a steep
slope (Type *a*) produces the same range
of prey sizes as a wider range of predator
sizes and a gentle slope on the prey size/

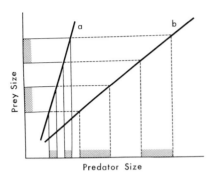

Figure 2 The effect of different relations of prey size to predator size on the relative sizes and relative variability of two types of predators (lines *a* and *b*). Shading indicates the range of prey sizes (ordinate) or range of predator sizes (abscissa). The relative sizes and variability of the prey are defined as identical for the pairs of species of each type of predator. To achieve the same variability and the same difference in prey size, predators of Type *a* (steep slope; projections on abscissa indicated by solid lines) require only a small difference and small range of predator sizes; those of Type *b* (gentle slope; projections on abscissa indicated by dashed lines) require a larger difference and larger range of predator sizes. Identical ecological consequences can thus be achieved by different morphological relationships among competitors (redrawn from Hespenheide, 1973a; cf. Figure 1 for empirical examples).

predator size graph (Type *b*). Conversely, the same range of predator sizes for both types of predators may lead to very different ranges of prey sizes, by a similar argument. However, empirical data seem to suggest that reality does not conform to these expectations (see Hespenheide, 1973a, for details), as shown in the following:

	Slope, prey size/ predator size	Mean variance in prey size	Coefficient of morphological variability of predator
Vireos	0.0275	0.384	0.055–0.070
Flycatchers	0.0088	0.385	0.021–0.045
Swallows	0.0060	0.505	0.027–0.034

Less variable predators with gentler slopes take the greatest range of prey sizes (swallows), and two types of predators with very different slopes and levels of morphological variability take identical ranges of prey sizes (flycatchers and vireos).

The solution to this apparent puzzle probably lies in the fact that the total range of prey sizes taken by a population of predators has two components: one represents the range of prey sizes taken by an individual predator of a given size; the other represents the increased range of prey sizes that results from considering predators of different sizes. The former has been termed the "within-phenotype" component of niche width; the latter, the "between-phenotype" component of niche width. Mathematically, the between-phenotype component is represented by the slope of the prey size/predator size relationship in the way we have discussed; the within-phenotype component is represented by the error variance associated with the regression of prey size on predator size (Hespenheide, 1973a). Thus, a small range of predator phenotypes, each selecting a wide range of prey items, can produce the same total range of prey sizes as a larger range of phenotypes, each selecting a narrower range of prey sizes. Until the relative importance of within- and between-phenotype components is determined for a wide range of types of birds, it will be difficult to make predictions or even analyses of the relationship between variability in diet and morphological variability of predators. The evidence from the flycatchers, swallows, and vireos, given above, suggests that the within- and between-phenotype aspects of the total niche width are inversely related. The between-phenotype component can be estimated by the product of the slope relating prey size to predator size and the midrange value for morphological variability; the within-phenotype component, by the quotient of the total niche width divided by the between-phenotype component:

	Total	Niche width Between-phenotype	Within-phenotype
Vireos	0.384	1.71	0.22
Flycatchers	0.385	0.27	1.42
Swallows	0.505	0.18	2.80

Although these data are suggestive, the matter needs much further study.

Prey Type

If predators choose prey entirely or even primarily on the basis of size characteristics, how can one explain differences in the taxonomic composition of bird diets? The orders of insects are known to differ among themselves in average size within a habitat (Schoener and Janzen, 1968). A bird that prefers an optimum size of prey would be expected to feed on taxa of prey whose members are closest in size to this preference, and to ignore those much larger or smaller, rather than to take all taxa in the proportions in which they occur in the habitat. Thus, if two birds differ in their preference for prey size, their preferences in prey type would also be expected to differ; birds similar in size should take similar sizes and types of prey, whereas birds dissimilar in size should take dissimilar sizes and types of prey. I have shown that a close relationship between similarity in prey size and similarity in prey type exists for forest flycatchers that live together in the same habitat, whereas no significant relationship could be shown for birds that occur in different habitats, including other flycatchers, swallows, and vireos (Hespenheide, 1971). The lack of relationship in the latter cases could be explained easily if different habitats had different relative frequencies and sizes of insect taxa present. It thus appeared possible that taxonomic composition of diets could be a simple consequence of size preferences. Figure 3a demonstrates that there is a relatively continuous change in the taxonomic composition of diets with increasing bird size among forest flycatchers. However, Figure 3b shows a more chaotic relation between diet composition and bird size within Root's (1967) foliage-gleaning guild, even though his birds also occurred together in the same habitat and were considered to forage in a similar manner (see below for further discussion of the foliage-gleaning guild).

To determine the relative importance of prey size and type, diets of two swifts and a swallow in Central America were compared with available flying insects (Hespenheide, 1975). The birds were found to select strongly certain insect taxa and to avoid others almost completely: Hymenoptera comprised 9 to 20% of sampled flying insects, but 44 to 82% of diets; Diptera comprised 66 to 75% of the flying insects, but only 3.7 to 4.3% of bird diets. Prey were selected only partly on the basis of their size. Beetles averaged closest among flying insects to the mean prey size

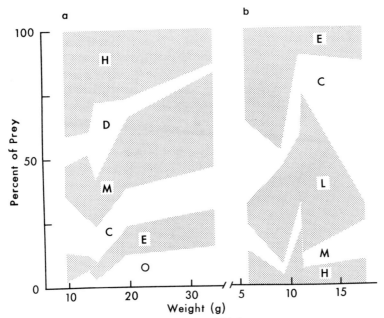

Figure 3 Composition of diet (ordinate) and bird size (abscissa) for two guilds: Figure 3a, left, five forest flycatchers of the eastern United States (Tyrannidae; data from Beal, 1912); Figure 3b, right, five foliage gleaners of California oak woodland (four families—see text; data from Root, 1967). Insect prey include C, Coleoptera; D, Diptera; E, Hemiptera (*sensu lato*); H, Hymenoptera; L, Lepiodoptera; M, miscellaneous other prey; O, Orthoptera. Diet composition is a regular function of bird size for the flycatchers, all of which forage in the same way. However, diet composition varies irregularly with bird size for the foliage gleaners, which suggests that they forage in different ways.

of the birds and were also more frequent in stomachs than in insect samples, but were not the most common prey. Available Diptera and Hymenoptera were about equal in size, but were eaten in inverse proportion to their abundance. The reason that smaller Diptera were avoided was probably their small size, but larger species seem to have been avoided because of their ability to escape. Larger flies of the housefly type are relatively fast and agile fliers. Indeed, their flight ability seems to be the basis for a Müllerian mimicry system in which beetles resemble flies; both are avoided not because they are distasteful, but because they are difficult to capture (Hespenheide, 1973b). The great preference for Hymenoptera was due to their tendency to occur in local concentrations as mating swarms, at nests, or at host plants. This bears out Emlen's (1966) prediction that frequency in diet is not a simple linear function of frequency in nature, but that rare species should be even less common in diets, and abundant species more common. (Similarly, Hutchinson, Chapter 17, cites several apparent instances among insects of uncommon morphs that are at a competitive disadvantage but are nevertheless maintained in the population, because rarity prevents their predators from forming a search image and thereby facilitates predator escape.) No attempt was made to determine whether qualitative differences in palatability or nutritional value (i.e., those independent of size) affected the preferences of these birds. Birds' recognition of differences in palatability among insect species has been demonstrated experimentally (Brower, Brower, and Collins, 1963), and nutritive value is known to affect preferences of mammals (Howell, 1974). The nutritive value of different types of insects is currently under study (S. White, personal communication). In summary, the taxonomic composition of diets of swallows and swifts is very sensitive to prey density and prey flight ability, and possibly other factors, and is not a simple reflection of size preferences per se.

Prey Type and Foraging Method

In the preceding example, swifts were found to avoid flies because they are difficult to capture; study of the diets of three species of swifts (Hespenheide, 1975, and unpublished) showed that flies comprised about 6% of all prey items. Published data on the prey of 17 flycatchers (Beal, 1912) and six swallows (Beal, 1918) show that their diets included an average of 14% and 25% fly prey, respectively. Swifts, swallows, and flycatchers all feed on flying insects, but take different proportions of the various insect groups.

These birds also forage in different places and in different ways: swifts fly rapidly, cannot change course easily to capture individual items, and forage high above the top of vegetation; swallows fly more slowly, can change course more easily, and forage near the top of vegetation or near water; flycatchers pursue individual prey from perches, often inside the vegetation cover. Thus, it appears that the birds that forage in different ways on the same insects take different proportions of them, and that the composition of diet must be related to foraging behavior. What is the basis for this relationship?

Only beetle prey were analyzed in my original study of the relation of prey size to predator size (Figure 1; Hespenheide, 1971). It was noted without explanation that inclusion of prey-size records of insects other than beetles tended to skew prey-size distributions away from normality. Subsequently, the swifts and swallows were found to take significantly different mean sizes of prey for different prey taxa. This can be shown to be true for a wide variety of predators and prey types (Table 1). Figure 4a shows that hawks of the genus *Accipiter* take bird prey of an average size that is smaller than the average size of mammal prey. The explanation for these observations lies in the cost-benefit relationships of prey capture. Individual prey of the same size but from different taxa will differ in flight or escape ability and/or ease of being handled (cost of capture) or in the proportion of digestible material (nutritive benefit expressed as number of calories or units of a particular nutrient). In the case of *Accipiter,* bird

Table 1. Mean prey size (mm) for different prey taxa

Predator	Prey											Source
	Coleoptera		Hymenoptera		Diptera		Lepidoptera		Misc. taxa			
	n	mean	n	mean	n	mean	n	mean	n	mean		
Birds												
Chaetura spinicauda	104	2.60	19	4.36	13	2.89		—	15	2.26		Hespenheide, 1974
Contopus sordidulus	28	5.10	38	6.74	43	7.26	29	11.59		—	⎱	Beaver and Baldwin,
Empidonax difficilis	6	3.66	6	6.50	23	6.04	17	9.82		—	⎰	1974 and unpublished
Polioptila caerulea	54	3.65	29	4.34		—	16	12.56	105	4.08		Root, 1967
Flies (Asilidae)												
Efferia helenae		—	7	9.00	13	9.22		—	13	17.06	⎱	Hespenheide, Lavigne,
E. frewingi		—	16	4.96	22	9.43	10	9.26	14	17.59	⎰	and Dennis, in preparation

"Mean" is the antilog of the average logarithm of prey size for *Chaetura*; all others are means of untransformed measurements of insects found in stomachs (birds) or being eaten (flies). Hymenoptera prey size for *Chaetura* are social species of Hymenoptera only. Miscellaneous prey taxa are, for *Chaetura*, parasitic Hymenoptera; for *Polioptila*, Hemiptera (*sensu lato*); and for Asilidae, Orthoptera. Predators do not prefer a single, overall prey size, but take different mean sizes of prey for different prey taxa.

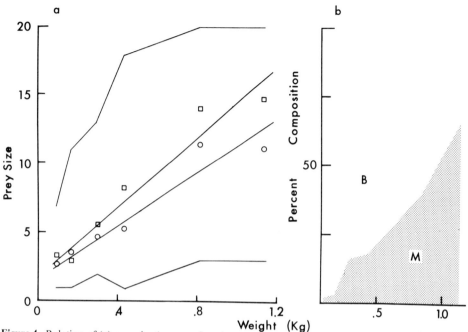

Figure 4 Relation of (a) prey size (expressed as size classes; see Storer, 1966, for definition) and (b) prey type to body size of hawks of the genus *Accipiter* (data from Storer, 1966 and unpublished). In order of increasing size the three species are *A. striatus, A. cooperi,* and *A. gentilis;* each is sexually dimorphic with males smaller, and data for the sexes are shown separately. Mean prey size on the ordinate of Figure 4a (left) is given separately for bird and for mammal prey (circles and squares, respectively); uppermost and lowermost lines show extremes of prey size. Composition of diet on the ordinate of Figure 4b (right) is given as the percent of birds (B) or mammals (M). Larger hawks take larger prey and a greater proportion of mammals, and exhibit a greater difference in size between the bird and mammal prey they select.

prey items are probably more difficult to capture than mammals of the same size, so that for a given expenditure of energy a larger mammal than a bird can be captured. Birds are "preferred" by small *Accipiter* only because small birds are so much more frequently encountered than the often-nocturnal small mammals such as shrews and mice; larger mammals, such as squirrels and rabbits, are more diurnal and comprise a larger proportion of the diets of larger *Accipiter*. Consistent differences in prey size between prey of different types are known for a variety of other predators in addition to *Accipiter* (Hespenheide, 1973a).

For insect prey, differences in escape mechanisms or in flying ability (speed or maneuverability) are likely to affect strongly the probability that an item will be included in the diet. The wing structure and ecology of insects, particularly the role of flight in the life history, differ radically among the orders. If we compare insects of the same size, for example, members of the order Diptera are probably more agile and faster, on the average, than Hymenoptera; Hymenoptera are more agile than Coleoptera; Coleoptera are more agile than Heteroptera; and so on. There are certainly great differences in flight or escape ability within each order, as well as between orders; for example, among beetles, members of the families Cicindelidae and Buprestidae are more difficult to capture than those of the Lampyridae and Scarabaeidae, which are slow and clumsy. However, if members of an insect group are relatively similar among themselves in flying or escape

ability, but are relatively different from other groups in these abilities, then a given type of insect will be captured most easily by a particular type of foraging maneuver or "tactic" (Root, 1967). If ease of capture depends on the foraging method of the bird, then birds that forage in different ways should take different types of prey. Table 2 gives a summary of the average taxonomic composition of diets of insectivorous birds that forage in a variety of ways. In general, aerial foragers (swifts, flycatchers, swallows) prefer Coleoptera and Hymenoptera; foliage gleaners (cuckoos, vireos) prefer Lepidoptera and Hemiptera or Orthoptera; and each specialist (hummingbirds, woodpeckers) takes its own characteristic type of prey. Among the aerial foragers, there are the differences already detailed in part above.

The relative variability in the types of prey taken by an insectivorous bird can be quantified for comparison with the relative variability in the sizes of prey taken and with the amount of morphological variability in the bird itself. The index of niche width $B = (\Sigma p_i^2)^{-1}$ (Levins, 1968) was calculated from the proportions p_i of prey among the categories in Table 2. The value of B varies from 1 to n, where n is the number of categories. If all prey categories are equally common in the diet (all $p_i = 1/n$), then $B = n$; if virtually all prey items are from one category and each of the other categories has only a trace amount, then $B \sim 1$. When n is large compared with the value of B, differences in n are relatively unimportant in their influence on B, and one may compare values

Table 2. Taxonomic composition and diversity of bird diets

Birds		Prey categories										No. taxa	Niche width		Source
Family	No. spp.	C	H	A	D	E	L	O	S	M	P		$\bar{B}$	B_{tot}	
Cuckoos	2	6	—	—	—	6	49	30	9	9	—	5	—	2.96	Beal and Judd, 1898
Swifts	3	23	35	19	6	13	T	—	T	3	—	12	3.74	4.36	Hespenheide, unpublished
Hummingbirds															
total	6	1	10	9	26	2	T	1	49	1	?	13	1.87	3.01	Stiles and Hespenheide, in preparation
hermits	3	T	1	2	2	T	—	2	92	1	?	10	1.17	1.17	
other	3	1	19	17	50	3	T	—	6	2	?	11	2.56	3.07	
Woodpeckers															
total	16	20	2	28	—	2	5	1	4	4	36	8	2.80	3.90	Beal, 1911
ant eaters	6	8	1	55	—	2	2	1	4	1	27	8	2.62	2.61	
generalists	5	46	1	12	—	5	14	1	4	4	17	8	3.69	3.58	
seed eaters	5	9	5	13	—	1	1	2	3	3	65	8	2.13	2.19	
Flycatchers	17	14	34		14	8	10	10	2	3	5	9	4.52	5.34	Beal, 1912
Swallows	6	16	18	9	25	19	3	T	—	9	T	10	5.01	5.74	Beal, 1918
Thrushes	5	20	5	13	3	2	10	4	6	2	36	11	4.67	4.89	Beal, 1915
Vireos	8	15	—	—	6	25	31	—	3	7	6	9	4.56	5.01	Chapin, 1925
Blackbirds	4	9	—	8	—	—	3	10	1	5	72	7	1.83	1.84	Beal, 1900

The diet composition is measured by percent of total number of items for swifts and hummingbirds, and by percent of total volume of stomach contents for other birds. In general, diets of all species treated in the cited references were used in calculating the averages; exceptions include the thrushes (members of the genus *Hylocichla* only used; *Myadestes* is largely frugivorous), swallows (*Iridoprocne bicolor* excluded because of winter frugivory), and blackbirds (Icteridae; only species for which a full year's diet is known were included, because of seasonal changes in diet). Data for the two species of cuckoos were not given separately in the original source. Prey types include Coleoptera (C), Hymenoptera other than ants (H), ants (A), Diptera (D), Hemiptera (*sensu lato*, E), Lepidoptera (L), Orthoptera (O), spiders (S), miscellaneous animal matter (M), and plant material (P). T indicates that prey were present in amounts less than 1%. Numbers placed between columns indicate that data were lumped for the two categories. Methods of calculating niche widths are given in the text. Each foraging type of bird has a characteristic set of prey taxa. This set does not normally vary widely among species within a given foraging type (as measured by the similarity of B and B_{tot}), except for hummingbirds and woodpeckers, which include distinct sets of "morphologically cryptic specialists."

of B even though n differs (see below for normalization procedures when n is small). The use of B as an index of dietary diversity rather than one derived from information theory (as in Hurtubia, 1973) has the advantage that (Σp_i^2) is also the denominator of the expression for niche overlap (see below).

In general, the prey categories given in Table 2 were defined as the orders of insects (viz., Coleoptera, Hymenoptera, Diptera, Hemiptera *sensu lato,* Lepidoptera, and Orthoptera); spiders were either considered separately or lumped with other animal material, and all forms of plant material were considered a single category. Ants were distinguished from other Hymenoptera in most cases because their very different ecology and behavior must require very different foraging methods. A biologically realistic value of B of course requires that the birds make the same or similar distinctions, or lack of distinctions, among insects that taxonomists do. Subdivision of taxa other than Hymenoptera was often possible (especially for beetle prey), but was not made because no obvious differences in foraging seemed necessary in order to exploit the subdivisions. Subdividing categories with large p_i causes large changes in B; subdividing or lumping categories with p_i less than about 0.05 has relatively little effect on B, especially if the largest value of p_i is greater than 0.5. The amount of nectar taken by the hummingbirds could not be estimated readily from stomach contents, but is not important for our discussion here.

B was calculated in two ways: by computing B for each species separately and then averaging over all species $(= \bar{B})$, and alternatively, by averaging p_i over all species for each prey category and then calculating B $(= B_{\text{tot}})$. If all species of birds of a given foraging type are taking the same proportions of prey categories, then $\bar{B} \sim B_{\text{tot}}$; if different species are taking rather different proportions of prey items, then $\bar{B} < B_{\text{tot}}$. Hurtubia (1973) presents a similar method of analyzing dietary similarity. For most groups of birds, the two values of B are quite similar, as are the compositions of individual species' diets. Two exceptions are hummingbirds and woodpeckers. Hummingbird diets and foraging behavior will be considered in detail elsewhere (Stiles and Hespenheide, in preparation), but the species studied appear to specialize either on spiders or flying insects, predominantly Diptera. Woodpeckers can be separated into three distinct groups on the basis of food: specialists on ants; specialists on plant material (pine seeds or acorns); and "true" woodpeckers whose prey is largely larvae of wood-boring beetles but includes a greater variety of other items. Both the hummingbirds and the woodpeckers are, in a sense, "morphologically cryptic specialists," in that striking differences in their prey are not accompanied by striking differences in morphology. This observation poses problems for studies that attempt to analyze community structure purely or primarily on morphological evidence (see below), or that analyze competitive relationships within guilds whose membership is at least partly defined on taxonomic grounds (Willson, 1970); in

both cases niche similarity is over-
estimated by morphological criteria. It
should be noted that lumping plant material into a single class obscures a variety
of food types that would differ considerably in a cost-benefit analysis of diet.
Plant material includes nectar, cambium,
and fruits; among fruits, birds may eat the
fleshy exocarp only, or swallow fleshy
fruits and defecate the seeds, or ingest
entire hard seeds, which are then ground
up in the gizzard, or husk seeds prior to
swallowing (Rosenzweig and Sterner,
1970). Of the intact dry seeds, some are
taken above ground (goldfinches, cross-
bills) and others on the ground (sparrows).

Prey and Foraging:
Analysis of a Guild

Relation of Prey Type
and Foraging

In the preceding sections we have demonstrated a relationship between foraging
method and prey size, and between prey

type and prey size, and we have suggested
a relationship between prey type and foraging method. The data of Root (1967)
allow us to investigate the relationships
among these factors for a guild of five
foliage-gleaning birds that occur in oak
woodlands in California. The birds studied were from four families and included
the plain titmouse (*Parus,* Paridae), the
blue-gray gnatcatcher (*Polioptila,*
Sylviidae), the warbling and Hutton's vireos (*Vireo,* Vireonidae), and the orange-
crowned warbler (*Vermivora,* Parulidae;
see Table 3). The birds occurred together
in the same habitat and took prey from
the same insect populations.

Root gives data on food size, food type,
foraging maneuvers (or "tactics"), and
vertical foraging zone for each bird species. For analysis here, prey items were
assigned to six "octaval" size classes; i.e.,
those whose limits are multiples of 2: 0.5,
1, 2, 4, 8, and 16 mm. This has the effect
of transforming the data to logarithms of
the base 2, and thereby normalizing the

Table 3.   Bird sizes, prey sizes, and niche widths of members of the California oak
foliage-gleaning guild

| Species | Bird | | Prey size (mm) | Niche width | | | |
| | Weight (g) | Bill (mm) | | Foraging | | Food | |
				Zone	Tactic	Type	Size
Polioptila caerulea	5.7	7.3	5.8	0.11	0.52	0.68	0.43
Vermivora celata	9.3	7.7	10.7	.18	.06	.42	.54
Vireo huttoni	11.2	6.5	10.6	.43	.62	.86	.66
Vireo gilvus	11.3	7.2	14.7	.11	.53	.34	.18
Parus inornatus	17.8	8.9	5.9	.70	.04	.44	.23

Bird and prey sizes are taken from, and niche widths are calculated from, the data of Root (1967). Note that prey size
is not closely related either to body size or to bill size. See text for method of calculating standardized niche widths (B_s),
which can vary from 0 to 1. Niche widths average slightly larger for food characteristics than for aspects of foraging behavior.

prey size distributions (see discussion above). Prey were also identified as belonging to one of five categories: Coleoptera, Hymenoptera, Hemiptera (*sensu lato*), Lepidoptera, or other. Foraging maneuvers were divided into three categories: gleaning insects from leaves or twigs while perched, hawking for flying insects, and hovering to pick items from leaf surfaces. Three broad foraging zones were defined: canopy oak foliage, subcanopy trunks or shrub foliage, and herbaceous ground vegetation.

Niche widths were calculated for each bird for each of the four niche aspects, or dimensions, according to the formula for B given above. Because the number of categories for each aspect of food or foraging was small and differed from 3 to 6 among the four, niche widths were standardized for comparison to fractions (0 to 1.0) of the maximum possible niche width, by the formula $B_s = (B - 1)/(n - 1)$. Niche overlaps (α) were measured by the formulae of MacArthur and Levins (1967),

$$\alpha_{ij} = \frac{\sum p_i p_j}{\sum p_i^2}$$

and

$$\alpha_{ji} = \frac{\sum p_i p_j}{\sum p_j^2}$$

where α_{ij} is the overlap of species j on species i and α_{ji} is the overlap of species

i on species j; p_i is the unweighted use of a particular resource category by species i relative to its use of the other categories of that resource or niche dimension (i.e., $\Sigma p_i = \Sigma p_j = 1.0$ for each dimension). The value of α varies from 0 with no overlap to 1 for complete overlap and may exceed 1 if niche widths are unequal. Values of α_{ij} and α_{ji} were not normalized to a single value (May, 1974), but calculated separately for analysis of covariance (Levins, 1968; Vandermeer, 1972). The table giving values of α for the effect of each species in a community or a guild upon all the others is called the α-matrix. Values of α were calculated for the members of the foliage-gleaning guild for each of the four dimensions (prey size, prey type, foraging method, and foraging zone) separately; the resulting α-matrices are given in Table 4. Similar guild or community matrices have been calculated by Brown and Liebermann (1973), Brown (Chapter 13), Levins (1968 and Chapter 1), and Pianka (1969 and Chapter 12) for other ends.

Although mean prey size is not correlated with bird size within the foliage-gleaning guild (Table 3), and although there is no orderly change in the taxonomic composition of diets with increasing bird size (Figure 3b), nevertheless, overlap in prey type correlates strongly with overlap in prey size (Figure 5; $r = 0.81$, probability that r is not different from zero < 0.001), as it does for forest flycatchers (Hespenheide, 1971). This means that one may estimate overlap in diet for these birds either from overlap in prey type or from overlap in prey size, or from a linear combination of the two ["summation α" of Cody (1974)], but not from their product [May (1974); my suggestion of 1971 that a full dietary overlap value was the product of size and compositional overlaps was incorrect]. Just as size and composition of prey items fail to correlate with bird size, so overlap in either prey size or diet composition is not significantly correlated with differences in predator bill or body size [measured as the ratio of larger to small species, after Hutchinson (1959); ϕ of MacArthur and Levins (1967)]. This means that the foliage-gleaning guild is a more heterogeneous group of birds ecologically, as well as taxonomically, than are the forest flycatchers.

The distinctions among members of the guild lie in differences in foraging

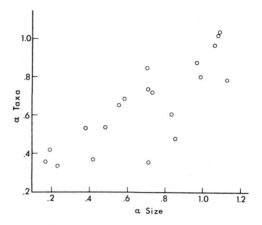

Figure 5 Overlap in composition of diet (α_{taxa}) compared with overlap in prey size (α_{size}) for members of the California oak foliage-gleaning guild (Root, 1967; data from Table 4). See text for method of calculation. The near-linear relationship shows that these are not independent dimensions of the niche.

Table 4. α-matrices of the foliage-gleaning guild

Foraging behavior					Prey characteristics				
Foraging tactics ($\bar{\alpha} = 0.85$, cov. $= -0.033$)					Prey taxa ($\bar{\alpha} = 0.65$, cov. $= 0.022$)				
P.c.	—	1.27	0.79	0.88	1.29	P.c.	—	0.85	0.74	0.53	0.96
V.c.	0.69	—	0.47	0.52	1.02	V.c.	0.61	—	0.48	0.78	0.35
V.h.	0.88	0.96	—	1.03	0.91	V.h.	0.88	0.80	—	1.02	1.04
V.g.	0.89	0.96	0.94	—	0.99	V.g.	0.34	0.69	0.54	—	0.35
P.i.	0.67	0.97	0.43	0.51	—	P.i.	0.72	0.37	0.65	0.42	—
	P.c.	V.c.	V.h.	V.g.	P.i.		P.c.	V.c.	V.h.	V.g.	P.i.
Foraging zone ($\bar{\alpha} = 0.94$, cov. $= -0.040$)					Prey size ($\bar{\alpha} = 0.69$, cov. $= 0.038$)				
P.c.	—	0.94	0.75	1.00	0.58	P.c.	—	0.70	0.70	0.38	1.06
V.c.	1.05	—	0.79	1.05	0.62	V.c.	0.83	—	0.85	1.13	0.71
V.h.	1.15	1.09	—	1.15	0.85	V.h.	0.96	0.98	—	1.08	1.08
V.g.	0.99	0.94	0.75	—	0.58	V.g.	0.23	0.58	0.47	—	0.16
P.i.	1.13	1.09	1.09	1.14	—	P.i.	0.73	0.41	0.55	0.19	—
	P.c.	V.c.	V.h.	V.g.	P.i.		P.c.	V.c.	V.h.	V.g.	P.i.

Values of niche overlap, α, are calculated from the data of Root (1967) for each of the four niche dimensions (foraging tactics, foraging zone, prey taxa, and prey size) by the formulae given in the text. The guild members are indicated by the initials at the row and column headings (*P.c.* = *Polioptila caerulea, V.c.* = *Vermivora celata, V.h.* = *Vireo huttoni, V.g.* = *Vireo gilvus, P.i.* = *Parus inornatus*). Average overlaps ($\bar{\alpha}$) and the covariance of $\alpha_{row, column}$ and $\alpha_{column, row}$ are given for each dimension (Levins, 1968; Vandermeer, 1972). Although average values of overlap are greater for foraging behavior than for food (prey) characteristics, analysis of covariance (discussed in text) suggests that species are more closely packed with respect to prey characteristics.

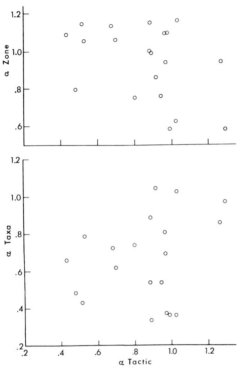

Figure 6 Overlap in vertical foraging zone (α_{zone}, upper figure) and overlap in composition of diet (α_{taxa}, lower figure) compared with overlap in foraging techniques (α_{tactic}) for members of the California oak foliage-gleaning guild (Root, 1967; data from Table 4). Foraging zone is unrelated to foraging techniques; similarity in diet is roughly but not significantly correlated with similarity in foraging techniques. Close correlation of prey type and foraging methods would have suggested that diet composition is determined by foraging behavior (see also Table 2).

methods. Close correlation of prey size and prey type independent of bird size may arise in two alternative ways: the birds prefer different types of prey because of their different foraging techniques, and prey size is a reflection of preferences for these prey types; or the birds prefer different sizes of prey because of their different foraging techniques, and type of prey is a reflection of prey size preferences. In both cases, foraging behavior determines both prey size and prey type, but in which order?

By the first hypothesis, foraging repertoire defines the general set of taxa taken, in the sense of Table 2; the particular proportions are determined in turn by the bird's size, as for the flycatchers, Figure 3a, or for *Accipiter,* Figure 4b. However, if overlap in prey type is compared with overlap in foraging methods (lower part in Figure 6), the correlation is very poor ($r = 0.26$; the probability p that r is not significantly different from zero is $0.3 > p > 0.2$). Moreover, if residuals from the plot of overlap in prey type against that in prey size (Figure 5) are compared with overlap in foraging method, there is no significant correlation ($r = 0.15$; probability that r is not significantly different from zero ~ 0.5); that is, differences in foraging method among the five species do not strongly affect the correlation of prey type with prey size. The reason may be that mean sizes of prey for different foraging methods tend to converge in the range of sizes of birds in the foliage-gleaning guild (i.e., lie near the intersections of the lines in Figure 1) and thereby minimize the effect of foraging differences on optimum prey size. Greater differences in foraging methods should alter the relation of prey type to prey size, however, because of the dependence of prey size/predator size relationships on foraging method as discussed above.

The alternative hypothesis would be that a bird's preferences for prey size are set by the cost-benefit aspects of prey capture, and that composition of diet reflects these size preferences. By this hypothesis, overlap of prey sizes ought to be correlated with overlap in foraging techniques. In fact, the correlation is less than that between prey type and foraging behavior ($r = 0.09$; probability that r is not different from zero ~ 0.7). Although neither prey type nor prey size alone is more important than the other, foraging differences that affect the choice of both seem the most likely explanation for the close correlation of prey size and prey type in the absence of the correlation of either with bird size.

Information about foraging zone does not increase understanding of the relationships between food and foraging method for Root's foliage-gleaning guild. Values of α_{zone} are high and seem unrelated to the other niche dimensions (Figure 6, upper part). Foraging height distributions are probably more critical in structurally more complicated habitats and are probably an independent niche dimension (Cody, 1974).

Niche Width, Overlap, and Species Packing

From Table 3 we see that niches are broadest with respect to prey type, somewhat less broad for prey size and foraging method, and narrowest for foraging zone. Except for one species in each case, niche breadths for prey size and prey type are correlated, as are those for foraging method and prey type; this again suggests

the interdependence of these three aspects of the niche. Average overlaps (Table 4) are largest for foraging zone, somewhat less for foraging method, and least for prey size and type. From these last observations we might expect that competition was greater and species more closely packed for niche characteristics related to foraging behavior and less for those related to food characteristics. However, analysis of covariance for each niche characteristic separately (Vandermeer, 1972) shows that, for the calculated overlaps and mean values of α, one would expect 10 species in the guild on the basis of foraging behavior, but only 4 or 5 on the basis of food characteristics; that is, the species are more closely packed with respect to food than to foraging behavior, even though the average overlaps for food characteristics are smaller. This implies that it is easier to adjust behavior (in this case, foraging zone or method) to avoid competition than to change food habits. This conclusion is consistent with the prediction of MacArthur and Pianka (1966) and MacArthur and Wilson (1967), who deduce from strategic arguments that predators should not avoid any prey items they encounter, but should instead restrict foraging to the richest parts of the environment. As species are added to communities, diets will remain as wide as possible, but foraging zones should contract; in sum, species become habitat and/or foraging specialists, but remain food generalists. This prediction is supported by comparisons of island and mainland bird communities (Crowell, 1962; Diamond, 1970, and Chapter 14). However, specialization in foraging method is also partly a way of dividing space; for example, foliage gleaners are in foliage, woodpeckers and creepers are on tree trunks, thrushes are on the ground, and flycatchers and others are in the air. This sort of spatial separation implies restriction in diet in the sense that foliage, tree trunks, and litter have different insect communities, but diets are otherwise so wide that division of diet independent of foraging method (i.e., within a guild) or habitat selection is difficult. At any one point in habitat space there may be a limit of two (Hespenheide, 1971) to four (Diamond, 1973) for the number of species dividing food by means of size differences alone.

The use of morphological data and only a general knowledge of food habits to discuss the composition and packing of entire communities (Findley, 1973; Karr and James, Chapter 11; Keast, 1972) involves several major difficulties. Birds that are quite similar morphologically for taxonomic reasons may be quite different in food for behavioral reasons, as in the cases of the hummingbirds and the woodpeckers discussed above, as well as in the foliage-gleaning guild. Birds that are morphologically quite different for taxonomic reasons may be very similar in food, also for behavioral (foraging) reasons [again, the foliage-gleaning guild; also gnatcatchers, *Polioptila,* and redstarts, *Setophaga,* foraging like flycatchers (Root, 1967; Williamson, 1971); also sunbirds and hummingbirds (Karr and James, Chapter 11)]. The same size differences may have quite different ecological meanings for birds that forage in different ways (Fig-

ures 1 and 2 above, and Hespenheide, 1973a). Size differences are important and interpretable within single taxonomic groups which are also similar in food; for example, the forest flycatchers and the genus *Accipiter* treated here, and the fruit pigeons and cuckoo-doves studied by Diamond (1973 and Chapter 14). They become more difficult to interpret at the community or faunal level where many taxonomic groups are included, each with its own prey-size/predator-size relationships. A community for which morphology may be a sufficient estimate of niche relationships are the moths whose diversity of appearance is critical to escaping their predators (Ricklefs and O'Rourke, in preparation), although there is a behavioral component of predator escape even in that system. Difficulty in relating differences in morphology to differences in niches and the way species are packed is matched by the difficulty in relating variability in morphology with variability in diet or the size of niches, as we shall now see.

Food and Morphology: Increasing Niche Breadth

The Roles of Food and Behavior

The amount of resources available to an organism are set by the interaction of resource productivity and the intensity of its use by other organisms. Broader niches would appear to be favored at the two extremes of food availability (Hespenheide, 1973a): high availability because of high productivity and low use, as in competitive release on tropical islands

(Crowell, 1962; Diamond, 1970; MacArthur and Wilson, 1967; Terborgh and Faaborg, 1973); low availability because of low productivity or high use, as in hawks and owls that feed on vertebrates (Earhart and Johnson, 1970). How does one achieve a broader niche? There has been considerable recent interest in the "niche-variation hypothesis" of Van Valen (1965; review in Hespenheide, 1973a; also Fretwell, 1972; Rothstein, 1973; Hutchinson, Chapter 17), which states that the morphological (and/or genetic) variability of an organism is related to its niche width. Interest in the hypothesis has concentrated on morphological evidence for broader niches, usually variability in bill characteristics of birds as evidence of broader diets, although behavioral changes in niche width are also known (Crowell, 1962; Diamond, 1970).

Our discussion above indicated that for theoretical reasons space and behavior are more easily subdivided than food, and that data for foliage-gleaning species and for birds in general support this theory. Morphological adaptations can be classed as related to foraging (wing dimensions, tarsal length, etc.) or to handling food (bill characteristics). If changes in niche optima and niche widths occur because of changes in competitive environments, and if these changes are largely changes in behavior in order to use space differently (habitat selection, foraging zone or methods), then we would expect major changes in size or in levels of morphological variability of foraging adaptations (Fretwell, 1972) rather than of bill characteristics, although the majority of studies

of morphological variability have concentrated on the latter (Rothstein, 1973; Soulé and Stewart, 1970; Van Valen, 1965; Willson, 1969). Because the taxonomic diversity of diets is determined in part by foraging method (above), even differences in dietary variability are likely to be reflected in morphological characters related to foraging. Table 2 shows that niche width with respect to prey type varies considerably among birds that forage in different ways. It points out, somewhat surprisingly, that birds that are thought of as foraging specialists are in some cases prey generalists (flycatchers, swallows) and that birds that are thought of as foraging generalists are food specialists (blackbirds). Fretwell (1972) has considered the relations among variability, habitat use, and foraging methods, as did my discussion of within- and between-phenotype components of niche width, above. Because the within-phenotype component of niche width is large and shows great seasonal changes (Hespenheide, 1975; Zaret and Rand, 1971), small differences in morphological variability would seem to have little effect on niche width. This leads us to ask under what conditions, if any, we should expect differences in levels of morphological variability because of differences in competitive environments; in particular, if dietary variability does not encourage more or less morphological variability, what does?

Niche Breadth and Social Structure
    Discussion of the niche variation hypothesis assumes that under normal circumstances *inter*specific competition limits

how variable a population can be, and that when interspecific competition is removed, *intra*specific competition gives an advantage to individuals very different from the population mode in that they are not competing with most other members of the population (Fretwell, 1972). This view of intraspecific competition fails to take into account the normal means by which this competition is alleviated, namely, by the social structure of the population. Just as in competition between species, competition within species is more easily solved by behavioral partitioning of space than by division of food resources on the basis of their size or type.

For territorial species, individuals or pairs are spatially isolated from other members of the population and are relatively uniformly spaced over the environment. A given individual is in competition for food with only one other individual of his species, who is also his mate. There is therefore no need for the population as a whole to be morphologically variable in body or bill size to avoid competition for food, since an individual is not competing with the rest of the population for that requisite. Because of the ease of dividing space, there may be morphological variation in the population in order to use different types of habitats, as Fretwell (1972) has shown in his discussions of "ecotypic variation," but this should primarily affect foraging rather than food-gathering characters. If available resources are either very common or very rare, sexual dimorphism may be advantageous in using a wider range of resources, but even competition between the sexes

can be solved by behavioral-spatial means rather than morphologically. For example, Robins (1971) has shown that male Henslow's sparrows (*Passerherbulus henslowii*) use the peripheral parts of the pair's territory while the females use interior portions ($\alpha_{\text{female, male}}$, the effect of the male on the female, $= 0.64$, and $\alpha_{\text{male, female}} = 0.75$); Ricklefs (1971) has discussed the advantages of this pattern of territory use by members of a pair. Vertical differences in foraging zones between males and females have been demonstrated for the following foliage-gleaning species by Morse (1968) and Williamson (1971):

Bird species	Overlap in vertical foraging zone	
	$\alpha_{\text{female, male}}$	$\alpha_{\text{male, female}}$
Vireo olivaceus	0.26	0.43
Dendroica magnolia	0.46	0.82
D. coronata	0.82	1.07
D. virens	0.89	0.96
D. fusca	0.82	0.93

The sexes of *Vireo olivaceus* overlap little in vertical foraging zone; those of *Dendroica magnolia,* to a somewhat greater extent; the other three species, very broadly. In each case the vertical range of the female is smaller than that of the male of the same species, so that the effect of overlap is greater on the male (Hespenheide, 1973a). Therefore one need not expect morphological differences between sexes in territorial species, although they are regular in hawks, owls, and woodpeckers, all of which tend to have very low-density prey. In woodpeckers both morphological and behavioral differences in foraging occur (Selander, 1972).

Birds with group social systems concentrate conspecifics and, therefore, increase intraspecific competition. Although groups are often seasonal, they tend to be formed at the worst season, when competition would be strongest and selection therefore greatest whatever the social structure. Morphological variability, including sexual dimorphism, should be favored under these circumstances; and, in fact, most examples of increased morphological variability have come from species or species groups whose members occur in flocks: crows and jays (Rothstein, 1973; Soulé and Stewart, 1970) and finches (Fretwell, 1972; Rothstein, 1973; Van Valen, 1965). Cody has pointed out (1971, and personal communication) that even flocks have mechanisms such as "rolling" to reduce competition within the flock by spatial means, but competition must still be greater than for solitary-territorial situations. In the cases where flocks exploit local, superabundant food resources, such as fruiting trees, competition may be negligible and increased morphological variability unnecessary.

In summary, even if we leave aside the difficult question of how food niche width is related to morphological variability, we find that we would not normally expect a priori differences in the levels of variability of morphological characteristics involved in food capture, partly because diets are always as wide as possible and are rarely altered in inter- or intraspecific competition, and partly because of the way social behavior structures intraspecific competition. In territorial species, sexual dimorphism in behavior or morphology is sufficient to avoid intraspecific

competition. Only in species that feed in social groups on relatively rare food would differences in feeding (bill) morphology be useful in reducing intraspecific competition. Ecotypic variation might be expected in either territorial or social species for characters related to foraging (wings, feet, tail) as hypothesized by Fretwell (1972), but not for those involved only in food capture.

The Food Niche in Retrospect

Among insectivorous birds the size and type of their prey and the foraging method used to obtain them are strongly interrelated. Mean size of prey increases with predator size, but in different ways for different foraging types of predators and for different types of prey. Type of prey is determined by the bird's foraging repertoire in two senses: foraging method restricts foraging to certain parts of the environment (air, foliage, ground), each of which has a characteristic fauna; and each group of insects has a characteristic ease of being captured and a characteristic value as food, such that different foraging methods make capture of some types of prey more profitable than others. For birds that forage in the same way and in the same place, similarity in prey size for two predators is correlated with similarity in their prey type. The difference in prey size and type may or may not be a function of the difference in the size of two predators; similarity of prey between birds of the same size, which forage in different ways, is related to the similarity in their foraging behavior. Coexistence depends on maintaining minimum differences between species and, for strategic reasons, space and behavior are more easily divided than food directly in competitive situations.

The width of the food niche of a population of insectivorous birds consists of two components: one reflects the variability in prey taken by a single individual, the other reflects differences between individuals. Because the relative importance of these two varies—perhaps inversely—among birds that forage in different ways, it is impossible to specify the ecological meaning of different levels of morphological variability of birds without data on their food. Again from strategic arguments, organisms should adjust niche widths in competitive situations by restricting or expanding habitat preferences or foraging behavior rather than diet, and should change morphology of characters related to foraging rather than those for gathering food. For characters related to food-gathering, different levels of morphological variability should be expected under different levels of interspecific competition only in organisms that have group social systems, because of the way social structure affects intraspecific competition.

This outline of the food niche and the relation between niche width and morphological variability is assembled from a minimum of empirical data. The relation of prey size to predator size, and particularly the aspect of the relative variability in diet within and between predator individuals, is critical in assessing the importance of morphological variability of phenotypes, and needs further study. The relationship between foraging behavior of predators and the relative ease of capture of their prey, which together appear to determine the type of prey taken by the predator, is possibly the most interesting aspect of the predator-prey interaction at the community level, and requires integrated study of behavior and diet. Finally, the suggestion that morphological variability is more likely to reflect species' differences in niche width for social birds than for territorial ones remains to be tested in a systematic fashion.

Acknowledgments

I would like to thank Donald L. Beaver and Robert W. Storer for providing unpublished data as indicated in the text. Conversations with Martin Cody, Stephen Fretwell, Robert May, and Robert Ricklefs have helped clarify my thinking. The comments of Martin Cody, Egbert Leigh, Jr., and, especially, Jared Diamond have greatly helped in the preparation of the manuscript. The types of questions asked and approaches used were stimulated by Robert MacArthur.

References

Beal, F. E. L. 1900. Food of the bobolink, blackbirds, and grackles. *Bull. U. S. Dep. Agric.* (*Bur. Biol. Surv.*) 13:1–77.

Beal, F. E. L. 1911. Food of the woodpeckers of the United States. *Bull. U. S. Dep. Agric.* (*Bur. Biol. Surv.*) 37:1–64.

Beal, F. E. L. 1912. Food of our more important flycatchers. *Bull. U. S. Dep. Agric.* (*Bur. Biol. Surv.*) 44:1–67.

Beal, F. E. L. 1915. Food habits of the thrushes of the United States. *Bull. U. S. Dep. Agric.* 280:1–23.

Beal, F. E. L. 1918. Food habits of the swallows, a family of valuable native birds.
Bull. U. S. Dep. Agric. 619:1–27.

Beal, F. E. L., and S. D. Judd. 1898. Cuckoos
and shrikes in relation to agriculture. *Bull.
U. S. Dep. Agric.* (*Bur. Biol. Surv.*) 9:1–26.

Beaver, D. L., and P. H. Baldwin, 1975. Ecological overlap and the problem of competition and sympatry in the Western and
Hammond's flycatchers. *Condor* 77:in
press.

Brower, L. P., J. V. Z. Brower, and C. T.
Collins. 1963. Experimental studies of
mimicry. 7. Relative palatability and
Mullerian mimicry among Neotropical
butterflies of the subfamily Heliconiinae.
Zoologica 48:65–84.

Brown, J. H., and G. A. Liebermann. 1973.
Resource utilization and coexistence of
seed-eating desert rodents in sand dune
habitats. *Ecology* 54:188–197.

Chapin, E. A. 1925. Food habits of the vireos,
a family of insectivorous birds. *Bull.
U. S. Dep. Agric.* 1355:1–42.

Cody, M. L. 1971. Finch flocks in the Mojave
Desert. *Theoret. Pop. Biol.* 2:142–158.

Cody, M. L. 1974. *Competition and the Structure
of Bird Communities.* Princeton University
Press, Princeton.

Crowell, K. L. 1962. Reduced interspecific
competition among the birds of Bermuda.
Ecology 43:75–88.

Diamond, J. M. 1970. Ecological consequences
of island colonization by southwest Pacific
birds. I. Types of niche shifts. *Proc. Nat.
Acad. Sci. U. S. A.* 67:529–536.

Diamond, J. M. 1973. Distributional ecology
of New Guinea birds. *Science* 179:
759–769.

Earhart, C. M., and N. K. Johnson. 1970. Size
dimorphism and food habits of North
American owls. *Condor* 72:251–264.

Emlen, J. M. 1966. The role of time and energy in food preference. *Amer. Natur.*

100:611–617.

Findley, J. S. 1973. Phenetic packing as a
measure of faunal diversity. *Amer. Natur.*
107:580–584.

Fretwell, S. D. 1972. *Populations in a Seasonal
Environment.* Princeton University Press,
Princeton.

Hespenheide, H. A. 1971. Food preference
and the extent of overlap in some insectivorous birds, with special reference to the
Tyrannidae. *Ibis* 113:59–72.

Hespenheide, H. A. 1973a. Ecological inferences from morphological data. *Ann. Rev.
Ecol. Syst.* 4:213–229.

Hespenheide, H. A. 1973b. A novel mimicry
complex: beetles and flies. *J. Entomol.* (A)
48:49–56.

Hespenheide, H. A. 1975. Selective predation
by two swifts and a swallow in Central
America. *Ibis* 117:82–99.

Howell, D. J. 1974. Bats and pollen: physiological aspects of the syndrome of chirop-
terophily. *Comp. Biochem. Physiol.* 48:
263–276.

Hurtubia, J. 1973. Trophic diversity measurement in sympatric predatory species.
Ecology 54:885–890.

Hutchinson, G. E. 1959. Homage to Santa
Rosalia, or Why are there so many different kinds of animals? *Amer. Natur.*
93:145–159.

Keast, A. 1972. Ecological opportunities and
dominant families, as illustrated by the
neotropical Tyrannidae (Aves). *Evol. Biol.*
5:229–277.

Levins, R. 1968. *Evolution in Changing Environments.* Princeton University Press,
Princeton.

MacArthur, R. H. 1972. *Geographical Ecology.*
Harper and Row, New York.

MacArthur, R. H., and R. Levins. 1967. The
limiting similarity, convergence, and divergence of coexisting species. *Amer.
Natur.* 101:377–385.

MacArthur, R. H., and E. R. Pianka. 1966. On optimal use of a patchy environment. *Amer. Natur.* 100:603–609.

MacArthur, R. H., and E. O. Wilson. 1967. *The Theory of Island Biogeography.* Princeton University Press, Princeton.

May, R. M. 1974. Some notes on measurements of the competition matrix, α. Submitted to *Ecology.*

Morse, D. H. 1968. A quantitative study of foraging of male and female spruce-woods warblers. *Ecology* 49:779–784.

Morse, D. H. 1971. The insectivorous bird as an adaptive strategy. *Ann. Rev. Ecol. Syst.* 2:177–200.

Orians, G. H., and H. S. Horn. 1969. Overlap in foods of four species of blackbirds in the potholes of central Washington. *Ecology* 50:930–938.

Pianka, E. R. 1969. Sympatry of desert lizards (*Ctenotus*) in western Australia. *Ecology* 50:1012–1030.

Ricklefs, R. E. 1971. Foraging behavior of mangrove swallows of Barro Colorado Island. *Auk* 88:635–651.

Robins, J. D. 1971. Differential niche utilization in a grassland sparrow. *Ecology* 52:1065–1070.

Root, R. B. 1967. The niche exploitation pattern of the blue-gray gnatcatcher. *Ecol. Monogr.* 37:317–350.

Rosenzweig, M. L., and P. W. Sterner. 1970. Population ecology of desert rodent communities: body size and seed husking as bases for heteromyid coexistence. *Ecology* 51:217–224.

Rothstein, S. I. 1973. The niche-variation model—is it valid? *Amer. Natur.* 107:598–620.

Schoener, T. W. 1965. The evolution of bill size differences among sympatric congeneric species of birds. *Evolution* 19:189–213.

Schoener, T. W. 1969. Models of optimum size for solitary predators. *Amer. Natur.* 103:277–313.

Schoener, T. W. 1971. Theory of feeding strategies. *Ann. Rev. Ecol. Syst.* 2:369–404.

Schoener, T. W., and D. H. Janzen. 1968. Notes on environmental determinants of tropical versus temperate insect size patterns. *Amer. Natur.* 102:207–224.

Selander, R. K. 1972. Sexual selection and dimorphism in birds. *In* B. Campbell, ed., *Sexual Selection and the Descent of Man 1871–1971,* pp. 180–230. Aldine Publishing Co., Chicago.

Soulé, M., and B. R. Stewart. 1970. The "niche variation" hypothesis: a test and alternatives. *Amer. Natur.* 104:85–97.

Storer, R. W. 1966. Sexual dimorphism and food habits in three North American accipiters. *Auk* 83:423–436.

Terborgh, J., and J. Faaborg. 1973. Turnover and ecological release in the avifauna of Mona Island, Puerto Rico. *Auk* 90:759–779.

Vandermeer, J. H. 1972. On the covariance of the community matrix. *Ecology* 53:187–189.

Van Valen, L. 1965. Morphological variation and width of ecological niche. *Amer. Natur.* 99:377–389.

Williamson, P. 1971. Feeding ecology of the red-eyed vireo (*Vireo olivaceus*) and associated foliage-gleaning birds. *Ecol. Monogr.* 41:129–152.

Willson, M. F. 1969. Avian niche size and morphological variation. *Amer. Natur.* 103:531–542.

Willson, M. F. 1970. Foraging behavior of some winter birds of deciduous woods. *Condor* 72:169–174.

Zaret, T. M., and A. S. Rand. 1971. Competition in tropical stream fishes: support for the competitive exclusion principle. *Ecology* 52:336–342.

8 The Temporal Component of Butterfly Species Diversity

Arthur M. Shapiro

Theoretical ecologists interested in the determinants of biotic diversity have tended to concentrate on birds and mammals. The activities of these animals are so nearly synchronous that time can be almost ruled out as a resource to be subdivided among them. Insects are small enough so that their numbers of generations per year are subject to natural selection and are potentially adjustable within fairly wide limits. If we examine an insect fauna over a complete year, we find internal patterns of species diversity, which presumably reflect the solutions found by the component species to their seasonal problems, including each other. For the past several years I have been studying butterfly faunas from this standpoint. The work is still in progress, so that this chapter will consist of a certain amount of data, a number of truisms, and a lot of speculation—a characteristic mixture in theoretical ecology.

It is part of the folk wisdom of butterfly collecting that there are good times and bad times of the year to go out with a net. On the other hand, very few collectors go to the same places throughout the season (however unproductive they might be), or record everything they see (however ubiquitous). The literature of butterfly biology, although massive, provides scant information concerning seasonal patterns of species diversity. The most thorough

faunistic study done in the United States, from a phenological standpoint, admits to the traditional bias in selecting localities to visit (Opler and Langston, 1968). The most tantalizing paper along these lines is that by Clench (1967), who demonstrates the existence of temporal sets of skippers (Hesperiidae) which replace one another through the season at a locality in western Pennsylvania. Clench suggests adult competition for nectar sources as an ultimate basis for the partitioning, but does not fully explain why discrete, synchronized groups of species would be more desirable than a continuous turnover in the Hesperiid fauna. His sets, moreover, have not been reproducible in other localities, although other groupings may occur.

Butterflies are holometabolous (have complete metamorphosis). Their larvae are specialized for feeding, whereas the adults function primarily for dispersal and reproduction. The seasonal characteristics of butterfly life cycles would be expected to reflect these specializations. Adult butterflies, being heliotherms, are able to fly only within a certain range of temperatures, and except in the warmest weather are restricted to flying in sunshine. Flight is essential for feeding, mating, and oviposition. The timing of adult emergence should depend on the availability of sunshine, nectar, and oviposition substrates; beyond these essentials predation and competition might exert significant pressures. A butterfly fauna may be analyzed as to the voltinism—the number and timing of flights—of its species: they may be univoltine (one flight per season), bivoltine, or multivoltine (several consecutive flights per season), and these characteristics may be facultative or obligate. Widespread species normally vary in voltinism over their geographic or altitudinal ranges. During the adverse season (e.g., winter in temperate zones) butterflies enter a state of arrested development, or diapause; this commonly occurs in the egg or pupa stages, which are quiescent anyway, but may occur in the larva (young to full-grown) or the adult (in which the diapause shuts down just the gonads). In most species only one stage can survive the adverse season, and this particular aspect of the seasonal history tends to be extremely conservative evolutionarily, even at the subfamily or family level. Thus, all of the large fritillaries (tribe Argynnini, Nymphalidae), world-wide, overwinter as young larvae; most Pieridae, as pupae; hairstreaks (Theclinae, Lycaenidae) as eggs, etc.

Multivoltine Strategies

Voltinism is correlated with the set of life-history adaptations resulting from what MacArthur and Wilson (1967) call r- and K-selection [reformulated by Hairston, Tinkle, and Wilbur (1970) as b- and d-selection]. Species living in ephemeral habitats would be expected to undergo selection for the following characteristics: ability to discover the habitat quickly; rapid reproduction in order to "use up the

habitat" before other, competing species arrive; and dispersal in search of new habitats as the old one begins to deteriorate. The overall effect is to increase r, the intrinsic rate of natural increase. In long-lived habitats most populations are usually at or near their saturation levels, the carrying capacity K. Such populations should be under intense selection for ability to compete with their own and other species (cf. the discussion and contrast of "supertramp" versus "high-S species" in Chapter 14, by Diamond). K-selected species might be expected to evolve narrow, specialized niches in response to interspecific competition, and territoriality in response to intraspecific competition. Slobodkin and Sanders (1969) arrive at similar conclusions by a slightly different argument, using a different terminology.

Multivoltinism in butterflies is usually (but not inevitably) correlated with high r, high vagility, association with disturbed habitats, and overall "colonizing ability" or "weediness."[1] Multivoltinism contributes to r not only by shortening the generation time but also by preventing the

---

[1] The contrapositive—that K-species are univoltines—seems to hold in middle-latitude faunas. Tropical r-species resemble their temperate counterparts in being multivoltine, with relatively rapid adult maturation and reproduction (cf. Owen, 1971). An increasing list of seeming K-species in the tropics demonstrates very long adult life, up to several months, e.g., *Marpesia berania* (Benson and Emmel, 1973). Little is known about the reproductive biology of these animals, and in particular the number of generations per year and the amount of generation overlap need to be investigated. Intuitively one may hypothesize that mortality due to predators is less at the adult than at the early stages, and that consequent selection for prolongation of the adult stage may increase mean generation length quite significantly.

abrupt truncation of population growth at the onset of adverse weather. For multivoltine species, the basic phenological problem is to produce as many generations as possible in a given season but still minimize the likelihood of catastrophic weather-induced mortality. The length of the butterfly season in a given locality depends basically on climate. In temperate latitudes it is not—as popularly imagined even by most field collectors—determined in normal years by the actual dates of lethal extremes of temperature. The adaptations that induce or terminate diapause in a species are its responses to high-probability events; abnormal weather may produce catastrophic disruptions of normal season patterns, but only with such low probability that selection cannot possibly buffer populations against it (compare Ehrlich et al., 1972). Diapause is normally keyed to photoperiod, the most reliable seasonal indicator and one that circumvents the variance of any meteorological variable. Most species, in an "average" year, are more conservative than they "need" to be in shutting down for the adverse season. The reproductive potential they sacrifice in this way can be viewed as the price they pay to avoid being wiped out in the frequent worse-than-average years. In Levins's (1968) fitness-set treatment of diapause the determinants of diapause strategy are essentially the fitness values in different weather occurrences, and the probabilities of those occurrences. The time at which a multivoltine species shuts down, then, reflects the variance as well as the means in the local climatology.

A few species, especially those tolerant of adverse conditions in more than one developmental stage (e.g., the West Coast lady, *Cynthia annabella*, Nymphalidae), continue breeding until deteriorating weather absolutely interdicts their activities; in exceptional years they may be able to breed continuously. A different sort of risk-spreading occurs in the pipevine swallowtail, *Battus philenor* (Papilionidae), in California. The host plant of this species, *Aristolochia*, is a vine that grows in foothill canyons, and its availability to larvae depends on how long the streams run and thus on the seasonal rainfall, which is extremely variable. In each brood of *B. philenor* part of the population diapauses and part does not, making the same population simultaneously one-, two-, and multiple-brooded. The spring azure, *Lycaenopsis argiolus pseudargiolus* (Lycaenidae), in the northeastern United States has a similar system. This insect oviposits on the flower buds of various trees, shrubs, and perennials and has a seasonal succession of host plants. These often do not all occur together; in local populations particular broods may be largely suppressed. Thus in upland woods *L. a. pseudargiolus* is mostly spring-univoltine, but in bottomlands nearby may be multivoltine, or may even largely lack a spring flight.

Multivoltine species are usually wide-ranging geographically and ecologically. In mountainous terrain they may show great variability with altitude and exposure in number of broods per season, a variability that may be confused even further by altitudinal migration (Shapiro, 1974c). In general, multivoltines are unable to exist permanently in sites where

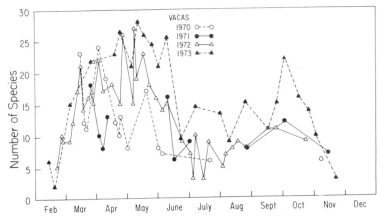

Figure 1 Number of species of butterflies recorded flying in Mix, Gates, and Thompson Canyons, east slope Vaca Mountains, California (inner coast ranges: elevation 175–1750 feet), during the years 1970 through 1973. Note that the peak flying time is in the spring, corresponding to the season of maximum vegetation growth here.

they can rear, on the average, only one generation a season. Unlike true univoltines closely adapted to the statistical properties of the climate, they are not buffered sufficiently to avoid local extinction in the worst or shortest years, and their adaptation is hindered by the regular input of genes from lower-altitude populations.

Phenology of Long-Summer Faunas

Many species do not produce a succession of broods over a long season, even where climate would seemingly allow them to. Every butterfly fauna contains a percentage of univoltine species, and their seasonal distribution may or may not reflect dominant climatic conditions. Uni-

voltines could presumably occur at random through the season; in fact, a major unsolved problem is that they do not, and one is forced to seek explanations of the non-random patterns one observes.

Figure 1 shows one pattern which is easy to interpret. This and the illustrations that follow plot the number of species flying (by weeks or on sampling dates) over the season, for the locality and year(s) indicated. No indication of the distribution of the individuals among species is given, for reasons that should be clear enough to field lepidopterists if no one else.[2] Figure 1 is a plot for three east-

[2] This point requires some elaboration. In 1964, when I was Robert MacArthur's student, I undertook a study in Arizona in the hope of identifying factors contributing to "butterfly species diversity." I

slope canyons in the Vaca Mountains, part of the inner coast ranges in central California. It would be equally representative of the fauna at comparable elevations in the western foothills of the Sierra Nevada. By far the greatest numbers of both species and individuals occur in spring; in midsummer only a few species—all multivoltines—are flying. The total Vaca fauna is about 40% univoltine, with all but one of these species flying in spring. In early years (1972) the univoltines are divisible into an early and a late spring component; in retarded years (1973) the two groups overlap broadly. There is a secondary peak in species number in autumn; this is composed largely of lowland species that wander up into the canyons at the time of their own population peak, but do not breed there. The phenology of the Vacas is obviously a function of climate. The vegetation includes many annuals that grow lushly in spring but are dead during the hot, rainless summer; there are virtually no nectar sources in July and

August. Many of the perennials are inedible or dormant in summer, unavailable to larvae.[3]

The Sacramento Valley has the same climate as the Vacas, but its vegetation is very different and so is the phenology of its butterfly fauna (Figure 2). The indigenous perennial grassland has been eradicated by agriculture and replaced by annual grassland and ruderal vegetation, both dominated by cosmopolitan plants mostly of European origin. Sizable remnants of the riparian forests remain along the major streams, however. There is virtually no butterfly fauna in the annual grassland, but a large one where irrigation or floodplain conditions provide enough water to keep the vegetation green all summer. The breeding butterfly fauna is only 17% univoltine, and the species-per-week curve shows a very long season with

---

quickly learned that gathering data on the equitability component of diversity (for a calculation of *H*) was virtually futile. Because of great differences in population structure and vagility, no single plot size was satisfactory for all the species in any area. The problem is analogous to sampling both canopy and understory vegetation with the same size quadrat. Moreover, the accurate censusing of all the species in any good-sized plot required a prohibitive amount of work and itself threatened serious perturbation of the system. The population characteristics of butterflies are so conspicuous that incommensurability becomes evident in dealing with whole faunas; there is too much information to be dealt with by the techniques one applies to such unstructured (artifactual?) pieces of data as light-trap collections.

[3] On Barro Colorado Island, Panama, Emmel and Leck (1970) found that the maximum number of butterfly species in nonforested environments flew at the end of the rainy season and beginning of the dry season (as in lowland California). In the forest the maximum was reached during the dry season, presumably reflecting the year-round higher humidities there, which would be inhibitory in the wet season but congenial in the dry. Fox et al. (1965) report that in humid-tropical Liberia 90% of the species fly only in the "dry" season. Here moisture in the "dry" season is ample, but in the "wet" season precipitation is so constant that suitable conditions for flight do not occur with any reliability.

At first glance it seems odd that the spring univoltines in the Vacas do not produce an autumn brood, which would take advantage of the large bloom of winter annuals. The answer is evident to anyone who has lived through a few California autumns: the timing of the onset of weather too cold, cloudy, or wet for flight is far too variable from year to year, and the margin for error too small for a butterfly to try to predict it.

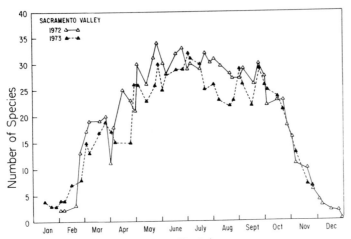

Figure 2 Number of species of butterflies flying on the floor of the Sacramento Valley, California (elevation 10–100 feet), in 1972 and 1973. The butterflies are restricted to areas near water, where vegetation remains green throughout the summer. Note the broader spread of flying times compared with those in Figure 1.

a shallow peak in late spring but essentially constant numbers of species for much of the year. The 1972 season was unusually hot and dry and followed an unusually dry winter; 1973 was cool and moist after one of the wettest winters of record. The highly multivoltine fauna responded by producing extraordinarily similar species-per-week curves despite the very different weather—reflecting the flexibility of these broadly-adapted animals.

The 44,000-acre Suisun Marsh is a tidal wetland nearly enclosed by the coast ranges, just east of San Francisco-San Pablo Bays. Its climate is more maritime than in the Sacramento Valley and more continental than in the valleys open di-

rectly to the coast. Its vegetation consists almost exclusively of late summer-blooming plants, mostly perennial. The role of vernal annuals in tidal marsh is the smallest for any lowland community in California. The species-per-week curve (Figure 3) shows a peak in late May, composed almost entirely of strays from the hills, and a higher peak in October made up of resident species. The resident Suisun fauna is only 8% univoltine, and the timing of flight periods there is obviously correlated with the condition of the vegetation, including the availability of nectar sources. In the field it is evident that not only the largest number of species but also by far the greatest number of individuals is to be found in autumn.

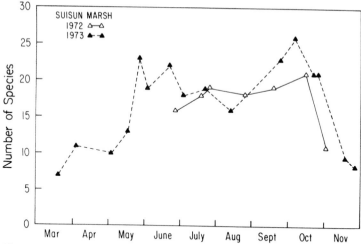

Figure 3 Number of species of butterflies flying in
the Suisun Marsh, California (sea level), in 1972 and
1973. Here vegetation consists mainly of late-
summer flowering plants, which is reflected by the
September-October flying peak.

The long-summer faunas are thus divisible into those dominated by univoltine
species (in summer-drought areas such as
the Vaca Mountains) and by multivoltines
(where summer water is available, as in
the Sacramento Valley and Suisun
Marsh). As noted above, there is a general
correlation of uni- and multivoltinism
with K- and r-selection. The ecologies of
the species in these Californian faunas
broadly support this correlation in both
the highly disturbed Sacramento Valley
and the much less disturbed Suisun
Marsh.

It must of course, be stressed that K-
and r-selection are not mutually exclusive
over the great majority of real habitats,
and are meaningful only relative to other

species. Apparent exceptions to the overall
pattern are neither surprising nor disquiet-
ing. For example, the West Coast lady
(*Cynthia annabella*) is one of the weediest
species in California and occurs in a wide
variety of disturbed habitats, but the
males are highly territorial—an attribute
classically associated with K-selection.
Territoriality in butterflies reflects competition among males for mates—specifically
for sites in which the probability of en-
countering a receptive female is highest.
Such sites are characteristic of each species, and are in short supply—an excellent
epigamic device, but one that necessitates
competition when males are at high density. Male territoriality cannot regulate
population size in butterflies; all the fe-

males are inseminated regardless of whether 1% or 100% of the males obtain territories. Nor is there any evidence to suggest that the number of different inseminating males affects the fitness of a female (although the number of actual *matings* may do so). If territoriality has nothing to do with population regulation in butterflies, it should be able to evolve independently of *K*- and *r*-selection as it apparently has done here.

Phenology of Short-Summer Faunas

In addition to these lowland localities, data are available for three stations with short summers, located in the middle ele-

vations of the west slope of the Sierra Nevada. Each includes a wide range of forest and meadow habitats within a small area.

At the Marin-Sierra Camp at 5000 feet (Figure 4), there is a well-defined peak in early June consisting of early spring univoltines. After the species of late spring and early summer appear in early July, there is a gradual decline in number of species per week, much as in the Sacramento Valley. Although the fauna is 60% univoltine, only one species (*Apodemia mormo*, Riodinidae), first appears in the last third of the season.

Boreal Ridge and Donner Pass (Figure 5) are at the same elevation (7000 feet) and only about four miles apart. The Donner fauna (see Emmel and Emmel, 1962, 1963) includes a number of east-slope species that occur as residents or frequent transients; a few of these reach Boreal Ridge. Both locations receive heavy snow pack and have quite short butterfly seasons. At Boreal Ridge in 1972 the early- and mid-season univoltine species overlapped rather broadly; at Donner in 1973 the same species were more disjunct, giving a clearly bimodal curve. Univoltinism is high (Donner > 65%; Boreal Ridge, > 67%), and several multivoltine or partially multivoltine species may be dependent on immigration from lower elevations. Univoltinism in the climate at 7000 feet may be accounted a virtual necessity, due to the uncertainties in the timing of snowfall, snowmelt, and low temperatures at both ends of the season. Among resident species only one (*Pieris occidentalis*, Pieridae) is fully bi-

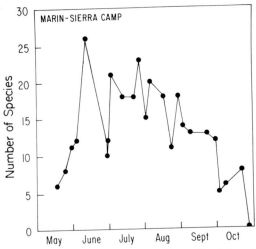

Figure 4 Number of species of butterflies flying at the Marin-Sierra Boy Scout Camp, Sierra Nevada, California (elevation 5000 feet), in 1972. The short summers considerably restrict the butterfly flying season. Note the distinction between the early-June and the mid-summer peaks.

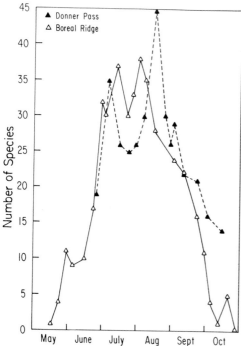

Figure 5 Number of species of butterflies flying at Boreal Ridge (1972) and Donner Pass (1973) in the high Sierra of California (elevations ca. 7000 feet). As in Figure 4, flying times are restricted to the short summer season. Note the two distinct peaks of Donner Pass, but the single peak at Boreal Ridge.

recognized peak of spring univoltines; this is best seen in the 1968 curve. In 1967 the spring was extremely wet and cold, and the spring and early summer species emerged together in June, eliminating the early peak but producing a noticeable June bulge in the species-per-week curve. The main peak in midsummer includes a large number of univoltines that emerge at the end of June or beginning of July; these represent four of the six largest butterfly families. The small autumn peak is composed mostly of immigrant species; there is only one autumn univoltine (*Hes-*

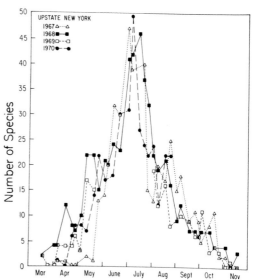

Figure 6 Number of species of butterflies flying in the vicinity of Cornell University on the Allegany Plateau in central New York, 1967 through 1970 (elevations 500–1800 feet). 1968 had a mild spring; 1967 a very cold one. The sharp summer peak is reminiscent of those at high-elevation California sites (Figure 4 and 5). Faunistic data from Shapiro (1974b).

voltine, and it is the most *r*-selected species in the fauna (Shapiro, unpublished).

Figure 6 gives the data for one very well-collected locality in the northeastern United States, the vicinity of Cornell University. At least partial data are on hand for the years 1967–1970 inclusive. Again, although the seasons differed markedly among themselves in weather, the curves match fairly well. In mild or "normal" years in the northeast there is an easily

peria leonardus, Hesperiidae) in the northeast, as well. The same curve, only somewhat broadened and more clearly trimodal, occurs further south at Philadelphia.

Central New York (50% univoltine) and the high Sierras have sufficiently similar climates and vegetation to justify asking whether they have similarities in their butterfly faunas. Both areas have well-defined groups of spring and summer univoltines, forming sets like those described by Clench. The difference between the Donner and Boreal Ridge curves, which share most of their species, seems to be due to the heavier snow pack at the former and lower temperature maxima in summer at the latter. If the high Sierran curves are superimposed on the New York ones, a greater similarity becomes apparent than might at first be noted. The slow drop-off in the Sierran curves may be due to lower night temperatures than in New York, making for a shorter daily activity period for the animals and a longer life-span.

In short-summer faunas the uncertainties at the ends of the season limit the opportunity for *r*-selection to operate by shortening the generation time; yet, intuitively, we might predict a predominance of *r*-strategists in the climatic regime at 7000 feet. Presumably *r*-selection could still operate by increasing vagility, fecundity, and the range of plants acceptable as larval hosts. Such a trend, if present, is certainly not conspicuous. The uncertainties of the climate have largely been eliminated by the evolution of long, secure diapause. The remainder of the life cycle—

the active part—has retreated into the benign part of the year where *K*-selection can and apparently does occur, perhaps even abetted by the competition implicit in the enforced synchrony of large parts of a fauna.

We have seen how species-per-week curves vary among selected communities and how this may be related to extrinsic factors. We have assumed that the habit of a species—univoltine vs. multivoltine, and if univoltine its season—is fixed and that climate, acting directly and through the host plants and nectar sources, selects a fauna from a pool of geographically available species. Now let us inquire briefly into univoltinism as a way of life.

Univoltine Strategies

Studies of the genus *Pieris,* the whites, have demonstrated that voltinism is evolutionarily flexible. *Pieris napi,* for example, is a very widespread holarctic species (or species complex), multivoltine in the southern half of its range, with two seasonal phenotypes under photoperiodic control. It feeds on a wide variety of annual and perennial Cruciferae. *Pieris virginiensis* has a much more restricted range, mostly south of *P. napi* in the northeastern United States. It is "obligately" univoltine in spring, with a single phenotype like the spring phenotype of *napi.* When *P. virginiensis* is reared under continuous light, a "nonsense" photoperiod, it does not diapause but develops directly as an animal phenotypically indistinguishable from the nondiapause, summer phenotype of *P. napi.* In this par-

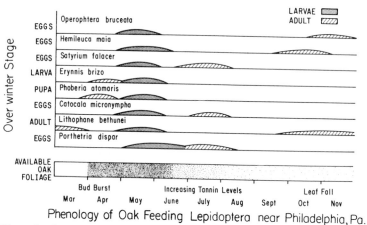

Phenology of Oak Feeding Lepidoptera near Philadelphia, Pa.

Figure 7 Phenology of univoltine oak-feeding lepidopterans near Philadelphia, Pennsylvania. Note
that although larval feeding periods are synchronous, both the adult flight period and the part of
the life cycle that overwinters vary greatly among
the species.

ticular case the phenology of *P. virginiensis* is tightly coupled to that of its only
natural host plant, the climax-understory
vernal Crucifer *Dentaria*. In adapting to
this plant and adopting a *K*-strategy, *P.
virginiensis* merely changed its diapause-
induction response, submerging not only
its potential for multivoltinism but its
phenotypic switch as well (Shapiro, 1971).
The same phenomenon has occurred in *P.
napi* itself in the east-slope canyons of the
Vaca Mountains in California and in the
belt of incense cedar and yellow pine on
the lower west slope of the Sierra. In both
places *P. napi* becomes just like *P. virginiensis* in behaving as a vernal univoltine, and in both cases its host is an
ephemeral vernal *Dentaria!* Coastwise in
central California *P. napi* is facultatively

bivoltine, and northward in the Pacific
Northwest it becomes fully multivoltine.

Although the phenology of *Dentaria*
determines the seasonal pattern of these
Pieris, not all univoltines are explicable in
host plant terms. Explanations of voltinism tend to be preoccupied with host plant
phenology, and certainly cruciferous, eri-
caceous, and certain other feeders do tend
to be spring-univoltine. But Figure 7
shows that adaptation to host phenology
need not say anything about when the
adults appear. It is a plot of the seasonal
appearance of larvae and adults of eight
univoltine oak-feeding lepidopterans in
the New Jersey pine barrens, representing
six families. The larvae of all of them feed
at essentially the same time in spring, but
the adults appear at widely differing sea-

sons. Among them these species exhaust the life-history stages in which a holometabolous insect can overwinter! The synchronization of the larval feeding periods is presumably explicable in terms of a limitation of larval feeding by the increasing tannin content of the oak leaves, as admirably worked out in England by Feeny (1970). Once that adaptation to host phenology has been achieved, other selective forces are quite free to work on other components of the life cycle, especially the quiescent stages. Studies of facultative diapause in butterflies (Shapiro, unpublished) make it abundantly clear that the length of time necessary to make a caterpillar from an egg or an adult from a pupa is very negotiable.[4] The timing of adult emergence in univoltines is in itself an adaptive characteristic.

Some Arguments From Competition

If univoltines actually occur in sets, negative (competitive) interactions within the sets must be minimized or fully offset by the advantages of the association. If adult competition is important, being associated with the same set of species every year would allow for more coevolution (and therefore greater fitness all around)

than could occur if species' emergence dates were independent of one another and each species faced a changing spectrum of competitors within its flight period within and between years. The very existence of sets would imply temporal displacement as an escape from competition, as Clench hypothesized, but only if the component species of each set were thus selected to minimize interference among them.

Early spring univoltines are notoriously at the mercy of the elements; if widely different years are compared, they show several weeks' difference in flight dates (as implied by Figure 6). As we have seen (e.g., at Boreal Ridge in Figure 8.5), in very late seasons they overlap the late spring or early summer species broadly, presumably increasing the intensity of competition. (Unfortunately we have no idea what to look for in the hope of demonstrating this.) Summer univoltines, with a long period of developmental buffering after diapause, are very resistant to temporal change (Shapiro, 1974a), although they may not differ significantly from spring species in susceptibility to weather-induced mortality. Such mortality must occur mostly in the developmental stages, since (as Levins's model predicts) selection would retard any species from emerging in spring until the odds against a dramatic climatic reversal were decidedly in its favor.[5]

---

[4] I noted earlier that the stage in which the adverse season is passed seems to be resistant to change. Presumably this is true because of the difficulties of evolving suitable physiological devices for getting through the adversity, and is a limiting factor on the flexibility of life histories. Pierids, for example, secrete alcohols into the haemolymph of their diapausing pupae, thereby lowering the freezing point. Every known *Pieris* overwinters as a pupa, and this would clearly help shape the phenological evolution of the *napi* group.

[5] In addition to the *K*-selected univoltines and *r*-selected multivoltines, a third category occurs in lowland California, the seasonal bivoltines. These are double-brooded species both of whose generations occur in the same season: spring (*Lycaenopsis argiolus echo*, Lycaenidae; *Anthocharis sara*, Euchloe

Explanations of the timing of univoltines grounded in interspecific competition may all be wide of the mark.[6] Competition for nectar sources, although an obvious hypothesis, is not an especially strong one: it implies that clusters of species are timed to take advantage of seasonal maxima in nectar sources—which broader experience does not suggest is the case. (Clench did find such a pattern in western Pennsylvania; it may be real, or it may be an artifact of the clumped butterfly emergences themselves.) At any rate, this hypothesis fails to account for the rarity of autumn univoltines everywhere, although in many habitats the sea-

sonal maximum of flowering occurs then.[7] Forty years after Saunders (1932) recorded flower preferences in the field and correlated them with proboscis length, we still know next to nothing about competition—if any—for adult food. A second candidate as an object of competition, male territories, is difficult to visualize as a seasonally regulated variable at all.

This leaves the matter of nonrandom univoltine seasonality in an intellectually unsatisfying state—which is where it must remain until many more data can be accumulated on the competition hypothesis. Univoltinism itself is on a firmer footing theoretically: it can be seen to be either a necessary outcome of selection in a short growing season at high altitudes or latitudes (in which case it enables a species to escape its seasonal problems in diapause and proceed with adapting to its biotic environment); or as part of an overall K-strategy evolved in a long-growing-season environment. In either case the association of univoltinism and K-selection seems inescapable. To find butterflies that are simultaneously univoltine and strongly r-selected we will probably have to look at arctic and arctic alpine faunas, for which environmental unpredictability is so pervasive that it cannot be avoided—but only minimized—by the timing of life-history phenomena.

One final point: the homeostasis implied by the similarity of the 1972 and 1973 species-per-week curves for the Sac-

---

ausonides, Pieridae) or late-summer–early-fall (*Ochlodes sylvanoides, Lerodea eufala,* Hesperiidae). (Normal double-brooded species have widely spaced generations that span the season: *Papilio multicaudatus,* Papilionidae; *Ochlodes yuma, Hesperia columbia,* and *H. juba,* Hesperiidae). Seasonal bivoltines appear to represent a transitional stage from uni- to multivoltinism or the other way around. All but *L. eufala* occur up to the high Sierra, where they are late spring (*E. ausonides, A. sara, L. a. echo*) or late summer (*O. sylvanoides*) univoltines. Only *L. eufala* is fully multivoltine anywhere in its range.
[6]Some inferences about competition can be drawn from my recent paper (Shapiro, 1973) on the New York butterfly fauna. In toto, the New York data on host selection, voltinism, and species/genus ratios support the idea of tighter species packing in richer faunas—an idea derived from competition theory. We are thus left with the usual paradox: competition is theoretically the most important organizing force in communities; empirically, it is exceedingly difficult to demonstrate at all. The usual escape from this paradox is the hypothesis that competition is a transient phase of interspecific association, evolving itself out of existence by character displacement. If so, the new, artificial fauna of the Sacramento Valley should have significantly greater niche overlaps than older, better integrated ones. My very preliminary data suggest this is true.

[7]Temperate butterfly faunas invariably show a migratory effect in autumn. The strength of this influx varies greatly from year to year but could conceivably be a selective factor opposing autumn univoltinism.

ramento Valley does not require coevolution. The largely multivoltine lowland California faunas are composed primarily of adventive species whose original ranges and habitats are unknown and whose northern California ranges are in some cases quite recent (tens of years). Theoretical ecology often seems to assume that structural aspects of communities are predetermined by climatic factors independently of the taxa involved. It is a great pity that the early Spanish explorers in the Sacramento Valley did not number in their party a lepidopterist, especially one with a penchant for collecting at odd places and seasons.

Acknowledgments

My interest in this problem became greatly strengthened about ten years ago in conversations with Robert MacArthur and correspondence with Harry Clench. Since then it has profited by discussions with dozens of persons, including theoreticians, climatologists, and butterfly collectors. Part of the data for Figure 1 was kindly provided by Oakley Shields, Entomology Department, University of California, Davis. The other California data are my own and were obtained through the generous support of the Committee on Research, University of California, Davis, in grant D-804.

References

Benson, W. W., and T. C. Emmel. 1973. Gregariously roosting populations of the Nymphaline butterfly *Marpesia berania* in Costa Rica. *Ecology* 54:326–335.

Clench, H. K. 1967. Temporal dissociation and population regulation in certain Hesperiine butterflies. *Ecology* 48:1000–1006.

Ehrlich, P. R., D. E. Breedlove, P. F. Brussard, and M. A. Sharp. 1972. Weather and the "regulation" of subalpine population. *Ecology* 53:243–247.

Emmel, T. C., and J. F. Emmel. 1962. Ecological studies of Rhopalocera at Donner Pass, California. I. Butterfly associations and distributional factors. *J. Lepid. Soc.* 16:23–44.

Emmel, T. C., and J. F. Emmel. 1963. Ecological studies of Rhopalocera at Donner Pass, California. II. Meterological influences of flight activity. *J. Lepid. Soc.* 17:7–20.

Emmel, T. F., and C. F. Leck. 1970. Seasonal changes in organization of tropical rain forest butterfly populations in Panama. *J. Res. Lepid.* 8:133–152.

Feeny, P. P. 1970. Seasonal changes in oak leaf tannins and nutrients as a cause of spring feeding by winter moth caterpillars. *Ecology* 51:565–581.

Fox, R. M., A. W. Lindsey, Jr., H. K. Clench, and L. D. Miller. 1965. The butterflies of Liberia. *Mem. Amer. Entomol. Soc.* 19:1–438.

Hairston, N. G., D. W. Tinkle, and H. M. Wilbur. 1970. Natural selection and the parameters of population growth. *J. Wildlife Mgt.* 34:681–690.

Levins, R. 1968. *Evolution in Changing Environments.* Princeton University Press, Princeton.

MacArthur, R. H., and E. O. Wilson. 1967. *The Theory of Island Biogeography.* Princeton University Press, Princeton.

Opler, P. A., and R. L. Langston. 1968. A distributional analysis of the butterflies of Contra Costa County, California. *J.*

Lepid. Soc. 22:89–107.

Owen, D. F. 1971. *Tropical Butterflies: The Ecology and Behavior of Butterflies in the Tropics with Special Reference to African Species.* Clarendon Press, Oxford.

Saunders, A. S. 1932. Butterflies of the Allegany State Park. New York State Museum Handbook 13. Albany.

Shapiro, A. M. 1971. Occurrence of a latent polyphenism in *Pieris virginiensis* (Lepidoptera: Pieridae). *Entomol. News* 82:13–16.

Shapiro, A. M. 1973. Ecological characteristics of the New York State butterfly fauna. *J. New York Entomol. Soc.* 81(4):201–209.

Shapiro, A. M. 1974a. The butterfly associations of Staten Island (Richmond County, New York). *J. Res. Lepid.* 12:65–128.

Shapiro, A. M. 1974b. The butterflies and skippers of New York (Lepidoptera: Papilionoidea, Hesperioidea). *Search: Agriculture, Entomol.* 12(4):1–60.

Shapiro, A. M. 1974c. Altitudinal migration of butterflies in the central Sierra Nevada. *J. Res. Lepid.* 12:231–235.

Slobodkin, L. B., and H. W. Sanders. 1969. On the contribution of environmental predictability to species diversity. *In* G. M. Woodwell and H. H. Smith, eds., *Diversity and Stability in Ecological Systems,* Brookhaven Symposium in Biology, pp. 82–95. U. S. Department of Commerce, Springfield, Va.

9 Markovian Properties of Forest Succession

Henry S. Horn

If a forest suffers a catastrophe, its place is invaded by a community of pioneering plants, which then undergoes a succession of changes in structure and specific composition that slowly rebuilds a semblance of the original forest. This succession has several properties that have intrigued and employed botanists since the birth of ecology as a science. Similar pioneer communities follow nearly the same course, and hence a predictable course, to the same final state. Strikingly different pioneer communities may converge on the same final state, which, in a given region, may closely resemble extant patches of virgin forest. Traditional analyses of these patterns have attributed them to uniquely biological causes, in some cases to an almost mystical organic integrity of the final or "climax" community. This vitalistic interpretation has been toppled by the fact that real climax communities are elusive, indistinct, and variable from place to place (Drury and Nisbet, 1973). The many fragments of the traditional analysis of succession have been christened and described in detail, but the resulting jargon has added more to Freudian imagery than it has to a genuine understanding of successional patterns.

The most dramatic property of succession is its repeatable convergence on the same climax community from any of many different starting points. This property is shared by a class of statistical processes known as "regular Markov chains." A Markov chain is a stochastic process in which transitions among various "states" occur with characteristic probabilities that depend only on the current state and not on any previous state (Kemeny and Snell, 1960). Applied to succession, the "states" may be patches of forest of a given composition; or they may be different species of trees, in which case the forest is represented by a collection of independent Markov chains. A Markov chain is "regular" if any state can be reached from any other state in a finite number of steps, and if it is not cyclic. A chain is cyclic if every state necessarily returns to itself after the same fixed number, greater than one, of steps or an integral multiple of that number.

The fundamental property of a regular Markov chain is that eventually it settles into a pattern in which the various states occur more or less randomly with characteristic frequencies that are independent of the initial state. This final "stationary distribution" of states is the analog of the climax community, and different climax communities imply different probabilities of transition among the states, rather than different initial communities. The models of this chapter are basically regular Markov chains, though my data include some "transient" states that can be left but never returned to, and one model adds some "absorbing" states that cannot be departed from once they are entered.

I should like to examine forest succession as a tree-by-tree replacement process. For each tree in the forest, I estimate the probability that it will be replaced by another of its kind or by another species. From a matrix of these probabilities, I can calculate how many trees of each species should be found at any stage of succession. Even in its rudimentary form, this model makes accurate predictions for the forests near Princeton, New Jersey. With Robert May's help, I have analyzed a generalized successional matrix, examining the relations among: the degree and pattern of exploitation or of natural disturbance, the pattern of reseeding, the diversity of species in the late stages, and the dynamic stability; that is, the speed with which the forest recovers from temporary indignities imposed by man or by nature.

The representation of succession as a plant-by-plant replacement process is not original. MacArthur (1958, 1961) recognized that many of the supposedly biological surprises of succession were routine statistical properties of Markov processes. John Vandermeer and Henry Wilbur (personal communication) use a plant-by-plant replacement process as a parable to introduce their students to succession and to matrix algebra. Anderson (1966) used stochastic theory ingeniously in a discussion of local groupings of plants and their stability. Stephens and Waggoner (1970) have characterized forest successions in New England by the transition probabilities among stands of different composition, and have tested the Markovian assumption (Waggoner and Stephens 1970). Leak (1970) found that short-term changes in species composition could be predicted with reasonable accuracy from measurements of the birth and death rates of each species, even though his model included no explicit interactions between different species. Botkin, Janak, and Wallis (1972) simulated the succession of a New Hampshire forest with great accuracy, using a Markovian model in which each species is characterized by a particular pattern of deterministic growth, stochastic birth and death, and response to the local physical environment, with interspecific interactions resulting from the effect of each species on the environment of the others. I have no doubt that there are many other instances hiding from me in the Russian literature and in the all-noun cryptology of compartment model systems analysis.

My models differ from those of previous authors in that they are specific enough to confront real data, general enough to apply to a wide variety of successions, and compact enough to produce analytical results. I shall routinely make outrageous assumptions, but I shall defend them in several ways. Some are needed only for analytic convenience and may be relaxed with no major effect on the result. In some cases a redefinition of a measurement is all that is needed to bring theory into line with fact. Astoundingly, some of the assumptions are even true.

I shall first confront a rudimentary model with data from nature, and evaluate the empirical validity of a host of theoretical assumptions. I shall then compare predictions with reality, and illustrate some of the general statistical properties of succession. After generalizing the rudimentary model, I shall analyze a generalized successional matrix, interpreting dynamic stability intuitively as well as analytically. All the properties of the rudimentary model will be shared by a more realistic model that allows overlapping generations and diverse rates of survival for different species at each of several stages in their life histories. Striking new properties will emerge from a third model that allows the rain of seeds from each species to be proportional to its local abundance. Finally, I shall analyze a linear, unbranched succession, producing simple expressions for the stationary distribution, its degree of disturbance, its diversity, and its dynamic stability, all independently defined in biologically realistic ways.

Study Site

My data were gathered in the woods behind the Institute for Advanced Study in Princeton, New Jersey. These woods have several stands, differing in composition from each other but each uniform within itself. One of them was apparently never farmed, and several were abandoned as field, pasture, or managed woodlot at times estimated as between 30 and at least 150 years ago. I have given a sketchy account of the succession in these woods as part of a strategic analysis

of the geometric distribution of leaves in different species of trees (Horn, 1971). I have documented thoroughly the forest succession in the Institute Woods, in a separate publication (Horn, 1975).

Rudimentary Model and Data

I shall first pretend that the forest is a honeycomb of independent cells, each occupied by a tree; that all trees are replaced synchronously by a new generation that arises from their understory; and that the probability that a given species will be replaced by another given species is proportional to the number of saplings of the latter in the understory of the former. Table 1 is a matrix of these transition probabilities, given as percentages. For example, among a total of 837 saplings found underneath gray birches scattered throughout the Institute Woods, I found no gray birch saplings, 142 red maples, and 25 beeches. Hence for the probability that a given gray birch will be replaced by another gray birch I have $0/837 = 0$, by a red maple $142/837 = 0.17 = 17\%$, and by beech $25/837 = 0.03 = 3\%$.

I can now calculate the number of gray birch in the next "generation" by finding all species in the current canopy that have some gray birch saplings in their understory, namely big-toothed aspen, sassafras, and blackgum, and summing their current abundances times the probability that each will be replaced by gray birch. From Table 1 we see that GB (next generation) $= 0.05$ BTA $+ 0.01$ $SF + 0.01$ BG.

I can do the same for each species: e.g.,

SF (next generation) = 0.09 BTA
+ 0.47 GB + 0.10 SF + 0.03 BG
+ 0.16 SG + 0.06 WO + 0.02 OK
+ 0.01 HI + 0.02 TU + 0.13 RM.

Finally, I take the predicted abundances in the next generation and combine them once more with the transition matrix according to the same recipe, to calculate the composition of the forest two generations later, and so on, for as many generations as I can endure. The matrix of Table 1 is applied for five generations in Table 2. After several generations the abundances of the various species settle down to a stationary distribution, for reasons that I shall discuss later.

For a similar description of the replacement process, see Stephens and Waggoner (1970), who also give a graphic representation of the approach to the stationary distribution.

The generations of trees are of course not synchronous; therefore the sequence of specific compositions obtained by successive multiplications of an initial distribution by the transition matrix does not accurately represent the actual course of succession. However, the stationary distribution should accurately represent the proportions of occurrences of individuals

Table 1. Transition matrix for Institute Woods in Princeton: percent saplings under various species of trees

	Sapling species (%)											Total
	BTA	GB	SF	BG	SG	WO	OK	HI	TU	RM	BE	
Canopy species												
Big-toothed aspen	3	5	9	6	6	—	2	4	2	60	3	104
Gray birch	—·	—	47	12	8	2	8	0	3	17	3	837
Sassafras	3	1	**10**	3	6	3	10	12	—	37	15	68
Blackgum	1	1	3	**20**	9	1	7	6	10	25	17	80
Sweetgum	—	—	16	0	**31**	0	7	7	5	27	7	662
White oak	—	—	6	7	4	**10**	7	3	14	32	17	71
Red oaks	—	—	2	11	7	6	**8**	8	8	33	17	266
Hickories	—	—	1	3	1	3	13	**4**	9	49	17	223
Tuliptree	—	—	2	4	4	—	11	7	**9**	29	34	81
Red maple	—	—	13	10	9	2	8	19	3	**13**	23	489
Beech	—	—	—	2	1	1	1	1	8	6	**80**	405

The number of saplings of each species listed in the row at the top, where the abbreviations are self-explanatory, is expressed as a percentage of the total number of saplings (last column) found under individuals of the species listed in the first column. The entries are interpreted as the percentages of individuals of species listed on the left that will be replaced one generation hence by species listed at the top. The percentage of "self-replacements" is shown in boldface.

The total number of recorded saplings is 3,286. A dash implies that no saplings of that species were found beneath that canopy; a zero, that the percentage was less than 0.5%.

The species are: BTA = *Populus grandidentata*, GB = *Betula populifolia*, SF = *Sassafras albidum*, BG = *Nyssa sylvatica*, SG = *Liquidambar styraciflua*, WO = *Quercus alba*, OK = Section *Erythrobalanus* of *Quercus*, HI = *Carya* spp., TU = *Liriodendron tulipifera*, RM = *Acer rubrum*, BE = *Fagus grandifolia*. Species mentioned only in the text are ash (*Fraxinus* spp.), dogwood (*Cornus florida*), and chestnut (*Castanea dentata*).

Table 2. Theoretical and empirical approach to stationary distribution: from the initial composition of a 25-year-old stand

| | Species (%) | | | | | | | | | | |
	BTA	GB	SF	BG	SG	WO	OK	HI	TU	RM	BE
Theoretical											
Generation: 0	0	49	2	7	18	0	3	0	0	20	1
1	0	0	29	10	12	2	8	6	4	19	10
2	1	1	8	6	9	2	8	10	5	27	23
3	0	0	7	6	8	2	7	9	6	22	33
4	0	0	6	6	7	2	6	8	6	20	39
5	0	0	5	5	6	2	6	7	7	19	43
⋮	⋮	⋮	⋮	⋮	⋮	⋮	⋮	⋮	⋮	⋮	⋮
Stationary distribution (%)	0	0	4	5	5	2	5	6	7	16	50
Longevity (years)	80	50	100	150	200	300	200	250	200	150	300
Age-corrected stationary distribution (%)	0	0	2	3	4	2	4	6	6	10	63
Empirical											
Years fallow: 25	0	49	2	7	18	0	3	0	0	20	1
65	26	6	0	45	0	0	12	1	4	6	0
150	—	—	0	1	5	0	22	0	0	70	2
350	—	—	—	6	—	3	—	0	14	1	76

Abbreviations are as in Table 1. The theoretical approach results from taking a row vector of species proportions in a 25-year-old stand, expressed above as percentages, and multiplying it once for each generation by the transition matrix of Table 1. The stationary distribution is the solution of eq. 4, given in the text. Note that initial changes are rapid, but that a close approximation to the stationary distribution is reached in only five generations. Several species (e.g., RM = red maple) show damped oscillations. The age-corrected stationary distribution is simply the stationary distribution, weighted and normalized by the rough approximations to the specific longevities listed above it.

The empirical approach results from independent measurements of 639 trees in stands that have been fallow for at least the number of years indicated. The percentages are of total basal area, calculated from diameters measured at breast height. Note the similarity between the age-corrected stationary distribution and the composition of the oldest forest. Other patterns in Tables 1 and 2 are discussed in the text.

of each species in a single cell through the course of time. If I take a synchronous sample of many cells I should encounter each species in proportion to the number of times that it occurs in the temporal sequence for each cell, weighted by the average amount of time that each occurrence spans; that is, by the lifespan of individuals of the species. Thus the stationary distribution, when weighted by the longevity of each species, should represent the actual distribution of species in the climax, and takes account of the facts that the lifespans of trees vary and that their generations are not synchronous. For the stationary distribution, the interval between generations is about equal to the lifespan for each species, because before a tree reproduces it must replace another tree that has died. I can and shall add varied lifespans and asynchrony to the next model, but I shall leave the fiendish empirical computations of such a model for a later paper.

I have assumed that abundance in the understory aids competition for space in the canopy. With two exceptions, this assumption is probably close to the truth in the upland parts of the Institute Woods, where the overriding cause of succession is the concomitant increase in shade tolerance and in amount of shade shed by the dominant species (Horn, 1971). I have simply ignored the two most prominent exceptions. Ash persists abundantly as a seedling in deep shade, but cannot rise to the canopy through that shade, and hence is not an important component of the upland Institute Woods, though it is common in the patchily disturbed flood plain, where it quickly rises to fill gaps in the canopy. Dogwood, an archetypal understory species, is abundant throughout the drier soils of the Institute Woods, but of course it never reaches the canopy. Since almost nothing grows in the compound shade of both canopy and dogwood, dogwood has little effect on the relative success of different species in the understory.

In any other environment—for example, even in the flood plain of the Institute Woods—I should have to visit dead and dying trees to find out what is actually replacing them in order to estimate the probabilities in the transition matrix. Indeed, the challenge in applying these models to any community is the development of a simple and biologically realistic measurement of the transition matrix.

Since I have chosen simplicity, I have assumed that abundance as a sapling is a reasonable predictor of survival to reach the canopy. With the restrictions given above, this assumption may approximate the truth for the comparative survival of individuals of one species on different sites, though of course saplings can grow in many places where adult trees cannot. However, for interspecific comparisons, this assumption is likely to be false since different species have different rates of survival when each is grown under its own optimal conditions. Hence my measurement of the transition probabilities is indirect and crude, which makes the agreement between theory and reality still more surprising. The theoretical part of Table 2 should not be mistaken for an empirical analysis of succession, but rather should be taken for an heuristic justification of the simple model of succession.

I have assumed that the transition probabilities do not change with forest composition. This is likely to be true if all measurements are made near the stationary state. It is fortuitously true for most successional patches in the Institute Woods, which are small enough for their rain of seeds to be dominated by the distribution of tree species over the whole woods, rather than by the local numbers of each species in each patch. The conspicuous exceptions are sweetgum, white oak, and beech. Sweetgum can establish itself as a pioneer on appropriate soil, but it will beat other species to a field only if there is a large and prolific sweetgum nearby; this behavior is partly represented in Table 1 by the lavish self-replacement of sweetgum, and its sparse invasion beneath other species. However, were sweetgum commoner throughout the woods, it would probably be more abundant beneath other species. White oaks are seriously underrepresented in the understory, in view of their dominance and fecundity in the canopy on well-drained soils. I suspect that a fluctuating and currently high population of gray squirrels has dispatched their acorns. Beech reproduces mainly by root sprouts; in fact I have yet to find a palpable seedling in the Institute Woods. However, large beeches are scattered throughout the woods, and I have traced lines of successive sons of beeches for nearly a quarter of a mile. Hence, I doubt that beech's invasiveness is underestimated.

Finally, I assume that the transition probabilities do not change with successional stage or with local edaphic conditions. This assumption is generally true, since the presence of a particular species in the canopy is itself a fairly accurate indicator of local conditions of shade, drainage, soil texture, nutrients, and associated plants. In fact, I can specify a narrow range of these conditions for all of the trees in the Institute Woods except the red oaks, sassafras, and red maple. I suspect that the first two have narrow requirements of which I am currently ignorant. Red maple occupies a wide variety of sites, but I have not found any systematic variation in the distribution of species in red maple's understory at different sites.

I shall now use Table 1 and Table 2 to illustrate several general properties of succession as a Markovian replacement process. General proofs of the following assertions can be found in Kemeny and Snell (1960). After a long enough time, the composition of the forest settles into either a cyclic pattern or a stationary distribution. In the latter case the stationary distribution is the same no matter what the starting distribution, and in my example the stationary distribution is closely approached after only five generations. The stationary distribution, especially when weighted by the lifespans of the species, resembles the actual specific composition of a forest that has not had a major disturbance in at least 350 years, except for the loss of its chestnuts to the blight early in this century. The initial changes in specific composition, shown in Table 2, are much more rapid than the later changes. This effect would be even greater if the calculations included the short lifespans of the early successional species and the hoary age of the later species. Thus "stability" in the naive sense of "absence of change" increases tautologically as succession proceeds.

Pioneers such as big-toothed aspen and gray birch are lost completely if they do not invade later stages of succession. Conversely, a beech forest could be monospecific only if its understory were completely uninvadable by any other species. Beech has three characteristics that assure its dominance in the stationary distribution: ability to invade under other species, resistance to invasion by other species, and copious self-replacement. The subdominant species in the stationary distribution, red maple, is a prolific and tenacious invader under other species, although its understory is diverse and its self-replacement is only average. However, self-replacement alone is not enough to ensure abundance in the stationary distribution, as sweetgum and blackgum testify.

During the approach to the stationary distribution, red maple shows damped cyclic behavior. Such cycles will invariably be damped unless the least common denominator of all possible cycles of replacement is more than one generation. Therefore if any species can replace itself, or if generations are not equally long for all species, intrinsic cyclic behavior will not persist. Hence a persistent and stably cyclic succession implies either a cycle in one of the dominant species, or an extrinsic environmental cycle. A dominant species might cycle if the control of its population involved time lags or a predator-prey system. An environmental cycle may be represented as an alternation between two different matrices of transition probabilities, e.g., **WDWDWD** for a cycle of wet and dry decades. The resulting cycles in the community would damp only if the matrices were peculiar enough to commute, i.e., only if **WD = DW**, which is an exceedingly rare property among matrices. Furthermore, if the environmental cycle were irregular (e.g., **WWDDDWDWW**), the community's cycle would damp only if **W = D**; that is, if both matrices were the same, in which case the environmental cycle would not exist from the community's point of view.

Generalized Discrete Model

The rudimentary model is generalized as follows. Let $N_j(t)$ be the proportion of species j in generation t. Let p_{ij} be the probability that an individual of species j replaces a given individual of species i. Let s be the number of species. Then the model is simply:

$$N_j(t + 1) = \sum_i N_i(t)p_{ij} \qquad (1)$$

Putting this into matrix notation with $\mathbf{n}$ a row vector of N_j and $\mathbf{P}$ an s by s matrix of p_{ij} we have

$$\mathbf{n}(t + 1) = \mathbf{n}(t)\mathbf{P} \qquad (2)$$

and after m generations:

$$\mathbf{n}(t + m) = \mathbf{n}(t)\mathbf{P}^m \qquad (3)$$

As m gets large, $\mathbf{n}$ settles down to a stationary distribution, $\mathbf{n}^*$, which is the solution of s linear equations:

$$\mathbf{n}^* = \mathbf{n}^*\mathbf{P} \qquad (4)$$

The proof of this assertion follows. First expand $\mathbf{n}$ as a linear combination of eigenvectors ($\mathbf{v}_i$) with distinct eigenvalues ($\lambda_1 > \lambda_2 > \cdots > \lambda_s$), and then:

$$\mathbf{n}(t)\mathbf{P}^m = \lambda_1^m\mathbf{v}_1 + a_2\lambda_2^m\mathbf{v}_2 \\ + \cdots + a_s\lambda_s^m\mathbf{v}_s \quad (5)$$

where the a's are constants. As will be proved later, λ_1 is 1 for inclusive probability matrices, and hence $\mathbf{v}_1 = \mathbf{n}^*$.

The second largest eigenvalue, λ_2, is a negative measure of dynamic stability, since if it is large, it takes a long time for the contribution of $\mathbf{v}_2$ to be overwhelmed by that of $\mathbf{v}_1$. Similarly, May (1973) measures the stability near equilibrium in a continuous model by the smallness of the largest real part among eigenvalues of a matrix with entries:

$$a_{ij} = \frac{\partial}{\partial N_j}\left[\frac{dN_i(t)}{dt}\right] \qquad (6)$$

measured at equilibrium. The matrix of a_{ij} is precisely the matrix used by Levins (Chapter 1) to analyze the dynamics of communities near equilibrium.

Eigenvalues of a Successional Matrix

I start by assuming a pattern of transitions such that early successional species are replaced only by later species. Furthermore, all species are crushed by the hand of fate with uniform probability d per generation, and the openings thus created are invaded by a single pioneer species. Species i is succeeded by later species with probability e_i. For notational convenience, the transition matrix is represented as:

$$\mathbf{P} = \begin{bmatrix} 1-e_1 & e_1 & 0 & \cdots & 0 \\ d & 1-e_2-d & e_2 & \cdots & 0 \\ \vdots & \vdots & \vdots & \vdots\vdots\vdots & \vdots \\ d & 0 & 0 & \cdots & e_{s-1} \\ d & 0 & 0 & \cdots & 1-d \end{bmatrix} \qquad (7)$$

The following calculations are also accurate for a more generalized matrix with the e_i distributed within the ith row to any columns numbered $> i$, that is, anywhere above the diagonal. Take away λ from each diagonal entry. Replace the first column by the sum of all columns, which equals $1 - \lambda$. Subtract the last row from all other rows. Now expand the determinant by the first column, last row $(1 - \lambda)$. The first eigenvalue is $\lambda = 1$.

Now remove the first column and the last row and treat the rest of the matrix as follows. Replace the first column by the sum of all columns, replace the second column by the sum of the second through the last columns, and so on. The result is a matrix with $(1 - e_i - d - \lambda)$ as its ith diagonal element, 0s below the diagonal, and garbage above. Expand the determinant by diagonal entries, and the other eigenvalues are $\lambda_i = 1 - e_i - d$. The eigenvalues for the more generalized matrix are also as follows: 1, the first diagonal entry minus d, and the second through the $(s - 1)^{th}$ diagonal entries of the original matrix.

The assumptions that allow explicit calculation of the eigenvalues are not as restrictive as they first appear to be. The constancy of d—that is, a constant sensitivity of each species to chronic disturbance—could be assured by letting the first "species" actually represent openings in the canopy that are invaded by pioneers, and measuring the abundance of each species in a weighted fashion. The weighting is simply arranged so that the sensitivity for the weighted abundances is indeed constant. The subdiagonal zeros represent the assumption that when succession "slides back" it returns to the first stage and not to any intermediate stage. To the extent to which a transition matrix departs from this assumption, either there is a mutualistic interaction between two species, or their interaction is less biologically determinate than it is random or historical. In principle, one could pool species into associations within which these assumptions are violated, creating a matrix of transitions among associations, for which these assumptions are fulfilled. This pooling may be ambiguous or impossible in practice, and I am left with the pious hope that real matrices will not be far different from the presumptive one, and with the somewhat priggish defense that the properties of the imagined matrix of eq. 7 are more interesting because they derive from rigid biological determinism.

Since most of the eigenvalues of the assumed successional matrix of eq. 7 are diagonal entries, or probabilities of self-replacement, the dynamic stability of a succession is easily interpreted. The higher the probabilities of self-replacement among successional species, the larger is the subdominant eigenvalue of the matrix, and the slower is the return to the stationary state after a perturbation. Thus copious self-replacement among successional species lowers stability. Intuitively, when a climax forest is disturbed, the return to a stationary distribution will be slowed if the early successional stages efficiently reproduce themselves rather than give way to later stages.

This interpretation is strictly applicable only to siblings of the peculiar matrix of eq. 7, but of course the method applies to any matrix whose eigenvalues can be calculated numerically by magic or by brute force.

Continuous Model with Fixed Rain of Seeds

This model is appropriate when departures from the stationary distribution are either small enough or localized enough so that different species of seeds rain down in fixed relative proportions throughout succession.

Let each tree species j have a characteristic and constant death rate d_j. Other symbols are as in the previous model. The rate of increase of each species j has a negative term $(-d_j N_j)$ due to deaths, and a positive term that sums the number of deaths among other trees times the probability that species j will replace the dying species. Symbolically:

$$\frac{dN_j}{dt} = -d_j N_j + \sum_i d_i N_i p_{ij} \qquad (8)$$

The matrix of eq. 6 for analysis of stability after the fashion of May (1973) has terms

$$a_{ji} = d_i p_{ij} (i \neq j)$$

and

$$a_{ii} = d_i (p_{ii} - 1) \qquad (9)$$

The same matrix can be derived more rigorously and compulsively from the Chapman-Kolmogorov diffusion equations (see, e.g., Bailey, 1964).

Note that the eigenvalues for the simplified successional matrix of eq. 7 are now 0, and $-e_i - d$ for i from 1 to $s - 1$. The stationary distribution is stable in the sense that the community returns to it after perturbations. However, as before, the return to the stationary distribution is slowed by copious self-replacement or marked longevity among the successional species.

Note also that this model is applicable to the realistic case where each species has different rates of survival as seedling, sapling, and adult in the canopy. The different life stages are simply treated as additional species.

Continuous Model with Seeding Proportional to Local Abundance

This model is appropriate when a large enough area is clear-cut, so that the rain of seeds at any stage of succession depends on what species of trees are present and on how abundant they are. It would also apply to intertidal communities whose successional patterns depend on which species' "seeds" are available at the times when vacant patches appear (Connell, Chapter 16).

The symbols are as in the previous models, except that p_{ij} is now the probability that species j replaces species i, weighted by the rain of seeds that each individual of species j produces. When this probability is weighted by the abundance of species j, the probability that species j replaces species i is $N_j p_{ij} / (\Sigma_k N_k p_{ik})$. Then following the same derivation as eq. 8 of the last model, we have:

$$\frac{dN_j}{dt} = -d_j N_j + \sum_i d_i N_i \left(\frac{N_j p_{ij}}{\sum_k N_k p_{ik}} \right) \quad (10)$$

The matrix for analysis of stability about the equilibrium abundances N_j^* is easier to interpret if I adopt a notation that puts it in the same form as the stability matrix of the previous model, namely:

$$\mathcal{P}_{ij} = \frac{N_j^* p_{ij}}{\sum_k N_k^* p_{ik}} \quad (11)$$

$\mathcal{P}_{ij}$ is then the probability that an individual of species i will be replaced by one of species j, measured at the stationary distribution.

An obscene amount of algebra produces the matrix for analysis of stability, eq. 6, which has entries:

$$a_{ji} = d_i \mathcal{P}_{ij}$$
$$- \frac{1}{N_i^*} \sum_k d_k N_k^* \mathcal{P}_{ki} \mathcal{P}_{kj} (i \neq j) \quad (12)$$

and

$$a_{ii} = d_i (\mathcal{P}_{ii} - 1)$$
$$+ \frac{1}{N_i^*} \sum_k d_k N_k^* \mathcal{P}_{ki} (1 - \mathcal{P}_{ki}) \quad (13)$$

The diagonal terms a_{ii} of this matrix are the same as those of the stability matrix of the previous model (eq. 9) with the addition of a term that is always positive. Hence, if the original transition matrix takes the form of the simplified successional matrix (eq. 7) so that some of the a_{ii} are eigenvalues, clear-cutting will be less stable than patch-cutting. More generally, if $\mathcal{P}_{ii}$ is large enough, a_{ii} will be positive; if enough of the a_{ii} are positive, their sum, which is also the sum of the eigenvalues, will be positive; hence at least one of the eigenvalues will be positive. Therefore if self-replacement is sufficiently copious, the equilibrium that contains all species is not only less stable, it is unstable so that even small disturbances result in the loss of one or more species. This instability should be expected intuitively, since if the replacement of each species is proportional to its abundance, then a secular change in the abundance of a given species will tend to multiply itself in each generation. Of course, if self-replacements are infrequent enough among all species, this tendency to instability will be balanced by the numerical responses of other species.

Note that the term $\mathcal{P}_{ki}\mathcal{P}_{kj}$ in eq. 12 for a_{ji} is in some sense a coefficient of competition between species i and j for openings under species k. Hence, changes in specific composition are slowed by interspecific competition for openings in the forest. Intraspecific competition enters a_{ii} as a negative term, and thus helps to stabilize the equilibrium that contains all species.

The sum of all species' abundances is soothingly constant, since $\Sigma_j(dN_j/dt) = 0$. Tree-by-tree replacements are no longer Markovian since the transition probabilities for any given tree depend upon many of its neighbors. However, the process as a whole is Markovian since the transition to a new specific composition depends on tree-by-tree replacements, which in turn depend only on the current composition of the forest, and not on any previous composition. There are several alternative equilibria, some of which are stable. In particular, the initial absence of any species precludes its appearance because there is no source of seeds. Hence a very efficient high-grade lumbering is trivially unstable. This kind of equilibrium is only of historical interest unless some species are prevented from invading, given the opportunity.

I shall examine one such equilibrium

that is not trivial. Let one species, say species i, be the only one present. We then ask under what conditions can species j invade; that is, when is $dN_j/dt > 0$, with $N_i \gg N_j$? The dynamic equation (eq. 10) for the invader is reduced to:

$$\frac{dN_j}{dt} = N_j\left(-d_j + \frac{d_i N_i^* p_{ij}}{N_i^* p_{ii}}\right) \quad (14)$$

dN_j/dt is positive when the parenthetic term is positive; that is, when

$$\frac{p_{ij}}{d_j} > \frac{p_{ii}}{d_i} \quad (15)$$

In words, a species can invade when its ratio of colonization of new openings to its death rate is higher than that of the current monopolist. By a similar argument, Horn and MacArthur (1972) have shown that competition favors a high ratio of migration rates to death rates in beasts that inhabit a diverse and patchy environment. A monopolist with a very high self-replacement and low death rate cannot be easily displaced. In fact, these are the characteristics of the competitors who eventually replace Diamond's early successional "supertramp" bird species (Chapter 14). Hence competition between species favors those characteristics that decrease the stability of an equilibrium with many species. Once again, for a community whose species interact solely via competition, diversity begets the seed of its own destruction (see Levins, Chapter 1).

This model is amenable to a more detailed and less compact analysis, but the result so far is a bewildering set of special cases with no qualitatively new properties.

The result that patch-cutting is more stable than either clear-cutting or high-grading is well known to foresters (e.g., Gifford, 1902) and nicely documented (Twight and Minckler, 1972). However, it is useful to have a dispassionate analysis that is independent of the immediate aesthetic and biological effects.

Diversity, Stability, and Disturbance in an Unbranched Succession

Let the succession matrix be:

$$\mathbf{P} = \begin{bmatrix} 1-e & e & 0 & 0 \\ d & 1-d-e & e & 0 \\ d & 0 & 1-d-e & e \\ d & 0 & 0 & 1-d \end{bmatrix} \quad (16)$$

or a similar s by s matrix. Its eigenvalues are 1 and $1 - d - e$ with multiplicity $s - 1$ (0 and $-d - e$ in the continuous case). Hence the stability about equilibrium is measured by $d + e$. (Note that technically this is not true since the eigenvalues are multiple, but I can deviously define the matrix so that it has $s - 1$ distinct eigenvalues that are arbitrarily close to each other.) To further simplify expressions, I define:

$$\delta = \frac{d}{d + e} \quad (17)$$

the ratio of disturbance to dynamic stability.

The stationary distribution, found by solving $\mathbf{n}^* = \mathbf{n}^*\mathbf{p}$ (eq. 4), is:

$$N_i^* = \delta(1 - \delta)^{i-1}(i = 1, 2, \ldots, s - 1)$$
$$N_s^* = (1 - \delta)^{s-1} \quad (18)$$

Note that this is a strict geometric series. A more realistic matrix with varied e's distributed above the diagonal would produce a somewhat sloppier geometric series. Hence in a community that is an equilibrial patchwork of a few successional species, the relative abundances of species might be expected to follow a geometric series, or its statistically fuzzy equivalent, the log series (May, Chapter 4).

The diversity of the stationary distribution is easily calculated, by using Simpson's (1949) index of diversity:

Diversity

$$= \frac{1}{\sum_i N_i^{*2}} = \frac{2 - \delta}{\delta + 2(1 - \delta)^{2s-1}} \quad (19)$$

In this calculation, I use a standard formula for the sum of a geometric series, obtained by dividing $1 - x$ into 1 and subtracting the fractional remainder once I have enough terms in the series.

The diversity is plotted as a function of δ for different total numbers of species in Figure 1. The chief results are lower diversity at high and low degrees of disturbance than at intermediate disturbance, less evenness in the community of maximum diversity as the number of species increases, and the identification of the ratio diversity/stability as a useful metric. All of these properties are shared by some more general matrices with varied e's distributed above the diagonal, which I have forced our local computer to analyze numerically, though the resulting curves are variously skewed.

In his celebrated experiments, Paine

(1966) found that a rocky intertidal shore supported a more diverse assemblage of large beasts in the presence of predatory starfish than in the absence of starfish. Figure 1 provides a conceptual basis for Paine's original contention that predation allows a successional patchwork to replace a community dominated by mussels, who most effectively monopolize the limited space. Subsequent analysis of Paine's ideas has been dominated by a search for the peculiar properties that make the predator a "keystone" (cf. Levins, Chapter 1, Figure 6b), though Dayton (1971) has shown that the effect of starfish is mimicked by several other chronic disturbances, including wave exposure and battering by gigantic driftwood. Figure 1 produces a further result documented by Dayton: low levels of disturbance may increase diversity, but high levels decrease it again. A similar peak in diversity at intermediate disturbance was found by K. Harper (personal communication) when he plotted plant species diversity over a gradient of grazing pressure.

Summary of Results

Several properties of succession are direct statistical consequences of a plant-by-plant replacement process, and have no uniquely biological basis. Something like the original community returns after a temporary disturbance. Rapid changes are followed by undetectably slow changes. Hence "stability," in the naive sense of "absence of change," increases tautologically as succession proceeds. Extrinsically caused cycles persist; intrinsic

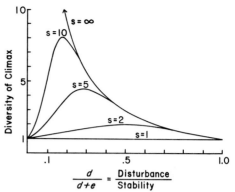

Figure 1 The diversity of the stationary distribution or "climax" is plotted against the ratio of disturbance to stability for the linear, unbranched succession of eq. 16. Curves are given for different numbers of species from one to an effectively infinite number. Diversity is measured by the index of Simpson (1949), which is the number of equally abundant species that would have the same probability that two randomly drawn individuals are conspecific as does the actual stationary distribution. Stability is measured by the negative of the subdominant eigenvalue of the matrix of eq. 16.

Note that intermediate disturbances produce higher diversity than either very high or very low levels. Even the curve for an infinite number of species plunges back to one for no disturbance. Do not make much of the fact that the curves are skewed to the left; this particular skew is not shared by models more general than eq. 16.

cycles are generally damped unless the cyclic organization of the community is unusually blatant. Pioneer species are eventually lost completely in undisturbed successions unless they are capable of invading later stages. To be abundant at climax, a species must be both invasive and noninvadable.

There is a further statistical consequence that often pretends to have biological significance. Different pioneer communities may converge to the same climax.

Some subtler properties of a plant-by-plant replacement process help to clarify arguments based on biological intuition. Copious self-replacement does not guarantee a species' abundance or even its persistence in late stages of succession. Hence the "stability" conferred by self-replacement is illusory, and self-replacement shows no consistent pattern with succession. In fact, dynamic stability, the speed with which a stationary community recovers from a temporary disturbance, is actually decreased by copious self-replacement of successional species. As a direct consequence, applicable to so-called human exploitation, patch-cutting should be more stable than either clear-cutting or high-grading.

Interspecific competition favors abundant self-replacement and a low death rate. Hence the dynamic stability of a complex community is lowered or even destroyed by natural selection on the scales of both ecological and evolutionary time.

A model of linear, unbranched succession allows simple expressions for the stationary distribution, its degree of disturbance, its diversity, and its dynamic stability, all independently defined in biologically realistic ways. The stationary distribution is a geometric series. Dynamic stability has no simple relation to succession, disturbance, or diversity. The ratio of disturbance to stability is identified as an intriguing metric. Diversity is higher for intermediate values of this ratio than for either high or low values.

Acknowledgments

I am indebted to Robert MacArthur for suggesting this project and for penetrating discussions in its early stages. Many of my theoretical insights, and most of my analytical results, were pilfered from Robert May. I had useful discussions with almost everyone at the Memorial Symposium, and I have made specific use of comments by John Endler, Thomas Givnish, Egbert Leigh, Richard Levins, and Robert Whittaker.

References

Anderson, M. C. 1966. Ecological groupings of plants. *Nature* (London) 212:54–56.

Bailey, N. T. J. 1964. *The Elements of Stochastic Processes with Applications to the Natural Sciences.* Wiley, New York.

Botkin, D. B., J. F. Janak, and J. R. Wallis. 1972. Some ecological consequences of a computer model of forest growth. *J. Ecol.* 60:849–872.

Dayton, P. K. 1971. Competition, disturbance, and community organization: the provision and subsequent utilization of space in a rocky intertidal community. *Ecol. Monogr.* 41:351–389.

Drury, W. H., and I. C. T. Nisbet. 1973. Succession. *J. Arnold Arboretum* 54:331–368.

Gifford, J. 1902. *Practical Forestry for Beginners in Forestry, Agricultural Students, Woodland Owners, and Others Desiring a General Knowledge of the Nature of the Art.* Appletons, New York.

Horn, H. S. 1971. *The Adaptive Geometry of Trees.* Princeton University Press, Princeton.

Horn, H. S. 1975. Forest succession. *Scientific American,* in press.

Horn, H. S., and R. H. MacArthur. 1972. Competition among fugitive species in a Harlequin environment. *Ecology* 53:749–752.

Kemeny, J. G., and J. L. Snell. 1960. *Finite Markov Chains.* Van Nostrand, New York.

Leak, W. B. 1970. Successional change in northern hardwoods predicted by birth and death simulation. *Ecology* 51:794–801.

MacArthur, R. H. 1958. A note on stationary age distributions in single-species populations and stationary species populations in a community. *Ecology* 39:146–147.

MacArthur, R. H. 1961. Community. *In* P. Gray, ed., *The Encyclopedia of the Biological Sciences,* pp. 262–264. Reinhold, New York.

May, R. M. 1973. *Stability and Complexity in Model Ecosystems.* Princeton University Press, Princeton.

Paine, R. T. 1966. Food web complexity and species diversity. *Amer. Natur.* 100:65–75.

Simpson, E. H. 1949. Measurement of diversity. *Nature* (London) 163:688.

Stephens, G. R., and P. E. Waggoner. 1970. The forests anticipated from 40 years of natural transitions in mixed hardwoods. *Bull. Conn. Agric. Exp. Stn.* (New Haven), No. 707:1–58.

Twight, P. A., and L. S. Minckler. 1972. Ecological forestry for the central hardwood forest. 12 pp. National Parks and Conservation Association, Washington.

Waggoner, P. E., and G. R. Stephens. 1970. Transition probabilities for a forest. *Nature* (London) 255:1160–1161.

III

Community Structure

10 Towards a Theory of Continental Species Diversities: Bird Distributions Over Mediterranean Habitat Gradients

Martin L. Cody

Introduction

In 1963 Robert MacArthur and E. O. Wilson published a landmark paper in biogeography in which they suggested that species diversity on islands is determined by a dynamic balance between species immigration and extinction rates. The following decade has produced great advances in our understanding of island patterns, including confirmation and elaboration of the predictions of their equilibrium model. Diversity patterns on continents are not so easily studied, and we do not understand them nearly as well. Whereas a large part of the theory of island species diversity is developed by simply using the total number of species recorded on islands of various sizes and distances from a source fauna, on mainlands such simple indices of diversity, area, and isolation are lacking. We are forced to consider different habitat types as equivalents of islands (cf. discussion of natural preserves as habitat islands: Chapter 18, Wilson and Willis), and must therefore measure habitat similarity in an objective and meaningful way. Different habitats are spread over continents in various complicated patchworks, and their configurations and distributions in a particular locale are such that "island" effects and the relative accessibility of each type

of habitat are at best obscure. The isolation of patches of each habitat type is no longer clear-cut, but depends on the extent and nature of surrounding or intervening habitat which must be variously inhospitable depending on the identity of the would-be colonist. Further, the extent and distribution of mainland habitat "islands" have been altered by climatic history and the effects of man and management; islands are relatively permanent and stationary.

In this chapter I draw attention to some distributional patterns of mainland species, show how they can be measured and depicted, and attempt to explain their evolution. In place of the species total that suffices to characterize an island, I use three diversity measures. A small patch of habitat within an extensive and uniform stand of the same type supports a number of species referred to as the α-diversity or species-packing level of that habitat. When one moves between different habitats in a systematic way, some species are lost and others gained at rates that are measured relative to the difference between the habitat types. These rates measure the β-diversity or the species-turnover rates among the habitats concerned. And lastly, similar habitats in different zoogeographic areas support different species for historical reasons. Such geographic species turnover is termed γ-diversity. I will provide simple measures for each aspect of species diversity, contrast the values obtained for these measures in similar habitat sets or gradients on three different continents, and discuss interactions among the diversity measures. In addition, some intercontinental contrasts and similarities will be illustrated at the level of ecologically-similar species sets, or guilds, that have to various degrees evolved in parallel on the three continents.

I chose to collect data on bird distributions over habitat gradients in three localities with Mediterranean climates: southern California, central Chile, and southwest Africa. These are three of the five areas in the world with the particular combination of cool, wet winters and hot, dry summers that makes for such delightful working conditions for both ecologists and non-ecologists. The three areas I studied are alike also in being located on the west- to southwest-facing coasts of major continental land masses, in having mountains close to the coasts, and in producing some fine wines. These similarities in climate, latitude, aspect, and topography have produced striking parallels in the range and types of habitats found in each area, from grasslands through a variety of scrub types of increasing heights and densities to woodlands of evergreen, broadleaf trees and to introduced or native pine forest. The data I collected in each geographic area consist of habitat measurements and bird censuses at 10–14 separate sites, each consisting of 4–10 acres in the most homogeneous, extensive,

and undisturbed vegetation stands I could find. The data presented were assembled for Chile in 1971, for California in 1972, and for Africa in 1973.

Observations on the bird communities of these three Mediterranean-like sites can be used to assess convergence or parallel evolution in the composition and organization of the communities, or in guilds (subsets of the communities). This is possible because sites on the three continents can be matched remarkably closely in the structure of the vegetation, and it is the vegetation structure that provides a major component of natural selection for bird community organization; Cody (1973) provides a review of this topic. In addition, the bird faunas of the three areas are but distantly related by taxonomy or recent ancestry, so that such similarities in species morphologies and community structure as are observed become all the more interesting and informative. Intercontinental convergences are discussed elsewhere in this volume in Chapters 11 (Karr and James) and 12 (Pianka), with respect to bird morphology and lizard community structure, respectively.

The Habitat Gradient

In the Mediterranean zones, type of habitat is strongly associated with exposure, slope, and elevation. Grasslands with a high proportion of annuals occur on coastal benches and inland valleys. A progression of increasingly tall and dense scrub follows with increasing elevation, or from seaward to inland facing slopes, or from equatorial-facing to polar-facing slopes. Patches of evergreen forest occur in riparian or flatter sites on the inland slopes, and the same trees occur in savannah grasslands in the interior valley floors. I tried to spread my attentions uniformly over this habitat range and selected a total of 31 study sites, which are listed in Appendix A.

If one begins looking for bird species in the grasslands, then visits sequentially more and more complex habitats, new species will be accumulated at the same time as earlier-encountered species are lost. Clearly the bird species are recognizing and selecting habitats on the basis of many of the same criteria that we use to classify these vegetation types. It appears that as simple an index as vegetation height suffices to rank habitats in a way such that, with few exceptions, bird species enter the cumulative list when a certain value of height is reached, occur predictably over a limited range of heights thereafter, and then drop out, not to be seen again on the habitat gradient. Bird species are also somewhat sensitive to the distribution of vegetation within the preferred height span of the species. A simple index of this distribution is obtained from the foliage profile of the habitat, which is a plot of vegetation density (in a horizontal plane) against height above the ground; then the vegetation "half-height" is defined as the height that divides the area under the foliage profile into two equal sections, and is used as the desired index. Thus, both vegetation height and vegetation half-height are derived from the foliage profile and can be used to rank

habitats on a gradient. A third variable, the total area under the foliage profile, which is proportional to the total leaf area per unit ground area in the habitat, appears to be less independent and less important, and ranking habitats by taking this area into consideration introduces rather than reduces disorderliness in the bird rankings. I therefore confined myself to the two variables mentioned in order to quantify position on the habitat gradient for each habitat studied. I used the principal component of the above two variables, which is the linear combination $(PC) = k_1(\text{height}) + k_2(\text{half-height})$ with the ks adjusted so that the variance of points, one from each of the 31 study sites, is maximized along this principal axis. In this way the habitat gradient is reduced to a single dimension.

Having selected simple habitat metrics that ensure as far as possible that species enter and exit the ranking in an orderly fashion, we are now faced with a different choice. We could take a simple logarithmic scaling of the height/half-height combination PC, so that the habitat gradient is defined as $H = \log(PC)$. This scaling appears here a posteriori, and elsewhere in general (Cody, 1974), to render bird species distributions as uniformly normal along H as habitat scaling can make them. Alternatively, we could adjust the scaling of PC so that the rates of gain and loss of species over habitats fit some criterion, such as linearity. I have chosen the former possibility, and will feel free to compare fitted gain and loss rate curves for the three continents and to draw biological conclusions from them.

The Birds Encountered

Birds were censused at the height of the breeding season in study areas that were mostly between 4 and 6 acres in size (23 sites; 6 were above and 2 below this range). I was concerned that the census area be large enough to include all species characteristic of the censused habitat type, and small enough so that its sample of the vegetation type would be homogeneous. I wished to ensure that no bird species turnovers or replacements occurred within the site between, for example, taller and shorter sections of the study area. This latter consideration is an important one, as I wanted species totals at a site to represent species packing levels (α-diversity) without contagion by species turnover (β-diversity). Thus when I found that two pipit species (*Anthus*) and two larks (*Mirafra* and *Calandrella*) appeared to occupy mutually exclusive parts of Gordon's Bay Flats in South Africa, a selection correlated with vegetation height, I split the 9-acre site into two sections, each more uniform in vegetation height than the original.

I revisited the study sites until I could be sure that all species had been listed and that my estimates of bird population densities were reasonable. This usually entails from five to twenty visits of several hours each, spread over two to four weeks, although I have a much more intimate acquaintance with thirteen of the sites, in which I conducted detailed analyses of bird community structure. The densities of some species could not be accurately measured with my techniques, for reasons associated with their rarity, social behavior, or very large territory sizes.

The census results are given in Appendices B, C, and D. All of the bird species encountered are listed there except for the following ecological groups: a) aerial feeders, which tend to be specific to nest site rather than to feeding habitat; b) raptors; c) nocturnal and crepuscular species; and d) nest parasites. The species are ranked, obviously, in a way that reflects the order in which they would most likely be encountered if habitats were sampled sequentially from low to high values of H, and thus the ranking reflects the preferred habitats of the birds. In addition and where satisfaction of the first condition permits, species which drop out first from the gradient are listed before species that drop out later. One can liken each numerical entry in these appendices, a species density at a site in pairs per acre, to a weight which is precisely fixed in a plane with respect to the two axes, habitat gradient and species rank. Thus, each row in an appendix has a "center of gravity" which reflects the center of distribution for that species on the habitat gradient, and each column has a "center of gravity" which reflects the rank of the species around which the census is centered. I attempted to position species so that the variances around these two centers of species distributions over habitats and census distributions over ranked species are minimized. Thus, the final ranking is a compromise between the two constraints of species entries into and exits from the charts, and the aim of keeping the highest weights (densities) closest to the center of the charts. To illustrate these compromises, note that *Anthus* in Appendix B ranks second because, although three species that are listed later first occur on the gradient slightly earlier, all three exit from the gradient much later than *Anthus*. Also, the species ranked 16 and 17, *Turdus* and *Elaenia*, reach high densities on the gradient earlier than species 15, *Zenaidura;* but *Zenaidura* was listed first because it does not reach the last census site on the gradient, whereas the other two are common there. Perhaps the neatest of all possible data would produce normal-looking distributions in both the horizontal and the vertical directions in the three charts, but such hopes are only approximately realized here.

Species Gain and Loss Rates: Diversity Measures

The data on the number of species gained and the number lost over the three parallel habitat gradients are summarized in Figure 1. By the time a certain point on the habitat gradient is reached, a certain number of species have been recorded. This is considered the number of species gained up to a point H, a number that increases smoothly and unambiguously with increasing H. But species are lost with increasing H somewhat irregularly, because species distributions over H terminate in only approximately the same order in which they are started, or entered. Thus the number of species lost by some point H on the habitat gradient is not simply read from Appendices B–D, but must be conventionalized. It is calculated

in the following manner. Each species in a census has a rank number i and a density d_i, and has "moments" around the H axis of $(i) \cdot (d_i)$, just as a weight on a lever has moments around the fulcrum equal to its product with the distance it is placed from that fulcrum. The total census at a point H has "moments" $\Sigma_i \ (i) \cdot (d_i)$ summed over all species i in the census. As a convenient measure of species number lost at H, I took the rank of the species reached (when reading down from early to later rankings) by the time 10% of the total census moments $\Sigma_i \ (i) \cdot (d_i)$ had been accumulated. Thus, at Chilean site 2, where $H = -0.071$, the census moments are $(2 \times 0.45 + 4 \times 0.22 + 6 \times 1.16) = 8.74$; 10% of this figure is 0.87, and this much has been accumulated at the species ranked 2 $(2 \times 0.45 = 0.9)$. Thus, the number of species S lost at the H value of -0.071 is $1 + (2 - 1)0.87/0.90 = 1.97$.

As mentioned above, species gain and loss functions can now be fitted to the points of Figure 1. Three sorts of functions were fitted: linear relations where S (number of species gained or lost) $= A + BH$ and where gain and loss rates are assumed constant over H; exponential relations $S = Ae^{BH}$, where gain or loss rates increase with increasing H; and logistic curves $S = A/(1 + Be^{-CH})$, which are sigmoidal or S-shaped curves that approach upper and lower S values asymptotically and have the highest gain or loss rates at intermediate H. Table 1 gives the results of fitting each of these three types of curves to the species gain and loss rates of Figure 1, where the parameters A, B, and C are chosen to give "least-squares"

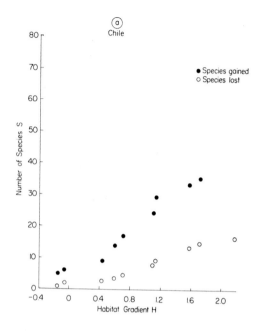

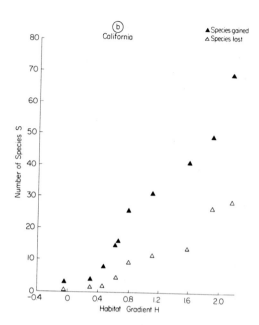

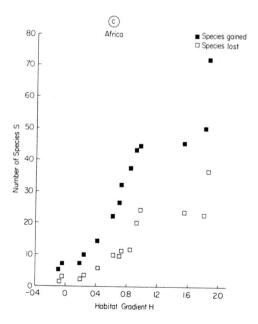

Figure 1 Species distributions are shown over structurally similar habitat gradients in the Mediterranean zones of three continents: a) in Chile; b) in California; c) in Africa. Position on the habitat gradient H (abscissa) is measured by $\log_{10}$ of the principal component of the height and the "half-height" of the vegetation, against which is plotted the numbers of species S (accumulated totals) gained (solid symbols) and the numbers lost (open symbols) up to certain positions on the habitat gradient.

Table 1. Parameter values and goodness-of-fit estimates for three types of functions that describe species gain and loss rates, $g(H)$ and $l(H)$, over Mediterranean habitat gradients on three continents

Locality		Model												
		Linear: $S = A + BH$				Exponential: $S = Ae^{BH}$				Logistic: $S = A/(1 + Be^{-CH})$				
		A	B	RSS	F	A	B	RSS	F	A	B	C	RSS	F
Chile	$g(H)$	5.70	17.23	42.75	168.37	6.65	1.09	0.21	136.77	40.2	6.8	2.25	19.94	361.01
	$l(H)$	0.73	7.32	15.72	82.51	1.58	1.38	0.29	161.32	21.5	16.2	2.07	1.75	744.18
California	$g(H)$	0.41	25.75	78.87	183.84	4.49	1.47	1.20	39.64	47.8	16.7	3.27	51.33	282.47
	$l(H)$	−2.23	12.53	64.02	53.60	0.01	1.97	1.70	50.05	20.0	25.0	3.19	82.10	41.79
Africa	$g(H)$	7.79	27.12	427.52	73.52	9.19	1.29	1.62	44.17	49.4	12.6	4.24	86.17	369.68
	$l(H)$	2.33	13.47	157.95	49.09	2.85	1.51	2.32	42.22	24.5	24.0	4.47	85.02	424.22

In each case the fitted parameters A, B, and C are chosen by least-squares analysis. In the linear and exponential models A is a position index and B is the slope or rate constant of the fitted curve. In the logistic model, A is the value of the upper asymptote to S which is approached at high H values, B is a position index that shifts the curve to the right or left, and C is the rate constant that measures the rate at which the asymptote is gained. RSS gives the residual sums of squares, and F measures the goodness of fit, high when F is high.

deviations of observed from predicted S values. A measure of the goodness of fit is the F-statistic, the ratio of explained sum-of-squares to the residual or error sum-of-squares, each corrected for the number of degrees of freedom. The table shows that all data sets except the California loss rate[1] are best fitted (highest F-value) by logistic curves. Because of decreasing gain and loss rates in taller habitats, linear curves fit the data rather better than exponential curves. It is worth pointing out, however, that gain and loss points are more orderly over the early part of the habitat gradient, and these orderly points are well fitted by exponential functions of habitat H. The first 11 points of the African set, for example, given an

[1] In fact, a logistic curve with $F = 73$ can be drawn through the California loss points, but this curve has a very high asymptote and treats all points as lying on the early part of the curve where $d^2S/dH^2 > 0$, and is biologically meaningless.

F-value of 330 in the case of an exponential model.

Tall forest, the right-hand extreme of the habitat gradient, is not found in Mediterranean climate zones, but generally occurs in the same geographic area. This habitat extreme is *Nothofagus* forest in south-central Chile and *Sequoia* forest in central California, found by a polewards jump of several hundred miles. In South Africa such a move is not possible, but evergreen forest reaches down the east coast from the extensive subtropical forests of Mozambique, and patches can be found several hundred miles east of the Mediterranean Cape region. These censuses constitute the last data set in Appendices A–D and Figure 1, but are not included in the regression analyses. The best-fit logistic curves are shown in Figure 2, which enables easy comparison between the three continents. Notice that the inflexion points occur earliest and that the asymptotes are highest in the African data, and that asymptotes are lowest and are reached later at the Chilean sites; California is intermediate on both counts.

From Figure 2 we can derive convenient indices of α-diversity and β-diversity, which will characterize broad sections of mainland rather than specific habitat types and local habitat assemblages. If we write the species gain curve as $S = g(H)$ and the species loss curve as $S = l(H)$, the number of species at some point H is given simply by $g(H) - l(H)$. Thus the difference $g(H) - l(H)$ gives the α-diversity or species-packing level as a function of H. α-diversity curves are given in Figure 3, from which it appears that Chilean habitats are rather more closely packed at the grassland extreme, African packing levels exceed those at other sites over a broad range of intermediate habitats, but Californian woodlands and forests support more species than their Chilean or African counterparts.

The curve midway between those of species gain and loss rates describes how successive censuses shift over species ranks; it might be called the "species accumulation curve" (Figure 4a), and its derivative, $S = d/dH[g(H) + l(H)]/2$, gives species turnover, the rate at which earlier species are being replaced in censuses by later ones at each point on the habitat gradient. This turnover rate defines β-diversity, the diversity component attributable to habitat change between census sites. It peaks high and early in African habitats (at $S = 38$, $H = 0.65$), and the peak comes successively later and lower in California ($S = 27$, $H = 0.91$)

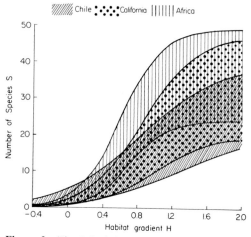

Figure 2 Fitted logistic curves for accumulated numbers of species gained or lost (ordinate) as a function of position on the habitat gradient H (abscissa). As described in Table 1, the curves were fitted through the observed values of Figure 1 representing the accumulated species gained or lost on each continent as a function of H. For each continent the upper curve is the fitted gain curve, the lower is the fitted loss curve, and the difference between the curves at any H value is the number of species present at that H value or position on the gradient (replotted in Figure 3).

and Chile ($S = 16$, $H = 1.02$). Figure 4b shows these β-diversity curves for the three continents.

Analysis of the Bird Distributions

We can note first that the shaded areas between gain and loss curves in Figure 2, or the numbers of species these areas represent in Figure 3, are not dissimilar in the three sites. In spite of the rather striking differences in β-diversity, as shown in Figure 4b, α-diversity changes with H in a similar manner and extent on the three

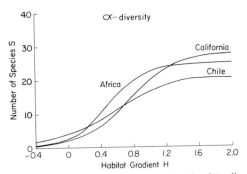

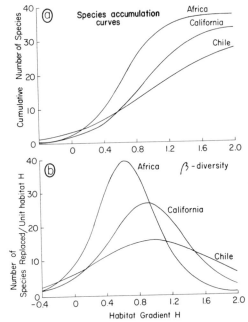

Figure 3 α-diversity or species-packing level (ordinate) as a function of position on the habitat gradient *H* (abscissa). The curve for each continent is the difference between the gain and loss curves of Figure 2. Note that each continent is richest in species over a different portion of the gradient (e.g., Africa is richest around $H = 0.3-1.1$).

Figure 4 The curves of Figure 4a (above) are those lying halfway between the species gain and loss curves of Figure 2 for each continent, and can be designated the "species accumulation curves." The differentials of these curves are plotted below (Figure 4b); these give the rate of species turnover with respect to *H*, or β-diversity as a function of *H*.

continents. Consider the fitted parameters of the logistic model in Table 1. The value of the upper asymptote is given by *A*; *B* sets the position of the curve in a right-left alignment relative to the *S*-axis; and *C* measures the rate at which the asymptote *A* is approached. On all three continents, at any given site the rate constant for gain of species is very similar to the rate constant for loss of species, although these constants vary widely, from 2 to 4.5, between sites. Thus, wherever gain rates of species over habitats are high, loss rates are similarly high, and likewise low gain

and loss rates are found together (Figure 2). This feature of course tends to equalize α-diversity values over each of the three habitat gradients. Furthermore, while California has the highest *A* (asymptote) value for α-diversity, it has the highest *B* value (Table 1), which shifts the α-diversity curve to the right (Figure 3) and reduces the area under the curve over the range of *H* illustrated. This area is roughly equivalent to the total or accumulated species-packing over the gradient. Chile, on the other hand, has the lowest upper asymptote *A*, and has the lowest B values; this lower B tends to maximize the area under the curve for a given A value. So we can see that the total area under the α-diversity curves tends to remain constant (Figure 3) despite very different rate constants C, asymptote heights A, and curve positions B.

The similarities in α-diversity at any point on the gradient and in sum over the whole gradient are striking (Figure 3), and are presumably produced by natural selection via the competitive interactions that determine the limiting similarity of coexisting species and the stability of the resulting community. These processes evidently occur in parallel over the three habitat gradients and produce similar results in a given habitat type on any of the three continents; we can conclude that parallel gradients of resource type and span are exploited by bird species that assort into communities of parallel size that vary over the gradients in comparable ways.

Nevertheless, there are some differences between continents in the packing levels of species at any gradient point. African scrub habitats, for example, in the H range of 0.4 to 1.0, support from one to five more bird species than do the equivalent habitats in California and Chile. We expect that such differences will alter species densities and distributions in predictable ways, and these expectations are examined next.

Density Compensation for Low α-Diversity

As we move from left to right with increasing H over the habitat gradient, the vegetation becomes taller and denser and presumably adds new resources, which are exploited by additional species. This results in increasing numbers of species with increasing H, as shown by the fitted curves of Figure 3 and by the crude data on species numbers given in Figure 5a. The rate of accumulation of extra species with increasing H is given by the rate constant of the best-fit logistic curve for the pooled sites, and is 4.9. With the addition of extra species come additional individuals, so that the plot of total bird individuals against H mirrors the plot of total species number against H, as given in Figure 5b. The logistic equation that describes the relation is: total density $D = 6.2/(1 + 19.3\ e^{-5.4H})$. If for any reason such as chance, history, or the configuration of surrounding habitats, the species total at any particular site is relatively impoverished, the existing species there can often use at least a part of those resources which would have gone to the missing species. The densities of such opportunistic species would thereby increase because they have access to additional resources (assuming that density is governed by resource availability). This process is called density compensation (MacArthur, Diamond, and Karr, 1972). The result of density compensation is that two sites with different numbers of species may be more similar in total bird population densities (of all species combined) than in number of species.

Whereas total bird density increases rather more rapidly than total species number over the earlier and middle section of the habitat gradient (rate constant 5.4 versus 4.9), densities and species counts are quite directly proportional.

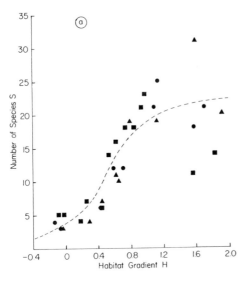

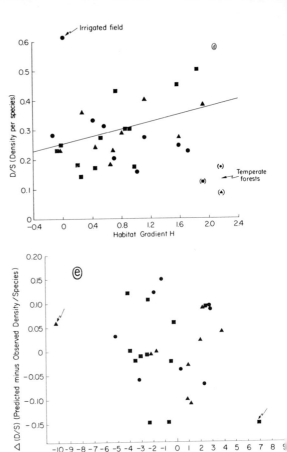

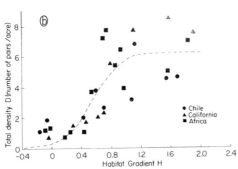

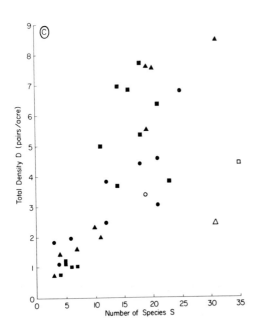

Figure 5 Figure 5a gives the observed values of α-diversity or numbers of species S as a function of H for each continent (see Figure 1 and Figure 5b for the meaning of the symbols, which is consistent throughout the chapter). Figure 5b gives total bird density D (pairs per acre of all species combined) as a function of H. Figures 5a and 5b are approximately parallel, so that the relation between D and S is approximately (Figure 5c) but not perfectly (Figure 5d) linear. The nonlinearity is best seen in Figure 5d, which shows that D/S increases slightly with H. Note that D/S deviates strikingly upward in an irrigated Chilean field, and somewhat downward in the tall forests lying outside the Mediterranean zone on each continent. In Figure 5e, which plots the residuals from Figure 5d (ordinate) against the residuals from Figure 5a (abscissa), a negative correlation is expected if density compensation is occurring, as discussed more fully in the text. This correlation is apparent in the extreme values: the high point marked by an arrow at the extreme left → ▲ represents supersaturated California woodlands (see text for discussion of supersaturation); the low points marked by arrows at the extreme right → ■ represent impoverished African woodlands. Note further in Figure 5c the three open symbols, which represent relatively impoverished forests outside the Mediterranean zone.

This relation is given in Figure 5c, and is expressed as $D = 0.31\ S$. Thus, on the average every new species adds 0.31 pairs per acre to the total density of breeding birds. The density of bird pairs that can be supported per acre must be an intrinsic property of the habitat gradient and its climate. The gradient produces a certain variety of bird food types at certain abundances at each point on H. Resource types increase with H as seen by increasing S (Figure 5a); total resource abundance increases with H as expressed by increasing D (Figure 5b); productivity per resource remains roughly constant, as shown by the approximately linear relation between D and S (Figure 5c).

But D and S do not increase at quite the same rate over H, as is shown by Figure 5d. Although D/S remains approximately constant, some of its variation is correlated with variation in H, so that each species is on average slightly more common at high H than at low H (D/S increases slightly with H). From Figure 5d I conclude that although overall productivity per resource does vary somewhat within these Mediterranean habitats, these variations in productivity over H within the Mediterranean zone are judged to be small, compared with the differences between the habitats within the zone and those produced by a different climatic regime. Two types of exceptional values, both marked on the figure, help to prove the rule. First, the point in the upper left-hand corner represents an exceptionally high D/S value at low H. It was measured in a Chilean field in the Mediterranean zone, but the field was irrigated! In species

number and in the appropriateness of the species present, this field appeared to be normal, but the additional moisture availability has presumably resulted in exceptionally high food levels for the three resident birds there. Second, three low points are seen bracketed in the lower right corner. All three represent censuses from the forests mentioned earlier that are outside the Mediterranean zone, one each from California, Chile, and Africa. The three clearly represent resource production regimes that differ from those of the remaining 31 study areas. When these four points are omitted, the regression equation is $D/S = 0.232 + (0.062 \pm 0.058)H$, where the figures in parentheses are the slope and its 95% confidence limits.

We are now in a position to discuss what could happen if a species that is normally found at some point H is locally absent or is removed from the community. There are clearly two extremes. On the one hand, the subtraction of that species could result in a reduction of the total species density there by an average factor of $(n - 1)/n$ in a community with a predicted α-density of n, or a reduction by a factor of $(n - 2)/n$ if two species are absent, etc. Thus, each species removed should subtract an average of 0.31 pairs per acre from the total bird density. This would mean that the remaining species do not utilize any of the resources vacated by the absentee, and therefore do not *compensate* for its absence. On the other hand, the other extreme is represented by species that utilize the vacated resources just as efficiently as the absentee, and increase their own densities in proportion to their share of the extra available resources. In this case the total bird density remains constant, and we would have perfect *density compensation*. Normally we would expect something intermediate, where the remaining species in the community are able to use just some of the extra resources, or use all of the extra resources but less efficiently than the species to which they would normally go; this would result in a partial density compensation. Both extremes occur in nature, depending on the relative abilities of the missing species and their replacements to harvest the available resources. For instance, there is complete density compensation in the species-poor avifauna of Bermuda compared with the North American mainland (Crowell, 1962), and of Puercos Island compared with the Panama mainland (MacArthur et al., 1972), but little or no compensation if one compares New Guinea with various satellite islands (Diamond, 1970), or different islands of the Pearl Archipelago with each other (MacArthur et al., 1973).

As shown in Figure 5a, some communities are rather impoverished and others rather rich relative to an average level of species packing. Has this uneven packing resulted in any density compensation, then? I tested this by plotting the residuals $\Delta(D/S)$, deviations of observations predicted from D/S in Figure 5d, against ΔS, the residuals from the plot in Figure 5a, which are the deviations of predicted from observed species number at each H. If some community at a particular H value has fewer species (relative to the average packing level at that H amongst the three

localities), we obtain a positive ΔS. And if this reduced species number has resulted in density compensation by the species that are present, the density per species will be unusually high and will result in a negative $\Delta(D/S)$. But if there is no density compensation, $\Delta(D/S)$ will bear no relation to deviations ΔS in species number, as density is added directly in proportion to species additions and is subtracted likewise as species are removed, leaving $\Delta(D/S)$ unchanged. Figure 5e gives the results of this test. It shows that for small deviations ΔS there appears to be little or no detectable trend in density compensation, as the central cluster of points shows at the very best a slight negative correlation; the scatter could easily be due to stochastic population fluctuations combined with errors of sampling. But it is informative to look at the points that represent much larger deviations in S from average levels. Take, for example, the upper left point indicated by an arrow on the figure. This represents a California oak woodland site, at $H = 1.587$. Thus the predicted D/S from Figure 5d is 0.330, and with no density compensation the 31 species present would yield a total density given by $D/31 = 0.330$, or $D = 10.23$ pairs per acre. But the predicted number of species at $H = 1.587$ from Figure 5a is 20.76, and so $\Delta S = 10.24$ species extra. Thus, if there were perfect density compensation, each of the 31 species would on the average be rarer than the species of a normal 31-species community, and would achieve a total density appropriate to these 20.76 species and the predicted value of D/S. That is, with perfect density compensation, $D/20.76 = 0.330$, or $D = 6.85$ pairs per acre. What in fact happens in the oak woodland is that the observed density is 8.45 pairs per acre. This means that the 31 species achieve a partial density compensation, where each is rarer than it would be in a normal, supportable 31-species community, but the total density of birds is larger than it would be in a 21-species community.

Two African woodland points at the lower right in Figure 5e represent the opposite situation, for each is impoverished in species, owing most likely to island effects in the isolated woodland patches of the Cape region. One of the sites is at $H = 1.560$, predicting D/S to be 0.328. With only 11 species actually present this would amount to a total density D given by $D/11 = 0.328$, or $D = 3.61$ pairs per acre. But since the predicted species number for this habitat type is $S = 20.74$, it is about 10 species short of saturation. With 20.74 species, an overall density of $D = 6.80$ pairs per acre would be expected $(6.80/20.74 = 0.328)$. But the eleven species yield a total density of 4.98 pairs per acre, a 43% level of density compensation compared with the 52% in the California woodland. At a second African woodland 14 species were censused, and at the measured H of 1.834, 20.84 were expected; D/S is predicted to be $0.232 + 0.0616 \times 1.834$, or 0.345. Thus, 14 species with no density compensation would produce 14×0.345 or 4.82 pairs per acre, and with perfect density compensation would yield 20.84×0.345 or 7.19 pairs per acre. The observed density

of 6.93 pairs per acre amounts to a near-perfect density compensation of 89%. Thus, D/S is reduced at the species-rich California sites, and increased at the species-poor African sites, from the value appropriate to the actual number of species towards the value appropriate to the number of species expected for the H value of the site.

I conclude that density compensation takes place, and is readily identifiable in exceptionally poor or exceptionally rich habitats. The level of this density compensation is such that deviations of actual $\Delta S/S$ from predicted values by around 50% produce about 50% density compensation, whereas smaller deviations in species number produce much more perfect density compensation. Thus $\Delta S/S$ is inversely proportional to the degree of density compensation, and we can guess at a constant of proportionality of around 25% (or 0.25, using fractions rather than percentages).

Niche-Breadth Compensation
for Low α-Diversity

There is a second way by which low α-diversity in a community may be compensated, and this constitutes a clear alternative to density compensation. This is achieved through extension of the habitat niche breadth B_H of species in adjacent habitats, so that their ranges on H expand into the habitat that is species-poor. Low α-diversity at a given site will permit this expansion, and high α-diversity will restrict it. This is a form of niche-breadth expansion, as was the density compensation just discussed, but now it is additional habitat types rather than additional food resources that are being exploited. The reader will have realized already that it is this form of compensation for missing species that accounts for the relatively constant α-diversities over H, despite rather striking differences in β-diversities.

To test for habitat compensation, we need to measure habitat niche-breadth B_H for each species in Appendices B–D, and plot this against the relative packing level in the habitats over which the species is distributed. I measure B_H in a simple, convenient way by using probability paper. This paper is designed for easy plotting of frequency distributions, such as ours where we plot $\%_i D_H$, the percentage of the total density of a particular species i that has been accumulated by a certain point H, against habitat gradient H. One axis of the probability paper will be H, the other percent. This second axis is scaled so that the cumulative frequency, or percentage, under a normal, gaussian curve results in a straight line. By convention, I take zero density for a species as $0.1\%_i D_H$, and obtain a series of more or less straight lines that represent the frequency distributions of species on habitats. Some of these are illustrated in Figure 7, for those who wish to look ahead. Now, the slope of the line is an index of variance in the frequency distribution it represents. Thus B_H can be measured as the range of H covered by the line between cumulative percentages of 10% and 90%; the value of H at the 50% mark gives the center or position of the habitat niche of the species.

Niche breadths are generally largest in Chile (mean $B_H = 0.63$), narrowest in Africa (mean $B_H = 0.38$), and intermediate in California (mean $B_H = 0.47$). But within each continent, species-packing levels vary on different parts of the gradient in different ways (see Figure 3). Thus I divided the H range into three sections: those that correspond to larger α-diversities in Chile (low $H < 0.2$), Africa (intermediate H), and California (high $H > 1.2$). I then took the smallest area under the α-diversity curves of Figure 3 over each of the three sections of H as unity, and obtained the values of the larger areas relative to the smallest. These areas reflect the relative packing-levels of species over each of the three sections of H, and are plotted against the average B_H of those species whose centers of distribution lie within those sections of H in Figure 6. As we expected, the figure shows that bird ranges over habitats are restricted where species are densely packed, but are allowed to expand over H in relatively impoverished habitats.

It is interesting though anecdotal to note that although conifer plantations of *Pinus radiata* have been introduced into Mediterranean South Africa, and flourish there in extensive stands, they are virtually an ornithological desert, being practically ignored by the local birds. Introduction of the same conifers into Chile has resulted in their colonization by large numbers of Chilean bird species, which are as unfamiliar with the plant and its form and structure as are African bird species. In fact, the Chilean pines now support the same number of bird species

as are supported in Californian habitats
where the pines are native. Although the
density of bird individuals in pines is
lower in Chile than in California, the
difference is no greater than the Chile-
California difference in bird densities in
native habitats where Chile is relatively
species-poor. Part of the explanation for
this different response by African and
Chilean birds to Californian pines may be
in the different response of the shrubby
plants to pines, because several plant species in Chile invade pine plantations and
can form quite a thick understory,
whereas in Africa the pine habitats have
no such understory. Furthermore, Chilean
birds breed and feed in the introduced
groves of tall *Eucalyptus* that are now
abundant in that country, whereas similar
stands of these trees in California are by
comparison very much underutilized.
There is an obvious correspondence between the degree to which exotic plant
forms are used and the species turnover
rate amongst the local, native vegetation
types; apparently the birds of high
β-diversity faunas, notably Africa, have
become stereotyped to a narrow habitat
range, but the birds of low β-diversity
faunas, notably Chile, maintain a behavioral plasticity that enables them to utilize
freely the novel habitat forms provided by
plant introductions.

Convergence and Contrast between Species and Guilds on Three Continents

In this section I shall make comparisons
among some ecologically isolated species,
and among sets of ecologically-related

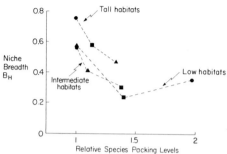

Figure 6 Compensation for low species-packing
levels can take the form of habitat expansion by
species in neighboring habitats, resulting in a negative association between habitat niche-breadth and
species-packing levels at each point on the gradient.
Each continent is averaged for the calculation of
mean niche-breadth for each of the three sections
of the habitat gradient over which a different continent is richer than the others (see Figure 3). Niche
breadth for each species is calculated as the number
of log cycles of habitat covered by the distribution
of the species so that from 10% to 90% (total 80%)
of the individuals of the species are accounted for
(cf. Figure 7). Symbols for each continent as in
previous figures, e.g., Figure 5b. Note that in each
section of the habitat gradient, the continent with
the poorest fauna has the broadest niches, and the
continent with the richest fauna has on average the
narrowest niches.

species ("guilds") that provide counterparts to each other on the three continents. Sometimes these counterparts are
one-for-one at the species level, but more
often the guilds match one or two species
on one continent to three or four or even
six species on another continent. In the
comparisons that match single species,
and in the small guilds of two or three
species, the one-to-one correspondence in
morphology and ecology between species
counterparts can be extremely exact and
dramatic, but these exact matches become
obscure among guilds of unequal number
of species and among larger guilds even

though they are of similar species numbers. With equivalent guilds of unequal species number, the species-rich guilds may show a higher α-diversity over H, a higher β-diversity, or both. Furthermore, the richer guild as a whole may occupy a greater range on H, or perhaps use more foraging methods or food types without violating the guild notion or transgressing its boundaries. Where appropriate, I shall indicate to what extent coexistence mechanisms other than habitat separation characterize the species within a guild, and I shall provide general comments on the importance of these alternatives in relation to foraging strategy. In these comparisons, I select merely those examples that are particularly striking, either in the perfections or in the inequities of species matches. A more inclusive community view, and a far more thorough morphological treatment of convergent evolution and its ecological correlates, is provided by Karr and James in Chapter 11. These intercontinental comparisons of ecologically equivalent guilds formed from taxonomically different species pools are analogous to interisland comparisons of guild subsets drawn from the same species pool. In the latter situation, too, species are generally matched on a more complex basis than one-for-one (see Diamond, Chapter 14, especially the discussion there of "permissible combinations").

There are only a few striking examples of convergent evolution at the single species level, but one of these is the classical case of the "meadowlarks," *Sturnella, Pezites,* and *Macronyx* (see Cody, 1974). The first two genera are in the family Icteridae, North and South American respectively, and the third genus includes the African pipits, family Motacillidae. The list of previously recognized convergent characteristics is impressive, and includes size (body lengths 236, 264, and 200 mm), shape (all are "chunky," stout-bodied and short-tailed), bill proportions (lengths 32.1, 33.3, and 17.8 mm, and depth/length ratio 0.36, 0.40, and 0.38) as well as feeding and breeding behavior patterns and incredibly similar plumage and coloration. We can now add to this list the fact that the three Mediterranean-zone representatives of the genera use habitats in a very similar way (Figure 7a). The species enter the gradient H at the same point, increase in density at the same rate, reach median abundance in the same habitat type, and drop out of the gradient at the same point on H, about as close a correspondence as could be hoped for. Note that the African *M. capensis* is not the *Macronyx* species (*croceus*) that shows such striking similarity in coloration to both *Sturnella* species *magna* and *neglecta,* yellow and black, nor the one (*ameliae*) that is much more like both *Pezites* species *militaris* and *defilippi,* red and black. But in the more fundamental ecological characteristic of habitat use, the three species of Figure 7a prove to be virtually identical.

The second single-species contrast, if we again use a reasonably well-defined and discrete niche, is that of a small, long-tailed insectivore foraging high on the tips of twigs on the outside of bushes. Here again each continent provides a single species, and each belongs to a different family: the Chilean *Leptasthenura* is an ovenbird (Furnariidae), the Californian *Psaltriparus* is a long-tailed titmouse (Aegithalidae), and the African *Prinia* is an old-world warbler (Sylviidae). The three are small-sized (wing lengths 57, 51, and 51 mm, respectively) and all have long tails, excessively so in *Prinia* (70 mm) and the spinetail *Leptasthenura* (91 mm). Although the habitat distributions are similar (Figure 7b), they are not nearly so coincident as in the meadowlarks. Both niche breadths and niche positions are somewhat different. *Prinia* is commoner in lower habitats than its counterparts, whereas *Psaltriparus* reaches highest densities in taller habitats. Moreover, *Psaltriparus,* which is in the richest insectivore fauna, has the broadest niche, and *Leptasthenura,* which is in the most impoverished insectivore fauna, has the narrowest niche,—just the opposite of simple expectations.

On all three continents there are "garden sparrows," so called because they are abundant in suburban gardens as well as in natural vegetation; these birds are present at both ends of the habitat gradient, but are absent from the dense scrub at the center (Figure 7c). California has two species, *Melospiza melodia* (Emberizidae) and *Carpodacus mexicanus* (Carduelidae), and both occur at both ends of the habitat gradient (*Carpodacus* favors drier and more open sites than *Melospiza*). Chile has just one species, the emberizine finch *Zonotrichia capensis,* with the same bimodal distribution over *H* as each California species. Africa is like California in supporting two species (both are weaver-

birds, Ploceidae), but because one species is restricted to one end of the habitat gradient and the second to the other end, Africa is like Chile in supporting just one species at any single point on *H*.

The "wren" guilds are pictured next, in Figures 7d–f. On the three continents these belong to two or three unrelated families, but all species are uniformly small (wing 40–60 mm), dull-colored with cocked tails, and forage low in thick cover. In Chile a single species occurs, the house wren *Troglodytes aedon* (family Troglodytidae), and occupies a broad range of hab-

Figure 7 Comparison of species distributions among continents in single-species to multi-species guilds of close competitors. The vertical axes represent the habitat gradient, the principal component of vegetation height and "half-height," plotted on a logarithmic scale. The horizontal axis is cumulative frequency, in this case the cumulative frequency of individuals of a species distributed over the habitat gradient. This sort of graph is known as a "probability plot," and frequency distributions that are log-normal will result in straight lines when plotted on this log-frequency × probability paper. Note that many species distributions are approximately linear in these graphs, meaning that the species are normally distributed over *H*, or log-normally distributed over the principal component of habitat structure (*PC*). To illustrate this with the bushtit *Psaltriparus minimus* in Figure 7b, the graph shows that 10% of all the bushtits censused had been counted by the point on the habitat gradient where PC = 6.6, 30% by PC = 11.6, 50% by PC = 17.2, 70% by PC = 26.5, and 90% by the point at which PC = 42.2. Each plot in Figures 7a–7c represents a simple guild, with intercontinent counterparts on the one figure; Figures 7d–7r show more complex guilds where the representatives on each continent are shown on separate figures. Examples are thus ordered from single to multispecies guilds, from close one-to-one correspondence to guilds with quite different numbers of representatives among continents, and from guilds that rely on habitat separation to those in which size and foraging-site differences become increasingly important.

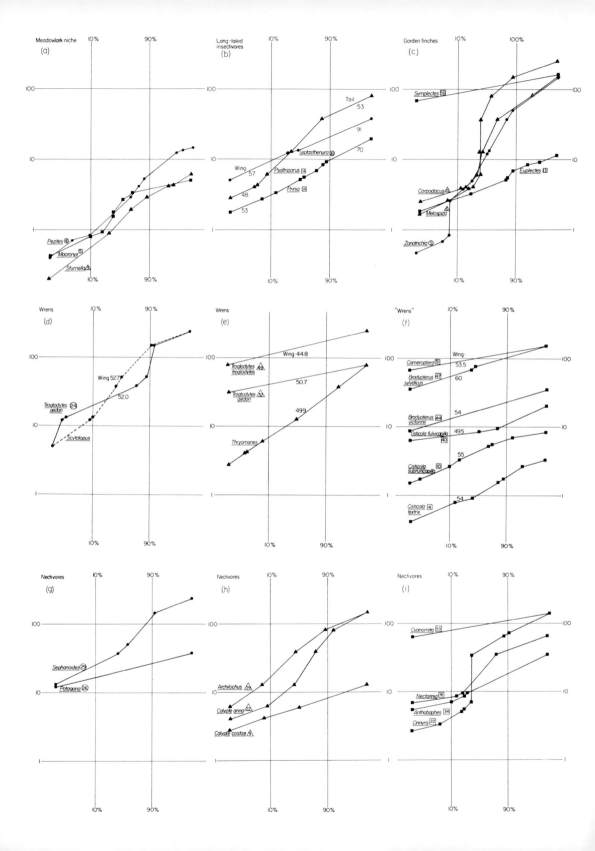

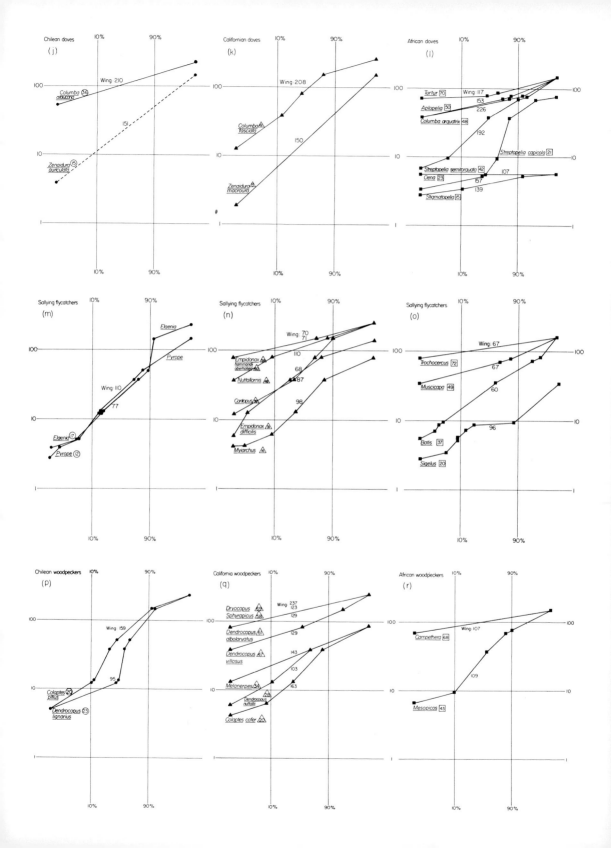

itats (Figure 7d). The only other species in the Chilean Mediterranean zone that is at all close in ecology to the wren is the smallest member of the family Rhinocryptidae, *Scytalopus magellanicus,* which feeds largely on the ground and is thus only marginally a member of the wren guild. California has three Mediterranean zone wrens (all Troglodytidae), all similarly-sized and all situated at different positions on the habitat gradient with but marginal overlap between them. South Africa has no true wrens, but several genera of warblers Sylviidae, in particular *Cisticola, Bradypterus,* and *Cameroptera,* are their substitutes. These share the typical wren characteristics, including uniform small size, similar proportions, and discrete habitat selection. Here too the six species are arranged so that they replace each other with bewildering rapidity as habitat changes.

Where specific habitats of different wren species merge, the species are usually interspecifically territorial (Cody, 1974), and thus the rule of one wren species per habitat patch is maintained. Some *Cisticola* warblers likewise show indications of interspecific territoriality. In the tall, rank vegetation around marshes and irrigation ditches in southwestern Africa, for example, one may find both *C. tinniens* and *C. juncidis,* whose similar songs elicit responses from conspecific and allospecific territorial neighbors alike.

Like wrens, the species of the nectivore guild (Figures 7g–i) sort out by habitat and are even more renowned for their interspecific aggression and spacing. The Chile and California representatives are

hummingbirds (Trochilidae), and the African representatives are sunbirds (Nectariniidae). The moderate convergence in morphology and in iridescent plumage between hummingbirds and sunbirds has often been remarked on (e.g., Karr and James, Chapter 11, Figure 1). The two Chilean species (Figure 7g) breed in largely different habitats. The four African species (Figure 7i) overlap broadly in their positions on the habitat gradient, but strong interspecific interactions appear to keep their territories spatially separate. In particular, in African chaparral *Nectarinia famosa* and *Anthobaphes violacea* defend territories that exclude each other quite precisely. Of the three Californian species (Figure 7h), *Calypte costae* and *Archilodrus alexandri* separate by habitat, whereas *Calypte anna* overlaps in habitat with both of these species but breeds earlier in the season.

In contrast to the preceding examples, the remaining guilds are those in which species coexist chiefly by dint of size differences (and therefore, *fide* in particular Chapter 7 and earlier papers of Hespenheide, by differences in their prey size distributions; cf. also Figures 30 and 31 of Chapter 14 by Diamond, and Figures 6 and 7 of Chapter 13 by Brown). Perhaps the simplest example is that of the doves (all species censused are in the family Columbidae), a homogeneous group of vegetarians with wide size range, which also provide the examples of size segregation discussed in Diamond's chapter. California and Chile are similar, with two species each: a small species appearing early in the habitat gradient (*Zenaidura auriculata,* wing 151 mm, in Chile; *Zenai-dura macroura,* wing 150 mm, in California) and a large species appearing only when woodland and forest are reached (*Columba araucana,* wing 210 mm, in Chile; *Columba fasciata,* wing 208 mm, in California). There is little habitat overlap, and doves are scarcest in thick brush (center of the gradient), which thus serves to isolate these species pairs. Africa provides a startling contrast, for it is very rich in doves (Figure 7j–l). Here the importance of size differences shows up clearly; each of the two ends of the gradient, which provide the best dove habitat, has four species rather than one, and a total of seven species was censused. The only two species of the same size, 107 mm, do not meet on the habitat gradient. Interestingly enough, the commonest species in Africa were the 157-mm *S. capicola* in the lower, more open brush, and the large 226-mm *C. arquatrix,* characteristic of woodland; these two species, of the seven African possibilities, give the closest matches in size and appearance to the Chilean and Californian pairs!

The sallying flycatchers provide a slightly more complex illustration (Figure 7m–o). Mediterranean Chile has two species (family Tyrannidae), one large and one small, with very considerable habitat overlap even though the larger species enters the habitat gradient earlier and the smaller persists longer. Similarly in California, in which the flycatchers are also tyrannids, a large species enters first and is soon followed by a small species, but thereafter the parallel breaks down. These first two California species are joined by

a medium-sized species in woodland, the earlier large species is replaced by a second large species, and the earlier small species is replaced by two congeners in the taller habitat. These latter two are identical in size (and in just about every other characteristic), but feed at rather different foraging levels in the *Sequoia* forest and supplement this difference with interspecific territoriality. The first three steps, an early large species followed by a small species and later by a medium-sized species, are repeated in Africa with flycatchers of the family Muscicapidae, but then a second medium-sized species is added, which feeds on the average higher in the canopy than the first. In this guild, then, primarily size and secondarily foraging height serve to separate members ecologically.

As with the flycatchers, body size is a primary ecological variable in the woodpecker guilds. All species are in the family Picidae, and species distributions are illustrated in Figure 7p–r. All three continents have a small, trunk-feeding species, which appears early in the habitat gradient (*Dendrocopos lignarius,* wing 95 mm, in Chile; *D. nuttallii,* wing 103 mm, in California; *Mesopicos griseocephalus,* wing 109 mm, in Africa). All three also have a larger ground-feeding species, which occupies similar positions in the habitat gradient in Chile (*Colaptes pitius,* wing 159 mm) and in California (*C. cafer,* wing 163 mm), but which in Africa is curiously restricted to rockier locations (*Geocolaptes olivaceus,* wing 129 mm) and did not occur in any of my study areas. The remaining Chilean species (the Magellanic woodpecker *Phloeoceastes magellanicus,* wing 215 mm), would have provided a very close counterpart to the largest Californian species (the pileated woodpecker *Dryocopus pileatus,* wing 237 mm), but the Chilean species now occurs only sporadically in its preferred habitat of *Nothofagus* forest and may be heading for extinction. These two large species are without parallel in Africa; in turn, the remaining African species, *Campethera notata* (wing 107 mm), has no equivalent in Chile, but parallels the Californian species *Dendrocopos villosus* (wing 127 mm) quite closely, both species being similarly sized and centered in the thicker woodland. Thus, California provides close equivalents to each Chilean species and to each African species, but California provides a host of additional species that segregate quite neatly by size. A size and foraging-site sequence of four species occurs in heavy forest in California (wing lengths 123, 129, 129, and 237 mm). Woodpecker species are abundant in California, less so in Chile, and inexplicably scarce in Africa.

As we proceeded through Figure 7, we have progressed from single- to multi-species guilds; and from quite exact one-to-one correspondence, to correspondence between one species and several or few species, to correspondence between many species. Examples were also ordered to show an increasing elaboration of coexistence techniques within guilds, beginning with simple habitat displacement as exemplified by wrens and ending with guilds that employ, besides habitat differences, differences in size, feeding height, and feeding technique. We showed that these techniques are used in various combina-

tions in different guilds. We also showed that habitat displacement can be modified by interspecific behavioral interactions, which occur especially in groups of species that differ only in their respective position on the habitat gradient and presumably serve to sharpen these habitat differences where one species meets another. In these Mediterranean habitats, two of these resource axes or "niche axes" prove to be of primary importance: habitat type, or position on H, and body size of species, an index of food size or type. Interspecific overlap within guilds in these two resources is largely complementary, as shown in Figure 8: guilds whose species show large habitat overlaps exhibit a size segregation, and only in these guilds with low habitat overlap are species of similar sizes.

As a final example in this series of deteriorating parallels, and one reflecting a dismaying though fascinating confusion of factors, I mention those deep-billed and stout-legged species that comprise a very loose guild of "ground-feeders" (not illustrated in the figure). In Chile these are drawn from the endemic South American family Rhinocryptidae, and pack up to three or four species per habitat. They segregate ecologically, chiefly by size differences [e.g., the wing sizes of the four species on Mocha Island are 53, 61, 77, and 106 mm (Cody, 1974)]. However, the species (six in four genera) are involved in a "character convergence" complex (see Cody, 1973) that employs interspecific territoriality as a backup strategy for ecological segregation. For reasons associated with chance, history, or habitat configuration, the number of species that occur together in a given section of the habitat gradient exceeds the number that can be supported by the variety of available resources. In this case, two or more of the species involved may converge in some morphological attribute such as bill length and perhaps also general appearance, defend territories interspecifically, and behave ecologically as a single species. There are also quail in Chile (introduced from California) and a native tinamou (Tinamidae, one species per habitat type) at the open, brushy end of the gradient. The closest California counterparts to the Rhinocryptidae are the towhees *Pipilo,* with two similarly-sized species that co-occur over most of the habitat gradient (see Appendix C). I strongly suspect that, although these two towhees appear to co-exist peacefully, they are also involved in interspecific behavioral interactions that reduce niche overlap between them. The same quail occurs in California as in Chile; it is replaced by the similarly-sized mountain quail *Oreortyx pictus* in montane forest, and primaevally a large forest grouse would have been found. In Africa the closest counterparts again differ, in virtually all criteria in which they can be compared. They are the extremely large francolins (Phasianidae, 220-mm and 162-mm wing sizes) and guinea fowl (Numidae, wing 282 mm), plus a tiny grassland quail (Turnicidae, wing 76 mm), and unlike the American species they occur with just one taxon to a habitat. Here the size differences seem not at all related to ecological coexistence in multi-species assemblages, but size is instead inversely related to the vegetation density of the preferred ground cover of the spe-

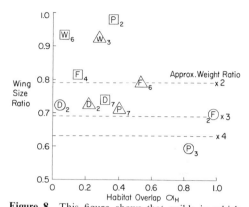

Figure 8 This figure shows that guilds in which species are primarily segregated by size are those in which the habitat distributions of their members overlap widely, and that guilds in which members differ in habitat distribution may also comprise species that are similarly sized. As before, the squares refer to African guilds, the triangles to Californian guilds, and the circles to Chilean guilds, all of which are depicted in Figure 7. P = the woodpecker guild; W = the "wren" guild, D = the dove guilds, and F = the flycatcher guild; these are represented in Figures 7p–r, 7d–f, 7j–l, and 7m–o respectively. The suffix to the lower right of each symbol gives the number of species in that guild on that continent. Wing size ratio is the average ratio of the wing sizes of species that are adjacent in the size series that forms the guild, and can be approximately translated into a weight ratio of adjacent species as indicated. Habitat overlap α_H is again an average for the guild, the average of each possible combination of species pairs in the guild. The pair-wise values for α_H are computed from the distributions given in Figure 7, since the coincident area of partially overlapping normal distributions is given by $e^{-(d_1-d_2)^2/2(\sigma_1^2+\sigma_2^2)}$, where d_1 is the center of one distribution and d_2 the center of the other, and the σ^2 are the variances of each of the distributions. Note that d_i come directly from Figure 7, as the value of H at which 50% of the distribution is accumulated, and that variances are equally easily obtained from Figure 7, since $\sigma^2 = [(H_{16\%} - H_{84\%})/2]^2$. The figure shows that, while guilds composed of very similarly-sized species (high wing-size ratio) are always habitat-segregated, and guilds of species with coincident habitat requirements (high α_H) are always of different-sized species, guilds may in fact comprise species that differ in both habitat and size, indicating that some interdependence exists between habitat and size constraints.

cies: dense grass has the smallest species, thick chaparral the second smallest, and the largest occurs in the low, open brush, which supports only the sparsest cover of annual weeds at ground level.

The Origin of The Species Diversity Patterns

This concluding section will necessarily be much shorter than its importance warrants, for there is little but anecdotal comment that I can offer on the origins and evolution of the diversity patterns described above. As a first-order generalization, the α-diversity patterns among the three Mediterranean localities appear to be very similar. These species-packing levels can be adequately explained by competition theory and by the constraint of a limiting similarity, between species on a resource gradient, that cannot be exceeded (May and MacArthur, 1972). By choosing all habitats from within a Mediterranean climatic regime we have minimized differences in resource production, and the number of species presents a fairly simple reflection of the span of resources as measured by H. Where chance has reduced the predicted species number over some limited range of H, expansion over a greater range of habitats by the species that are present compensates for such low numbers. In addition, there is density compensation from within the existing impoverished community, so that the density per species tends to increase in impoverished communities and to decrease in exceptionally rich communities. Thus patterns in α-diversity conform well

with simple theoretical expectations that ignore history.

However, diversity differences, and in particular the marked differences in β-diversity between continents, must have explanations with a strong component of history and chance, both in the production and distribution of various habitat types and in the buildup through time of their respective bird faunas. Bird censuses reflect in a sense a snapshot of the kaleidoscopic processes of radiation, invasion, competition, and extinction in bird species groups, processes that are mediated by just the same processes in the vegetation groups and types in which the bird species live. I believe that the censuses must reflect an equilibrium solution to the present ecological conditions; however, this equilibrium may be a complex function of the invasion and extinction possibilities peculiar to each site, may not yet be closely approximated in some faunas, and therefore a full explanation of β-diversity patterns may prove elusive.

But we can guess the identity of some of the important factors in this explanation, and one may be the relative accessibility of each habitat to a species pool of more or less appropriate colonists from habitats elsewhere on the continent. I shall next discuss in a qualitative way how this explanation might operate, and provide preliminary data for the three continents to defend its relevance.

In the species charts of Appendices B–D we really have pictures of the niche space available on the gradient to bird species. The charts can be "compacted" by putting all bird species that replace each other with changing habitat (i.e., species that contribute to β-diversity) on the same line in the ordinate; then the ordinate represents food/feeding type and is one resource or niche dimension, and the abscissa is habitat type and constitutes a second dimension in a simple two-dimensional scheme (Figure 9 below). Niche space is then some roughly triangular area on the plane of H (habitat) and F (food or foraging site). It is subdivided by birds whose α-diversity is measured by reading vertically in the plane, and whose β-diversity is read by viewing the niche space horizontally.[2] Starting, then, from the basic niche space, a zone of F-H space that must be very similar in the three continents, how are the different diversity patterns generated?

It may be supposed that the compacted species charts reflect a more or less steady state between the production or immigration of species and their ultimate extinction in the niche space, between expansions and contractions of species in each of the two niche dimensions. Rosenzweig (Chapter 5) gives an elaboration of this concept. After the local extinction of a

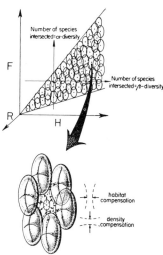

Figure 9 A pictoral representation of niche space and its subdivision by bird species. There are two primary niche dimensions, H (habitat type) and F (food, feeding behavior, foraging site). Niche space occupies a triangular area in the F-H plane, and can be estimated by charts such as those that produced Figure 2 by putting all species that replace each other by habitat type on the same F-line. This means that, when species niches are represented by elliptical areas in the plane, α-diversity is read by counting species niches vertically above some H value, and β-diversity is read by counting species niches horizontally at some F-value. The figure shows (below) what happens after an extinction in the niche space. Neighboring species may usurp some of the vacant niche space, by density compensation and by niche-breadth compensation in both the F- and H-dimensions but particularly in the H-dimension. The new colonist must arrive before this compensation has greatly reduced the resource availability, if it is to be successful. The dimension R represents resource level, which must be high over some contiguous area of the F-H plane for successful invasion.

[2] It can be seen from Figure 9 that a symmetrical version of this argument casts α-diversity as a function of H, and β-diversity as a function of F. This means that β-diversity can be understood in terms of H, as it is pictured in Figure 4b, only in so far as the resources of the F-axis can be mapped onto the H-axis. Notice how the shape of the niche space puts constraints on the value of β-diversity. Whereas the number of food niches and hence the number of species that can coexist at a point (α-diversity) increases smoothly with H, the food niches with the broadest distribution over H, and therefore the food niches that might support several different species with changing H (β-diversity), are centered at intermediate H.

species, an immigrant arrives whose identity is statistically predictable from a probability distribution of likely immigrants; this probability distribution is generated from the amounts, proximities, and similarities of the habitats nearby that supply potential colonists. Thus, we might be able to generate the most likely arrival sequence and schedule. Potential immigrants arrive early and often if the depauperate patch of habitat is not isolated, and these immigrants are appropriate consumers for the unused resources if the habitat in which the extinction occurred is well represented in the surrounding environment. The more appropriate and rapid the colonization, the less likely that these unused resources will be accounted for by density compensation or habitat-breadth compensation by the neighboring species in niche space. The populations of colonists with low efficiencies on the unused resources would grow slowly; if the immigration schedule is rapid, they will soon be replaced by more appropriate consumers, but if it is slow they might possibly withstand later invasions by better adapted species, as in the interim their populations will have built up.

After an extinction in some part of the niche space, expansion by species in neighboring niches is slow in the F-dimension and relatively rapid in the H-dimension. We know this from the constancy of α-diversity and from studies such as those by Crowell (1962), Diamond (1970), and Yeaton (1972), which show that island birds can readily use a variety of habitats from which they were competitively excluded in rich, mainland faunas, but rarely (or slowly) do they change their feeding habits. Then niches are akin to contour ellipses (sometimes called confidence or probability ellipses) in two dimensions, much more expansible in one dimension (habitat) than in the other (see

Figure 9), areas that delimit with a certain degree of confidence the resources going to the niche occupant. Early colonists can preempt vacant niche space, but later colonists find much less unused resource to control, partly because of density compensation (an increase of niche breadth in the F-dimension[3]) and particularly because of habitat compensation (expansion in the H-dimension). There must of course be limits to the maintainable niche breadth in either dimension, limits that are probably a function of competitor pressure. It is reasonable to assume that the degree to which unused resource levels remain high, and therefore the degree to which successful invasion by a competitor is likely, decreases from the niche center of a species to its periphery and sets a maximum to niche width, whereas a minimum is determined by extinction rates in narrower niches that permit only low densities and small population sizes. To further embroider this picture, then, we can imagine a third dimension, resource level, coming out of the F-H plane, and picture species-niches as elongated bowl-shaped depressions in resource levels (Figure 9). Only those parts of niche space with contiguous areas of high resource levels are open to invasion, areas where the combined utilization densities of the species present remain low.

Rather than attempt any further elabo-

---

[3]Note that the increase in niche breadth will affect the range of foraging sites much more than it will the range of food types. We suspect that genetic changes, not just behavioral changes, are required if the mean value or variance of food items consumed is to change significantly.

ration of this scheme in a general fashion, I will try to make some specific predictions. If appropriate colonists are slow in arriving at the scene of a recent extinction, whether for reasons of their scarcity in the surrounding countryside or because of the isolation of the habitat patch suffering the loss, compensation will occur; we predict that diversity in the patch will be low, and particularly low in β-diversity for that part of the habitat gradient. A corollary of the same slow colonization process is that the same habitat in different geographic areas will be occupied by a partially different species set (cf. discussion of "patchiness" by Diamond, Chapter 14; e.g., Figures 34–38). This occurs because of the chance element in colonization in determining which resources will eventually go to which consumer, as opposed to the increased predictability of this allocation if the total species-pool is always at hand to compete freely for the resources. Such species turnovers between geographic areas independently of habitat change measure a third type of diversity, γ-diversity, and may be attributable to the same governing factors as is β-diversity and in this instance may be negatively associated with β-diversity.

I made a simple test of a part of this scheme by obtaining from vegetation maps the relative and absolute amounts of the habitat types censused on the gradient, within a thousand-mile radius of the census sites. This test is motivated by the possible analogy between mainland habitat patches and islands. If the analogy holds, then the positive correlation between island size and species number

should have a parallel on continents, where habitat patches associated with larger source areas are expected to hold larger numbers of species. These data are shown in Figure 10, which gives amounts

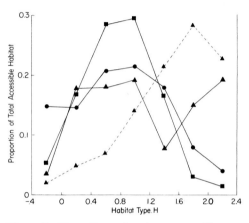

Figure 10 The ordinate gives the amount of "accessible" habitat of type H (abscissa) expressed as a function of the total accessible habitat of all types. "Accessible" habitat is defined as habitat within 1000 miles of the Mediterranean sites censused; this distance is measured on journeys, not necessarily on a straight line, that entail elevational shifts of less than 5000 vertical feet. The figure was constructed from vegetation maps by Evans (1935) for South Africa, Küchler (1964) for North America, and Sahab (1966) for Chile. The dashed line is the estimated vegetation distribution for the southwestern United States in late Pleistocene times. Compare the dashed and solid North American curves, ▲; note that there was less low habitat and more woodland in the late Pleistocene than at present. Note also that at present Africa has the largest proportion of accessible habitat in the form of intermediate-height scrub ($H = 0.3$–1.4), California in the form of forest ($H = 1.6$–2.4), Chile in the form of low habitat ($H = -0.2$–0). Correspondingly, these are the ranges over which each continent exceeds the other two continents in α-diversity (Figure 3), as further shown by Figure 11a.

and distributions (in terms of H) of the habitat types in the potential source habitat from which colonists might be derived. I allowed these potential colonists to fly a thousand miles or less, but this distance might be flown either in a straight line or in a jagged course around barriers. For this purpose, I defined a barrier to be a mountain block that constitutes an elevational change of at least 5000 feet. Thus the colonist could fly around such barriers from a source habitat, as long as the total distance would not exceed 1000 miles. Such mountain barriers form "shadows" over the potential source habitats within the 1000-mile radius, and eliminate portions of the source area from contribution to the colonization of habitats at the census sites.

In Figure 11a, I plot α-diversity on the ordinate against the cumulative area under the curve of source habitats in Figure 10. The α-diversity values are taken from the fitted curves for each continent as shown in Figure 3. This plot produces two rather striking results, both of which are direct continental analogues of familiar island-biogeography patterns. First, the continent with the most source area over some range of H has the highest α-diversity values there. This is seen from the fact that, whereas the three continents show different rates of increasing α-diversity with increasing H (Figure 3), in Figure 11a these differences have been submerged, at least over the lower H range. In other words, if we take into account the differences between continents in source areas at different H, the observed differences in α-diversity reduce to a common

relation between species-packing level and availability of source habitat, as if habitats over a limited range of H were akin to islands and as if islands of similar habitat supported more species with increasing island area. Second, the asymptote reached by α-diversity bears a direct relation to the total area of putative source habitats. Thus in California, with a source area of 9.5×10^5 square miles, the asymptote of α-diversity is 27.8 species, and in Chile with a source area of 1.3×10^5 square miles the asymptote is around 20 species. Africa has a source area just a little smaller than California's, 8.7×10^5 square miles, and its α-diversity asymptote is likewise just below that of California at 24.9 species. This result seems to be another manifestation of the island-like pattern of more species in larger areas.

Furthermore, β-diversity at a given H correlates fairly well with $p(A)$, the proportion that a given habitat contributes to the total source area (Figure 11b). This means that if a certain type of habitat at some H value is well represented in the source habitat distribution [$p(A)$ is high], β-diversity or turnover rate of bird species at that point on H is also high. There are easily recognized manifestations of this phenomenon. Africa, for example, has vast areas of savannah, and these savannahs are extremely rich in species restricted to particular types of savannah. Southern Africa has vast areas of short scrub habitats (see Figure 10), and the Mediterranean zone supports a large number of scrub insectivores such as the *Cisticola* warblers and their relatives,

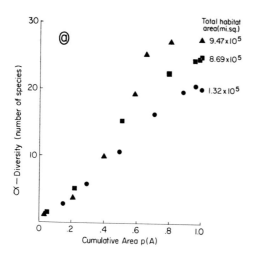

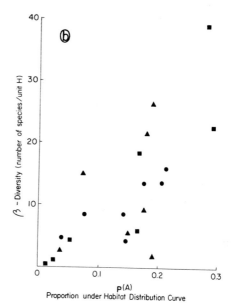

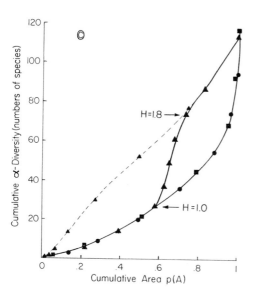

Figure 11 Figure 11a gives α-diversity (ordinate, from Figure 3 in any habitat type H on each continent, plotted against the cumulative proportion of accessible habitat reached by that point H (abscissa), accessible habitat being defined as in Figure 10. In effect, integration of the curves of Figure 10 is used to transform the abscissa of Figure 3 from H to cumulated area. Note that this correction for intercontinental differences in the distribution of accessible habitat eliminates the intercontinental differences for α-diversity for $H < 0.5$ (Figure 3), and that the intercontinental differences in the asymptote of α-diversity at high H correspond to differences in absolute totals of accessible habitat (numbers to the right of each curve). Figure 11b plots β-diversity in any habitat type H on each continent (ordinate, from Figure 4a, below) against the proportion of accessible area at each point H (abscissa, from Figure 10). This correction for area reduces but does not eliminate intercontinental differences in β-diversity (Figure 4b, below). Figure 11c differs from 11a only in that the ordinate is cumulative α-diversity up to habitat type H. From this plot it appears that Chile and Africa have reached similar equilibria over all H, and that California has reached the same equilibrium at low H, but at high H California is supersaturated. This supersaturation in Californian woodlands and forests is presumably a carryover from Pleistocene times, when woodlands and forests were more extensive than they are today. The dashed line is the relation predicted for California from Pleistocene vegetation distributions; the other lines are those actually observed on each continent. See text for discussion.

243

which are serially arrayed over habitats and thus show high β-diversity. Chile and California with less scrub have many fewer of these habitat-specialist insectivores in ecologically equivalent roles. California, on the other hand, is the only continental area of the three with extensive woodlands and forests, and California is the area so rich in flycatchers and woodpeckers, two guilds that are dependent on the taller habitats. The correlation of Figure 11b, although far from perfect, shows that we have in fact accounted for around 50% of the variation in β-diversity by using this index of colonist accessibility. Quite reasonably, then, species turnover between habitats in the Mediterranean zone is highest in those habitats for which there are large and accessible areas of related habitat to provide appropriate colonists.

In Figure 11c, I look at one further aspect of the relation between source area for colonists and the degree of species-packing in the Mediterranean zones. In this graph I plot cumulative α-diversity, a measure of the rate at which total species-packing is accumulated with increasing H, against as before the cumulative area $p(A)$ under the source habitat curves. This plot yields a particularly notable result: African and Chilean points fall on exactly the same line, whereas Californian points follow this line for a while and then diverge dramatically from it. The divergence of the Californian points indicates that, after H values of around 1.0, which correspond to habitats between tall chaparral and low woodland, species are accumulated much faster than

the representation of the taller habitats in the source area would predict. I take this to mean that for some reason the oak and pine woodlands and montane forests of California are "supersaturated" with bird species compared with the base line provided by all shorter vegetation types and by the taller habitats of both Chile and Africa.

This situation is strikingly reminiscent of results from some island studies, and proves to be yet another continental analogue of a phenomenon previously recognized only on islands. I refer to the studies of "relaxation," particularly by Diamond (1972 and Chapter 14). These studies demonstrate that when island species diversities are displaced either above or below their equilibrium values, the species count relaxes back to the equilibrium value at a rate predictable from island size and the direction and distance of the displacement. The main results of these studies are a) that relaxation times to equilibrium are proportional to the island size, and b) that relaxation times are faster by an order of magnitude when the equilibrium is approached from below than when it is regained from above.

I am tempted to attribute the striking deviation of bird species-packing in California woodlands to the relaxation of the fauna following the effects of Pleistocene glaciations and pluvials, which had drastic consequences for the biogeography of the native vegetation types. We are fortunate in knowing a good deal about what habitat distributions were like in the southwestern United States between 12,000 and 20,000 B.P. This information comes from pollen analysis (e.g., Axelrod and Ting, 1961; Martin and Mehringer, 1965) and from the radio-carbon dating of plant remains taken from desert woodrat dens (Wells, 1966; Wells and Berger, 1967). At the height of the glacial period, the southwest looked very different, and was much more heavily wooded than it is today. The elevational ranges of plants were shifted down in response to a cooler ($-5°$) and wetter ($+15''$ of precipitation) climate, by as much as 3000–4000 ft. Most of the Mohave and Great Basin deserts was covered with woodland of pinyon pine, juniper, and oak, and much of the Lower Sonoran Desert supported sagebrush and oak-chaparral. The extent of the montane forests, particularly of open yellow-pine woodland, was greatly increased, and at the southern end of the Sierra Nevada grew a plant association that is typical today of forests 800 miles farther north. Could the present California bird distribution, then, be a carryover from the time, which ended 12,000 B.P., when the area of woodland was much more extensive? To test this hypothesis, I plotted cumulative α-diversity against the cumulative $p(A)$ for the source area as it would have looked about 15,000 B.P. (the lighter line in Figure 10). The results are shown in Figure 11c, from which it appears that the current distribution of Californian bird species fits the present vegetation distribution for lower habitats up to tall chaparral, deviates upward between low oak woodland and the lower montane forest habitats, and thence matches the curve predicted for Pleistocene forest distribution. After the sudden climatic change

around 12,000 B.P., the bird fauna must have undergone a gradual reorganization towards a new equilibrium between local extinctions and the immigration of new colonists at altered rates consonant with the new habitat distributions of the post-Pleistocene climate.

The biogeographies of both African and Chilean vegetation types were presumably also affected by Pleistocene climates, but these effects were most likely not nearly as pronounced as they were in the southwestern United States. Although nothing is known of these changes from Chile, except the extent of glaciation in the southern Andes (Antevs, 1929), it would appear that nothing on the same scale as in North America could occur in the narrow region between the perennially formidable Andes and the ocean-tempered coastal zone. In Africa, it is known from limited pollen studies in the northeast Cape Province that around 12,000 B.P. the vegetation types supported there oscillated between the present Karroo vegetation and a grassveld vegetation; the latter can now be found to the northeast and at slightly higher elevations from the sampled site (Coetzee, 1967). There was a limited expansion of the evergreen woodlands along the southeast coast, but there could not have been any appreciable increase of woodland or forest types over the Mediterranean source area. Thus, the current Chilean and African bird patterns probably reflect an equilibrium, one that has been reached since the Pleistocene and reflects a conformation of bird distributions to the present habitat circumstances. California, on the other hand, is at the same equilibrium over only the lower habitat types, and the faunas of woodland and forest are apparently still undergoing relaxation from high Pleistocene levels set by the greater extent of woodland and forest at that time. I can suggest two reasons why relaxation in woodland and forest has been much slower than in lower habitat types. First, the lower grassland, sage, and low chaparral vegetation types were surely those least affected by Pleistocene glaciations, and therefore their bird diversities had to make only small shifts in order to reach new equilibria. Second, these low habitats are those that are the most fragmented into "small islands" and had to approach post-Pleistocene equilibrium species-diversity from below, since grassland and sagebrush were underrepresented in the Pleistocene vegetation. In contrast, the forest and woodland habitats occur in large blocks (associated chiefly with the Sierra Nevada and Rocky Mountain ranges) and approached equilibrium from above, since their Pleistocene areas (hence species diversities) were greater than at present. Island studies (Diamond, 1972) show that avifaunal relaxation times increase with island size, are much longer from above than from below, and are of the order of 10,000 or more years for supersaturated large islands. Thus, one would expect the low habitats to have reached equilibrium but woodland and forest still to be supersaturated, in agreement with Figure 11c.

I have summarized the relation between α-diversity and β-diversity for the three Mediterranean regions in Figure 12, in which each study site is represented by a point. The sequence of censuses on each continent is represented by a sweep of points for which β-diversity increases and then decreases with increasing α-diversity (cf. footnote 2). Each study site falls into a certain range of H values, and the way in which H increases along the sweep of censuses is shown by a series of rays (the dashed lines), which originate close to the origin of the graph and divide the plane into zones of equal H values. Each series of censuses, then, falls on a line or sweep that characterizes in a specific way the bird diversities over the habitat gradient sampled by the censuses. The rate at which β-diversity increases with increasing α-diversity matches the proportional representation of habitat types $p(A)$ over various H within the source area, and may be interpreted in terms of the availability of suitable colonists for the habitats on the gradient. The "breadth" of the sweep appears to correspond to the total area of the source habitats, and broader sweeps correspond to larger source areas and indicate higher diversity values. Clearly these patterns are not yet fully understood, but in view of the rather simplistic analysis of source habitats attempted here, this sort of treatment appears profitable and informative.

Finally, the differences among continents, including differences in α-, β-, and γ-diversity, are represented as sections of the species-area curves for the three localities in Figure 13. This figure shows careful plots of the numbers of species accumulated with increasing area, beginning in each case with tiny $50' \times 50'$ plots in tall

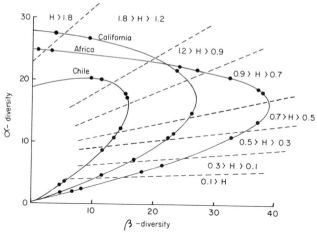

Figure 12 α-diversity is plotted against β-diversity for each census site on each continent. The areas swept out by these curves bear some resemblance to the absolute areas under the habitat distribution curves given in Figure 10, and the proportions of each habitat type correspond to distances along the radii that separate different sections of the habitat gradient in the figure. Thus, α-diversity and β-diversity are reasonably predictable from the information on habitat distribution given in Figure 10.

chaparral, and expanding the area sampled until geographic regions the size of the state of California are counted. I believe that these species-area plots can yield much more than an exponent (slope) of uncertain biological meaning (cf. May, Chapter 4). When compiled from a detailed knowledge of bird distributions, the curves can show successively measures of point diversity, α-, β-, and γ-diversity for the habitats and areas studied. Normally such curves are plotted merely to extract the value of the slope z; in this case the slope is 0.13, a typical value for continental surveys; MacArthur and Wilson

(1967, p. 9) give the range of mainland z values as 0.12-0.17.

But of more interest is the fact that, although all three localities reach endpoints that are within just a few species of each other, these species totals are accumulated in quite different ways in different continents. That is, at any point on the abscissa, the slopes can differ, and these differences can tell us quite a bit about the structure of the bird faunas, and in particular about the various sorts of diversities we have been discussing. Take, for example, the species numbers found in areas around 10^{-2} square miles in tall

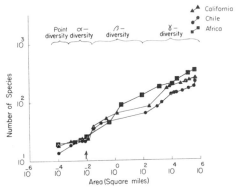

Figure 13 The species-area curves for three continents, centered in thick chaparral in the Mediterranean zone. Each point gives the number of species (ordinate) encountered in a circle or in the most rounded area possible of a given area (abscissa) centered in habitat $H = 1.11$. For larger sample areas to the right on the abscissa I used data from Roberts (1972) for South Africa, Grinnell and Miller (1944) for California, and Johnson, Goodall, and Philippi (1957) for Chile. For the left-hand section of small-area resolution I used my own detailed census maps of chaparral sites and their surrounding regions. See text for discussion of how various diversity measures can be read from the ordinate values and slopes of these curves.

interspecific behavioral interactions that prevent territory overlap there. These reflect difference in "point diversity."

Proceeding to the right, as area increases above 10^0 square miles we start to get into habitat types different from the tall chaparral in which we started. One can encounter patches of (Californian) oak or riparian woodland (or their ecological counterparts on the other continents), or perhaps one begins to find coastal sage vegetation types. This means that we begin to pick up species that are habitat replacements, and so we start to measure β-diversity. If the accumulation of species is rapid, β-diversity is high, and vice versa. So the *slope* of the curve, between around 10^0 and 10^3 or 10^4, measures β-diversity. These slopes are highest in Africa, with high β-diversity, and lowest in Chile, which has the bird fauna with the lowest β-diversity. At this point we have accumulated a fairly complete sample of all the habitat types we can find in the region. Yet we still continue to accumulate species, because species may replace each other not just between habitats but also between geographic areas independent of habitat change. These species are geographical replacements, and the rate at which they are accumulated is a measure of γ-diversity. Thus the slope of the last part of the species-area curve measures γ-diversity, and we can now compare the three continents in this characteristic. Notice that the slope of the African species-area curve in the terminal section is lower than that for either California or Chile (compare where the sections begin at around 10^3 and where they end at 10^6),

chaparral. This is close to the area of the study sites we have described in this paper, and this is just where all three curves are most coincident. This then shows that the α-diversities on the three continents are very similar; indeed, species-packing levels are similar among the three continents in the area range between about one-half-acre plots and about 200-acre plots (10^{-3} to $10^{-0.5}$ square miles). But notice that at the extreme left of Figure 13 Chile differs from Africa and California, for there are fewer species with overlapping territories in Chile and more

and that the greater slope is found in the Chilean curve. These differences are just the opposite of the differences in β-diversity for the three continents, and are just those we predicted from the presumed slower faunal buildup in Chile. A concluding scenario: it seems that localities that are relatively isolated from source faunas present immigration problems to colonists and these result in low β-diversity. The same isolation must produce a "faunal gateway" effect, such that new immigrants cannot reach sections of the habitat complex without passing through a "gateway," where fully saturated communities present a barrier for the further dispersal of colonists into the less saturated interior habitats. But pressure builds up at these gateways, and is relieved as close ecological counterparts of incoming colonists are forced into the interior of the habitat set, where they finally end up as geographic counterparts to the species responsible for the "pushing." There is clearly a wealth of information in species-area curves, and although a start has been made in their interpretation, their further elaboration awaits future work.

Acknowledgments

The field work described in this study has been generously supported by the National Science Foundation grants GB-35,896 and GB-27,154 and by the University of California-Universidad de Chile Convenio. Studies in South Africa were greatly facilitated by the Percy Fitzpatrick Institute of African Ornithology and its director W. R. Siegfried and his staff, and by the South African Department of Forestry. The manuscript has benefited from critical reading by friends and colleagues, in particular by my coeditor Jared Diamond.

References

Antevs, E. 1929. Maps of Pleistocene glaciations. *Bull. Geol. Soc. Amer.* 40:631–720.

Axelrod, D. I., and W. S. Ting. 1961. Early Pleistocene floras from the Chagoopa surface, south Sierra Nevada. *Univ. Calif. Publ. Geol. Sci.* 39:119–194.

Cody, M. L. 1968. On the methods of resource division in grassland bird communities. *Amer. Natur.* 102:107–137.

Cody, M. L. 1970. Chilean bird distributions. *Ecology* 51:455–464.

Cody, M. L. 1973. Parallel evolution and bird niches. *In* F. DiCastri and H. Mooney, eds., *Ecological Studies No. 7.* Springer-Verlag, Wien.

Cody, M. L. 1974. *Competition and the Structure of Bird Communities.* Princeton University Press, Princeton.

Coetzee, J. A. 1967. Pollen analytical studies in East and Southern Africa. *In* E. M. van Zinderen Bakker, ed., *Palaeoecology of Africa.* Balkema, Cape Town.

Crowell, K. 1962. Reduced interspecific competition among the birds of Bermuda. *Ecology* 43:75–88.

Diamond, J. M. 1970. Ecological consequences of island colonization by southwest Pacific birds. II. The effect of species diversity on total population density. *Proc. Nat. Acad. Sci. U.S.A.* 67:1715–1721.

Diamond, J. M. 1972. Biogeographic kinetics: Estimation of relaxation times for avifaunas of southwest Pacific islands. *Proc. Acad. Nat. Sci. U.S.A.* 69:3199–3203.

Evans, I. B. P. 1935. Vegetation map of South Africa. *Bot. Survey Mem.* 15:1–23.

Grinnell, J., and A. H. Miller. 1944. *Distribution of the Birds of California.* Publ. Cooper Ornith. Soc. No. 27.

Johnson, A. W., J. D. Goodall, and R. A. Philippi. 1957. *Las Aves de Chile.* Platt Estab. Graf. S. A., Buenos Aires.

Kilgore, B. M. 1971. Responses of breeding bird populations to habitat changes in a giant Sequoia forest. *Amer. Midl. Natur.* 85:135–152.

Küchler, A. W. 1964. *Potential Natural Vegetation of the Coterminus United States.* American Geographic Society, New York.

MacArthur, R. H., J. M. Diamond, and J. Karr. 1972. Density compensation in island faunas. *Ecology* 53:330–342.

MacArthur, R. H., J. MacArthur, D. MacArthur, and A. MacArthur. 1973. The effect of island area on population densities. *Ecology* 54:657–658.

MacArthur, R. H., and E. O. Wilson. 1963. An equilibrium theory of insular zoogeography. *Evolution* 17:373–387.

Martin, P. S., and P. J. Mehringer. 1965. Pleistocene pollen analysis and biogeography of the Southwest. *In* H. E. Wright and D. G. Frey, eds., *The Quaternary of the United States.* Princeton University Press, Princeton.

May, R. M., and R. H. MacArthur. 1972. Niche overlap as a function of environmental variability. *Proc. Nat. Acad. Sci. U.S.A.* 69:1109–1113.

Roberts, A. 1972. *The Birds of South Africa.* 3rd. ed., revised. MacLaughlin and Liversidge. Cape Times, Cape Town, R.S.A.

Sahab, 1966. *South America: Vegetation and Land Use.* Sahab Drafting Institute, Teheran, Iran. Library of Congress No. G 5201/D2/11m.

Wells, P. V. 1966. Late Pleistocene vegetation and degree of pluvial climatic change in the Chihuahuan desert. *Science* 153:970–975.

Wells, P. V., and R. Berger. 1967. Late Pleistocene history of coniferous woodland in the Mohave Desert. *Science* 155:1640–1647.

Yeaton, R. I. 1972. *A Comparison of Chaparral Bird Communities between Mainland California and Santa Cruz Island.* Ph.D. thesis, University of California at Los Angeles, Los Angeles.

Appendix A Study sites and their characteristics.

The first column of words names 31 Mediterranean study sites on three continents (plus 3 non-Mediterranean sites), and the second column lists their dominant plant species. The last named site in each geographic locality represents the tall-forest site closest to the Mediterranean zone on that continent. For each study site I plotted foliage profile, which is a graph of vegetation density against height above the ground. From this plot were derived the two basic habitat metrics "height" and "half-height," in feet, which are indices of the span and distribution of the vegetation (columns 3 and 4). Column 5 gives the principal component combination of these two habitat variables, the logarithm of which is a measure of position on the habitat gradient H. Columns 6 and 7 give numbers of bird species and the total bird species density (summed over all species) in pairs per acre. Chilean sites 1 and 2 are study areas H and I respectively of Cody, 1968. Chilean sites 3, 6, and 7 are study areas 3, 1, and 2 respectively of Cody, 1970. California site 10 is that of Kilgore (1971), from whom the bird census data are taken, and African site 14 uses bird census data compiled with the generous and gracious assistance of John Harkus.

Appendix B Chilean Bird Distributions and Densities

Data from bird censuses in Chile at 10 sites, 9 of which are in the Mediterranean zone. The axis across the top of the figure represents a habitat gradient H, and census sites are located at a point on this gradient, as indicated by the vertical lines. Bird species are ranked in the chart according to when they are first encountered on the habitat gradient, when they reach maximal abundance, and when they exit from the censuses. The values in the chart are bird densities, in pairs per acre. A + means that a species was present, but too rare to be accurately estimated. The symbol + + means that the species was present and common, but again no accurate estimate of density could be obtained. The symbol * means that I expected to find a particular species in the census but did not do so.

Appendix C Californian Bird Distributions and Densities

Data from bird censuses in California at 10 sites, 9 of which are in the Mediterranean zone. Entries in the chart are bird densities. See introduction to Appendix B for further information.

Appendix D African Bird Distributions and Densities

Data from bird censuses in Africa at 14 sites, 13 of which are in the Mediterranean zone. Entries in the chart are bird densities. See introduction to Appendix B for further information.

Appendix A. Study sites and characteristics

Site Habitat	Plant species	Habitat measures			Bird censuses	
		Height	Half-Height	Log (PC)	No. species	Pairs/acre
Chile						
1. Short grassland, Maipu	*Phalaris, Erodium*	0.7	0.3	−0.148	4	1.13
2. Mid-grassland, Maipu	*Various introduced species*	0.9	0.3	−0.071	3	1.83
3. Low coastal scrub, Pichidangui	*Bahia, Happlopappus*	3.0	0.8	0.043	6	1.95
4. Coastal scrub, La Laguna	*Bahia, Muhlenbeckia*	4.4	1.1	0.590	12	3.76
5. Successional scrub, Las Docas	*Muhlenbeckia*	5.9	1.2	0.701	12	2.45
6. Savannah, Melipilla	*Acacia, Prosopis*	12.0	5.4	1.090	21	3.05
7. Matorral, Puchuncavi	*Lithraea, Quillaja*	16.2	2.5	1.121	25	6.74
8. Evergreen woodland, Concepcion	*Cryptocarya*	39.0	14.5	1.578	18	4.39
9. Pine forest, Valparaiso	*Pinus radiata*	49.8	22.5	1.709	21	4.55
10. Temperate forest, Temuco	*Nothofagus*	142.0	66.7	2.169	19	3.31
California						
1. Short grassland, Oceanside	*Bromus, Erodium*	1.0	0.3	−0.054	3	0.71
2. Tall grassland, Oceanside	*Bromus, Avena*	2.3	0.4	0.282	4	1.45
3. Grass-sagebrush, Oceanside	*Artemesia, Salvia*	3.5	0.5	0.452	7	1.65
4. Coastal sage, San Onofre	*Artemesia, Eriogonum*	4.7	1.2	0.620	11	2.00
5. Sage-chaparral, Oceanside	*Artemesia, Rhus*	5.6	0.8	0.650	10	2.32
6. "South-slope" chaparral, Topanga	*Adenostoma, Rhus, Arctostaphyllos*	7.0	1.8	0.794	19	5.51
7. "North-slope" chaparral, Topanga	*Quercus, Ceanothus, Heteromeles*	13.8	4.6	1.114	19	7.63
8. Oak woodland, Topanga	*Quercus agrifolia, Heteromeles*	39.8	14.8	1.587	31	8.45
9. Pine forest, Point Lobos	*Pinus radiata*	68.8	46.5	1.911	20	7.54
10. Montane forest, Sierra Nevada	*Sequoia, Pinus, Acer*	138	73	2.170	31	2.40
South Africa						
1. Coastal flats, Gordon's Bay	*Many Restios, Compositae*	1.0	0.2	−0.090	5	1.17
2. Coastal flats, Gordon's Bay	*Many Compositae, Iridaceae*	1.1	0.2	−0.030	5	1.27
3. Open heath, Cape peninsula	*Erica, Restios*	1.9	0.4	0.198	4	0.70
4. Low renosterbos, Gordon's Bay	*Elytropappus*	2.1	0.4	0.247	7	1.01
5. Tall restio, Cape peninsula	*Restio*	3.4	0.4	0.431	6	1.00
6. Renosterbosveld, Devon Valley	*Elytropappus*	3.9	0.8	0.527	14	3.72
7. Tall renosterbos, Devon Valley	*Elytropappus*	5.5	1.9	0.719	16	6.86
8. Coastal macchia, Swartklip	*Metalasia, Euclea, Acacia*	5.9	1.9	0.739	18	7.72
9. "South-slope" macchia, Jonkershoek	*Protea repens, Cliffortia, Rhus*	7.8	2.1	0.845	18	7.72
10. Open Protea macchia, Jonkershoek	*Protea arborea, Restio, Asparagus*	10.9	1.1	0.929	21	6.35
11. "North-slope" macchia, Jonkershoek	*Protea, Cliffortia, Myrsine, Rhus*	11.6	1.6	0.970	23	3.80
12. Kloop woodland, Jonkershoek	*Brabeium, Cunonia, Podalyria*	36.0	15.3	1.560	11	4.98
13. Evergreen forest, Kirstenbosch	*Virgilia, Dyaspyrum, Cunonia, Cassive*	77.3	19.2	1.834	14	6.93
14. Evergreen forest, Alexandria	*Virgilia, Pytosporum, Dyaspyrum*	83	24	1.879	35	4.34

Appendix B. Chilean bird distributions and densities

Species ranking	1	2		3	4	5	6	7		8	9	10
Habitat gradient H:	−0.2	0	0.2	0.4	0.6	0.8	1.0	1.2	1.4	1.6	1.8	2.0
1. *Geositta cunicularia*	1.54											
2. *Anthus correndera*	*	0.45										
3. *Phrygilus alaudinus*	2.05	0		0.63	0.67	0	0.03					
4. *Pezites militaris*	1.03	0.22		0.38	0.42	0.30	0.19	+				
5. *Zonotrichia capensis*	0.51	0		+	1.17	0.70	0.48	0.83		0.28	0.50	
6. *Sicalis luteola*		1.16		0	0.17	0.20	0.03	*		*	*	
7. *Agriornis livida*				0.09	0.08	0	0.05	0.11		*	*	
8. *Mimus tenca*				0.25	0.17	0.10	0.13	0.44		*	*	
9. *Diuca diuca*				0.50	0.50	0.30	0.32	0.33		+	0.10	
10. *Phytotoma rara*					+	0	+	0.33		*	*	
11. *Asthenes humicola*					0.33	0.20	0.16	0.33		0	+	
12. *Pyrope pyrope*					+	0.10	0.05	0.28		0.18	0.15	
13. *Lophortyx californica*					+	++	++	++		+	+	
14. *Curaeus curaeus*					+	0.05	0.26	0.22		0.28	0.10	0.27
15. *Zenaidura auriculata*						+	0.22	+		+	+	*
16. *Turdus falklandii*						0.20	0	0.33		0.14	0.15	0.15
17. *Elaenia albiceps*						0.10	0.06	0.33		0.73	0.40	0.56
18. *Leptasthenura aegithaloides*							0.19	0.22		*	*	*
19. *Pteroptochos megapodius*							+	0.22		*	*	*
20. *Colaptes pitius*							0.03	+		0	0.10	0.03
21. *Anaeretes parulus*							0.26	1.11		0.46	0.35	0.18
22. *Scytalopus magellanicus*							0.13	0.22		0.18	0.30	0.24
23. *Dendrocopos lignarius*							0.10	+		+	0.05	+
24. *Troglodytes aedon*							0.06	0.67		0.37	+	0.18
25. *Nothoprocta perdicaria*								0.06		*	*	*
26. *Patagona gigas*								0.44		*	*	*
27. *Scelorchilus albicollis*								0.11		*	*	*
28. *Spinus barbatus*								0.11		0.09	0.10	0.23
29. *Sephanoides sephanoides*								+		0.28	0.15	0.12
30. *Eugralla paradoxa*										0.55	0	0.18
31. *Pteroptochos tarnii*										0.09	0	0.18
32. *Aphrastura spinicauda*										0.28	0.30	0.44
33. *Sylviornithorhynchus desmurii*										0.18	0.75	0.15
34. *Phrygilus gayi*											0.55	*
35. *Pygarrhichas albogularis*											+	0.06
36. *Columba araucana*												0.12
37. *Microsittace ferruginea*												0.06
38. *Colorhamphus parvirostris*												0.06
39. *Scelorchilus rubecula*												+

Appendix C. Californian bird distributions and densities

Study area gradient H: −0.2, 0, 0.2, 0.4, 0.6, 0.8, 1.0, 1.2, 1.4, 1.6, 1.8, 2.0 (study areas numbered 1–10)

Species ranking	1	2	3	4	5	6	7	8	9	10
1. Eremophila alpestris	0.31									
2. Ammodramus savannarum	+	0.10	0.40							
3. Sturnella neglecta	0.30	0.75	0.30	0.09	0.05					
4. Passerculus sandwichensis		0.40	0.40	*	*					
5. Guiraca caerulea			0.20	*	*					
6. Melospiza melodia			0.15	0.50	0.25	0.65	0	0.40	0.54	
7. Pipilo fuscus			+	0.25	0.40	0.52	0.27	0.40	*	
8. Zenaidura macroura			+	+	0.40	0.35	*	+	+	+
9. Calypte costae				+	0.40	0.35		+		
10. Toxostoma redivivum				0.05	0.05	0.64	0.66	0.25		
11. Chamaea fasciata				0.25	0.10	0.44	1.72	0.10		
12. Thryomanes bewickii				0.06	0	0.35	1.06	0.30		
13. Lophortyx californica				+	0	0.41	0.33	0.60		
14. Psaltriparus minimus				0.19			0.73	0.30		
15. Carpodacus mexicanus				0.31	0.40	0.18		0.20		
16. Myiarchus cinerascens					0.07	0.18	0.20			
17. Amphispiza belli						0.29	*	*		
18. Spizella atrogularis						0.29		*		
19. Aimophila ruficeps						0.18	*			
20. Colaptes cafer						+	0.07	+		
21. Polioptila caerulea						0.12	+	0.25	0.22	
22. Calypte anna						0.47	0.53	0.20	+	
23. Aphelocoma caerulea						0.18	0.33	0.30		
24. Parus inornatus							0.20	0.50	0.10	0.09
25. Pipilo erythrophthalmus						0.35	0.53	0.60	0	0.25
26. Pheuticus melanocephalus						0.09	0.30	0.15	0	+
27. Vermivora celata							0.20	0.40	+	*
28. Dendrocopos nuttalli							0.10	0.25	0.20	*
29. Archilochus alexandri							0.07	0.30	0.38	*
30. Vireo huttoni							+	0.40	0.65	*
31. Empidonax difficilis							+	0.30	*	*
32. Troglodytes aedon								0.40	*	*
33. Phainopepla nitens								0.10	*	*
34. Melanerpes formicivorus								+	*	*
35. Spinus lawrencei							+	+	*	*
36. Spinus psaltria						+	+	+	*	*

254

#	Species			
37.	Icterus galbula	0.15	*	*
38.	Vireo gilvus	0.20	0	0.02
39.	Contopus sordidulus	0.30	+	0.23
40.	Carpodacus purpureus	0.15	0.87	0.07
41.	Columba fasciata		+	0.03
42.	Sitta pygmaea	+	1.25	*
43.	Spinus pinus		0.87	*
44.	Parus rufescens		0.54	
45.	Junco oreganus		0.87	0.32
46.	Certhia familiaris		0.22	0.10
47.	Dendrocopos villosus		0.22	0.02
48.	Nuttallornis borealis		0.05	0.03
49.	Troglodytes troglodytes			0.01
50.	Piranga ludoviciana		+	0.24
51.	Turdus migratorius			0.21
52.	Cyanositta stelleri			0.17
53.	Parus gambeli			0.11
54.	Sitta canadensis			0.09
55.	Regulus satrapa			0.06
56.	Hesperiphona vespertina			0.06
57.	Dendroica occidentalis			0.05
58.	Sphyrapicus varius			0.04
59.	Empidonax hammondii			0.03
60.	Empidonax oberholseri			
61.	Dendrocopos albolarvatus			0.03
62.	Oreortyx pictus			0.03
63.	Vireo solitarius			0.03
64.	Vermivora ruficapilla			0.02
65.	Myadestes townsendi			0.02
66.	Catharus guttatus			0.02
67.	Carpodacus cassinii			0.02
68.	Dendroica coronata			+
69.	Dryocopus pileatus			+

255

Appendix D. African bird distributions and densities

Study area ranking: 1–14. Habitat gradient H (axis): $-0.2,\ 0,\ 0.2,\ 0.4,\ 0.6,\ 0.8,\ 1.0,\ 1.2,\ 1.4,\ 1.6,\ 1.8,\ 2.0$

Species	1	2	3	4	5	6	7	8	9	10	11	12	13	14
1. *Anthus novaeseelandiae*	0.25													
2. *Calandrella cinerea*	0.12	0	0.12											
3. *Calandrella magnirostris*	0.14	0.14	0.12	0.16										
4. *Cisticola textrix*	0.60	0.60	0.36	0.16	0.10									
5. *Macronyx capensis*	0.06	0.06	0	0.06	0	0.12								
6. *Anthus leucophrys*		0.34	0.10	*	*	*								
7. *Mirafra apiata*		0.13	0	0.31	*	*								
8. *Coturnix coturnix*				0.08	*	0.12	0	0		*	*			
9. *Telephorus zeylonicus*				0.24	0	0.64	1.61	1.18	0.64					
10. *Cisticola subruficapilla*				0.16	0.16	0.64	0.16	0.10	0.18					
11. *Francolinus capensis*				0.12	0.12	0.11	0.47	0.73	0.54					
12. *Sphenoeacus afer*				0.24	0.24	0.32					0.05			
13. *Euplectes capensis*				0.05	0.05		0.23	0	0.09	0.11	0.19			
14. *Prinia maculata*				0.24		0.64	1.03	1.06	0.45	0.43	0.19			
15. *Stigmatopelia senegalensis*						0.16	0.46	*	*	*	*			
16. *Numida meleagris*						0.08	0.23	*	*	*				
17. *Crithagra flaviventris*						0.40	0.92	0.71	*	*	*			
18. *Erythropygia coryphaeus*						0.28	0.57	0.47	0.47	*	*			
19. *Fringillaria capensis*						0.16	0.23	0.35	*	*	*			
20. *Sigelus silens*						0.08	0	0	0	0.16	0.05	0.27	0.21	
21. *Streptopelia capicola*						0.40	0.46	0.12	0.36	0.33	0.05	0.10	0.83	+
22. *Cinnyris chalybeus*						0.16	0.23	0.47	0	0	0	0	*	*
23. *Oena capensis*							0.11	0			0.15	*		*
24. *Serinus canicollis*							0.11	0	0.36	0.43	0.15	0.39	*	*
25. *Cossypha caffra*							0.34	0.47	0.45	0.16	0.31	0.63	0.41	*
26. *Zosterops pallidus*							0.34	0.47	0.18	0.33	0.34	0.55	1.04	0.30
27. *Parisoma subcaeruleum*								0.24	0.06	*	*	*	*	*
28. *Saxicola torquata*								0.06	0.18	0.16	0.24	*	*	*
29. *Pycnonotus capensis*								0.18	0.24	0.10	0.43	*	*	*
30. *Onychognathus morio*								0.24	0.10	0.43	0.15	*	*	*
31. *Francolinus africanus*								0.24	0	0.33	0.19	0.55	0	0.30
32. *Apalis thoracica*								0.24	0.05	0	0.58	*	*	*
33. *Promerops cafer*									0.64	0.33	0.44	*	*	*
34. *Anthobaphes violacea*									0.36	0.33	0.15	*	*	*
35. *Poliospiza leucoptera*									0.18	0.33	0.27	*	*	*
36. *Laniarius ferrugineus*									0.09	0.10	0.27	0.63	0.62	0.05

256

#	Species						
37.	*Batis capensis*	0.09	0	0.27	1.09	0.62	0.05
38.	*Monticola rupestris*		0.33	0.08	*	*	*
39.	*Crithagra sulphurata*		0.43	0.08	*	*	*
40.	*Cisticola fulvicapilla*		0.11	0.06	*	*	*
41.	*Nectarinia famosa*		0.65	0.06	0.27	0.21	+
42.	*Streptopelia semitorquata*		0.05	0.05	0.11	0.10	0.05
43.	*Mesopicos griseocephalus*		0.05	0.05	0	*	0.05
44.	*Bradypterus victorinii*			0.08	*	*	*
45.	*Turdus olivaceus*			0.10	0.39	0.41	0.10
46.	*Andropadus importunus*					1.04	0.84
47.	*Bradypterus sylvaticus*					0.41	+
48.	*Columba arquatrix*					0.41	0.05
49.	*Muscicapa adusta*					0.41	0.15
50.	*Aplopelia larvata*					0.21	+
51.	*Camaroptera brachyura*						0.45
52.	*Tauraco corythaix*						0.15
53.	*Cyanomitra veroxii*						0.10
54.	*Dryoscopus cubla*						0.10
55.	*Tychaedon signata*						0.10
56.	*Symplectes bicolor*						0.05
57.	*Phyllastrephus terrestris*						0.05
58.	*Apalis flavida*						0.05
59.	*Cossypha dichroa*						0.05
60.	*Apaloderma narina*						0.05
61.	*Dicrurus adsimilis*						+
62.	*Phoeniculus purpureus*						+
63.	*Oriolus larvatus*						+
64.	*Pogoniulus pusillus*						+
65.	*Indicator variegatus*						+
66.	*Bycanistes bucinator*						+
67.	*Pogonocichla stellata*						+
68.	*Campethera notata*						+
69.	*Coracina caesia*						+
70.	*Turtur tympanistra*						+
71.	*Serinus scotops*						+
72.	*Trochocercus cyanomelas*						+

11 Eco-Morphological Configurations and Convergent Evolution in Species and Communities

James R. Karr and Frances C. James

Introduction

That structure and function vary together in predictable ways among sets of organisms is fundamental to all of biology. D'Arcy Thompson (1917) pioneered the graphic analysis of allometric morphological adaptations in relation to function and evolution. Frazzetta (1962, 1970) and Bock (1966) have emphasized the importance of adaptive precision in the morphological characteristics of organisms. The objective of this chapter is to consider this phenomenon in the framework of the ecological configurations of species that coexist in nature.

Classical examples of the evolutionary process are often cast in the framework of the adaptive radiation of an ancestral genotype into a set of species that are morphologically specialized to exploit a set of ecological opportunities. One can reverse this approach, which was first used by Darwin (1859) to describe the geospizid finches of the Galapagos Islands, to discuss cases of convergent evolution: animals or plants with similar ecology developing similar morphology despite different phylogenies. Examples of convergence range over a wide spectrum of taxa and habitats. Frequently-cited examples include the "fish-like" form of sharks, ichthyosaurs, and whales, or the common properties of distantly related rainforest tree species such as "drip-tip" leaves, buttresses, and thin, smooth bark. Another way to look at convergent evolution is to consider ecologically equivalent species occurring in similar environments on different continents. Among mammals, pairs of morphological types come to mind, such as the duikers of Africa and the agouti of the South American forests, flying phalangers and flying squirrels, pangolins and armadillos, elephant shrews and tree shrews. Outstanding examples among birds are the toucans and hornbills, or the hummingbirds, sunbirds, and meliphagids (Figure 1).

If environments impose constraints on the adaptations of organisms and these are expressed simultaneously in their morphology and their ecology, then sets of coexisting species should also have predictable properties. This question has been addressed to varying degrees by Harrison (1962), Lein (1972), Karr (1975), and elsewhere in this volume by Cody (Chapter 10), Pianka (Chapter 12, Table 3), and Brown (Chapter 13, Figure 5). These and other studies demonstrate striking similarities in ecology and behavior of unrelated species and communities in different parts of the world.

Another approach, the result of early studies by MacArthur and his associates, asks how the structural characteristics of habitats can be used to predict the prop-

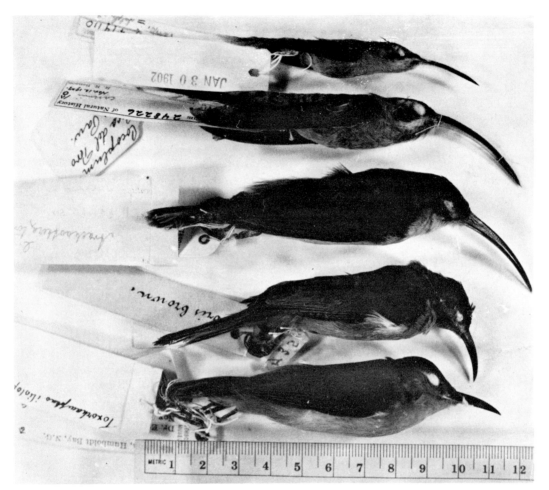

Figure 1 Five species of nectarivorous birds of the
forest undergrowth from four continents; from top:
Central America (two species), Southeast Asia,
Africa, New Guinea. They represent two orders and
three families: from top: two species of Trochilidae
(Apodiformes), two species of Nectariniidae (Passeri-
formes), and one species of Meliphagidae (Passeri-
formes). Note convergence on relatively long, down-
curved bill, small size, and drab brown or green
plumage (not shown). Species from top are *Phae-
thornis longuemareus, P. superciliosus, Arachnothera
longirostra, Nectarinia olivacea,* and *Toxorhamphus
iliolophus.*

erties of their corresponding communities of birds (MacArthur and MacArthur, 1961; MacArthur, Recher, and Cody, 1966; Recher, 1969; Karr and Roth, 1971; Blondel, Ferry, and Frochot, 1973), lizards (Pianka, 1967; Schoener, 1968), fish (Sheldon, 1968), or mammals (Rosenzweig and Winakur, 1969). Beginning with Harrison (1962), several studies have considered trophic relationships among sets of coexisting species (Orians, 1969; Blondel, 1969; Karr, 1971; McNab, 1971a; Lein, 1972; Fleming, 1973; Wilson, 1973).

Many biologists have noted that within families there are predictable relationships between morphology and ecology. Birds that are adept at feeding from the ground generally have longer legs than species that feed from the vegetation (Dilger, 1956; Newton, 1967). Hespenheide (1971) demonstrated that body size in tyrannid flycatchers and the size of insects utilized for food are positively correlated. More recently (Chapter 7) Hespenheide has expanded his studies to show that the relation between bird size and food varies with bird taxon, bird foraging behavior, and also food taxon.

Lack (1947) has shown that bill structure varies among species of geospizid finches, with insectivores having longer, thinner bills and seed eaters having shorter, stouter bills. In addition, Lack demonstrated that presence or absence of related species on specific islands affects the bill dimensions and the food of certain Galapagos finches. Clearly, complex interactions mold the morphological patterns observed within and between species.

Other studies (Lack, 1954; Hutchinson and MacArthur, 1959; Schoener, 1968; MacArthur, 1972; Diamond, 1973) have dealt with selected species or species groups in an effort to determine: 1) how available resources are partitioned to minimize competition for limited resources, and 2) how morphology is tuned with ecology and behavior to maximize the efficiency of each species. Selander (1966) accounts for sexual differences in the morphology of woodpeckers by invoking intraspecific competition. Van Valen (1965), Willson (1969), Soulé and Stewart (1970), Grant (1968, 1971), McNaughton and Wolf (1970), Rothstein (1973), Willson, Karr, and Roth (1975), Hespenheide (Chapter 7), and Hutchinson (Chapter 17) ask whether niche width may be related to the amount of morphological variation within and between species.

The only previous studies integrating both ecological and morphological data quantitatively within communities are those by Keast, which consider morphological patterns of sets of sympatric species in fishes and birds. The fish study (Keast and Webb, 1966) examined mouth and fin structure and body shape of 14 species of fish cohabiting fresh water areas in eastern Ontario, Canada. This study attempted to determine the extent to which the competitive exclusion principle is sustained in fishes with broadly overlapping food habits. More recently, an expansion of the study showed how factors such as different combinations of species and different age distributions interact with food organisms and relative abundances of predator and prey (Keast,

1970). In studies of bird morphologies, Keast (1972) explored the ecological and morphological diversification involved in the proliferation of the family Tyrannidae in the neotropics, and related that diversification to patterns observed in species exploiting the same adaptive zone in other regions.

Using information on the morphology and the ecology of the birds occurring in tropical lowland forests of Panama and Liberia, plus the breeding birds of a bottomland forest in Illinois, we will apply graphical and mathematical methods to explore the eco-morphological configurations of the avifaunas and the extent of convergence among them. We present plots of a variety of combinations of morphological variables that convey insight into morphological consequences of known ecological differences. In some cases these consequences are obvious and are used to standardize or calibrate our methods. In other cases they are less obvious. Sometimes ratios of the original variables and transformations such as cube roots and logarithms of weights are used when they seem biologically meaningful. In addition we present two examples of applications of multivariate analysis to ecological data. The first is a principal components analysis of a set of ratios of the morphological measurements. This procedure has been applied previously to problems in avian systematics (e.g., Johnston and Selander, 1973; Niles, 1973; Schnell, 1970) and to considerations of avian behavior and distribution by James (1971) and James and Shugart (1974). Discriminant function analysis was used

to separate birds according to aspects of their preferred habitats by Cody (1968) and James (1971). The second multivariate method used here is a canonical correlation analysis of all of the morphological and ecological data for 196 species. This method has been applied to ecological problems previously by Jameson and associates in their studies of the relationship between climatic variation and morphological variation in tree frogs (Calhoon and Jameson, 1970; Vogt and Jameson, 1970; Jameson, Mackey, and Anderson, 1973). See also Austin (1968) and Webb et al. (1973).

Study Areas and Methods

For our ecological data we draw on field work conducted over the past seven years, including an eight-month study of several habitats in east-central Illinois (Karr, 1968), a thirteen-month study in Panama (Karr, 1971), and a three-month study in Liberia (Karr, 1975; unpublished). This chapter will deal primarily with the forest areas involved in these studies. Initially, we shall discuss the ecological and morphological patterns of many of the 184 forest birds in the vicinity of Mount Nimba, Liberia, West Africa. We use the expression "forest bird" in the sense of Moreau (1966): forest birds are species dependent on forest vegetation. Later, we concentrate on the resident avifaunas of the three forest study areas surveyed in the studies mentioned above. The "resident avifauna" of an area consists of the resident species during the season of highest breeding activity; a resident is a species "that could be seen or netted almost daily on the study area" (Karr, 1971). The problems of applying that definition in tropical faunas are discussed in another paper (Karr, 1975). All species were categorized according to their ecologies, as described in Karr (1971, 1975) with some modifications. Ecological categories are listed in Table 5.

Two of the three forest study areas (Panama and Liberia) are lowland tropical forests that receive 2600 to 2800 mm of rainfall per year, and the third (Illinois) is a temperate bottomland forest receiving only about 1000 mm per year (Karr, 1975). Since both of the tropical forests are in the lowlands (<600 m), they have high mean annual temperatures (Panama, 27°C; Liberia, 23°C). The mean annual temperature in Illinois is 11°C, but, of course, the breeding season temperature is higher (19°C). Trees commonly exceed 100 feet in the tropical forests but are less tall in the temperate forest. The foliage height diversities (MacArthur and Mac-Arthur 1961) as measured by Karr (1968) for three vegetation strata are 1.07, 1.09, and 1.04 for the Illinois, Liberia, and Panama forest study areas, respectively. The volume occupied by foliage is greater in the tropical forests because of the taller trees.

All three forest areas were disturbed to varying degrees. The Illinois forest was strip-mined for coal in the late 1800s but has been undisturbed since that time. The effects of that mining on the avifauna, if any effects remained into the 1960s, were probably to increase the species diversity of birds (Karr, 1968). The Panama study area was adjacent to a hunting camp, so

that some disturbance to both the vegetation and the birds was inevitable. A few large trees had been removed from the Liberian forest by loggers about eight years before the study, and heavy hunting pressure had reduced, or exterminated, populations of several large species. Similarly, hunting pressure had exterminated macaws, currasows, and other groups in the vicinity of the Panama study area.

Specimens of the species observed in the three study areas were measured from study skins either in the National Museum of Kenya, the National Museum of Natural History in Washington, or the University of Kansas Museum of Natural History. The following characters were measured for one male and one female of each species: culmen (bill) length, bill width, bill depth, gape, wing length, tail length, tarsal length, and length of central toe. The data reported in this chapter are for males, except in a few species for which we were unable to obtain male specimens.

Throughout our discussion, when we refer to the diversity or species diversity of our avifaunas, we will be concerned merely with the number of species present, not with the information-theoretic or some other measure affected by both the number of species and their relative abundance.

Segregation Within a Lowland Forest Avifauna

Several families of birds are known for their extreme morphological and ecological diversification. Classical examples include the Hawaiian honeycreepers (Ama-

don, 1950) and the Galapagos finches
(Lack, 1947). Other less diverse families
also exhibit structural variations correlated with feeding ecology. Let us consider
several examples.

On the basis of food and feeding site,
two major groups of pigeons (Columbidae) occur in the forest fauna of Liberia.
The first group feeds primarily on seeds
or fruits, or both, that have fallen to the
ground, whereas the second gleans fruit
in the forest canopy. These two groups
also segregate by morphology, the canopy
feeders being larger and having relatively
longer wings (Figure 2a). Furthermore,
the pattern is consistent for both the African and neotropical columbids in our
study areas (compare triangles L and P in
Figure 2a). Presumably the large size and
relatively long wings of the frugivores are
adaptations that enable the birds to make
the long flights between the widely dispersed patches of food associated with
fruiting trees. The smaller sized terrestrial
pigeons have longer legs. This is more
suitable for species that walk a great deal.
Their short wings allow quick escape, like
the explosive take-off and short flight of
quails and pheasants. In fact, the similarity of one genus of neotropical forest pigeons to the quails gives them their common English name, quail-dove.

Liberian hornbills illustrate a similar
segregation in morphology and feeding
ecology (Figure 2b). Frugivorous species
are larger and relatively longer-winged
than the species that feed on arthropods
and lizards. One omnivorous species
(symbol O in Figure 2b) is intermediate.
Again, the long flights from fruiting tree
to fruiting tree require the greater effi-

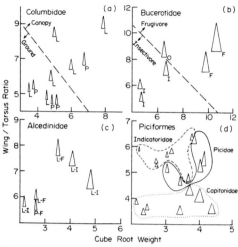

Figure 2 Segregation by morphological characters
in several Liberian (L) bird families. Each triangle
refers to one bird species. The position of the triangle gives the wing/tarsus ratio and the cube root of
weight; the base and height of the triangle are proportional to bill width and bill length, respectively.
(a) Columbidae (pigeons) segregate morphologically
by feeding location (canopy vs. ground). (b) Bucerotidae (hornbills) segregate morphologically by food
habits. I = Insectivore; F = frugivore; O = omnivore. (c) Alcedinidae (kingfisher) do not segregate
by food habits for these morphological characters.
I = insectivore; F = fish-eater. (d) Piciformes
(woodpeckers and allies). Three families segregate
morphologically by wing-to-tarsus ratios, and, within
families, by size. Note similarities between Panama
(P) and Liberia (L) species in Figures 2a and 2c.

ciency of relatively longer wings. The insectivores have shorter stouter wings for
more lift as they hover briefly to snatch
a large insect from its hiding place. These
two families exploit resources that are either of the same type in two different
locations (pigeons), or else two different
types of resources (hornbills). In both
cases the morphology of two groups
within the family is adapted to different
foraging behaviors.

The kingfishers do not segregate by food habits according to these morphological characters of wing/tarsus ratio versus size (Figure 2c). Although two types of food are exploited by the kingfishers, the similarity in foraging methods required to obtain insects and fishes has resulted in similar size and similar wing-tarsus ratios. Instead of segregating by food type there are two small species—one feeding on insects and the other on fish—and also several large-bodied species feeding on each resource. The single kingfisher known from the Panama forest study area is a small fish-eater with about the same morphology as the small Liberian piscivore (compare triangles L-F and P-F in Figure 2c). The kingfishers do, however, segregate according to food habits by bill ratios. The insectivores have culmen-length/bill-width ratios of 3.5 to 5.0, whereas the culmen-length/bill-width ratios of piscivores range from 6.4 to 6.7 (compare proportions of the triangles F and I in Figure 2c). With the same wing and size a fisheater needs a relatively narrower bill than does a kingfisher that feeds on insects.

Within the Piciformes (woodpeckers and allies) three families are represented in the forest birds of Liberia. The woodpeckers (Picidae) are mostly bark gleaners and drillers, the barbets (Capitonidae) are fruit or insect gleaners, or both, and the honeyguides (Indicatoridae) are specialized insectivores. All three families cover the same range of sizes, as measured by weights, but the relative lengths of wings and legs vary among the families (Figure 2d). The wide-ranging honeyguides have relatively the longest wings, and the

barbets the shortest. The broader bills of the barbets and narrower bills of the woodpeckers can be seen in the proportions of the triangles (Figure 2d). Within the families of the Piciformes there is little morphological segregation correlated with ecological differences, but significant segregation exists between families.

Many other examples of morphological segregation are obvious. For example, most sunbirds in the genus *Nectarinia* have longer thinner bills than members of the genus *Anthreptes*, because most species of *Nectarinia* feed in association with flowers. Two groups of sunbirds, the nectarivores and insectivores, segregate well by bill proportions (Figure 3). The nectarivorous group are all *Nectarinia;* the insectivorous group is composed of three species of *Anthreptes* and one of *Nectarinia*. Similarly, the relative shapes of wing, bill, and tail in the flycatcher family Muscicapidae suggest a variety of foraging techniques (Figure 3). Here we see that the foliage gleaning flycatchers have

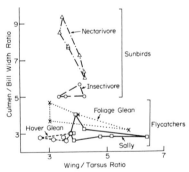

Figure 3 Segregation by morphological character ratios in Liberian Muscicapidae (flycatchers) and Nectariniidae (sunbirds). There is no overlap between the two groups of sunbirds, but moderate overlap among the three groups of flycatchers.

relatively thinner bills than their hover-gleaning or sallying associates. Salliers are longer winged (or with shorter tarsi) than hover-gleaners. However, the separation of these three groups is not as clear as in the two groups of sunbirds.

Principal Components Analysis of Morphological Variation Within Communities

The usual role of statistics in data analysis is that of testing hypotheses, which have been deduced from theory, against empirical results. But there are heuristic uses that are potentially equally important. Cattell (1966) argued for this approach, emphasizing the importance of the free observation of complex systems as opposed to the manipulative experimental approach. By means of multivariate analysis we can view such systems with non-interfering statistical finesse. Although there are no experimentals and controls in the classical sense. this type of analysis is "experimental" in the sense of an exploratory survey (Cooley and Lohnes, 1971). Since one of the goals of ecology is to describe the interrelationships between organisms and their environment, the use of multivariate analysis would appear to be a powerful and appropriate tool.

Of the various multivariate methods available, the principal components and canonical correlation models are particularly appropriate to ecological problems (Pielou, 1969). First, the assumption of normally distributed data is not a prerequisite. Second, the results are suitable for graphic display. Since sets of variables

are often dependent among themselves, they should be considered together as a system, rather than be split off from one another (Kendall, 1957). The procedures may indicate refinements of constructs that would be impossible with manipulative experimental techniques (Tiedeman and Tatsuoka, 1963).

Thus far in this study we have presented graphic displays of a variety of morphological relationships among species of birds. Transformations and ratios of the originally measured variables were used as axes of two-dimensional plots to view the patterns of within-family and between-family segregation (Figures 2,3). These graphs are useful, but are unsatisfactory for two reasons. First, the axes are more or less arbitrarily selected, and may inadequately represent the major ways in which the morphologies differ in the original sense of D'Arcy Thompson (1917). To display combinations of linear measurements for species that in fact vary in complex allometric ways is limiting. It would be preferable to represent the variation in a space determined by the total morphologies of the species in question. The second drawback of the plots of within- and between-family variation is that they overemphasize phylogenetic relationships at the expense of some potentially important ecological considerations. For instance, the extent of convergence or divergence among many of the remaining coexisting species is overlooked. The only methods available for this expanded complex problem are those of multivariate analysis.

Principal components analysis, a form of factor analysis, was applied to a covari-

Table 1. Correlations of principal components I and II constructed by analysis of seven primary morphological ratios

Morphological ratio	Principal component		
	I	II	
Wing/tail	0.37	0.03	
Wing/culmen	.98*	− .14	
Wing/tarsus	.75*	.61*	
Tarsus/toe	− .38	− .53	
Culmen/gape	− .51	.63*	
Culmen/bill width	− .45	.60*	
Bill width/bill depth	− .05	− .23	
Per cent of			Total
total variance	71.5	16.3	[87.8]

Principal components I and II were extracted for morphological characters of 184 lowland forest bird species of Mount Nimba, Liberia. Asterisks indicate ratios most significantly correlated with the principal components.

ance matrix of ratios of the morphological measurements for bird species that coexist in the lowland forest area of Mount Nimba, Liberia (Table 1). This procedure extracts the combination of original variables that accounts for the greatest amount of variation in the original data matrix. This combination of original variables is called the major axis, principal axis, or principal component I. Subsequent principal components (II, III, etc.) account for progressively smaller amounts of variation. With this multivariate technique it is possible to consider simultaneously all of the original variables.

The ratios were chosen to include those aspects of morphology that we intuitively felt would best represent evolutionarily significant morphological relationships in birds. For this analysis we pooled the data

from all 184 species of Liberian forest birds, totalling 335 individuals. Both males and females were included in the analysis. Only males are shown in the following discussion, except in a few species for which male specimens were unavailable.

Correlations with original variables (Table 1) indicate that PC I is most strongly correlated with wing-to-culmen ratio and secondly with wing-to-tarsus ratio, while PC II is highly correlated with culmen-to-bill width and/or gape ratios and also with wing-to-tarsus ratio. PC I is separating relatively long-winged species from relatively short-winged species, in terms of the length of wing relative to culmen and tarsal length. PC II is influenced especially by the shape of the bill. Loadings of correlations between the ratios and the first and second principal components are given in Table 1. We should stress that size variation is not expressed in this analysis as it was in Figures 2–3, which used the cube root of weight.

Although we have seen that a plot of the Muscicapidae (flycatchers) in a space determined by two ratios of measurements does not show complete segregation of the three types of foraging behavior (Figure 3), a striking segregation of the three groups of flycatchers becomes evident in a space determined by PC I and PC II (Figure 4, symbols □, X, ◇). In addition, the new plot preserves the segregation of the nectar-feeding sunbirds (symbol △) in the upper left of the plot due to their relatively long, thin bills. The sallying muscicapids (□) are segregated by their

relatively long wings. The short, broad-
billed, hover-gleaning flycatchers (◇) are
segregated from the sunbirds and from the
sallying (☐) and foliage-gleaning (X) flycatchers. The foliage-gleaning muscicapids (X) and the foliage-gleaning (insectivorous) sunbirds (○) are similar. The
Sylviidae (warblers) overlap the morphological space of the foliage-gleaning sunbirds and flycatchers. The latter two
groups are similar to the warblers in
resource-exploitation pattern, demonstrating convergence on a similar morphology by ecologically similar species of
three different phylogenetic stocks. Within
the Muscicapidae we can see three major
foraging behaviors: sallying, hover-gleaning, and foliage-gleaning. Salliers are
longer-winged than hover-gleaners (Mann
Whitney U test significant, p < 0.05) and
foliage-gleaners (MWU not significant,
p < 0.10). The foliage-gleaners have
larger culmen/bill-width ratios than the
hover-gleaners (MWU significant,
p < 0.02). The latter group includes a
number of species with very broad bills,
useful for snapping food from a substrate
without alighting.

The Muscicapidae have diversified be-
haviorally and morphologically. They
could be considered plastic in their behavioral responses to selection pressure.
By comparison, the warblers are more
conservative for variation in morphology
and foraging behavior. Within the sets of
species depicted in Figure 4, segregation
occurs by different habitat selection patterns that are not associated with major
shifts in foraging behavior or morphology
(Table 2). Some species are found in forest

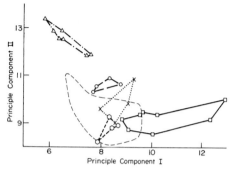

Figure 4 Segregation of Liberian Muscicapidae
(flycatchers) and Nectariniidae (sunbirds) on principal components I and II constructed by analysis of
seven primary morphological ratios (Table 1). Note
the almost complete segregation of two groups of
sunbirds and three groups of flycatchers (symbols as
in Figure 3). Warblers (Sylviidae) fall within the area
enclosed by the dashed lines and overlap extensively
with ecologically similar groups of sunbirds and
flycatchers.

edge while others are more at home in
forest. In addition, species sort out in
space by using different strata of the vegetation in both habitats. A few species
found to some extent in both forest and
forest-border situations have well-defined
stratal preferences. Since Liberia, and, in
fact, most of the Upper Guinea forest in
West Africa, has only isolated low mountains, no distinct altitudinal segregation of
the sort well documented for areas with
greater topographic complexity (Diamond, 1973) has taken place in Liberia.

Like the Sylviidae (warblers), the bulbuls (Pycnonotidae) are fairly conservative in size and shape, although they
are the richest Liberian family in number
of species. Most bulbuls are small to medium sized gleaners with drab plumage of
greens, grays, and browns. The 20 species
segregate on the basis of food into 10

Table 2.   Habitat segregation in two groups of Muscicapidae (flycatchers) and the Sylviidae (warblers)

Habitat	Stratum		
	Low	Medium	High
Sallying Muscicapidae			
Forest interior	—	1	1
Forest edge	—	*	—
Both	1	2	2
Stream edges	1	—	—
Hover-gleaning Muscicapidae			
Forest interior	1	—	1
Forest edge	1	*	1
Both	—	1	—
Sylviidae			
Forest interior	—	2	1
Forest edge	4	*	1
Both	1	2	2

Each entry gives the number of species in the given group coexisting in the given vegetational stratum of the given habitat type. (The Sylviidae and Muscicapidae are sometimes considered subfamilies of the same family.) Asterisks indicate that no medium stratum is recognized in forest edge habitat by the classificatory procedures of Karr (1971). Note that it is rare for more than two species in the same group to coexist in the same stratum.

insectivores, 8 frugivores, and 2 omnivores. Attempts to discover morphological complexes that might serve to segregate the species of bulbuls were not satisfactory when original variables were utilized. The best segregation was obtained with the principal components analysis of seven primary ratios (Figure 5). There is a distinct separation of insectivorous and frugivorous bulbuls. In addition, two species that feed on both insects and fruits are intermediate between the insectivore and frugivore groups.

Since PC I is highly correlated with relatively long wings and PC II is associated with long culmen in relation to bill-width or gape, we might guess from Figure 5 that frugivorous bulbuls have long wings and similar culmen lengths when compared with insectivores from the same family. Surprisingly, the Mann Whitney U test for differences between wing and culmen lengths indicates that the wings of frugivores and insectivores are not significantly different ($p > 0.10$) but the culmen lengths of the two groups are significantly different (insectivores 22.5 mm; frugivores 17.3 mm; $p < 0.01$). Segregation based on PC I is not due to varying wing lengths but to varying culmen lengths with wing length held constant. The fundamental bill differences in the two groups of bulbuls cannot be discerned in PC II, as it is influenced primarily by a *ratio* of bill dimensions.

Simultaneous consideration of large arrays of morphological data by principal components analysis has produced insight into the morphological characters associated with various ecologies. In a later section we expand our

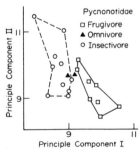

Figure 5 Segregation by food habits among Liberian Pycnonotidae (bulbuls) on principal components I and II constructed by analysis of seven primary morphological ratios (Table 1).

analysis to the simultaneous consideration of matrices of both morphological and ecological variables by means of canonical correlation analysis.

Congeners and Character Displacement

These general patterns notwithstanding, there is still substantial variation within some families, such as the warblers, that cannot be accounted for by consideration of just one type of ecological variation. Sets of ecological variables covary in complex ways. One way to investigate finer-scale tuning between and among species is to look at similarities and differences in patterns of congeners, which might behave in either of two opposite ways. On the one hand, congeners might exhibit greater similarity than non-congeners because of their presumed more recent common ancestry; that is, congeners are more likely to be nearest neighbors (in morphological space) than non-congeners when all members of a family are compared. On the other hand, the ecological similarity of sympatric congeners might result in divergence in their morphologies due to character displacement.

In order to operate from a conservative hypothesis, we suggest that the random distribution of species in morphological space will result in nearest neighbors being congeners in proportion to their frequency in the population. For example, consider a family with five Liberian forest species, three of which are congeners, like the shrikes (Laniidae) (Figure 6b). If the five species are distributed at random in space, the nearest neighbor over all spe-

cies will be a congener three in every ten possible occurrences. Table 3 summarizes, for seven families having significant numbers of congeners, the expected and actual frequencies of nearest neighbors that are congeners. In all families except the Muscicapidae the actual number of congeneric nearest neighbors is less than that predicted by the conservative, random-distribution hypothesis, implying that congeners are more different on the average than any random selection of two species (Figure 6). This suggests that morphological differences between the presumably more similar sympatric congeners may be exaggerated by character divergence through competitive displacement. The only exception in the seven groups analyzed is due to the extreme similarity of two species of *Fraseria* in the Muscicapidae. It is perhaps significant that these two species rarely encounter each other in the wild. *F. ocreata* is a forest-interior species, whereas *F. cinerascens* occurs along streams where thickets overhang the water surface.

An examination of the pattern of differences between the expected and actual frequencies is instructive (Table 3). The families with large differences (Picidae, Laniidae) segregate very little or not at all by foraging behavior and/or habitat as defined above. Most of the segregation in these families occurs in morphological space, as sequences of species with the same general shape but differing in size. In the Laniidae, for example, a three-species sequence of congeners (*Malaconotus*) have weights of 53, 75, and 96 gms. Two families with low differences in actual

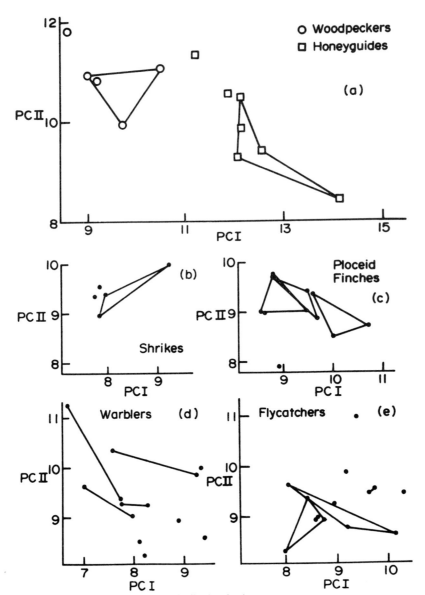

Figure 6 Segregation within several Liberian families on principal components I and II constructed by analysis of seven primary morphological ratios (Table 1). Note that congeners (connected by lines) are usually widely spaced. (See text and Table 3 for more complete discussion.)

Table 3. Actual and expected frequency of congeneric nearest neighbors for seven families

Family	No. species	No. genera	Frequencies of nearest neighbor congeners		Difference between expected and actual E − A
			Expected (E)	Actual(A)	
Picidae	5	3	0.30	0.00	+0.30
Indicatoridae	7	4	.29	.25	+ .04
Pycnonotidae	10	4	.20	.11	+ .09
Laniidae	5	3	.30	.00	+ .30
Sylviidae	13	9	.05	.00	+ .05
Muscicapidae	16	11	.06	.14	− .08
Ploceidae	11	5	.16	.06	+ .09

For each family the second and third columns give the number of species and of genera, respectively, that are represented in Liberian forest and for which we have male specimens. Column 4 gives the frequency with which nearest neighbors in morphological space (i.e., graphs like Figure 6) would be congeners if species were distributed at random, whereas column 5 gives the actual incidence of congeners. Values for the Pycnonotidae refer only to insectivorous species. Note that the actual incidence of congeneric nearest neighbors in almost all families is less than the incidence predicted for random distributions (column 6).

and expected differences have been shown to segregate to a great extent by habitat. In the Sylviidae, species share the same general size and shape. In the Muscicapidae, members of each foraging group segregate by habitat. The low value of the Indicatoridae may seem anomalous. However, an examination of Figure 2d shows that there is more variation in bill shapes in this family than in the other picid families, indicating greater behavioral or ecological segregation.

The intermediate values of the insectivorous Pycnonotidae and the Ploceidae (forest species only) underline an earlier frustration we encountered in trying to account for patterns of species diversity in these groups. In both cases there are frequently three congeneric species in the same mixed insectivorous flock. The patterns of segregation and overlap are not as clear as the patterns of segregation in the ant-wrens (Formicariidae) in Central America. The forest bulbuls and weaver-finches of African forests warrant more detailed study.

To review: with a limited series of families it appears that congeners are less likely to be close together than noncongeners in morphological space, supporting the suggestion that character displacement is operating to reduce competitive interactions. However, one must guard against interpreting results strictly on the basis of morphological configurations. Interactions of morphology, ecology, and behavior are optimized in complex ways.

Intercontinental Ecological and Morphological Equivalents

If, as we have seen, there are convergences on eco-morphological configurations under similar selective pressures between families, and divergences within families under varying selective pressures, we might expect to find convergences of eco-morphological configurations on different continents with similar selective pressures, as discussed for Mediterranean birds by Cody (Chapter 10) and for desert lizards by Pianka (Chapter 12). In this section, we shall compare the birds of the lowland forest in Liberia with the birds of similar forest in Panama.

A species-by-species analysis quickly demonstrates that precise correspondences between species of the two faunas are limited (cf. Figure 7 and text of Cody, Chapter 10). That many species cannot be matched on a one-to-one basis is due to a variety of reasons. In some cases the correspondences in morphological space are prevented by taxonomic correlates of morphological characters. For example, the sunbirds and hummingbirds are often

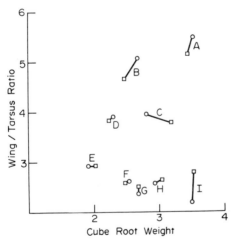

Figure 7 Morphological similarities of ecologically equivalent species in Panama (○) and Liberia (□). See Table 4 for species names, and text for discussion.

similar in ecology, behavior, and morphology. Any attempt to have them coincide in morphological space is doomed, however, if tarsal length is included as an original morphological variable. The very short tarsi of the hummingbirds separate them in morphological space from their nearest African equivalents, the sunbirds.

In addition, the assignment of one-to-one equivalents leaves a large pool of species that replace each other on 2-for-3, 3-for-5, or other complex bases (cf. Figure 7c–7r of Cody, Chapter 10). Given the variety of adaptations on behavioral, morphological, and ecological criteria illustrated by birds, these results should come as no surprise.

It is possible to describe a number of equivalents, however, similar to such widely known equivalents as the longclaw (*Macronyx*), meadowlark (*Sturnella*), and red-breasted meadowlark (*Pezites*) of Africa and North and South American grasslands, respectively (cf. Figure 7a of Cody, Chapter 10). Nine species pairs (Figure 7, Table 4) illustrate two apparently general phenomena. First, the probability of precise correspondence increases with decreasing bird size. Second, the frequency of morphological equivalents is more common in passerines than in non-passerines and more common in insectivores than in other trophic groups. Intercontinental variation in non-insect food resources and their availability reduces the frequency of precise morphological correspondence in sets of species exploiting these resources. (See Karr, 1975, for a comparison of resource-exploitation patterns among the three forests.) Simi-

Table 4.   Scientific names and ecological configurations of equivalent species in Panama and Liberia

Pair code	Species		Ecology
	Panama	Liberia	
A	*Tyrannus melancholicus*	*Meleanornis annamarulae*	High stratum, sallying, forest border
B	*Onychorhychus mexicanus*	*Terpsiphone rufiventer*	Low-to-medium stratum, sallying, forest
C	*Rhynchocyclus olivaceus*	*Fraseria ocreata*	Medium stratum, sallying, forest
D	*Myiobius sulphureipygius*	*Trochocercus nitens*	Low-to-medium stratum, sallying, forest
E	*Polioptila plumbea*	*Apalis nigriceps*	High stratum, foliage-gleaning, forest
F	*Henicorhina leucosticta*	*Stiphrornis erythrothorax*	Ground-gleaning, forest
G	*Microcerculus philomela*	*Sheppardia cyornithopsis*	Ground-gleaning, forest
H	*Cyphorhinus phaeocephalus*	*Trichastoma cleaveri*	Ground-gleaning, forest
I	*Grallaria perspicillata*	*Trichastoma puveli*	Ground-gleaning, forest

The table lists nine cases in which there is one-to-one equivalence between a Liberian species and a Panamanian species both in ecology (last column) and in morphology (displayed in Figure 7 with the code letters of the first column). All cases involve insectivorous passerines.

larly, phylogenetically conservative morphological characters (e.g., the short tarsi of hummingbirds cited above) reduce the frequency of morphological correspondence. An outstanding example of ecological constraints occurs in the bark feeders in the New World. Temperate bark-stratum feeders, primarily woodpeckers, harvest insects by drilling and/or scaling the rough bark of most temperate forest trees. Neotropical bark feeders (especially woodcreepers), on the other hand, have very different foraging strategies imposed on them by the thin smooth bark of tropical forest trees. The temperate woodpeckers have stout bills with a number of adaptations for "drilling," while the woodcreepers have long, narrow, often decurved bills for probing into crevices between branches, under vines, or associ-

ated with epiphytes. The two groups do have one common problem—hanging onto the trunk and larger branches of trees. Their morphological adaptations include, in parallel, large feet with impressive claws and elongated and/or strengthened tails for support.

Canonical Correlation Analysis

Commonly used methods of analysis, e.g., correlation, may be used to determine how morphological characters vary with one another. In the case of bird morphology, for example, we might ask how leg and tail lengths are correlated. Similar questions might be asked with respect to ecological variables. An equally interesting and rarely asked question in ecological research is: "How are morphology *and* ecology related?" With single morphological and ecological variables it may be possible to measure association of variables quantitatively, or perhaps associations may be obvious without extensive quantification. Significant problems may arise, however, when one attempts to depict the associations involved between two complex sets of variables. "How are the variables in one group correlated with the variables in the other group?" In our study we ask, "How are morphology and ecology related? What are the ecological correlates of patterns of morphological variation?"

Hotelling has developed a technique, the canonical correlation procedure, which is appropriate to this problem. The method was developed for studying the relationships between two batteries of psychological tests administered to the same subjects. It has been used widely in educational research and also in economics (Morrison, 1967). Basically, this is an extension of the principal components method whereby two multivariate sets of data that pertain to the same individuals or species are analyzed simultaneously. It is a way of comparing two matrices of information simultaneously and is thus ideally suited for the mathematical description of many kinds of ecological problems. We will compare a matrix of ecological information on bird diets, foraging behaviors, and preferred strata of the vegetation with a second matrix of measurements of bird morphologies.

The canonical correlation procedure is a form of generalized regression utilized to extract sets of factors that are uncor-related within each system but have maximum correlations of pairs of factors between systems. The first pair of variates exhibits maximum correlation between the two data sets. The resulting coefficient will be the largest product-moment correlation that can be developed between linear functions of the two systems. Then a second factor orthogonal to (uncorrelated with) the first is located; then a third orthogonal to the first two; and so on (Cooley and Lohnes, 1971).

Our data set consists of 17 morphological variables and 14 ecological variables for 196 species of birds. The morphological variables include the square roots of seven of the originally measured variables, nine ratios of these measurements, and $\log_{10}$ of the weight. The ecological variables include five categories of food, five of foraging behavior, and four of vegetation stratum (Table 5). The species consist of 50 resident species from the tropical lowland forest of Liberia, 100 resident species from the tropical lowland forest and the late shrub of Panama, and 46 resident species from the bottomland forest and late shrub of Illinois. The advantage of pooling the data from several areas is that subsequently one can draw out subsets and make valid comparisons among communities. The pooled canonical correlation analysis yields linear indices of morphology and ecology that exhibit the highest possible correlation. Subsets of the species occurring in the separate communities are plotted in a space defined by these indices.

In this system there are six highly significant relationships between morphol-

Table 5.   Thirty-one variables of the ecology and morphology of birds used in the canonical correlation analysis, and their correlations with the six significant canonical variates

Variables	Correlation with canonical variate					
	I	II	III	IV	V	VI
Ecology						
1. Food: fruit			−0.43			
2. Food: insects, fruit						
3. Food: insects, nectar	0.92					
4. Food: insects			.81	0.48		
5. Food: carnivore			− .56	.57		−0.42
6. Foraging behavior: bark-drilling						− .48
7. Foraging behavior: bark-gleaning						
8. Foraging behavior: ground-gleaning		0.55			0.40	
9. Foraging behavior: foliage-gleaning					− .59	
10. Foraging behavior: hover-gleaning	.38	− .43			.45	.40
11. Stratum: bark				.40		− .51
12. Stratum: ground		.53			.40	
13. Stratum: medium						
14. Stratum: high		− .44				
Morphology						
15. Culmen*						
16. Bill width*		− .39	− .39			
17. Bill depth*			− .43			
18. Wing*						− .63
19. Tail*						
20. Tarsus*		.42				
21. Central toe*						− .45
22. Culmen/bill width	.65					
23. Culmen/bill depth	.77		.36			
24. Culmen/wing	.56					
25. Culmen/tarsus	.77					
26. Bill width/bill depth		− .39		.31	.39	
27. Wing/tail					.55	
28. Wing/tarsus		− .62				
29. Wing/central toe	.50	− .64				
30. Wing/culmen			− .32			− .37
31. Weight**			− .33			− .50

The ecological data are multinomial. Since the analysis involves inverting the covariance matrix, one category each of food, foraging behavior, and stratum was defined as a base level and dropped from the list of variables to avoid singularities. These variables were: food—insects, seeds; foraging behavior—sallying; stratum—low. See Morrison (1967) for discussion of the statistical methods. *Square root of the measurement (mm). **log$_{10}$ of the measurement (g). Word descriptions of the six variates derived from these correlations are given in Table 6.

ogy and ecology (Tables 5 and 6). Word descriptions of these correlated canonical pairs, derived from their correlations with the originally measured variables, are presented in Table 6. Thus, birds with extremely long thin bills and small bodies tend to be hover-gleaners. Birds having relatively long legs tend to occupy the ground stratum. Small birds having long bills tend to be insectivorous. When the variation attributable to this relationship has been removed, other birds having relatively wide bills tend to be either insectivorous or carnivorous. Birds with short tails and flat bills tend to be either ground-gleaning or hover-gleaning. Small size is correlated with hover-gleaning

habits, rather than with bark-drilling or carnivorous habits. Ornithologists are familiar with all of these relationships and will immediately think of examples. What has been discovered is merely a mathematical way of expressing the relationships simultaneously.

To explore the usefulness of the method in greater detail, we have taken canonical variate II and plotted the species in the space determined by the correlation between morphology and ecology for that relationship (Figures 8 and 9). This variate expresses the significant relationship between the relative length of the leg and the stratum of the habitat that is occupied. Long-legged ground-dwelling birds hav-

Table 6. Results of a canonical correlation analysis of the relationships between morphology and ecology for 196 species of temperate and tropical birds

Canonical variate	Canonical correlation between morphology and ecology	Chi-square	Degrees of freedom	Word description	
				Morphology	Ecology
I	0.81	830.8	238	Relatively longer thinner bill; smaller body	Hover-gleaning
II	.79	635.4	208	Relatively longer leg, narrower bill	Ground stratum
III	.71	460.8	180	Longer bill, lower weight	Insectivore; bark stratum
IV	.64	336.3	154	Wider bill	Carnivore or insectivore
V	.57	242.0	130	Shorter tail, flatter bill	Ground-gleaning or hover-gleaning, not foliage-gleaning
VI	.51	170.2	108	Decreasing size	Hover-gleaning, not bark drilling or carnivore

The analysis demonstrates six canonical variates, each of which represents an independently significant relationship between the morphology and the behavior of the species in question. The ecological and morphological characters used are summarized in Table 5. With the null hypothesis that the remaining canonical correlations are zero, the probability of larger chi-squares by chance is less than 0.01 for all six roots.

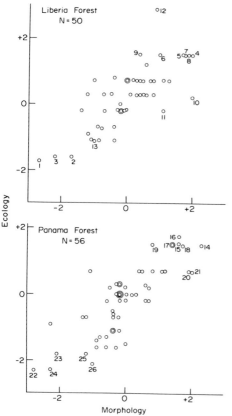

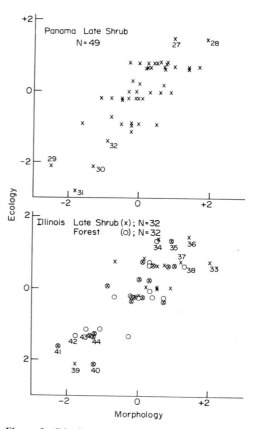

Figure 8 Distribution of Liberia and Panama forest
residents in a space determined by canonical variate
II, expressing the correlation between the relative
length of the leg (abscissa: long-legged birds to the
right, short-legged birds to the left) and stratum of
the habitat occupied (ordinate: ground birds towards
the top, canopy birds towards the bottom). Note that
ground birds tend to be long-legged, canopy birds
short-legged. Numbers refer to species named and
discussed in the text.

Figure 9 Distribution of Panama late shrub and
Illinois late shrub and forest residents in a space
determined by canonical variate II, expressing the
correlation between the relative length of the leg and
the stratum of the habitat occupied. See legend of
Figure 8 for explanation. Numbers refer to species
named and discussed in text.

ing high positive correlations for this variate would be expected to occur in the
upper right section of the plots. Short-
legged canopy-dwelling birds having high
negative correlations for this variate

should occur in the lower left section of
the plots. Species not occurring in the
direct regression between the two
extremes—i.e., species occurring in the
upper left or lower right sections—are

exceptions to the general relationship. Selected species are labelled in Figures 8 and 9 and identified by scientific name and number in the following text.

Taking first the plot from the Liberian forest (Figure 8, above), note the species that occur in the lower left of the plot. These three species—*Tropicranus albocristatus* (1), *Terpsiphone rufiventer* (2), and *Dicrurus ludwigii* (3)—are in fact, short-legged and forage in middle to high levels of the forest. At the other extreme (upper right) are the long-legged ground-gleaners such as *Francolinus lathami* (4), *Turtur brehmeri* (5), *Stiphrornis erythrothorax* (6), *Trichastoma cleaveri* (7), *T. rufescens* (8), and *Spermophaga haematina* (9). Not in the direct regression are a hawk, *Accipiter tousenellii* (10), and *Tauraco macrorhynchus* (11). The hawk presumably requires long legs for capturing prey. The touraco uses its long legs to forage in the canopy by walking on larger branches, while maintaining the long reach required to pluck canopy fruits. Another species that does not fall in the general pattern of this plot is the kingfisher *Alcedo leucogaster* (12). This species, located high in the right center, is specialized for eating small fish and, therefore, differs in food habits from any other resident species. The insectivorous kingfisher, *Halcyon malimbica* (13), is located closer to the general regression of the plot.

Similarly, in the Panama forest (Figure 8, below), a number of long-legged, ground-gleaning species occur in the upper right of the figure. These species include *Tinamus major* (14), *Geotrygon montana* (15), *Myrmornis torquata* (16),

Sclerurus guatamalensis (17), *Grallaria perspicillata* (18), *Formicarius analis* (19), *Cyphorhinus phaeocephalus* (20), and *Microcerculus philomela* (21). The lower left species include *Trogon massena* (22), *T. rufus* (23), *Electron platyrhynchum* (24), *Baryphthengus ruficapillus* (25), and *Myiarchus ferox* (26), a group of middle to high salliers and/or hover-gleaners. The same pattern occurs in the late shrub in Panama (Figure 9, above); the species in the upper right are the ground gleaners *Crypturellus soui* (27), and *Myrmeciza longipes* (28); the lower left species are the two nightjars *Chordeiles minor* (29) and *Nyctidromus albicollis* (30), plus the middle- to high-stratum species *Momotus momota* (21) and *Myiodynastes maculatus* (32). Finally, the same pattern can be seen in Illinois shrub and forest areas (Figure 9, below) with such long-legged, ground-dwelling species as *Colinus virginianus* (33), *Oporornis formosus* (34), *Hylocichla mustelina* (35), *Toxostoma rufum* (36), *Pipilo erythrophthalmus* (37), and *Thryothorus ludovicianus* (38) occupying the upper right; several short-legged, middle- to high-stratum species such as *Tyrannus tyrannus* (39), *Myiarchus crinitus* (40), *Contopus virens* (41), *Dendrocopos villosus* (42), *Dendrocopos pubescens* (43), and *Melanerpes erythrocephalus* (44) occupy the lower left. The power of this technique to extract correlated sets of ecological and morphological variables is obvious. Similar patterns of segregation of species can be demonstrated with the other significant pairs of canonical variates.

Anyone familiar with avian ecology is likely to read the suggested correlation of canonical pair I with skepticism. The generalization that hover-gleaners are small and with narrow bills seems easily challenged. Many hover-gleaning species (e.g., flycatchers, trogons, and broadbills) are not small and certainly do not have narrow bills. The high correlation of canonical pair I occurs because of the morphologically extreme hummingbirds and sunbirds. Later, in canonical variates V and VI, when variation attributable to earlier relationships has been extracted, we see the significant correlations between hover-gleaning and a broad, flat bill.

Finally, we should stress that these analyses are done to explore correlations over large sets of species from several continents. Patterns among small sets of species may be swamped by the patterns of variation in other groups. Also, correlations of ecology and behavior not represented in our species sample will obviously not be demonstrated in our results. If we had included waders, the results might have indicated correlations between mud probing and long narrow bills. We have tried to show the power of a multivariate method for analyzing complex phenomena. But we also feel the need to stress the use of caution in their application, at least equal to the caution required by less complex forms of statistical analysis.

Species-Packing Versus Expansion of Morphological Space

Comparisons of the morphological characteristics (Schoener, 1971), the foraging behavior (Orians, 1969), and the trophic exploitation patterns (Harrison, 1962; Karr, 1971, 1975; Fleming, 1973) of variously defined sets of coexisting species show that tropical faunas contain a wider range of ecological types than comparable temperate faunas. However, none of these earlier studies considers the morphological characteristics of all of the "resident" species of one vertebrate class. In order to assess the relative patterns of morphological space occupation by the resident avifaunas of the lowland forest study areas in Panama, Liberia, and Illinois, we have plotted the ratio of $\log_{10}$ of wing length divided by tarsal length on the abscissa and tarsal length on the ordinate (Figure 10). Triangles plotted in Figure 10 have bases proportional to bill width, and heights proportional to culmen length. The total morphological space occupied by the species, in terms of the distribution of the triangles and the range in size and shape of the triangles, is greater for the tropical than for the temperate avifaunas, as expected. This result is confirmed when the area of Figure 10 is divided into 24 equal quadrats (Table 7). Twelve and thirteen quadrats contain species occurring in Liberia and Panama, respectively; only seven quadrats contain species occurring in Illinois.

The axes of the morphological space in Figure 10 were chosen to represent ecologically significant aspects of morphology. Tarsal length gives a general representation of overall size. The ratio of the wing to the tarsus segregates similar-sized species by their relative proportions. For instance, since ground-dwelling species tend to have relatively longer legs than

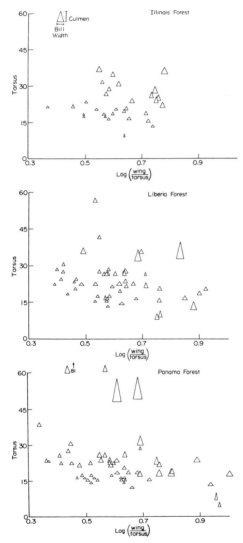

Figure 10 Configurations of wing, tarsus, and bill morphologies of all resident birds of forest study areas in Illinois, Liberia, and Panama. The ordinate is the tarsus length (mm), the abscissa is the $\log_{10}$ of wing length divided by tarsal length, and the bases and heights of the triangles are proportional to bill width and culmen length, respectively.

Table 7.   Number of species per quadrat in a morphological space

Tarsal length (mm)	\multicolumn{4}{c}{Log_{10} (wing/tarsus)}	Area
	0.30–0.475	0.475–0.650	0.650–0.825	0.825–1.000	
0–15	—	1	1	—	Illinois
15–30	2	14	9	—	
30–45	—	4	1	—	
45–60	—	—	—	—	
0–15	—	2	2	1	Liberia
15–30	8	21	6	3	
30–45	1	2	2	1	
45–60	—	1	—	—	
0–15	—	2	1	3	Panama
15–30	7	24	10	2	
30–45	2	—	1	—	
45–60	—	1	1	—	
60–75	—	1	—	—	
75–90	1	—	—	—	

The morphological space is that of Figure 10, in which tarsal length is plotted against $\log_{10}$ (wing/tarsus ratio) for the resident avifaunas of three forest-study areas. The quadrats are determined by the ranges of tarsal lengths at the left and by the ranges of $\log_{10}$ (wing/tarsus ratios) above. Note that more quadrats contain species in the tropical areas (Liberia and Panama) than in the temperate area (Illinois).

species in the canopy (e.g., Dilger, 1956), the wing-over-tarsus ratio tends to segregate ground-dwelling species toward the left of the plot. The triangles representing bill size and shape permit visual comparisons among species of this food-getting device (cf. also Figure 2).

Karr (1975) showed that much of the increase in resident species in tropical forests is attributable to changes in the complexity of exploitable food resources. At first, the expanded space occupied by the extreme morphological types in the tropics (Figure 10) might suggest that the species feeding on the "new" food resources (such as fruit, large insects, and nectar) have the extreme morphologies. (The expression "new resources" is a convenient way to refer to resources that are exploited in tropical forest but not in temperate forest. It should not be taken to mean they are not available in temperate areas; rather, they are not reliable resources throughout the year in temperate areas; see Karr, 1975.) However, a detailed examination of the plots (Figure 10) showed that there are more tropical than temperate species in central as well as peripheral areas.

To what extent is the increased number of species in the tropics due to greater species-packing? To answer this question, it is important to remember that since the total number of species varies, one must consider whether there is a concentration of either extreme or conservative morphologies that is independent of the differences attributable to sample size. In other words, is the increased range of morphological types in the tropics simply a result of the increase in the number of species (Figure 11)?

To resolve this question, we have constructed equal-frequency ellipses (Sokal and Rohlf, 1969) for the species in the morphological space of Figure 10. We use three frequency levels (0.50, 0.95, 0.99) to assess the relative areas of equal frequency ellipses in the total morphological space (0.95, 0.99) and central areas (0.50). This calculation yields the area of an ellipse containing 50, 95, and 99% of the observations in a given sample. These are largest in Panama and smallest in Illinois, Liberia being intermediate (Table 8). This substantiates the earlier conclusion that the morphological space of tropical avi-

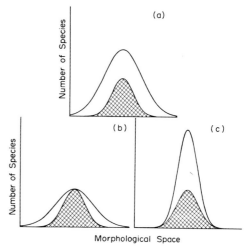

Figure 11 The upper curve in each figure gives the hypothetical distribution of species in morphological space when increased tropical species diversity is accommodated by (a) equal increases in the volume of morphological space (i.e., greater morphological variance, symbolized by a wider range of abscissa values) and in closeness of species-packing (symbolized by higher peak of ordinate values); (b) greater morphological space (greater variance), with no increase in species packing; (c) increased species-packing (higher peak) with no increase in morphological space occupied. The hatched areas under the lower curve in each figure are equal reference areas, corresponding to species distributions in a hypothetical temperate community.

faunas is expanded. To determine whether species-packing is also increased in the tropical faunas, we estimated the niche volume for morphological characters by determining the area within each equal-frequency ellipse and dividing the area by the number of species within the ellipse. The resultant index of species-packing (A/S, Table 8) allows comparisons of niche volume (in morphological space)

Table 8a. Comparisons of extent of morphological space and closeness of species-packing, defined by wing and tarsus measurements, in the avifaunas of three continents

Frequency level	Study area	Ellipse area (A)	Number of species (S)	Niche space (A/S)
0.50	Panama	6.78	39	0.174
	Liberia	4.31	29	.149
	Illinois	2.51	15	.167
.95	Panama	30.64	53	.578
	Liberia	19.50	49	.398
	Illinois	11.75	32	.367
.99	Panama	48.80	55	.887
	Liberia	31.25	49	.638
	Illinois	19.03	32	.595

Equal-frequency ellipses at frequency levels of 0.50, 0.95, or 0.99 were constructed as described in the text, for species distributions in Panama, Liberia, and Illinois in the morphological space of Figure 10 [tarsus length plotted against $\log_{10}$ (wing/tarsus ratio)]. Columns 3, 4, and 5 give, respectively, the ellipse area A, the number of species S contained within the ellipse, and the niche space per species A/S.

Table 8b. Percent change in niche space per species (A/S)

Areas compared	Frequency level	Percent change
Illinois to Panama	0.50	+4
	.95	+57
	.99	+49
Illinois to Liberia	.50	−9
	.95	+8
	.99	+7

Percent change A/S at each frequency level in Panama and Liberia is compared with A/S in Illinois. See text for discussion.

per species. For the 0.50 frequency ellipse there is only slight variation in the niche volume per species, showing that species-packing in central areas of the morphological space is approximately the same in the tropical and temperate areas. Illinois is intermediate with an A/S value of 0.167, and Panama and Liberia have the high and low values, 0.174 and 0.149, respectively. A shift occurs as 0.95 and 0.99 frequency ellipses are used. The niche volume per species is then about 50%

greater in Panama and 8% greater in Liberia than in the Illinois fauna (lower part of Table 8). This reflects the addition of several large birds such as the toucans, motmots, and others in the Panama fauna. Very intense hunting pressure for food in the vicinity of Mount Nimba in Liberia has reduced populations of many large species (hornbills, touracos, etc.). The loss of these species may be responsible for the smaller niche space per species in Liberia than in Panama.

We conclude that, for the data at hand, the increase in species diversity of the resident tropical avifaunas over their temperate counterpart appears to be due to an expansion of morphological space rather than to increased packing in the morphological space occupied by the temperate fauna.

A detailed analysis of Table 7 permits a closer examination of the morphological structure of the three forest avifaunas. The large number of species in the left-hand column of the tropical matrices mostly are relatively long-legged species associated with terrestrial habits. Nine or ten such species occur in each of the tropical areas but only two in Illinois (see also Karr, 1971). Most of the additional tropical species have tarsal lengths between 15 and 30 mm. The number of species having a tarsal length greater than 30 mm is approximately the same in all three areas. But this may be attributable to the loss of large species in the tropics due to hunting or other effects of man. Small species are more abundant in the tropical forests than in Illinois. This is probably due to

a release, in the warmer tropical lowlands, from selection pressure imposed by thermodynamic limitations in temperate areas; this release reduces the minimum size attainable in homeotherms (Karr, 1971; McNab, 1971b).

To evaluate the relative importance of morphological expansion versus closer packing in bill dimensions, we constructed equal-frequency ellipses on a morphological space determined by a plot of culmen length against the ratio of bill width to bill depth. As in the previous analysis of Table 8, equal-frequency ellipses are larger (column 3, "ellipse area," of Table 9) for the tropical than for the temperate avifaunas, although the differences are less than those for wing and tarsus morphologies.

A strikingly different pattern emerges, however, in niche space per species. At all frequency levels the niche space per species is larger for the temperate forest birds, indicating closer species-packing in the tropical fauna (last column, A/S, of Table 9). Furthermore, the tropical niche volume per species is proportionately much smaller in core areas (0.50 frequency ellipse) than in peripheral areas relative to temperate niche volume; Panama niche volume per species is 28% below that in Illinois for the 0.50 frequency ellipse. Panama niche volume is only 7 to 8% less per species than in Illinois for the 0.95 and 0.99 frequency ellipses (lower part of Table 9). We conclude, on this basis, that some increases in morphological space occur with respect to bill dimensions in the tropics but that

Table 9a.   Comparison of extent of morphological space and closeness of species-packing defined by bill dimensions, in the avifaunas of three continents

Frequency level	Study area	Ellipse area (A)	Number of species (S)	Niche space (A/S)
0.50	Panama	2.07	35	0.059
	Liberia	1.55	36	.043
	Illinois	1.32	16	.083
.95	Panama	9.47	52	.182
	Liberia	6.96	47	.148
	Illinois	6.09	31	.197
.99	Panama	15.09	53	.285
	Liberia	11.07	49	.226
	Illinois	9.94	32	.311

This table differs from Table 8 only in that morphological space was defined by bill-width/bill-depth ratio (ordinate) and culmen length (abscissa), instead of by the axes of Figure 10. See text for discussion.

Table 9b.   Percent change in niche space per species (A/S)

Areas compared	Frequency level	Percent change
Illinois with Panama	0.50	−28
	.95	−7
	.99	−8
Illinois with Liberia	.50	−48
	.95	−25
	.99	−27

the greatest proportion of the increased tropical diversity is due to increased species-packing.

Similarly, when the morphological space defined by the bills is divided into 16 quadrats, the number of quadrats occupied is similar for the three forests: 10 or 11 (Table 10). Closer examination reveals some interesting differences. For instance, the number of thin-billed species (bill-width/bill-depth ratio < 0.8) is higher in the tropical forests (10 and 13 compared with 6 for Illinois). In Liberia these birds include a hawk, doves, a cuckoo, hornbills, some bulbuls and a shrike. In Panama they include a hawk, a dove, a motmot, toucans, and several small insectivorous passerines. Species having broad bills (BW/BD > 1.2) are more numerous in Panama (13) than in

Liberia (6) or Illinois (7). Within these groups there are two sets of species. Those with ratios of 1.2 to 1.45 are generally types of flycatchers. In Illinois they include three tyrannids (*Contopus, Empidonax,* and *Myiarchus*); in Panama they include several flycatchers, a woodcreeper, and an ant-wren; in Liberia they include a kingfisher, a thrush, and two Old World flycatchers. Ratios larger than 1.45 characterize species that hover-glean—e.g., the broadbills (Eurylamidae) in Africa, and the Panamanian broad-billed flycatchers such as the spadebill (*Platyrinchus*), the flatbill (*Rhynchocyclus*), and *Tolmomyias.* There are twice as many hover-gleaners in Panama as in Liberia (15 to 7). Salliers are unrepresented on the Liberian resident list, whereas three occur on the Panama list.

In summary, when bird sizes are compared, the morphological configurations of the tropical faunas are more similar to each other than to those of the temperate forest. The increased tropical diversity in wing and tarsus morphology is accommodated by an enlargement of morphological space in the tropics rather than by increased species-packing (Figure 12a). When bill sizes and shapes are compared between faunas, tropical faunas show some expansion in morphological space but relatively greater species-packing in centrally located morphological configurations (Figure 12b). Intercontinental tropical comparisons of bill morphologies indicate fewer similarities than do leg and wing morphologies, because of the relative lack of certain kinds

Table 10. Number of species per quadrat in a morphological space determined by plotting culmen length against the ratio bill-width/bill-depth for the resident avifaunas of three forest study areas

BW/BD	Culmen (mm)				Study area
	10–17	17–31	31–59	>59	
0.4–0.8	4	1	1	—	Illinois
0.8–1.2	5	12	2	—	
1.2–1.6	2	2	1	—	
1.6–2.0	1	1	—	—	
0.4–0.8	1	8	2	2	Liberia
0.8–1.2	12	17	2	—	
1.2–1.6	2	2	—	—	
1.6–2.0	2	—	—	—	
0.4–0.8	3	4	3	—	Panama
0.8–1.2	9	18	6	—	
1.2–1.6	5	5	1	—	
1.6–2.0	2	—	—	—	

This table is similar to Table 8, except that the axes of morphological space are taken as the ratio bill-width/bill-depth (BW/BD) and as culmen length, as in Table 9. The quadrats are defined by the ranges of BW/BD ratios at the left and by the ranges of culmen length above.

of species—especially hover-gleaners and salliers—in Liberia.

Why do the relative influences of species-packing and expansion of morphological space vary between wing and tarsus, on the one hand, and bill characters, on the other hand? One possibility is that allometric growth patterns of bill characters preclude space expansion. This could be due to the smaller size of bird bills and the limits placed on magnitude of expansion. It may be easier evolutionally to produce extremes of wing and tail

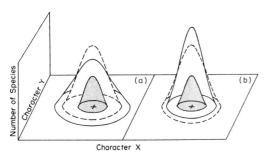

Figure 12   Three-dimensional figure showing the relative importance of closer species-packing and expansion in morphological space in tropical faunas. The shaded volume is equivalent to a temperate fauna; the volume outlined with dashed lines is the volume for a tropical fauna with increased species diversity scaled equally for increase in morphological space (basal area of cone) and in closeness of species packing (see Figure 11.11a). (a) The solid-line volume represents a proportionately greater increase in morphological space occupied than in species-packing. This is the pattern observed in wing and tarsus morphologies (Figure 10, Table 8) where Character *X* represents log (wing/tarsus) and Character *Y* represents tarsal length. (b) The solid-line volume represents greatly increased species-packing in central areas and a lower-than-proportionate increase in morphological space. This pattern is observed in bill morphologies (Table 9), where Character *X* is culmen (bill) length and Character *Y* is bill width/bill depth ratio.

morphology than of bill morphology. Alternatively, the relatively restricted use of bills—obtaining food—may set limits on the range of reasonable morphological options in birds. Wing and leg morphologies, on the other hand, may be less canalized since wings and legs are used in a context, locomotion, with relatively greater options. Flight options range from almost constant flight in such birds as swifts to no flying in the ratites such as the ostrich. Similarly, leg morphologies

vary from the short tarsi of hummingbirds to the long tarsi of ground-feeding species. This second suggestion implies that bill morphologies are conservative by comparison to more general size-parameters (wing and leg) for ecological and evolutionary reasons and not for allometric growth reasons.

Summary

The present analysis of correlations between morphology and ecology in sets of birds that coexist is based upon data from forest birds in Panama, Liberia, and Illinois. Graphical and mathematical methods are applied to explore the properties of the avifaunas and the extent of convergence among them. In the tropics, ground-feeding pigeons are smaller and have relatively shorter wings and longer legs than the canopy-feeding pigeons. Liberian hornbills that eat insects have relatively shorter wings than fruit-eating species. The insectivorous kingfishers have relatively narrower bills than the fish-eating ones. Within the Piciformes, the wide-ranging honeyguides have the longest wings, and the bark-gleaning woodpeckers have the narrowest bills. Several other similar examples of predictable relationships between morphology and behavior are discussed.

A pooled principal components analysis of a set of ratios of morphological measurements for 184 Liberian lowland forest birds shows that the shapes of these lowland species vary primarily in the ratios of wing to culmen and wing to tarsus. After that variation has been removed from the data set, the next principal component of variation is highly correlated with bill shape. Graphs having these two components as axes permit interesting insights into sets of morphological types within and between families of birds.

The following general patterns are demonstrated: 1) convergence between ecologically similar species of different phylogeny; 2) divergence between ecologically different species of the same phylogeny; and 3) in many cases, wider morphological spacing between congeners than non-congeners. The last suggests that competitive displacement may be a particularly important determinant of community structure. Demonstrations of convergences and divergences across generic and familial bounds indicate that organisms are adaptive complexes with high degrees of integration among their ecological, behavioral, and morphological characteristics. The relative influence of ecological, behavioral, and morphological segregation varies among the families: behavioral and morphological segregation characterizes the flycatchers (Muscicapidae), whereas morphological conservatism and ecological segregation (spatial by habitat) characterizes the warblers (Sylviidae). Other families (Columbidae, Bucerotidae, Alcedinidae) segregate by complex integrations of food habits, behavior, foraging location, and morphology.

A few examples of one-to-one convergences of species, producing ecological equivalents in different geographic areas, can be demonstrated. Many species cannot be matched on a one-to-one basis, however, because of phylogenetic correlates of morphological characters or variation in structural aspects of communities related to history. For example, variation in the availability of certain kinds of food resources may produce mosaic patterns in the distribution of ecological and morphological equivalents.

Canonical correlation analysis is a multivariate method that has rarely been applied to ecological problems. It is particularly appropriate in cases in which there are two types of information about one set of species. In the present study, for a set of 196 species representing the resident birds from five study areas, we have compared a matrix of 17 morphological variables with another matrix of 14 ecological variables. The analysis yields six significant relationships between the two matrices. Subsets of the species occurring in the separate communities have been plotted in a space determined by examples of these relationships. Canonical variate II, expressing primarily the relationship between ground-gleaning and long legs and narrow bills, is discussed in detail.

Earlier studies by a number of researchers have shown the presence of a wider range of ecological types in tropical than in temperate faunas. This is confirmed for both wing-tarsus and bill morphologies. However, without control for variation in species numbers, it is difficult to determine the relative influences of expansion of morphological space and increased species-packing. To sort out these two influences, we measure niche space per species by using equal-frequency ellipses. In the case of wing and tarsus morphology, increased species numbers in the tropics are due particularly to expansion of morphological space. For bill characters a striking increase in tropical species-packing in core morphologies is observed. These differences are probably due to ecological selective pressures rather than to allometric growth patterns.

Acknowledgments

Special thanks go to the Office of International Activities of the Smithsonian Institution, and the Smithsonian Tropical Research Institute, for their support of the field studies of the senior author. Alec Forbes-Watson, Stuart Keith, and Helen Lapham assisted with the field work at Mount Nimba, Liberia. For helpful discussions and access to museum collections we are indebted to Alec Forbes-Watson at the National Museum of Kenya, Nairobi; to Paul Slud and Alexander Wetmore at the National Museum of Natural History, Washington; and to Richard Johnston at the University of Kansas Museum of Natural History. The American Museum of Natural History provided specimens for Figure 1. The section of multivariate analysis was developed in consultation with James E. Dunn, University of Arkansas. K. Fujii, M. Levy, A. S. Rand, N. G. Smith, M. F. Willson, and the participants of the MacArthur Memorial Symposium commented on early versions of the manuscript. Sincere tribute is offered to Robert H. MacArthur, whose contributions to ecology catalyzed many of the thoughts in this paper.

References

Amadon, D. 1950. The Hawaiian honeycreepers (Aves, Drepaniidae). *Bull. Amer. Mus. Natur. Hist.* 95:151–262.

Austin, M. P. 1968. An ordination study of a chalk grassland community. *J. Ecol.* 56:739–757.

Blondel, J. 1969. Synécologie des passereaux résidents et migrateurs dans un echantillon de la region méditeranéen français. Thèse. Centre Regional de Documentation Pédagogique. Marseille.

Blondel, J., C. Ferry, and B. Frochot. 1973. Avifaune et végétation: essai d'analyse de la diversité. *Alauda* 41:63–84.

Bock, W. J. 1966. Functional analysis of bill shape. *Auk* 83:10–51.

Calhoon, R. E., and D. L. Jameson. 1970. Canonical correlation between variation in weather and variation in size in the Pacific tree frog, *Hyla regilla,* in southern California. *Copeia* 1970:124–134.

Cattell, R. B. 1966. Multivariate behavioral research and the integrative challenge. *Multivariate Behav. Res.* 1:4–23.

Cody, M. L. 1968. On the methods of resource division in grassland bird communities. *Amer. Natur.* 102:107–147.

Cooley, W. W., and P. R. Lohnes. 1971. *Multivariate data analysis.* Wiley, New York.

Darwin, C. 1859. *On the Origin of Species.* John Murray, London.

Diamond, J. M. 1973. Distributional ecology of New Guinea birds. *Science* 179:759–769.

Dilger, W. C. 1956. Adaptive modifications and ecological isolating mechanisms in the thrush genera *Catharus* and *Hylocichla. Wilson Bull.* 68:171–199.

Fleming, T. H. 1973. Numbers of mammal species in North and Central American forest communities. *Ecology* 54:555–563.

Frazzetta, T. H. 1962. A functional consideration of cranial kinesis in lizards. *J. Morphol.* 111:287–320.

Frazzetta, T. H. 1970. From hopeful monsters to bolyerine snakes? *Amer. Natur.* 104:55–72.

Grant, P. R. 1968. Bill size, body size and the ecological adaptations of bird species to competitive situations on islands. *Syst. Zool.* 17:319–333.

Grant, P. R. 1971. Variation in the tarsus lengths of birds in island and mainland regions. *Evolution* 25:599–614.

Harrison, J. L. 1962. The distribution of feeding habits among animals in a tropical rain forest. *J. Animal Ecol.* 31:53–63.

Hespenheide, H. A. 1971. Food preference and the extent of overlap in some insectivorous birds, with special reference to the Tyrannidae. *Ibis* 113:59–72.

Hutchinson, G. E., and R. H. MacArthur. 1959. A theoretical ecological model of size distributions among species of animals. *Amer. Natur.* 93:117–126.

James, F. C. 1971. Ordinations of habitat relationships among breeding birds. *Wilson Bull.* 83:215–236.

James, F. C., and H. H. Shugart, Jr. 1974. The phenology of the nesting season of the robin, *Turdus migratorius.* Condor 76:159–168.

Jameson, D. L., J. P. Mackey, and M. Anderson. 1973. Weather, climate, and the external morphology of Pacific tree toads. *Evolution* 27:285–302.

Johnston, R. F., and R. K. Selander. 1973. Evolution in the house sparrow. III. Variation in size and sexual dimorphism in Europe and North and South America. *Amer. Natur.* 107:373–390.

Karr, J. R. 1968. Habitat and avian diversity on strip-mined land in east-central Illinois. *Condor* 70:348–357.

Karr, J. R. 1971. Structure of avian communities in selected Panama and Illinois habitats. *Ecol. Monogr.* 41:207–233.

Karr, J. R. 1975. Production, energy pathways, and community diversity in forest birds. *In* F. B. Golley and E. Medina, eds., *Tropical Ecological Systems: Trends in Terrestrial and Aquatic Research,* pp. 161–176. Springer-Verlag, New York.

Karr, J. R., and R. R. Roth. 1971. Vegetation structure and avian diversity in several new world areas. *Amer. Natur.* 105:423–435.

Keast, A. 1970. Food specializations and bioenergetic interrelations in the fish faunas of some small Ontario waterways. *In* J. H. Steele, ed., *Marine Food Chains,* pp. 377–411, Oliver and Boyd, Edinburgh.

Keast, A. 1972. Ecological opportunities and dominant families, as illustrated by the neotropical Tyrannidae (Aves). *In* T. Dobzhansky, M. K. Hecht, and W. C. Steere, eds., *Evolutionary Biology.* 5:229–277.

Keast, A., and D. Webb. 1966. Mouth and body form relative to feeding ecology in the fish fauna of a small lake, Lake Opinicon, Ontario. *J. Fish. Res. Bd. Canada* 23:1845–1874.

Kendall, M. G. 1957. *A Course in Multivariate Analysis.* Hafner, New York.

Lack, D. 1947. *Darwin's Finches.* Cambridge University Press, London.

Lack, D. 1954. *The Natural Regulation of Animal Numbers.* Oxford University Press, London.

Lein, M. R. 1972. A trophic comparison of avifaunas. *Syst. Zool.* 21:135–150.

MacArthur, R. H. 1972. *Geographical Ecology.* Harper and Row, New York.

MacArthur, R. H., and J. W. MacArthur. 1961. On bird species diversity. *Ecology* 42:594–598.

MacArthur, R. H., H. Recher, and M. Cody. 1966. On the relation between habitat selection and species diversity. *Amer. Natur.* 100:319–332.

Margalef, D. R. 1968. *Perspectives in Ecological Theory*. University of Chicago Press, Chicago.

McNab, B. 1971a. The structure of tropical bat faunas. *Ecology* 52:352–358.

McNab, B. 1971b. On the ecological significance of Bergmann's Rule. *Ecology* 52:845–854.

McNaughton, S. G., and L. L. Wolf. 1970. Dominance and the niche in ecological systems. *Science* 167:131–139.

Moreau, R. E. 1966. *The Bird Faunas of Africa and its Islands*. Academic Press, London.

Morrison, D. F. 1967. *Multivariate Statistical Methods*. McGraw-Hill, New York.

Newton, I. 1967. The adaptive radiation and feeding ecology of some British finches. *Ibis* 109:33–98.

Niles, D. M. 1973. Adaptive variation in body size and skeletal proportions of horned larks of the southwestern United States. *Evolution* 27:405–426.

Orians, G. H. 1969. The number of bird species in some tropical forests. *Ecology* 50:783–801.

Recher, H. 1969. Bird species diversity and habitat diversity in Australia and North America. *Amer. Natur.* 103:75–80.

Pianka, E. R. 1967. On lizard species diversity: North American flatland deserts. *Ecology* 48:333–351.

Pielou, E. C. 1969. *An Introduction to Mathematical Ecology*. Wiley-Interscience, New York.

Rothstein, S. I. 1973. The niche-variation model—is it valid? *Amer. Natur.* 107:598–620.

Rosenzweig, M. L., and J. Winakur. 1969. Population ecology of desert rodent communities: habitat and environmental complexity. *Ecology* 50:558–572.

Schnell, G. D. 1970. A phenetic study of the suborder Lari (Aves). I. Methods and results of principal components analyses. *Syst. Zool.* 19:35–57.

Schoener, T. W. 1968. The *Anolis* lizards of Bimini: niche partitioning in a complex fauna. *Ecology* 49:704–726.

Schoener, T. W. 1971. Large billed insectivorous birds: a precipitous diversity gradient. *Condor* 73:154–161.

Selander, R. K. 1966. Sexual dimorphism and differential niche utilization in birds. *Condor* 68:113–151.

Sheldon, A. L. 1968. Species diversity and longitudinal succession in stream fishes. *Ecology* 49:193–198.

Sokal, R. R., and F. J. Rohlf. 1969. *Biometry: The Principles and Practice of Statistics in Biological Research*. W. H. Freeman, San Francisco.

Soulé, M., and B. R. Stewart. 1970. The "niche-variation" hypothesis: a test and alternatives. *Amer. Natur.* 104:85–97.

Thompson, W. D'Arcy. 1917. *On Growth and Form*. Cambridge University Press, London.

Tiedeman, D. V., and M. M. Tatsuoka. 1963. Statistics as an aspect of the scientific method in research on teaching. *In* N. L. Gage, ed., *Handbook of Research on Teaching*, pp. 142–170. Rand McNally, Chicago.

Van Valen, L. 1965. Morphological variation and width of ecological niche. *Amer. Natur.* 99:377–390.

Vogt, T., and D. L. Jameson. 1970. Chronological correlation between change in weather and change in morphology of the Pacific tree frog in southern California. *Copeia* 1970:135–144.

Webb, L. J., J. G. Tracy, J. Kikkawa, and

W. T. Williams. 1973. Techniques for selecting and allocating land for nature conservation in Australia. *In* A. B. Costin and R. H. Groves, eds., *Nature Conservation in the Pacific,* pp. 67–84. C.S.I.R.O., Melbourne.

Willson, M. F. 1969. Avian niche size and morphological variation. *Amer. Natur.* 103:531–542.

Willson, M. F., J. R. Karr, and R. R. Roth. 1975. Ecological aspects of avian bill size variation. *Wilson Bull.* 87:32–44.

Wilson, D. E. 1973. Bat faunas: A trophic comparison. *Syst. Zool.* 22:14–29.

12 Niche Relations of Desert Lizards

Eric R. Pianka

Lizards have proven to be especially suitable subjects for investigation of competition, community structure, and species diversity. Data on diets, microhabitats, reproductive tactics, and times of activity are readily obtained. Since much of the empirical base of modern ecology comes from data on birds, it may be judicious to examine other nonavian taxa. In many ways, lizards may be closer to a modal animal than birds; they are terrestrial, poikilothermic ("cold-blooded"), and generally lack parental care (Schoener, 1975). Moreover, most lizards are primary and/or secondary carnivores, relatively high in the trophic structure of a community, and should therefore often encounter relatively keen competition. Finally, lizards span a wide range of the *r-K*-selection continuum, which may make them particularly suitable for testing many current theoretical developments in population biology (Pianka, 1972).

Study Systems and Methods

I have devoted much of the past decade to studying lizard faunas on some 30-odd desert study areas at similar latitudes on three continents (Pianka, 1967, 1969a, 1971, 1973). These sites vary widely in total number of lizard species. For example, 14 areas in western North America (Great Basin, Mojave, and Sonoran des-

erts) support from 4 to 11 species, another 10 study areas in the Kalahari Desert of southern Africa support from 11 to 18 species of lizards, and 8 sites in the Great Victoria Desert of Western Australia support from 18 to 40 sympatric lizard species. In terms of the numbers of species they support, or species densities, the lizard communities of the Australian deserts are probably the richest on earth. I designed my research to elucidate the factors determining species diversity.

My assistants and I walked slowly through these desert habitats making observations on their saurofaunas. We spent more than 40 months in the field, observing lizards over their entire annual period of activity. Microhabitat and time of activity were recorded for the majority of undisturbed lizards encountered active above ground at their own volition. Most lizards were collected, and their stomach contents and reproductive condition analyzed.[1] These data were also augmented with both museum specimens and data from the literature. For this paper, I use 15 basic microhabitat categories: subterranean; sunny places in the open, in grass, in bushes, in trees, in other sunny loca-

[1] Resulting collections of some 5,000 North American lizards, over 6,000 Kalahari animals, and nearly 4,000 Australian ones, representing some 90-odd species, are now lodged with the Los Angeles County Museum of Natural History.

tions; shaded places in the open, in grass, in bushes, in trees, in other shaded locations; and four additional categories that describe whether lizards are found perched high or low in sun or in shade. Low perches are those within 30 centimeters of the ground, high perches are all those higher than this. Lizards at the interface of two or more microhabitats were assigned partial representation in each. Using just these 15 very crude categories allows separation of many species: for example, some lizard species frequent the open spaces between plants, whereas others tend to stay much closer to cover. Because time of activity shifts seasonally with changes in ambient temperature, all times of activity are expressed either as time since sunrise or time since sunset. I use 14 hourly categories for diurnal species here (limitations on human endurance dictated only 8 categories for nocturnal species). Any bias introduced through the shorter night-time sampling period is similar on different areas and among the three continental desert-lizard systems. In the following analysis, I use 20 very crude prey categories: spiders, scorpions, solpugids, centipedes, ants, wasps and other hymenopterans, grasshoppers and crickets, roaches (blattids), mantids and phasmids, ant lions, beetles, termites, bugs (Homoptera-Hemiptera), flies (Diptera), pupae, insect eggs, all insect larvae, miscellaneous insects not listed above including unidentified ones, all vertebrate material including sloughed skins, and plant materials (floral and vegetative). The proportional volumetric representation of each food category is used

in the following analysis. Even these very crude categories allow reasonably good separation of many lizard species by foods they eat. When prey items are analyzed by either numbers or size, separation is much less than when these 20 taxonomic volumetric categories are used.

This body of data from three basically rather similar, but independently evolved, subsets of natural communities allows detailed analysis of the niche relations and community structure of entire lizard faunas, including estimation of species diversities, diversities of resources actually exploited by entire lizard faunas, and both niche breadths and overlaps in saurofaunas that vary widely in species densities and diversities.

Species densities and diversities of the lizard faunas on 28 study sites are listed in Table 1. Lizard species diversities were calculated from the relative abundances of various species in the above-mentioned collections. These estimates are biased to the extent that different species may not have been collected in proportion to their true relative abundances; however, they are the best available estimates. In any case, lizard species diversity is very strongly correlated with the estimated number of lizard species ($r = 0.84$, $P < 0.001$).

Composition of Lizard Communities

Some major aspects of the niche relations and organization of lizard communities within each of the continental desert-lizard systems are summarized according to 5 niche categories in Table 2 and shown

Table 1. Desert lizard faunas at 28 study sites on three continents

Study site	(1) Estimated total number of lizard species	(2) Estimated lizard-species diversity	(3) Long-term mean	(4) Standard deviation	(5) Food	(6) Place	(7) Time	(8) Multiplicative (All)	(9) Summation (Nonzero)	(10) (All)
North America	4	1.4	18.4	5.6	0.49	0.80	0.58	0.20	0.20	0.63
	5	2.3	14.0	5.5	.75	.78	.53	.33	.36	.69
	5	1.8	9.6	4.4	.52	.92	.49	.25	.36	.64
	6	2.9	11.6	6.6	.55	.55	.47	.22	.37	.52
	6	1.9	18.7	10.5	.34	.55	.40	.12	.27	.43
	7	2.3	9.3	5.9	.39	.42	.31	.11	.39	.37
	8	2.1	12.7	9.2	.56	.31	.32	.10	.24	.40
	9	2.7	9.4	6.1	.28	.52	.58	.11	.23	.46
	9	2.8	19.2	8.0	.38	.32	.39	.06	.18	.36
	10	2.7	20.9	7.6	.37	.33	.50	.08	.25	.40
	Mean 6.9	2.3	14.4	6.9	.46	.55	.46	.16	.29	.49
Kalahari	11	2.6	22.7	11.3	.92	.35	.28	.18	.41	.52
	13	5.3	21.7	10.3	.36	.39	.30	.08	.26	.35
	13	6.2	22.7	11.3	.56	.47	.34	.13	.36	.46
	14	7.3	16.3	—	.56	.21	.15	.04	.27	.31
	15	7.7	14.5	8.6	.45	.23	.21	.04	.22	.30
	15	8.1	19.0	—	.56	.25	.24	.06	.23	.35
	16	7.0	16.7	7.8	.72	.22	.23	.09	.35	.39
	16	7.3	19.0	9.7	.44	.22	.24	.05	.22	.30
	16	8.3	28.6	9.3	.71	.28	.26	.11	.36	.42
	18	8.2	15.2	7.2	.51	.26	.27	.07	.24	.35
	Mean 14.7	6.8	19.6	9.4	.58	.29	.25	.08	.29	.37
Australia	18	8.3	16.0	—	.23	.16	.18	.01	.14	.19
	20	8.5	15.2	—	.18	.36	.13	.01	.02	.22
	28	6.3	21.5	13.4	.25	.32	.16	.03	.21	.24
	30	7.3	23.5	14.0	.23	.24	.19	.02	.19	.22
	31	7.0	15.2	—	.19	.28	.18	.02	.16	.22
	30	10.5	21.3	10.0	.37	.24	.27	.03	.18	.28
	29	9.7	21.9	9.3	.27	.30	.27	.04	.23	.28
	40	11.8	20.2	13.1	.23	.25	.22	.02	.15	.24
	Mean 28.3	8.7	19.4	12.0	.24	.27	.20	.02	.16	.24

Lizard species diversity is given in column 1 simply as the total number of species, and in column 2 as the number of species weighted by relative abundances by use of Simpson's Index $\Sigma\, 1/p_i^2$, where p_i is the proportion of the total number of individuals in species i and the summation is over all i species. Columns 3 and 4 are the mean and standard deviation (in centimeters) of annual rainfall, respectively (moisture availability is critical in these deserts). Three components of niche overlap are given separately in columns 5–7 (see text for formula), and the components are combined in three different ways in the last three columns: by multiplying all three together (column 8), by multiplying just the non-zero components together (column 9), and by averaging the three components (column 10). The values in the last 6 columns are community averages of all possible combinations of species pairs.

diagrammatically in Figure 1. Even these extremely crude categories reflect some of the differences between continents in the importance of various components of their lizard faunas. For example, fossorial (subterranean) lizards and arboreal nocturnal species are entirely absent from North American communities, but contribute an average of three to four species per area in the two deserts of the southern hemisphere. Moreover, although the average number of species of diurnal terrestrial lizards does not differ greatly in North America and the Kalahari (5.4 and 6.3 species, respectively), there are considerably more diurnal terrestrial species in Australia (14.4 species). However, when expressed as a percentage of the entire saurofauna, diurnal terrestrial species actually constitute a smaller fraction of the total saurofauna in the two deserts of the southern hemisphere. The percentage contribution of all diurnal species, both arboreal and terrestrial, to the total fauna decreases with lizard species density (Figure 1). The increased relative importance of nocturnality in the Kalahari and the Australian deserts probably stems largely from historical factors (see below), although my interpretation remains speculative. Among all arboreal lizards (nocturnal plus diurnal species), both the numbers of species and the percentage contribution to the total saurofauna tend to increase with lizard species density (Figure 1 and Table 2); however, even within the diverse Australian desert system, arboreal species are less well represented on areas with low diversity (Figure 1), which are structurally simple.

Table 2.   Intercontinental comparisons of desert saurofaunas by niche categories

Niche category	North America			Kalahari			Australia		
	$\bar{X}$	(range)	%	$\bar{X}$	(range)	%	$\bar{X}$	(range)	%
Diurnal	6.3	(4–9)	86	8.2	(7–10)	56	17.0	(9–25)	60
Terrestrial	5.4	(4–7)	74	6.3	(5.5–7.5)	43	14.4	(9–21.5)	51
Arboreal	0.9	(0–3)	12	1.9	(1.5–2.5)	13	2.6	(0–5.5)	9
Nocturnal	1.0	(0–2)	14	5.1	(4–6)	35	10.2	(8–13)	36
Terrestrial	1.0	(0–2)	14	3.5	(3–5)	24	7.6	(6–9)	27
Arboreal	0	—	0	1.6	(0.5–2.5)	11	2.6	(1–4)	9
Fossorial	0	—	0	1.4	(1–2)	10	1.1	(1–2)	4
All terrestrial	6.4	(4–8)	88	9.8	(9–11)	67	22.0	(15–30.5)	78
All arboreal	0.9	(0–3)	12	3.5	(2–5)	24	5.2	(1–9)	18
Totals	7.4	(4–11)	100	14.7	(11–18)	101	28.3	(18–40)	100

Lizard species fall into five crude categories in their foraging activities: diurnal and terrestrial, diurnal and arboreal, nocturnal and terrestrial, nocturnal and arboreal, and below ground (fossorial). The table gives the average number $\bar{X}$ and range of numbers of lizard species, and the percentage of the total saurofauna, in each category on each continent. Semiarboreal species are assigned half to arboreal and half to terrestrial categories. Notice that the proportion of species in each category differs among continents; that North America is poorest in all five categories; and that Australia is richest in all categories except for fossorial lizards, which reach slightly higher species densities in the Kalahari Desert.

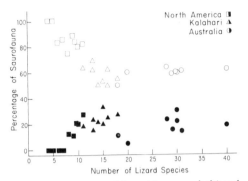

Figure 1 Percentages of diurnal (open symbols) and arboreal (solid symbols) lizard species plotted against the number of lizard species coexisting on various study areas in the three continental desert-lizard systems. Note changes in the proportions of diurnal and arboreal species with different numbers of lizard species.

Lizards are predominantly insectivorous. Most are rather opportunistic feeders and consume a wide variety of arthropods with little evident preference. However, differences in foraging technique and time and place of activity often result in exposure to different spectra of prey types. Some lizard species have evolved pronounced dietary restrictions; for example, *Phrynosoma* and *Moloch* are ant specialists in North America and Australia, respectively. Both the Kalahari and the Australian deserts support termite-specialized species; these include *Typhlosaurus, Rhynchoedura,* and certain species of *Diplodactylus* (Huey et al., 1974; Pianka and Pianka, 1976). The North American genus *Dipsosaurus* is almost entirely herbivorous, whereas *Crotaphytus* is a secondary carnivore, preying mostly on other lizards (*Varanus* in the Australian deserts also eat largely other lizards). Foods of all these dietary specialists are

at least temporarily very abundant, which makes specialization economically feasible (MacArthur and Pianka, 1966).

Temporal separation of activities may reduce competition between lizard species, provided that being active at different times leads to exploitation of different resources, such as basking sites or prey species. In addition to the conspicuous dichotomy of nocturnal versus diurnal lizards used in Table 2, a much more subtle temporal separation of daily and seasonal patterns of activity is widespread among lizards (Pianka, 1973; Schoener, 1974). In principle, niche separation resulting from temporal differences should be reflected in differential use of resources such as food and microhabitats; however, the crude prey and microhabitat categories employed here are most unlikely to subsume all such differences resulting from temporal separation of activity. Time of activity is therefore treated as a "resource" in the following analysis.

Ecological Equivalents

Animals that fill similar ecological niches in different, independently-evolved, faunas are termed ecological equivalents. Examples among birds are discussed elsewhere in this volume (Cody, Chapter 10; Karr and James, Chapter 11). Some such convergent evolutionary responses of lizards to the desert environment are evident among the three continents (Table 3). For example, as mentioned above, the North American and Australian deserts support a cryptically-colored, thornily-armored, ant-

Table 3.   Some approximate ecological equivalents in the three continental desert-lizard systems.

North America	Kalahari	Australia
Phrynosoma	—	*Moloch*
Callisaurus	—	*Amphibolurus scutulatus*
Dipsosaurus	—	*Amphibolurus inermis*
Crotaphytus	—	*Varanus eremius*
—	*Typhlosaurus*	*Rhodona*
—	*Mabuya occidentalis*	*Ctenotus pantherinus*
—	*Mabuya variegata*	*Ablepharus butleri, Ctenotus piankai*
—	*Chondrodactylus*	*Nephrurus*
—	*Colopus*	*Diplodactylus stenodactylus*
—	*Pachydactylus rugosus*	*Diplodactylus strophurus*
—	*Pachydactylus capensis*	*Heteronotia*
Sceloporus	*Agama*	*Amphibolurus barbatus*
Cnemidophorus	*Eremias lugubris, Nucras*	various *Ctenotus* spp.
Uta	*Eremias lineo-ocellata*	*Amphibolurus isolepis*

Species are aligned so that approximate ecological counterparts on different continents appear in the same row but in different columns. North America and Australia have four counterparts missing from the Kalahari. Seven niches are filled with roughly convergent species in Australia and the Kalahari, but only three have obvious counterparts on all three continents. See text for further details.

specialized species: *Phrynosoma* occupies this niche in the North American deserts (Pianka and Parker, 1975), and *Moloch* fills it in Australia (Pianka and Pianka, 1970). Similarly, each of these desert systems has a medium-sized lizard-eating lizard (*Crotaphytus wislizeni* in North American and *Varanus eremius* in Australia) and long-legged species that frequent the open spaces between plants (*Callisaurus* in North America, *Amphibolurus scutulatus, A. cristatus,* and *A. isolepis* in Australia). A number of species pairs in the Kalahari and Australia are also convergent: for example, the subterranean *Typhlosaurus* and *Rhodona* have somewhat similar ecologies, as do the agamids *Agama hispida* and *Amphibolurus barbatus* and several pairs of gecko species (*Chondrodactylus* and *Nephrurus, Colopus*

wahlbergi and *Diplodactylus stenodactylus,* and *Pachydactylus capensis* and *Heteronotia binoei*). However, few convergences are apparent among all three desert systems (Table 3). Even the above-mentioned convergent pairs of species exhibit marked ecological differences when subjected to close scrutiny (e.g., Pianka and Pianka, 1970). Indeed, there is considerable disparity among the ecologies of most species in the three continental desert-lizard systems.

Environmental Variability

Water is a master limiting factor in deserts. A convenient result is that long-term mean annual precipitation provides a reasonably good estimate of average annual productivity. Furthermore, year-to-year variation in annual precipitation should generate temporal variability in food availability, and standard deviation in annual precipitation should therefore reflect environmental variability. Brown (Chapter 13) uses the same climate index to predict the diversity of desert rodent species. Table 1 lists precipitation statistics and lizard species densities for the various study areas. Interestingly, the number of lizard species is positively correlated with both long-term mean precipitation and the standard deviation in annual precipitation ($r = 0.42$, $P < 0.05$ and $r = 0.68$, $P < 0.001$, respectively). Long-term mean and standard deviation in precipitation are themselves strongly correlated ($r = 0.70$, $P < 0.001$). To attempt to separate variables, I computed partial correlation coefficients. When effects of long-term mean precipitation are held constant, the correlation between lizard species density and the standard deviation in precipitation remains significant ($r = 0.54$, $P < 0.01$). However, when standard deviation in precipitation is partialled out, the number of lizard species is no longer significantly correlated with long-term mean annual precipitation ($r = 0.02$). This result suggests that productivity, per se, does not promote diversity, but rather that variability in productivity does. More productive areas might be expected to support a greater number of species; at first glance, it is more difficult to see how increased variability in itself might promote diversity. I reconsider this puzzling correlation later, after examining the niche relationships of the lizards.

Historical Factors

These lizard faunas have clearly been influenced profoundly by various historical factors, such as degree of isolation and available biotic stocks, particularly those of potential competitors and predators. One reason the Australian deserts support such very rich saurofaunas is that competition with, and perhaps predation pressures from, snakes, birds, and mammals are probably reduced on that continent. Climate doubtless shapes lizard faunas and regulates their species densities as well. Effects of other historical variables, such as the Pleistocene glaciations, on lizard communities are very difficult to evaluate, but could well be as considerable as they are in birds (as Cody, Chapter 10, and Diamond, Chapter 14, indicate; see also below).

An extremely powerful ecological technique is comparison of historically independent but otherwise basically comparable ecological systems. The degree of convergence between such independently-evolved systems reflects the extent to which evolutionary outcome is determined by the interaction between a given animal's body plan and a particular physical environment. Both Cody (Chapter 10) and Karr and James (Chapter 11) make use of the same type of comparisons. The degree to which evolutionary pathways are determinate is of considerable interest, since independent evolution of similar patterns strongly suggests common underlying selective forces. Moreover, such a convergence, if observed, indicates that general theories explaining niche relations and community structure can eventually be formulated (Cody, 1973; Recher, 1970). In appropriately selected situations, such "natural experiments" may actually allow some measure of control over historical factors such as the Pleistocene glaciations. Thus, differences between faunas of independently evolved areas with similar climates and vegetation structures presumably reflect their different histories.

Historical events have shaped lizard faunas in other ways as well. Australian desert lizards have clearly usurped some of the ecological roles occupied by other taxa in the other two desert systems. Thus pygopodid and varanid lizards in Australia replace certain snakes and mammal carnivores (Pianka, 1969a). The mammalian fauna is conspicuously impoverished in Australia, and the snake fauna less so. Clearly such usurpation of the ecological roles of other taxa has expanded the diversity of resources and the overall niche space exploited by Australian desert lizards. However, these snake-like and mammal-like lizards contribute only a very minor amount to the increased lizard species density of Australia, ranging from one to four species on various study areas. Hence there are many more "lizard-like" lizards in Australia than in the other two continental systems. In addition to such conspicuous usurpation of ecological roles, more elusive but important competitive interactions with bird faunas doubtless occur. For example, proportionately more species of ground-dwelling insectivorous birds exist in the Kalahari than in Australia, suggesting that competition between lizards and birds is keener in southern Africa (Pianka, 1971; Pianka and Huey, 1971). With increases in total community species density (birds plus lizards), the number of lizard species increases faster than bird species in Australia, whereas in North America and the Kalahari, bird species density increases faster than lizard species density (Figure 2). Reasons for this difference among the continental systems are elusive and interpretation must remain conjectural (Pianka, 1971, 1973). One relevant and salient fact is that there are very few migratory birds in Australia, whereas a

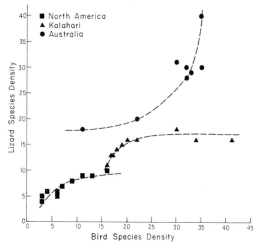

Figure 2 Lizard species density plotted against bird species density for 28 desert study sites on three continents. The number of bird species increases faster than the number of lizard species in North America and in the Kalahari, whereas in Australia the species density of lizards increases faster than that of birds.

fair number of migrant bird species periodically exploit the North American and
the Kalahari deserts. Clearly, competitive
pressures from migrants must influence
lizard faunas of the latter two desert systems in an adverse manner.

The most diverse, and perhaps the most
interesting, lizard communities presently
extant occur in deserts (a possible exception is the *Anolis* communities of the Caribbean islands). Indeed, lizards may actually capitalize on scant and variable
amounts of primary production, and this
could contribute to their relative success
over birds in arid regions (Pianka, 1967).

Still other differences in the composition of lizard communities of these three
continents presumably have a historical
basis as well. Southern Africa is renowned
for its diverse termite fauna, which in turn
has probably facilitated evolution of
termite-specialized subterranean lizards
(Huey and Pianka, 1974; Huey et al., 1974).
Similarly, the higher incidence of arboreal
and nocturnal lizard species in the Kalahari
and Australia, as compared with North
America, is probably related to fundamental differences in the niches occupied
by other members of these communities,
such as arthropods, snakes, birds, and
mammals. The high diversity of nocturnal
lizards in the two southern-hemisphere
deserts could be due to one or more of
the following factors: (1) The effects of the
Pleistocene glaciations are generally ac-
knowledged to have been stronger in the
northern hemisphere (cf. Cody's results in
Chapter 10). This could have had its effect
on the evolution of nocturnal lizards;

however, present-day climates in at least
the southern parts of the northern hemisphere's deserts seem to be quite adequate
for nocturnal lizards. Indeed, the euble-
pharine gecko *Coleonyx* has made a successful existence as a nocturnal terrestrial
lizard in these deserts. The absence of an
arboreal gecko from the flatland desert
habitats of the Sonoran desert is puzzling,
especially in view of the fact that rock-
dwelling geckos (*Phyllodactylus*) are found
within the system. Indeed, a successful
climbing gekkonid could probably invade
this desert system, given an opportunity
(Pianka, 1973). (2) Of course, the various
desert systems could differ in the diversity
and abundance of available nocturnal resources, such as nocturnal insects. (3) The
ecological role of arboreal nocturnal lizards could be filled in North America by
other taxa, say spiders. Differences in the
numbers and/or densities of insectivorous
and carnivorous nocturnal snakes, birds,
and mammals might also play a role.

Components of Species Diversity

Saturated communities can differ in
species diversity in only three ways, which
are not mutually exclusive (MacArthur,
1965, 1972). (1) The diversity of available
resources determines the variety of opportunities for ecological diversification
within a community. Communities with
fewer different resources will support
fewer species than those with a greater
variety of resources, all else being equal.
(This corresponds to a "smaller overall
niche space" or "fewer niches.") (2) As the

diversity of utilization of resources by an average species increases, the number of species that can coexist within a community must decrease. (This corresponds to "larger niches.") (3) Two communities similar in both the above respects can still differ in species diversity if they differ in the average extent to which resources are shared, or the amount of niche overlap. A community with greater overlap will support more species than one with less overlap simply because more species use *each* resource. (This corresponds to "smaller exclusive niches.")

Briefly, diversity should increase with the range of available resources and the extent of tolerable niche overlap, but decrease as niche breadths of component species become larger.

MacArthur (1972) derived the following intriguingly simple approximate equation for the species diversity of a community:

$$D_S \cong \frac{D_R}{D_U}(1 + C\overline{\alpha}) \qquad (1)$$

where D_S is the diversity of species, D_R is the diversity of resources used by the entire community, D_U represents the diversity of utilization (niche breadth) of each species, C measures the number of potential competitors or "neighbors in niche space" (a function that increases more or less geometrically with the number of subdivided niche dimensions), and $\overline{\alpha}$ is the "mean competition coefficient" (I prefer to consider this the mean niche overlap). The relevance of this equation to species diversity of desert rodents is discussed by Brown in Chapter 13.

Although his derivation is essentially tautological and requires the perhaps unrealistic assumption that all niche breadths are equal (identical D_U's for all species), MacArthur's community equation does focus attention on the importance of niche breadths and overlaps in considerations of species diversity. MacArthur (1972, p. 185) stated that "people have seldom even measured single components in these equations." It should prove profitable to estimate the various terms in eq. 1 for real communities. Here I present such estimates of D_S, D_R, D_U, and α for the above-mentioned lizard communities, and I examine interrelationships among these components and with the above climatological estimates of productivity and climatic variability. While I use MacArthur's elegant equation as a launch, my results do not depend upon its validity (though they may serve to test it).

Niche Dimensions

Like most animals, desert lizards subdivide resources in three major ways; they differ in what they eat, where they forage, and when they are active. Ecological differences in each of these three niche "dimensions" should reduce competition and facilitate coexistence of a variety of species.

With most lizard species, it would be virtually impossible to evaluate the degree of interdependence of these three niche dimensions (trophic, spatial, and temporal); however, in some relatively sedentary fossorial skinks (genus *Typhlosaurus*), we have attempted to assess the degree to which foods eaten depend upon microhabitat (Huey et al., 1974). These two niche dimensions appear to be largely independent in *Typhlosaurus*. Moreover, some diurnal and nocturnal species pairs consume many of the same prey types, sometimes the same prey species (though they are usually captured in different ways). Thus there appears to be a substantial degree of independence among the niche dimensions I have chosen to use here (see also Schoener, 1974, 1975). In still other cases, however, clear interactions among these three niche dimensions are apparent (Pianka, 1973). Because the vast majority of interspecific pairs of sympatric lizard species have substantial niche separation along one or more of these three dimensions, it is unnecessary to subdivide niche dimensions further.

Methods of Estimating Niche Parameters

Many different techniques of estimating niche breadths and niche overlaps have been suggested and used (Simpson, 1949; Horn, 1966; MacArthur and Levins, 1967; Levins, 1968; Schoener, 1968; Colwell and Futuyma, 1971; Pielou, 1972; Roughgarden, 1972; Vandermeer, 1972; Pianka, 1969b, 1970, 1973). Throughout this paper, I quantify diversity and niche breadth with the index proposed by Simpson (1949), $1/\Sigma p_i^2$, where p_i is the proportion of the ith species or resource category. Overlaps are computed with the following improved version of the equation proposed by MacArthur and Levins (1967) and Levins (1968) for estimating

"alpha" from field data:

$$a_{jk} = a_{kj} = \frac{\sum\limits_{i}^{n} p_{ij} p_{ik}}{\sqrt{\sum\limits_{i}^{n} p_{ij}^2 \sum\limits_{i}^{n} p_{ik}^2}} \quad (2)$$

where p_{ij} and p_{ik} are the proportions of the ith resource used by the jth and the kth species, respectively. May (1975) recently gave mathematical rationale for the superiority of this symmetric measure over the original nonsymmetric form. I do not consider values obtained from eq. 2 "competition coefficients," but merely measures of niche overlap (for discussion of the distinction between overlap and competition, see Colwell and Futuyma, 1971 and/or Pianka, 1974a).

In the following analyses I quantify the trophic dimension, using the volumetric representation of prey in the 20 different food categories listed earlier. Similarly, the 15 basic microhabitats listed earlier are recognized for analysis of the spatial dimension of the niche. Time dimension computations are based on 22 hourly time categories, expressed in hours since sunrise or sunset to help to correct for seasonal shifts in the time of activity, as noted above. Each niche dimension is given equal weight by dividing computed diversities by the total number of possible categories, which allows all diversities to be expressed as a proportion of their maximal possible value.

Ideally, a multidimensional analysis of resource utilization and niche separation along more than a single niche dimension

should proceed by considering all resources present as a simultaneous function
of all niche dimensions (May, 1975).
However, in practice it is extremely difficult or even impossible to obtain such
multidimensional utilization data, both
because animals move and because they
integrate over time. (Stomachs of most
lizards contain prey captured in a variety
of microhabitats.) To obtain true estimates
of multidimensional utilization, one would
have to follow an individual animal and
record the exact time and place of capture
of all prey items. Instead, I must work
with three separate unidimensional utilization distributions, for reasons indicated
above.

Provided that niche dimensions are
truly independent (orthogonal), with any
given prey item being equally likely to be
captured at any time and in any place,
overall multidimensional utilization is simply the product of the separate unidimensional p_i's (May, 1975). In this case, estimates of various niche parameters along
component dimensions can be multiplied
to obtain multidimensional estimates.
However, should niche dimensions be
entirely dependent upon one another,
resource utilization becomes additive, and
the appropriate procedure is to sum or
average the separate unidimensional estimates of various niche parameters. Since
real niche dimensions are presumably seldom, if ever, either perfectly dependent or
perfectly independent, neither technique is
entirely satisfactory. Recognizing these
very considerable difficulties, I estimate
multidimensional niche parameters along
the three niche dimensions, using both

multiplicative and summation multidimensional estimates for resource diversity, niche breadths, and niche overlaps.
May (1975) shows that summation niche
overlap actually constitutes an upper
bound on true multidimensional overlap;
moreover, he points out that multiplicative overall overlaps can both under-
estimate and overestimate the true multidimensional overlap.

Resource Diversity

There are many striking differences
among the three continental desert-lizard
systems in the relative importance of various resource categories. For example, termites comprise 41.3% of the diet of all
Kalahari lizards, but represent only 16.5%
and 15.9% of the saurian diet in North
America and Australia, respectively. As a
result, the diversity of foods eaten by
Kalahari lizards tends to be lower than in
the other two deserts. Ants increase in
importance from North America to the
Kalahari to Australia (9.7, 13.6, and
16.4%, respectively), whereas beetles decrease (18.5, 16.3, and 7.3%, respectively).
Vertebrates, largely lizards, constitute
24.8% of the diet by volume of Australian
desert lizards, but only 7.8% and 2.3% of
the food eaten by lizards in North America and the Kalahari, respectively.
Whereas 45.3% of all North American
lizards were first sighted in the open sun,
only about 19% of those in the two southern hemisphere deserts were in the open
sun (open shade percentages are, respectively, 1.7, 11.4, and 17.4%). This heavy
use of one microhabitat category in North

America results in low values for micro-habitat diversity on that continent. The percentage of arboreal animals above the ground increases from 4.2% in North America to 14.9% in the Kalahari to 18.2% in Australia. Many other interesting differences in resource utilization patterns are also evident among the continents (Pianka, 1973).

Diversities of resources actually used along each niche dimension by all the lizards (of all species) in the saurofaunas on the various study sites are plotted against lizard species densities in Figure 3. The diversity of foods eaten by all lizards does not correlate with the number of lizard species when all 28 areas are considered, although the correlation between these two measures *is* statistically significant *within* the North American deserts ($r = 0.77$, $P < 0.01$). When all 28 areas are grouped, the diversity of micro-habitats exploited by all lizards and the diversity of times of activity of all lizards are both strongly correlated positively with lizard species density (Figure 3). Trends within continental systems are usually less pronounced or nonexistent; lizard species density is significantly correlated with the diversity of microhabitats exploited in both North America ($r = 0.64$, $P < 0.05$) and the Kalahari ($r = 0.79$, $P < 0.01$). None of the correlations between resource diversity and lizard species density is significant for the Australian deserts. Multiplicative and summation estimates of the overall diversity of resources used by lizards are very strongly correlated with one another ($r = 0.95$, $P < 0.001$), and less strongly, but significantly, positively correlated with

lizard species density (Figure 3, lower right).

Thus the diversity of resources actually used by lizards, or the total volume of lizard niche space, tends to be greater in areas with more diverse saurofaunas (Figure 3). This result is not as circular as it might at first seem; the correlations often do not hold up within any single continental desert-lizard system, and thus high lizard diversity on any particular continent is not necessarily associated with high diversity of resource use there.

Niche Breadth

To maximize sample sizes and confidence in estimates of niche breadths, I group all individuals of each species from various study areas in the following analysis. (A few species exhibit distinct, usually relatively slight, niche shifts between study areas, but these constitute a definite minority.) Similar relative results are obtained in a more complex area-by-area analysis, except that niche breadths tend to be smaller and more variable, especially for uncommon species.

Niche breadths, standardized by dividing by the number of resource categories, were computed for 86 lizard species along with two estimates of overall niche breadth (summation and multiplicative). Niche breadths along particular dimensions and overall breadths vary considerably among species. Certain specialized species, described above, have very narrow niches along a particular niche dimension. Breadths along various dimensions appear to be independent (as judged by nonsignificant correlation coefficients

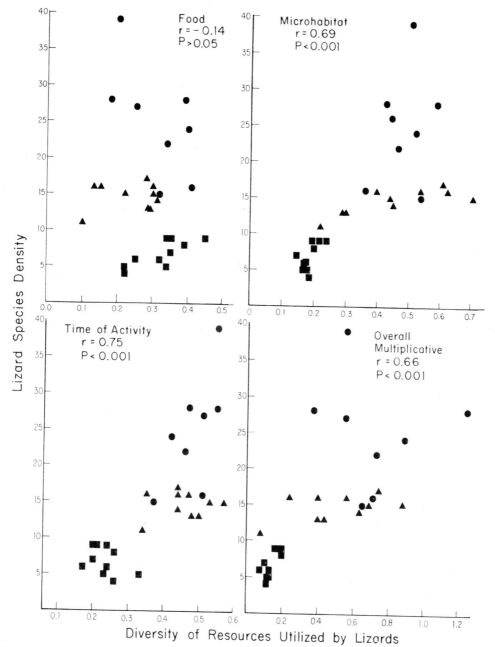

Figure 3 Four plots of the diversity of resources utilized by entire saurofaunas against their lizard species densities. Symbols coded by shape represent different continents, as in previous figures. Three niche dimensions (trophic, microhabitat, and time of activity) are shown, as well as their product, in-tended to reflect the overall diversity of resources exploited along all three niche dimensions. Note that, except for the trophic dimension, diversity of resources used by lizards tends to increase with the number of species. See text.

between dimensions), except that the diversity of microhabitats used by various species is weakly but significantly correlated with the diversity of time of activity ($r = 0.39$, $P < 0.001$).

The average diversity of use of foods and microhabitats by the component species varies among continental systems. For example, the diversity of use of the trophic dimension by an average species is low in the Kalahari where many species consume a lot of termites, while the spatial niche dimension is narrow in North America where many animals are first sighted in the open sun. Average diversities of utilization of the three niche dimensions are less variable in Australia. Confidence limits on means within continental desert-lizard systems overlap broadly (Table 4), however, so that no statistically significant variation in niche breadth is evident among continents. Because summation and multiplicative overall niche breadths are quite strongly correlated ($r = 0.88$, $P < 0.001$), I use multiplicative values hereafter.

Species diversity should increase with the ratio of the diversity of utilized resources (D_R in eq. 1) over the diversity of utilization by an average species. I calculated average niche breadths for the species actually occurring together on each study area ($\overline{D_U}$ for various subsets of species). Thus estimated, mean overall niche breadths *within* each continental system vary inversely with the number of lizard species, as might be expected, although these inverse relationships are not statistically significant. Somewhat surprisingly, however, when all 28 areas are grouped, the only evident relationship is a *positive* correlation of niche breadth with estimated lizard-species diversity ($r = 0.57$, $P < 0.01$). There is also a positive correlation between overall mean D_U and D_R ($r = 0.52$, $P < 0.05$), but this disappears when the effects of lizard species diversity are held constant by partial correlation.

Table 4. Average niche breadths of desert lizards on three continents

Niche dimension	North America	Kalahari	Australia
Trophic	0.232 (0.168–0.296)	0.198 (0.196–0.312)	0.214 (0.202–0.278)
Spatial	.146 (.108– .184)	.228 (.186– .270)	.201 (.177– .225)
Temporal	.241 (.167– .295)	.254 (.148– .258)	.240 (.180– .248)
Overall summation	.206 (.178– .234)	.237 (.205– .269)	.218 (.196– .240)
Overall multiplicative	.077 (.045– .109)	.138 (.078– .198)	.112 (.078– .146)

This table shows what proportion of the total range of a particular resource axis is used by the average lizard species in the three continental desert systems. The three main resource axes are food, space or microhabitat, and time of foraging activity. Each entry gives the mean and 95% confidence limits (in parentheses) of average niche breadth along the given axis on the given continents. Notice that lizards of all three continents have very similar temporal niche breadths, but that North American lizards have broader diets and narrower habitat ranges than lizards elsewhere, Kalahari lizards have the broadest habitat ranges and the narrowest diets, and Australian lizards are intermediate in both respects.

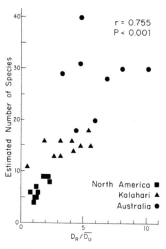

Figure 4 Ratios of the overall multiplicative diversity of resources used by all lizards D_R over the average niche breadth of the species on a given study area $\bar{D}_U$ are plotted against the number of lizard species. This ratio reduces variance in lizard species density by some 58%.

As expected, ratios of $D_R/\bar{D}_U$ are strongly positively correlated both with lizard species densities (Figure 4) and with lizard diversity ($r = 0.71$, $P < 0.001$). Overall D_R values alone account for 49% of the variation in lizard species density ($r^2 = 0.487$), whereas the ratio of overall D_R over mean overall niche breadth ($D_R/\bar{D}_U$) reduces variance in species densities by some 58% ($r = 0.76$, $r^2 = 0.578$). When the effects of niche breadth are removed by partial correlation, r remains similar (0.73), demonstrating that niche breadth contributes little further to the latter correlation, given knowledge of D_R. Hence, variations in niche breadth appear to be of relatively minor importance in considerations of diversity in these lizard communities. Increased lizard-species densi-

ties do not stem from conspicuously reduced niche breadths, but rather are closely associated with larger overall niche space (D_R).

Patterns of Niche Overlap

Species pairs with high overlap along one niche dimension often, though certainly by no means always, overlap little along another. Correlation coefficients of overlaps between dimensions are seldom significant, however. The average extent of overlap along the various dimensions varies among deserts (Table 1). For example, overlap in microhabitat is high in North America where many lizards frequent the open sun, whereas dietary overlap is extensive in the Kalahari where most lizards eat considerable numbers of termites. Overlap tends to be low along all three niche dimensions in Australia (Table 1).

May (1975) points out the great difficulties in obtaining reasonable estimates of overall multidimensional niche overlap from unidimensional patterns of utilization. In addition to average overlap values for each niche dimension on the various study areas, Table 1 lists the means of all multiplicative overlaps, average summation overlaps (upper bounds on the true multidimensional overlap; see also Figure 5) and means of all pairs of multiplicative (overall) overlap that are nonzero. The latter values might be expected to reflect maximal tolerable niche overlap. Fortunately, all these approaches to estimating multidimensional overlap and/or the upper limit on tolerable overlap yield the

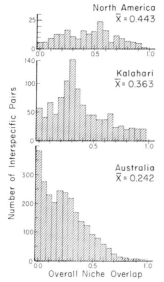

Figure 5 Frequency distributions of observed overall summation niche overlaps for all interspecific pairs on all study areas within each of the three continental desert systems. These values overestimate true multidimensional niche overlap, but they do constitute upper bounds on the true values. Note that overlap tends to decrease from North America to the Kalahari to Australia. Compare with Figures 6, 7, and 8.

same basic result: namely, overall niche overlap generally *decreases* with increasing lizard species density.

Average overall summation overlaps are also inversely correlated with standard deviation in annual precipitation (Figure 6). However, mean overall summation overlap is even more strongly correlated (again, negatively) with the number of lizard species (Figure 7). When effects of lizard species density are held constant by partial correlation, mean overall summation overlap and the standard deviation in precipitation are no longer significantly

correlated $(r = -0.28)$.[2] However, the correlation between lizard species density and overall summation overlap remains significant when the standard deviation in precipitation is held constant by partial correlation $(r = -0.56, P < 0.01)$. This result suggests that the extent of tolerable niche overlap is *not* a function of the degree of environmental variability, but rather that it is related to the number of competing species [MacArthur (1972) termed this "diffuse competition"]. Exactly comparable results are obtained with multiplicative overall overlap values. Elsewhere I have hypothesized that maximal tolerable niche overlap should vary

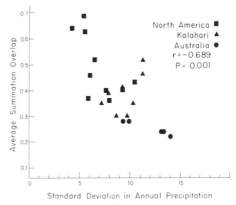

Figure 6 Average overall summation niche overlap plotted against the standard deviation in annual precipitation for 23 study sites, with climatic data. Although none of the correlations within continental desert systems is significant, the correlation coefficient for all areas is highly significant statistically. When the number of lizard species on the various areas is held constant by partial correlation, however, this correlation disappears.

[2] Averages of the largest tenth of all multiplicative and summation overlaps also decrease significantly with lizard species density (Pianka, 1974b).

inversely with the intensity of competition (Pianka, 1972); these results support that prediction.

The vast majority of interspecific pairs overlap very little or not at all when overlaps along the three dimensions are multiplied (Pianka, 1973, 1974b). The possible number of such nonoverlapping pairs increases markedly with the size of overall niche space, provided that niche breadths remain relatively constant (demonstrated above). Overlaps between those pairs with some overlap and those with greatest overlap are of most interest, as they are likely to reflect limiting similarity and/or maximal tolerable overlap. Such estimates behave similarly to average summation overlaps, and tend to decrease with increasing lizard species densities (Figures 7 and 8).

May and MacArthur (1972) developed an elegant analytic model of niche overlap as a function of environmental variability. Their theory predicts a distinct upper limit on the permissible degree of overlap; moreover, the derivation suggests that maximal tolerable overlap should be relatively insensitive to environmental variability. Leigh (Chapter 2) provides an alternative pathway to this same result. Although the May-MacArthur niche-overlap model assumes a one-dimensional resource spectrum, May (1974) recently expanded the argument without qualitative change to a multidimensional niche space. At first glance, Figure 6 seems at odds with the May-MacArthur prediction

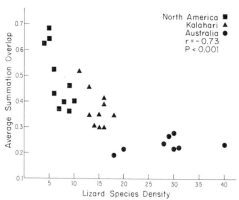

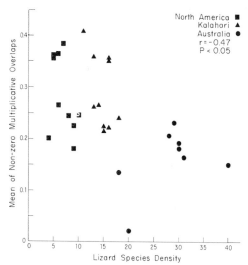

Figure 7 Average summation niche overlap plotted against the number of lizard species for the 28 study areas. The inverse correlation is highly significant ($r = -0.73$, $P < 0.001$). A similar inverse correlation exists between lizard species density and mean multiplicative overall overlap ($r = -0.67$, $P < 0.001$). Three estimates of maximal niche overlap are similarly correlated (see Figure 8).

Figure 8 Mean of all nonzero values of multiplicative overall overlaps plotted against lizard species densities. The averages of the largest tenths of all summation and multiplicative overall overlap values are even more strongly correlated with lizard species densities ($rs = -0.63$ and -0.62, respectively, $Ps < 0.001$).

that maximal overlap should be insensitive to environmental variability. However, overlap values used here are not entirely appropriate for testing the May-MacArthur theory, since this model is expressed in terms of the ratio of niche separation over niche breadth, effectively the inverse of niche overlap scaled by niche breadth. To approximate more closely the conditions of their model, I therefore estimated niche separation as one minus overlap for all interspecific pairs and then expressed these values as ratios of separation over standardized niche breadths. Means and confidence limits of these values are given in Table

5; many distinct differences are apparent both among niche dimensions and among the three desert systems. For example, the ratio of overall niche separation over niche breadth is essentially unity in Australia, but it is significantly greater than one in the Kalahari and significantly lower than unity in North America (Table 5). Plausible explanations for these differences are elusive, but the high overlap in North America could be a result of that continent's relatively low degree of environmental variability and/or reduced intensity of diffuse competition due to lower lizard species densities. Such speculation would be more plausible if the mean for

Table 5. Niche overlap, breadth, and separation of desert lizards on three continents

Niche parameter and dimension	North America	Kalahari	Australia
Overlap/breadth			
(1) Food	2.09 (1.81–2.37)	4.79 (4.37–5.21)	2.18 (2.06–2.30)
(2) Microhabitat	2.51 (1.87–3.15)	1.30 (1.12–1.48)	1.92 (1.84–2.00)
(3) Time	2.33 (1.97–2.69)	1.64 (1.38–1.90)	1.79 (1.69–1.89)
Niche separation			
(4) Food	0.54 (0.48–0.60)	0.36 (0.34–0.38)	0.68 (0.66–0.70)
(5) Microhabitat	.66 (.58– .74)	.71 (.67– .75)	.69 (.67– .71)
(6) Time	.42 (.34– .50)	.65 (.61– .69)	.68 (.66– .70)
(7) Overall (summation)	1.62 (1.48–1.76)	1.72 (1.66–1.78)	2.05 (2.03–2.07)
Separation/breadth			
(8) Food	3.73 (2.87–4.59)	2.51 (2.21–2.81)	5.18 (5.0–5.36)
(9) Microhabitat	5.79 (4.83–6.74)	4.23 (3.85–4.61)	4.60 (4.48–4.72)
(10) Time	2.78 (2.0–3.56)	5.08 (4.48–5.68)	5.11 (4.93–5.29)
(11) Overall	0.50 (0.38–0.62)	1.54 (1.16–1.92)	1.03 (1.00–1.06)

Each entry gives the mean and 95% confidence limits (in parentheses) for the indicated ratio or parameter, over all study sites on each continent. A simple hypothesis would be that species diversity is directly proportional to niche overlap and inversely proportional to niche breadth, so that the overlap/breadth ratio should be positively correlated with species diversity. Comparison of the first three rows of the table shows this naive deduction to be quite wrong, for North American deserts have high overall values of this ratio but support lizard faunas of low species diversity. Niche separation is estimated as 1 minus niche overlap (rows 4–7). This yields values that are very similar among the three niche dimensions in Australia, with the richest fauna. But niche separation is lowest in North America, again the opposite of simple expectations. The ratio of niche separation over niche breadth is given for each continent in rows 8–10, and the overall value is expressed in the last row. As expected, the Kalahari, with fewer lizard species than Australia, has a higher ratio. Contrary to expectations, North America, with the fewest species, has the lowest ratio. No simple hypothesis explains these results, as discussed in the text.

the Kalahari was intermediate between those for North America and Australia.

If the number of neighbors in niche space (C in eq. 1) could somehow be estimated independently, MacArthur's equation would generate predicted species diversities that could be compared with observed diversities. Unfortunately, I have been unable to find a satisfactory way to estimate C independently of the remaining parameters of eq. 1. It does seem likely, however, that the number of neighbors in niche space should be least in North America and greatest in Australia.

In sum, species diversity in rich saurofaunas is *not* facilitated by increased niche overlap; rather, quite the reverse is true: overlap tends to decrease with increasing lizard species density. I interpret this pattern as indicating that competitor species density or diversity influences tolerable niche overlap; a greater number of competing species, or stronger "diffuse competition," demands greater average niche separation among coexisting species. Similarly, Diamond (1973) finds precise altitudinal replacement among competing species only in the most species-rich bird communities. High overlap with fewer competitors could be equivalent to lower overlap with more competitors, since in the former case the coefficients a_{ij} in the competition expression $\Sigma\, a_{ij} X_j$ (cf. Levins, Chapter 1, eqs. 1–3) are larger and in the latter case the summation is over a larger number of competitors. If so, the actual intensity of interspecific competition per species, or the *total* of the interspecific inhibitory effects, could thus be similar in communities of widely divergent species densities. A first hypothesis might be that

total overlap among sympatric species remains constant; however, total overlap actually *increases* significantly with lizard species density, even though the average amount of overlap between pairs decreases (Pianka, 1974b). Indeed, it is intriguing to speculate that such adjustments of overlap with species density might actually result in a relatively constant level of interspecific competitive inhibition among a community's component species. If so, it is not overlap that remains constant, but rather the degree of competitive inhibition tolerated by the individuals comprising an average species.

Synthesis and Conclusions

Earlier I noted that lizard species density is positively correlated with standard deviation in annual precipitation, a measure of environmental variability. Hutchinson (1961) suggested that temporal heterogeneity might actually facilitate coexistence, both by continually altering the relative competitive abilities among members of a community, and by periodically reducing population sizes and thus the intensity of competition. Similar mechanisms might certainly be expected to operate in desert faunas. A difficulty with these interpretations, however, is that the above data on niche relationships make it difficult to avoid the conclusion that competition is in fact actually *keener* in the more diverse lizard communities. Clearly, future improvements in the theory of species diversity will have to include more sophisticated considerations of environmental variability, as Leigh, Chapter 2, shows; furthermore, overlap will have to be treated as a variable that varies inversely with diversity.

The diversity of resources actually exploited by lizards along various niche dimensions, and the extent of niche overlap along them, varies widely among the three continental desert-lizard systems. As a result, the relative importance of various niche dimensions in separating niches differs among continents. Food is a major dimension separating niches of North American lizards, whereas in the Kalahari niche separation is slight on the trophic dimension and differences in microhabitat and time of activity are considerable. All three dimensions separate niches more or less equally in Australia. Differences in diversity among continents stem largely from differences in the overall diversities of resources exploited by lizards or in the size of the lizard niche space, but are not due to conspicuous adjustments in overall niche breadths. Overall overlap decreases, rather than increases, as lizard diversity increases, so that niche overlap does not enhance diversity, but rather contributes negatively to it. Two of the four parameters in eq. 1 vary consistently with differences in lizard species density: D_R positively and $\bar{\alpha}$ inversely. The number of neighbors in niche space, C, probably influences maximal tolerable niche overlap and $\bar{\alpha}$. Mean niche breadth, $\overline{D_U}$, however, does not vary much and contributes little to observed differences in lizard diversity. This empirical finding could partially justify MacArthur's assumption that the D_U's among members of a community are identical. Moreover, if this result is generally true, niche breadths may not play an important role in future diversity

studies or theoretical developments. Since overlap varies inversely with lizard species density, whereas the number of competitors (or diffuse competition) varies positively, the term $C\bar{\alpha}$ in eq. 1 may change relatively little with diversity. If so, species diversity should be approximately proportional to the diversity of resources utilized (this implies that eq. 1 can be greatly simplified to $D_S \propto D_R$). In short, only one of the factors in eq. 1 is implicated as a major determinant of the number of species coexisting in these lizard communities: namely, the size of the lizard niche space as measured by the diversity of resources actually exploited by lizards, or D_R.

To my knowledge, this is the first empirical demonstration that species density is negatively correlated with the average extent of overlap among the members of a community. [Brown and Lieberman (1973) and Brown (Chapter 13, Figure 10) reported *positive* correlations between overlap and species densities of small mammals, but these could be artifacts of the increased numbers of interspecific pairs in more diverse communities.] I conclude that both species diversity theory and niche overlap theory need to be expanded to incorporate more fully the important phenomenon MacArthur (1972) termed "diffuse competition."

Acknowledgments

Robert MacArthur inspired this work and provided me with continual food for thought over the ten years I have been involved in it.

This research has benefited from contacts with so many other persons that it is impossible to list them all here. Some who deserve particular acknowledgment are my field assistants: Nicholas Pianka, William Shaneyfelt, Michael Thomas, Larry Coons, and Raymond Huey. My wife, Helen, has assisted in numerous ways. I thank Virginia Denniston and Glennis Kaufman for much help in data processing and analysis. Michael Egan painstakingly analyzed the contents of thousands of lizard stomachs. Participants in the MacArthur Memorial Symposium, especially the consulting editors, provided useful comment on a preliminary draft of the manuscript. The project was made possible by financial support from the National Science Foundation (grants GB-5216, GB-8727, GB-31006).

References

Brown, J. H., and G. A. Lieberman. 1973. Resource utilization and coexistence of seed-eating desert rodents in sand dune habitats. *Ecology* 54:788–797.

Cody, M. L. 1973. Parallel evolution and bird niches. *In* F. DiCastri and H. Mooney, eds., *Ecological Studies* No. 7, pp. 307–338. Springer-Verlag, Wien.

Colwell, R. K., and D. J. Futuyma. 1971. On the measurement of niche breadth and overlap. *Ecology* 52:567–576.

Diamond, J. M. 1973. Distributional ecology of New Guinea birds. *Science* 179:759–769.

Horn, H. S. 1966. Measurement of overlap in comparative ecological studies. *Amer. Natur.* 100:419–424.

Huey, R. B., and E. R. Pianka. 1974. Ecological character displacement in a lizard. *Amer. Zool.* 14:1025–1034.

Huey, R. B., E. R. Pianka, M. E. Egan, and L. W. Coons. 1974. Ecological shifts in sympatry: Kalahari fossorial lizards (*Typhlosaurus*). *Ecology* 55:304–316.

Hutchinson, G. E. 1961. The paradox of the plankton. *Amer. Natur.* 95:137–147.

Levins, R. 1968. *Evolution in changing environments.* Princeton University Press, Princeton.

MacArthur, R. H. 1965. Patterns of species diversity. *Biol. Rev.* 40:510–533.

MacArthur, R. H. 1972. *Geographical ecology: Patterns in the Distribution of Species.* Harper and Row, New York.

MacArthur, R. H., and R. Levins. 1967. The limiting similarity, convergence, and divergence of coexisting species. *Amer. Natur.* 101:377–385.

MacArthur, R. H., and E. R. Pianka. 1966. On optimal use of a patchy environment. *Amer. Natur.* 100:603–609.

May, R. M. 1974. On the theory of niche overlap. *Theoret. Pop. Biol.* 5:297–332.

May, R. M. 1975. Some notes on measurements of the competition matrix, α. *Ecology* 56:in press.

May, R. M., and R. H. MacArthur. 1972. Niche overlap as a function of environmental variability. *Proc. Nat. Acad. Sci. U.S.A.* 69:1109–1113.

Pianka, E. R. 1967. On lizard species diversity: North American flatland deserts. *Ecology* 48:333–351.

Pianka, E. R. 1969a. Habitat specificity, speciation, and species density in Australian desert lizards. *Ecology* 50:498–502.

Pianka, E. R. 1969b. Sympatry of desert lizards (*Ctenotus*) in western Australia. *Ecology* 50:1012–1030.

Pianka, E. R. 1970. Comparative autecology of the lizard *Cnemidophorus tigris* in different parts of its geographic range. *Ecology* 51:703–720.

Pianka, E. R. 1971. Lizard species density in

the Kalahari desert. *Ecology* 52:1024–1029.

Pianka, E. R. 1972. *r* and *K* selection or *b* and *d* selection? *Amer. Natur.* 106:581–588.

Pianka, E. R. 1973. The structure of lizard communities. *Ann. Rev. Ecol. Syst.* 4:53–74.

Pianka, E. R. 1974a. *Evolutionary Ecology.* Harper and Row, New York.

Pianka, E. R. 1974b. Niche overlap and diffuse competition. *Proc. Nat. Acad. Sci. U.S.A.* 71:2141–2145.

Pianka, E. R., and R. B. Huey. 1971. Bird species density in the Kalahari and the Australian deserts. *Koedoe* 14:123–130.

Pianka, E. R., and W. S. Parker. 1975. Ecology of horned lizards: a review with special reference to *Phrynosoma platyrhinos.* *Copeia* 1975:141–162.

Pianka, E. R., and H. D. Pianka. 1970. The ecology of *Moloch horridus* (Lacertilia: Agamidae) in Western Australia. *Copeia* 1970:90–103.

Pianka, E. R., and H. D. Pianka. 1976. Comparative ecology of twelve species of nocturnal lizards (Gekkonidae) in the Western Australian desert. *Copeia* 1976:in press.

Pielou, E. C. 1972. Niche width and niche overlap: a method of measuring them. *Ecology* 53:687–692.

Recher, H. 1970. Bird species diversity and habitat diversity in North America and Australia. *Amer. Natur.* 103:75–79.

Roughgarden, J. 1972. Evolution of niche width. *Amer. Natur.* 106:683–718.

Schoener, T. W. 1968. The *Anolis* lizards of Bimini: resource partitioning in a complex fauna. *Ecology* 49:704–726.

Schoener, T. W. 1974. Resource partitioning in ecological communities. *Science* 185:27–39.

Schoener, T. W. 1975. Competition and the niche. *In* D. W. Tinkle and W. W. Milstead for C. Gans, eds., *Biology of the Reptilia.* Academic Press, New York.

Simpson, E. H. 1949. Measurement of diversity. *Nature* 163:688.

Vandermeer, J. H. 1972. Niche theory. *Ann. Rev. Ecol. Syst.* 3:107–132.

13 Geographical Ecology of Desert Rodents

James H. Brown

In the introduction to *Geographical Ecology,* MacArthur (1972) wrote, "To do science is to search for repeated patterns, and to do the science of geographical ecology is to search for patterns of plant and animal life that can be put on a map." With a few notable exceptions (e.g., Connell, 1961 and Chapter 16; Patrick, Chapter 15; Simberloff and Wilson, 1969), geographical ecology has not been an experimental science. Usually it is either impractical or morally undesirable to transplant or exterminate organisms on a geographic scale. Most geographic patterns are detected by quantitative comparisons of the organisms or communities in different areas. For the past several years I have used geographic comparisons to study the distribution and community ecology of seed-eating rodents in the southwestern United States.

Desert rodents provide excellent material for research in geographical ecology. There are many species belonging to three major families: Heteromyidae (pocket mice and kangaroo rats), Cricetidae (deer mice and woodrats), and Sciuridae (ground squirrels). In most desert habitats these rodents are abundant and easily studied. In relatively simple desert ecosystems rodents play important roles as herbivores, particularly as consumers of seeds. Most of the specialized granivores belong to the family Heteromyidae, which have external cheek pouches that are used to collect and transport food. The diets of these species are readily sampled by dead trapping the animals and collecting the contents of their cheek pouches. Desert rodents are of particular interest because they are much less vagile than the birds which have been studied by most geographical ecologists. As a result, historical events and geographic barriers have had an important influence on their distribution and community structure.

The present chapter attempts to answer three primary questions about the ecology of desert rodents: 1) What determines the number of species that coexist in a habitat and occur together in larger geographic areas? 2) How are communities of coexisting species structured, and to what extent are communities in geographically isolated areas structured similarly? 3) How do coexisting species differ in their utilization of resources and thus avoid competitive exclusion? In order to answer these questions, I shall utilize both previously published work on sand dune habitats in the Mojave and Great Basin deserts (Brown, 1973; Brown and Lieberman, 1973) and more recent work on sandy flatland and rocky hillside habitats in the Sonoran Desert.

Methods

The present chapter is based on samples of the rodent communities in 31 geographically separated habitats in the southwestern United States. Eighteen of these habitats were sand dunes in the Mojave and Great Basin deserts of California, Nevada, and Utah that I surveyed prior to 1972. The results of that work have been published (Brown, 1973; Brown and Lieberman, 1973) and may be consulted for a detailed description of the methods. A standardized sampling procedure was developed that utilized dead trapping to provide comparable data on the relative abundances, foraging areas, and diets of the rodent species in each habitat.

Since 1972 I have sampled eight sandy flatland and four rocky hillside habitats in the Sonoran Desert of southern Arizona. The work in the Sonoran Desert was designed to test the generality and repeatability of the patterns described in the earlier papers. This study constituted an independent test of the earlier results in that it used habitats different from those sampled in the Great Basin and Mojave deserts and was not begun until the results of the earlier study had been submitted for publication. The location of the Sonoran Desert study sites and the methods used to sample them are described in the Appendix. In all important respects the methods were identical to those employed in my earlier work in the Mojave and Great Basin deserts.

The Patterns

Species Diversity

The number of species that coexist in a habitat is determined by both historical and ecological factors. The historical events of speciation and crossing of geographic barriers determine the supply of potential colonists to a habitat. The ecological requirements of these species and their interactions with each other determine which of the potential colonists can coexist within the habitat. For some extremely vagile organisms such as birds, it may often be possible to ignore the effects of historical factors on species diversity in continental habitats (MacArthur, 1972), but with careful study one can recognize these effects even for birds (Cody, Chapter 10). With more sedentary organisms such as desert rodents, it is imperative to consider historical factors. In order to evaluate the effects of ecological parameters on rodent species diversity, one must restrict the analysis to large, interconnected deserts to ensure that approximately equivalent numbers and kinds of species have had access to the habitats.

Species diversity increases with productivity. Where approximately equal numbers of rodent species have had opportunity to colonize sand dunes in the Mojave and Great Basin deserts, the number of common species varies from one to five. This variation should be attributable to ecological differences among the habitats. In fact, about 70% of the variation can be attributed to a single parameter, $\bar{x} - \sigma$ (mean minus one standard deviation) of annual precipitation (Figure 1). This is a measure of the predictable amount of rainfall[1] or, more precisely, the minimum

[1]Robert May (personal communication) has suggested that $\bar{x} - \sigma^2$ (mean minus the variance) may be a better statistic than $\bar{x} - \sigma$ for measuring pre-

amount that can be expected in $\frac{5}{6}$ of all years. The amount of precipitation in arid environments provides a good estimate of net primary productivity (see Rosenzweig, 1968) and seed production. The reason for this is that desert plants are almost totally dependent on the amount and timing of precipitation for germination, growth, and reproduction (Went, 1948, 1955; Went and Westergaard, 1949; Beatley, 1967, 1969; Schaffer and Gadgil, Chapter 6). Thus the close correlation between the number of common species and $\bar{x} - \sigma$ of annual precipitation indicates that rodent species diversity within habitats depends largely on the abundance and predictability of food resources. Dependence of species diversity on indices of precipitation is also observed for desert lizards (Pianka, Chapter 12).

The number of common species of seed-eating rodents in sandy flatland habitats in the Sonoran Desert also is well correlated with $\bar{x} - \sigma$ of annual precipitation (Figure 1). Although the slope of this relationship is less than for the sand dunes in the Mojave and Great Basin deserts, the similarity between the patterns is striking. In both cases the number of common species varies from one to approximately five, and the correlation coefficients are virtually identical. In each case $\bar{x} - \sigma$ of annual precipitation ac-

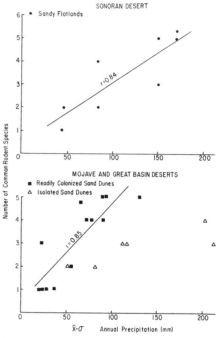

Figure 1 The relation between rodent species diversity in sandy soil habitats and the predictable amount ($\bar{x} - \sigma$, mean minus one standard deviation) of annual precipitation, an index of productivity. Note that species diversity increases with productivity in readily colonized habitats, and that geographically isolated habitats have lower diversity than expected on the basis of their productivity.

dictable productivity in ecological systems. Originally (Brown, 1973), I chose the latter on intuitive grounds, but I find that it gives clear, repeatable patterns and has the useful property that the curves for both species diversity and population density extrapolate approximately to zero when $\bar{x} - \sigma = 0$ (Figure 2).

counts for about 70% of the variation in rodent species diversity.

In contrast to the relationships for sandy habitats, the number of rodent species inhabiting rocky hillsides in the Sonoran Desert varies only from one to two (Table 1). Although the results are consistent with the hypothesis that species diversity is directly related to the predictable amount of precipitation, the data are too few to be convincing. However, it is

Table 1.   Location, sampling effort, and species diversity of study sites in the Sonoran Desert

Location	Date	Sampling effort (trap nights)	Seed-eating rodent species		$\bar{x} - \sigma$ annual precipitation (mm)
			Number (S)	Common (>5%)	
Sandy flatland habitats					
Arizona, Yuma Co., 11 km E Yuma, 75 m (Yuma A)	21–22 June 1973	200	1	1	46.7
Arizona, Yuma Co., 16 km E Yuma, 75 m (Yuma B)	22–23 June 1973	200	2	2	46.7
Arizona, Maricopa Co., 24 km W and 10 km N Gila Bend, 225 m (Gila Bend A)	24–25 June 1973	200	2	2	84.3
Arizona, Maricopa Co., 24 km W and 3 km N Gila Bend, 225 m (Gila Bend B)	25–26 June 1973	200	4	4	84.3
Arizona, Pinal Co., 11 km E Casa Grande, 425 m (Casa Grande A)	28–29 June 1973	200	3	3	150.6
Arizona, Pinal Co., 8 km S Casa Grande, 425 m (Casa Grande B)	27–28 June 1973	200	5	5	150.6
New Mexico, Hidalgo Co., 8 km N and 1.5 km E Rodeo, 1225 m (Rodeo A)	1–2 July 1973	200	5	5	170.4
Arizona, Cochise Co., 6.5 km E Portal, 1225 m	11–12 May 1972	300	6	5	170.4
(Rodeo B)	18–22 July 1972	600	6	5	170.4
	7–9 May 1973	600	7	6	170.4
Rocky hillside habitats					
Arizona, Yuma Co., 25 km E Yuma, 150 m (Yuma C)	23–24 June 1973	200	1	1	46.7
Arizona, Maricopa Co., 24 km SSW Gila Bend, 300 m (Gila Bend C)	26–27 June 1973	200	1	1	84.3
Arizona, Pinal Co., 18 km S Casa Grande, 450 m (Casa Grande C)	29–30 June 1973	200	1	1	150.6
Arizona, Cochise Co., 2.5 km NW Portal, 1425 m (Rodeo C)	2–3 July 1973	200	2	2	170.4

The table shows that rocky hillsides support fewer species than nearby sandy flatlands. S (column 4) is the total number of granivorous rodent species captured, "common" (column 5) is the number of species accounting for more than 5% of the catch (see Appendix), and $\bar{x} - \sigma$ of annual precipitation is an index of predictable productivity (see text).

apparent that rodent diversity is less dependent on precipitation in rocky than in nearby sandy habitats, and that rocky hillsides usually are inhabited by significantly fewer species than sandy flatlands that received similar rainfall.

The differences in rodent species diversity between readily colonized habitats are consistent with the hypothesis that diversity is determined primarily by the abundance and predictability of seeds. Rodent species diversity is greater and more dependent on variation in rainfall in sandy than in rocky habitats in the Sonoran Desert. Steep rocky habitats are less productive than flat sandy ones because they have little soil and much of the precipitation that falls is lost as runoff (Hillel and Tadmor, 1962). It is not obvious why rodent species diversity is more sensitive (steeper slope of Figure 1) to variation in rainfall in the sand dunes of the Mojave and Great Basin deserts than in the sandy flatlands of the Sonoran Desert. (The reason may be that the higher temperature of the Sonoran Desert makes productivity there less sensitive to rainfall.) The two kinds of sandy habitats probably differ either in seed production or in the presence of competing taxa that harvest seeds and make them unavailable to rodents. In this regard, it is likely that granivorous ants are more abundant and diverse and also active for a greater proportion of the year in the Sonoran than in the Mojave and Great Basin deserts (D. W. Davidson and G. A. Lieberman, personal communication).

Population density increases with productivity. The population density of rodents of all species within a readily colonized habitat increases with $\bar{x} - \sigma$ of annual precipitation in a manner similar to the increase in species diversity (Figure 2). This is not an artifact of the sampling method; the relative abundance of species within a habitat is remarkably uniform so that my measure of diversity is not influenced significantly by sample size. The correlation between total population density and $\bar{x} - \sigma$ of precipitation constitutes additional evidence that the latter provides an accurate estimate of the availability of seed resources. The relationships of both population density and species diversity to the predictable amount of precipitation are similar within desert regions but differ in the same way between

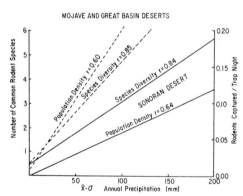

Figure 2 The relation of total population density (measured as the number of individual rodents captured per trap-night) and rodent species diversity (measured as the number of common species as defined in the Appendix) in sandy soil habitats to the predictable amount of annual precipitation, an index of productivity. Note that within deserts both population density and species diversity increase at comparable rates as productivity increases, but that the slopes of these relationships are lower for the Sonoran than for the Great Basin and Mojave deserts.

the Sonoran Desert on the one hand, and the Mojave and Great Basin Deserts on the other (i.e., that the dependences are less steep in the Sonoran Desert). This fact suggests that productivity determines both population size and species diversity through a common mechanism.

The production and availability of resources affect species diversity chiefly by limiting the sizes of populations that a habitat can support. Unpredictable fluctuations in the level of resources have a similar effect. When resources are so scarce or unpredictable that local populations of a species frequently go extinct, then selection will favor the development of habitat selection to prevent individuals from recolonizing those habitats even in times of plenty. Rodent species differ in the seeds they collect and in their methods of harvesting them. As deserts become drier and less productive, individual species tend to become restricted to progressively fewer kinds of habitats, specifically to those habitats where their foraging techniques yield sufficient returns to sustain individuals (and hence populations) even when seeds are scarce. For example, *Dipodomys merriami, Perognathus penicillatus, P. longimembris, Peromyscus maniculatus,* and *Reithrodontomys megalotis* are widely distributed in the deserts of North America. In areas where precipitation and seed production are high, these species are common in sandy flatland and dune habitats, but they are absent from the same kinds of habitats in the drier parts of the Mojave and Sonoran Deserts where productivity is low. Two of these species (*D. merriami* and *P. longimembris*) occur in

unproductive deserts, but they are restricted, perhaps in part by habitat selection, to other kinds of habitats that are often only a few meters from dunes or sandy flats. The other species are restricted to relatively productive habitats and resemble some Andean bird species studied by Terborgh (in preparation).

The geographic pattern of species diversity depends on the area sampled. The relationship between seed availability and habitat specificity produces a pattern of species diversity on a macrogeographic scale that differs from that observed within habitats (Figure 3). The largest number of granivorous rodent species (Figure 3, right) is found in the western Great Basin, eastern Sonoran, and southern Mojave deserts. The first two areas also have high species diversity within habitats, but the southern Mojave is extremely arid and unproductive, and its habitats characteristically support the lowest species diversity of all the North American deserts (Figure 3, left). Comparable levels of macrogeographic diversity in productive and unproductive deserts are achieved in different ways. Unproductive areas have only one or a few species in each type of habitat (low α-diversity), but these species tend to have highly restricted habitat distributions so that there is a high turnover of species between habitats (high β-diversity; see Cody, Chapter 10, for detailed discussion of the differences between α- and β-diversity). In contrast, in more productive deserts several species coexist in each kind of habitat, but there is less habitat specificity and turnover of species between habitats.

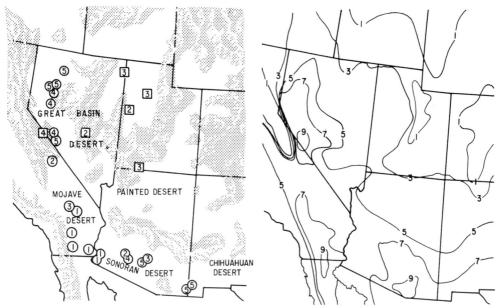

Figure 3 Comparison of species diversity in seed-eating desert rodents within habitats and on a larger scale. Left: numbers of common species that coexist in sandy soil habitats; circles indicate readily colonized habitats and squares denote geographically isolated habitats. Right: numbers of species that occur in geographic areas large enough to include several kinds of habitats. Note (map on left) that the greatest diversity within habitats is in the northwestern Great Basin and eastern Sonoran deserts. and the least is in the southern Mojave Desert. All three of these areas have equally high diversity on a macrogeographic scale (map on right). This pattern reflects an inverse relationship between diversity within habitats (α-diversity) and species turnover between habitats (β-diversity) in readily colonized deserts.

Thus, there tends to be an inverse relationship between α- and β-diversity in desert rodents. As productivity decreases, so does the range of habitats from which each species can harvest sufficient food to maintain local populations in times of seed scarcity, and species become habitat specialists.

Isolated deserts have low species diversity. Small mammals are such poor dispersers that climatic or habitat barriers only a few kilometers in extent may be sufficient to prevent several species from colonizing an area (Brown, 1971). This produces, in continental habitats, patterns of species distribution and community structure that usually are associated with islands. Such insular patterns are sometimes detectable in continental birds (Cody, Chapter 10), and are particularly apparent in desert rodents. Desert basins that have been isolated from the major centers of distribution and speciation by mountain ranges have fewer total species of rodents and lower species diversity within habitats than areas of similar productivity and habitat diversity in the major deserts (Figure 3; see also Figure 1, below, and Brown, 1973). The effects of such reduced species diversity on population density and community structure will be discussed later.

Body Size and Community Structure

Since the patterns of rodent species diversity in relation to productivity in the sandy flatlands of the Sonoran Desert and the dunes of the Mojave and Great Basin deserts are extremely similar, it is important to ask whether there are also similarities in the composition and functional organization of the rodent communities. This is of particular interest because the rodent fauna of the Sonoran Desert is quite distinct from that of the Mojave and Great Basin deserts. For example, the most thoroughly sampled sandy soil habitats are the dunes in Fishlake Valley, Nevada (Dune 7) and the flatlands near Portal, Arizona (Rodeo B). These two sites share only two common species, even though there are five common species

inhabiting the former and six inhabiting the latter (Figures 4 and 5).

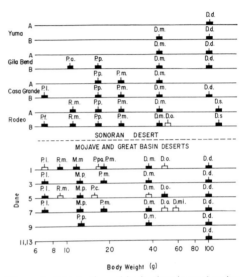

Figure 4 Distribution of body sizes in seed-eating rodent communities from sandy soil habitats in the Sonoran and Mojave-Great Basin deserts. Note that: 1) communities of coexisting species have distributions of body sizes with remarkably regular spacing; 2) species of similar size "replace" each other in different habitats (e.g., the 7-g species labeled *P.l.* and *P.f.,* or the 12-g species labeled *P.a., R.m., M.m.,* and *M.p.*); and 3) some species are displaced in size between the Sonoran and Great Basin-Mojave deserts in a manner that promotes regular spacing of sizes within communities (e.g., the species labeled *P.m.* and *D.m.*). Shaded symbols represent common species; unshaded symbols, rare species that probably are not permanent members of the community. Body weights of species indicated are the means for all adult individuals within that desert. The widths of the symbols are just for illustration and do not indicate variation in weight. Species are abbreviated as follows: *Perognathus longimembris (P.l.), P. flavus (P.f.), P. amplus (P.a.), P. penicillatus (P.p.), P. parvus (P.pa.), Reithrodontomys megalotis (R.m.), Peromyscus maniculatus (P.m.), Peromyscus crinitus (P.c.), Microdipodops megacephalus (M.m), M. pallidus (M.p.), Dipodomys merriami (D.m.), D. ordi (D.o.), D. microps (D.mi.), D. deserti (D.d.),* and *D. spectabilis (D.s.).*

SONORAN DESERT

GREAT BASIN DESERT

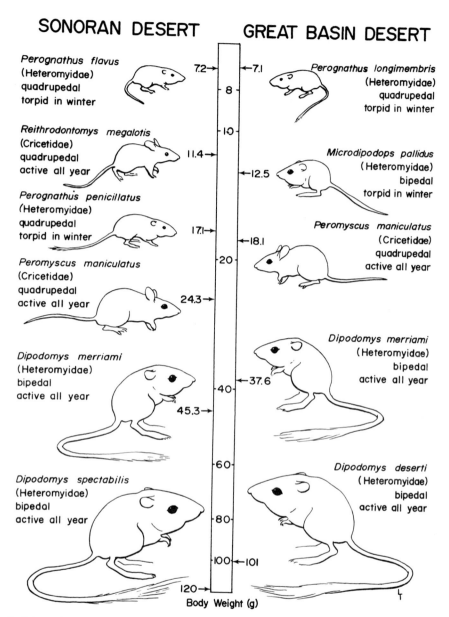

Perognathus flavus
(Heteromyidae)
quadrupedal
torpid in winter
7.2→

←7.1
Perognathus longimembris
(Heteromyidae)
quadrupedal
torpid in winter

8

Reithrodontomys megalotis
(Cricetidae)
quadrupedal
active all year
11.4→

10

←12.5
Microdipodops pallidus
(Heteromyidae)
bipedal
torpid in winter

Perognathus penicillatus
(Heteromyidae)
quadrupedal
torpid in winter
17.1→

←18.1
Peromyscus maniculatus
(Cricetidae)
quadrupedal
active all year

20

Peromyscus maniculatus
(Cricetidae)
quadrupedal
active all year
24.3→

Dipodomys merriami
(Heteromyidae)
bipedal
active all year

←37.6
Dipodomys merriami
(Heteromyidae)
bipedal
active all year
40

45.3→

60

Dipodomys spectabilis
(Heteromyidae)
bipedal
active all year

Dipodomys deserti
(Heteromyidae)
bipedal
active all year

80

100←101

120→

Body Weight (g)

Figure 5 Schematic representation of convergence in structure between a six-species community from the Sonoran Desert (Rodeo B, left) and a five-species community from the Great Basin Desert (Dunes 3 and 7, right). Numbers are average weights (grams). Note the similarities in body size, form, taxonomic affinity, and other characteristics between species occupying similar positions in each community. Also notice the displacement in body size in *Peromyscus maniculatus* and *D. merriami* (both are larger in the Sonoran Desert) to compensate for the different numbers and sizes of coexisting species.

Coexisting species differ in body size. One of the most striking characteristics of desert rodent communities is that the component species differ greatly in body size (Brown, 1973; Rosenzweig and Sterner, 1970). Similar patterns for other organisms in other habitats are discussed elsewhere in this volume by Cody (Chapter 10), Diamond (Chapter 14), Hespenheide (Chapter 7), and Karr and James (Chapter 11). In my 1973 paper I noted several cases in which species of similar size "replaced" each other on different dunes in the Mojave and Great Basin deserts. I concluded that the coexistence of seed-eating rodents in sand dune habitats depends more on the body sizes of species than on the identity of the species. Comparisons between the communities that occupy similar habitats (in this case sandy flatlands and dunes), in the Sonoran Desert on the one hand and the Mojave and Great Basin deserts on the other, support the generality of this conclusion (Figure 4). In each of these areas the most diverse communities consist of five or six species that show a remarkably regular spacing of body sizes. Also, as species diversity varies between habitats, the body sizes of the component species change in a quite regular manner. As species diversity decreases, the differential in size between pairs of coexisting species tends to increase (Figure 4; for best documentation, see Brown, 1973). Decreases in diversity along the gradient of productivity are accomplished by the gradual elimination of small and medium-sized species. Figure 4 is strikingly similar to Figure 32 of Chapter 14 (Diamond), which demonstrates regular spacing patterns in communities of New Guinea fruit pigeons.

Comparisons of five- and six-species communities between the Great Basin and Sonoran deserts provides additional evidence for the importance of body size in the structuring of rodent communities. The best sampled sites were Dune 7 in the Great Basin and the flatlands at Rodeo B in the Sonoran Desert. The rodents in each of these communities range in size from about 7 g to slightly more than 100 g (Figures 4 and 5). However, these habitats share only two species of intermediate size (*Peromyscus maniculatus* and *D. merriami*); the other species, including the largest and the smallest in each community, differ in the two sites. Intraspecific variation in body size between the Sonoran and Great Basin deserts also is related to community structure. In the Great Basin, *P. longimembris* and *R. megalotis* are similar in size (7.1 and 8.6 g respectively) and never are common on the same dune (Figure 4 and Brown, 1973). In the Sonoran Desert *P. flavus,* the pocket mouse ecologically equivalent to *P. longimembris,* is of similar size (7.2 g), but Sonoran *R. megalotis* is enough larger (11.4 g) so that the two species are able to coexist on the sandy flatlands at Rodeo B (Figure 4). To accommodate an additional small species, the other members of this Sonoran Desert community are displaced to larger sizes than conspecific populations or ecological counterparts in the Great Basin (e.g., compare the species labeled as *P.p.*, *P.m.*, *D.m.*, and, less striking, *D.o.* in Figure 4). These patterns of body size, which include intraspecific variation (character displacement) in response to community composition, are best interpreted in terms of selective pressures to maximize the differences in size and resource utilization among coexisting species. As yet unexplained evolutionary constraints seem to limit the minimum and maximum sizes (at approximately 7 and 120 g respectively) of seed-eating desert rodents, but within these limits there are striking parallels in the size structure of different communities despite differences (sometimes at the generic or even familial level) in the identity of the component species.

Geographically distant communities are structured similarly. The parallel structuring of desert rodent communities is not restricted to patterns of body size; it extends to similarities in taxonomic relationships and morphological, physiological, and behavioral characteristics (Figure 5). In communities of high species diversity occupying the sandy habitats of the Great Basin and Sonoran deserts, members of the family Heteromyidae usually span the entire range of body sizes. The smallest species are quadrupedal pocket mice of the genus *Perognathus,* and the largest are bipedal kangaroo rats of the genus *Dipodomys*. Additional species of *Perognathus* and *Dipodomys* and, in the Great Basin, species of *Microdipodops* (kangaroo mice) occupy intermediate positions in the spectrum of body sizes. The cricetid rodents found in these habitats are quadrupedal and occupy intermediate positions in the body size spectrum. Two cricetids, *Peromyscus maniculatus* and *R. megalotis,* commonly coexist

with the heteromyids in the most productive habitats, but granivorous cricetids usually are absent from readily colonized dunes or sandy flatlands which support three or fewer species. Another similarity between the species-rich communities of the Great Basin and Sonoran Deserts is that each usually contains two species of heteromyids, which readily enter torpor for extended periods, particularly during the winter when energetic costs are high and seeds are likely to be scarce. These are the smallest species of *Perognathus* and one of intermediate size of the genus *Perognathus* or *Microdipodops*.

Communities in isolated and rocky habitats have distinctive structure. Compared with the rodent communities of readily colonized sand dune and sandy flatland habitats that have just been discussed, the communities inhabiting dunes in isolated desert basins contain fewer species and differ markedly in structure. Although α-diversity is relatively low in isolated habitats, many of the species are habitat generalists, and β-diversity also is relatively low. The species that coexist in these habitats exhibit a regular spacing of body sizes, but the large kangaroo rats characteristic of almost all habitats with sandy soils are conspicuously absent. The cricetids, *Peromyscus maniculatus* and *R. megalotis,* are important constituents of two- and three-species communities in isolated habitats, but not of readily colonized dune or sandy flatland habitats with similar species diversity.

The structure of communities on rocky hillsides warrants brief comment. In the Sonoran Desert these habitats support one or two species. *Perognathus intermedius,* a quadrupedal heteromyid of intermediate size, occurs at all four sites, and a larger quadrupedal cricetid, *Peromyscus boylei,* is present on the most productive hillside. This seems to be a general pattern. My limited collecting on rocky hillsides in the Mojave and Great Basin deserts indicates that these habitats usually support one species of *Perognathus* (either *P. fallax* or *P. formosus*) and sometimes, in areas of higher rainfall, a species of *Peromyscus* (either *Peromyscus crinitus* or *Peromyscus maniculatus*). The bipedal kangaroo rats, which are important components of nearly all sandy soil and desert pavement habitats, are conspicuously absent from rocky hillsides throughout the North American deserts. Again there is a marked tendency for communities occupying the same kinds of habitats to be structured similarly, even though they are composed of different species.

The patterns of body size and community structure lead to two general conclusions. First, they indicate that competition plays a major role in determining the composition of rodent communities. The regular distribution of body sizes within communities, the interchangeability of species of similar size, and the geographic intraspecific variation in size to accommodate a species to different sets of competitors, all demonstrate that the ability of species to coexist is dependent largely on their being of different sizes. This implies that the important resources of these habitats are competed for and subdivided among species on the basis of size. Second, as indicated by the common patterns,

desert habitats set such severe constraints on the characteristics of the rodents that can inhabit them, that unrelated species have converged to fill similar ecological roles, just as has happened to birds and lizards on an intercontinental scale (Chapters 10, 11, and 12, by Cody, Karr and James, and Pianka). This has resulted in the parallel organization of rodent communities in different deserts, even though many of the species and even some of the genera are different. Finally, the predictable losses of species when one goes from species-rich to species-poor communities (e.g., from accessible to isolated deserts, or from productive to unproductive habitats) suggest the possibility of acquiring sufficient data to formulate "incidence functions" and "assembly rules" for communities of desert rodents as of New Guinea birds (cf. Diamond, Chapter 14).

Resource Utilization and Coexistence

One of MacArthur's major contributions was the suggestion that communities were structured primarily by competitive interactions among the component species. This viewpoint focused attention on mechanisms of coexistence and stimulated field ecologists to attempt to quantify the ways that limited resources are apportioned among species.

Coexisting species harvest different sizes of seeds and forage in different microhabitats. The diversity of body sizes among coexisting rodent species suggests that some resource (probably the seeds that comprise the majority of the diets of all species) is subdivided among species on the basis of size. This indeed proves to be the case, as also shown for birds (Hespenheide, Chapter 7, Figures 1 and 4; Diamond, Chapter 14, Figure 31). The sizes of seeds in the cheek pouches of heteromyid rodents from the sandy flat-land habitats of the Sonoran Desert are directly and strongly ($r = 0.94$) correlated with the body sizes of the species (Figure 6). This pattern is remarkably similar in slope and variability to the comparable relationship for the species inhabiting sand dunes in the Great Basin and Mojave deserts, but the rodents in the Sonoran Desert consistently select smaller seeds. Even when the same species occurs in both areas (e.g., species *D.m.*, *D.o.*, and *D.d.* in Figure 6), the population in the Sonoran Desert collects significantly smaller seeds.

The actual distribution of seed sizes collected by each species in each area is shown in Figure 7. It is apparent that each species takes a wide range of seed sizes, so that there is considerable overlap even among species that differed in size by a factor of 2 or 3 (cf. Figure 31 of Chapter 14 by Diamond, demonstrating overlap in fruit size taken by pigeons of different sizes). The variances in the sizes of seeds taken by species of comparable body size in the sandy flatlands of the Sonoran Desert and the dunes of the Mojave and Great Basin deserts are roughly similar. All of these species exhibit much less vari-ation in the sizes of seeds harvested than *P. intermedius*, which usually is the only species inhabiting rocky hillsides. *Perog-nathus intermedius* also selects larger seeds on the average ($\bar{x} = 2.2$ mm) than hetero-

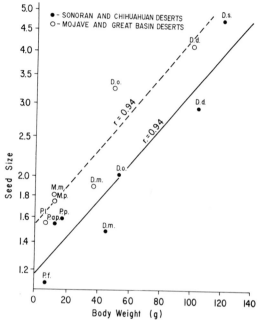

Figure 6 The relation between the size (length) of natural seeds in the cheek pouches and body size for heteromyid rodents from sandy soil habitats. Note that within both deserts the average size of seeds increases with body size, but that the rodents collect smaller seeds in the Sonoran than in the Mojave and Great Basin deserts. Species abbrevia-tions follow Figure 4 except for *Perognathus apache* (*P.ap.*).

myids of comparable size from any of the sandy habitats.

The large overlaps in seed-size utiliza-tion among coexisting species make it questionable whether subdivision of seeds on the basis of size is by itself sufficient to permit coexistence of four or five spe-cies in the same habitat. However, the species differ not only in the sizes of the seeds they harvest but also in the parts of the habitat where they forage for the seeds (Figure 8). This is quantified in Figure 9.

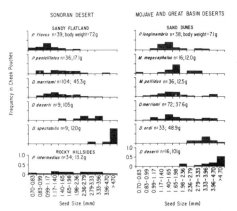

Figure 7 The frequency distributions of the sizes of natural seeds in the cheek pouches of heteromyid rodents. Note that the size of seed harvested is generally correlated with the size of the rodent, but that there is considerable overlap among species. *Perognathus intermedius,* the only heteromyid found in rocky hillside habitats, collects a wider range of seed sizes than any of the species that usually coexist in sandy soil habitats. To obtain the data on which this figure is based, size categories of seeds were determined by Tyler sieves, which form a geometric series. The mean size of all seeds in the pouches of each individual was determined, and then the values for all individuals were plotted; this was done to avoid bias by unduly weighting the samples from the few individuals whose pouches were filled with hundreds or even thousands of similar-sized seeds.

In the sandy flatlands of the Sonoran Desert some species (*R. megalotis* and *P. penicillatus*) forage largely under the cover of perennial shrubs, whereas others (all three species of *Dipodomys*) search primarily over the open ground between the shrubs. Although there are some differences between species of the same size (*P. flavus* and *P. longimembris*) and even between populations of the same species (*Peromyscus maniculatus*), rodents inhabiting the sand dunes of the Mojave and Great Basin deserts resemble those of

the sandy flatlands of the Sonoran Desert in their tendency to forage in different microhabitats as well as to harvest seeds of different sizes.

Overlap in resource utilization varies with species diversity. Similarities between species in the utilization of resources were quantified as described in Brown and Lieberman (1973) and in the legend of Table 2. The overlaps in both seed-size utilization and foraging area for the species that are common in the sandy flatland habitats of the Sonoran Desert are shown in Table 2. It is apparent that coexisting species that are similar in resource utilization in one dimension tend to differ in the other dimension. Thus, *D. merriami* and *P. penicillatus* feed on similar seeds but obtain them from different microhabitats, whereas *D. merriami* and *D. deserti* both forage in the open, but they harvest different sizes of seeds. Since it seems reasonable to assume that seed size selection and microhabitat selection are independent for each species, the values of overlap in the two dimensions have been multiplied to provide a measure of overall similarity in resource utilization (Table 2). These values of total overlap range from nearly zero to approximately 0.8. They are similar to those obtained for the rodents inhabiting sand dunes in the Mojave and Great Basin deserts (Brown and Lieberman, 1973). Although these overlap values are quantitative estimates of the intensity of competition between species, they should not be interpreted as equivalent to the alphas of Lotka-Volterra competition equations. Overlap between species with multi-

Figure 8 Schematic representation of horizontal foraging areas utilized by desert rodents. Equal numbers of traps set in each of four positions relative to shrub cover catch different proportions of coexisting species. Note that quadrupedal rodents forage predominantly under or close to shrubs, whereas bipedal species forage mostly in open areas (also see Figure 9). Abbreviations follow Figure 4.

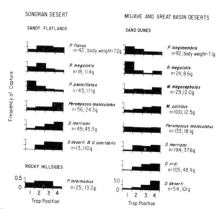

Figure 9 The relative utilization of horizontal foraging areas by granivorous rodents, measured by frequency of capture in equal numbers of traps set in four positions as follows: 1) under a shrub; 2) at the edge of a shrub; 3) on open ground 1 m from nearest shrub; and 4) on open ground at least 2 m from nearest shrub. Note that most species concentrate their foraging in particular microhabitats (e.g., *P. penicillatus* mainly under shrubs), but some utilize all areas indiscriminately (e.g., *Peromyscus maniculatus* in the Mojave and Great Basin deserts).

dimensional niches is discussed elsewhere in this volume by Hespenheide (Chapter 7, Table 4), Pianka (Chapter 12, Table 1), and Diamond (Chapter 14, Figure 45).

It is of interest to inquire how the competitive interactions within communities vary with species diversity. As shown in Figure 10, the mean values of overlap for all pairs of species increases with species diversity, the opposite of Pianka's finding for desert lizards (Chapter 12). The pattern for the sandy flatland communities of the Sonoran Desert is virtually identical to that for the sand dune communities of the Mojave and Great Basin deserts. These results suggest that increases in species diversity occur in part because closer competitors can coexist in more productive and predictable habitats. However, they also suggest that there is a limit to the overlap in resource utilization that can be tolerated by coexisting species and this limit is approached when these communities contain four or more species.

At this point it is necessary to insert a word of caution. Overlap has been meas-

ured in only two dimensions: seed size and foraging area. I am confident that these are the most important means of resource partitioning among desert rodents, but I am not sure they are the only ones. For example, the tendency of *Peromyscus maniculatus* to feed on insects during the summer months (O. J. Reichman, personal communication) may reduce its overlap with the more strictly granivorous heteromyids, but competition among desert rodents should be most intense during the winter when seeds are least abundant and alternative foods such as insects are unavailable. The fact that some species may be torpid throughout the winter suggests that this is a time when seeds are

Table 2. Overlaps in resource utilization between species of seed-eating desert rodents in the Sonoran Desert

Resource utilization	Species	P. long.[a] / P. flav.	P. ampl.[b]	R. meg.[c]	P. pen.	P. manic.[c]	D. mer.	D. des.[d] / D. spec.[d]
					Species			
Seed size	P. long., P. flav.		0.58	0.75	0.56	0.60	0.62	0.08
	P. ampl.			.81	.92	.84	.94	.36
	R. meg.				.78	.84	.83	.22
	P. pen.					.87	.89	.41
	P. manic.						.94	.32
	D. mer.							.32
Horizontal foraging area	P. long., P. flav.		1.0	.73	.61	.79	.81	.78
	P. ampl.			.83	.87	.66	.57	.54
	R. meg.				.75	.63	.54	.51
	P. pen.					.55	.46	.43
	P. manic.						.89	.80
	D. mer.							.88
Overall: seed size × horizontal foraging area	P. long., P. flav.		.58	.55	.34	.47	.50	.06
	P. ampl.			.67	.80	.62	.54	.19
	R. meg.				.59	.53	.45	.11
	P. pen.					.48	.41	.18
	P. manic.						.84	.26
	D. mer.							.28

Species that overlap greatly in one dimension (seed size or foraging area) usually show much less overlap in the other, so that overall overlaps are usually less than 0.80. Overlap is measured as the proportion of the area under the frequency distribution of one species that overlaps with that of the other (see Brown and Lieberman, 1973 for details). Overall overlaps are the products of the overlaps in seed size and foraging area.

[a] Based on data for P. flavus, which is very similar to P. longimembris in body size and general ecology.

[b] Assumed values calculated on the basis of body size and foraging areas from data on other Perognathus species.

[c] Since these cricetids do not have cheek pouches, assumed values of seed sizes were calculated on the basis of body size.

[d] Data on foraging areas were combined, data on seed sizes were for D. deserti because pouches of D. spectabilis contained mostly large fruits with many small seeds rather than individual seeds.

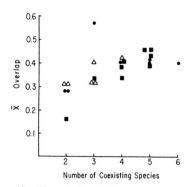

Figure 10 The relation between average overall overlap (seed size times foraging area) in resource utilization for all pairs of species (ordinate) and the number of coexisting species in a sandy soil habitat (abscissa). Note that as the number of species in a community increases, the mean overlap first increases and then approaches a constant value. Circles indicate communities from sandy flatland habitats in the Sonoran Desert; squares, readily colonized sand dunes in the Mojave and Great Basin deserts; triangles, geographically isolated sand dunes in the Great Basin Desert.

scarce. On the other hand, it raises the possibility that torpor itself serves to reduce competition. The diverse communities inhabiting sandy soils of the Great Basin and Sonoran deserts all have at least one hibernator and most have two. This suggests that these habitats contain seeds that are sufficiently abundant and predictable on a seasonal basis to support one or two part-time species as well as three or four that are active throughout the year.

Species utilize a wider range of resources and attain higher densities in the absence of competitors. In the absence of close competitors, rodents tend to utilize a wider range of resources than when other species with similar requirements are present. In the eastern Great Basin, *D.*

ordi is the only kangaroo rat common in sandy habitats, and it collects a wider range of seed sizes than it does farther west where it coexists with a larger (*D. deserti*) and a smaller (*D. merriami*) species (Brown and Lieberman, 1973). Similarly, *P. intermedius,* the only heteromyid found on most rocky hillsides in the Sonoran Desert, takes a much wider range of seed sizes than any of the species occur together in sandy flatland habitats (Figure 7).

The ability of rodent species to utilize a wider range of resources when close competitors are absent should enable them to attain higher population densities in habitats where geographic isolation has prevented colonization of appropriate competitors. Table 3 presents evidence confirming this expectation. The density of individuals per species does not differ

Table 3. Comparison of the density of rodents per species in readily colonized and isolated sandy soil habitats

Habitat	Relative density per species (rodents captured/trap night/species)
Readily colonized flatlands, Sonoran Desert	0.023 ± 0.008
Readily colonized dunes, Mojave and Great Basin deserts	0.033 ± 0.011
Isolated dunes, Great Basin Desert	0.049 ± 0.018

There is no significant difference between the densities on the readily colonized habitats in the different deserts ($P > 0.10$), but the isolated sand dunes support significantly higher rodent densities per species than the readily colonized habitats ($P < 0.025$).

significantly between the readily colonized dunes of the Mojave and Great Basin deserts and the sandy flatlands of the Sonoran Desert ($P > 0.1$). In fact, the two best and repeatedly sampled sites have almost identical densities (0.033 and 0.035 individuals/trap-night/species at Dune 7 and Rodeo B respectively). In contrast, the isolated sand dune habitats in the Great Basin support significantly higher densities per species than the readily colonized sandy soil habitats ($P < 0.05$). These data suggest that as productivity increases, accessible habitats are colonized by competing species at a rate that maintains a fairly constant density of individuals per species, as also suggested by Figure 2. On the other hand, equally productive, but isolated, habitats are not colonized by as many species as could utilize the available resources. As a result, some of these resources are utilized by the few species that are present, and these attain higher densities. It is apparent that these few species are less efficient at utilizing these resources than are their missing competitors, because total rodent densities in isolated habitats average 39% lower than in equally productive, but readily colonized, habitats. This corresponds to a level of density compensation of 44% when calculated as described by Cody (Chapter 10). Similar cases of incomplete density compensation in bird communities have been discussed by Cody (Chapter 10), Diamond (1970, and Chapter 14, Figure 18, solid points), and MacArthur et al. (1973).

At least three aspects of the relationship between seed-size utilization and foraging behavior in desert rodents require further study. First, it is interesting that it is one of the largest species, *D. deserti*, which characteristically inhabits the least productive sandy habitats (Figure 4; Brown and Lieberman, 1973). This suggests either that these habitats produce mostly large seeds or that the seeds are distributed in such a way that they can be harvested most efficiently by a species that can forage over a wide area. Second, it is interesting that the size of seeds collected by a rodent of a given size varies with the kind of habitat, even though those species that coexist within any particular kind of habitat also subdivide seeds on the basis of body size. Thus, rodents of the same size, and sometimes of the same species, collect smaller seeds on the sandy flatlands of the Sonoran Desert than on the dunes of the Mojave and Great Basin deserts. Similarly, *P. intermedius* collects larger seeds in its rocky habitats than any heteromyid of comparable size harvests from sandy soils. These observations suggest that foraging behavior depends on the size, abundance, and spatial distribution of available seeds, and at least some of these parameters vary between habitats. Third, I have estimated overall overlaps in resource utilization by multiplying observed overlaps in seed size and foraging area. Given the data available, these are reasonable approximations, but measurement of actual two-dimensional overlaps requires information on the sizes of seeds harvested by each species from each microhabitat (R. May and H. Horn, personal communication). More research on foraging strategies is necessary to solve all of these problems.

Discussion and Synthesis

The new data from the Sonoran Desert demonstrate the repeatability and generality of the geographic patterns of desert rodent distribution and ecology, which were described for the Mojave and Great Basin Deserts in earlier papers (Brown, 1973; Brown and Lieberman, 1973). The new data are also consistent with the mechanisms that have been proposed to account for these patterns. I shall try briefly to place these results in a more general and theoretical perspective.

Comparison with Related Studies of Desert Rodents

Rosenzweig and his collaborators also have been studying the population and community ecology of desert rodents. Rosenzweig and I have been interested in many of the same problems, but we have tended to approach them in different ways. For the most part, he has used mark-recapture techniques and field experiments to study local communities for extended periods, whereas I have utilized short-term sampling methods to compare geographically distant communities in similar habitats. In general, our work has not overlapped, but where it has we have obtained gratifyingly similar results.

Rosenzweig and his collaborators have studied species diversity and seed-husking behavior of desert rodents in southern Arizona. Rosenzweig and Winakur (1969) analyzed patterns and correlates of species diversity within small geographic areas. They concluded that variation in diversity between neighboring habitats was related

to characteristics of the vegetation and soil. My work has compared communities on a geographic scale, and indicates that when soil and vegetation parameters are deliberately kept relatively constant, diversity is related to productivity. Rosenzweig and Sterner's (1970) observation, that seed-husking efficiency is not a likely basis for the allocation of seeds among rodent species that differ in body size, is consistent with my observation that the sizes of seeds that rodents collect are related to habitat, and not directly to the sizes or other inherent characteristics of individual seeds.

More recently Rosenzweig's group has concentrated on the mechanisms of resource exploitation that permit two species, *D. merriami* and *P. penicillatus,* to coexist in many habitats in the Sonoran and Chihuahuan deserts. They found that the two species harvested an almost identical range of seed sizes (Smigel and Rosenzweig, 1974) but foraged in different microhabitats, *D. merriami* preferring open, bare ground and *P. penicillatus* favoring dense shrub cover (Rosenzweig, 1973). These results are almost identical to my findings for these species (Figures 7 and 9). The mechanisms of seed allocation and coexistence in communities of granivorous desert rodents warrant further study. It is apparent that some species reduce competition for seeds by foraging in different habitats and others do so by collecting different sizes of seeds, but the differences in foraging techniques, which vary with habitat and produce the apparent seed size selection, remain to be elucidated.

Species Diversity

MacArthur (1972, pp. 170–172) provided an elegant model that demonstrates how species diversity of a community utilizing a common resource depends on three parameters: the amount and distribution of resource (R), the average range of resource utilized by each species ($\bar{U}$), and the average overlap in utilization between species ($\bar{O}/\bar{H}$) (see Pianka, Chapter 12, for further discussion of this relation). Productivity (R) is the only parameter that is independent of the characteristics of the species comprising the community. Since $\bar{U}$ and $\bar{O}/\bar{H}$ can be reduced by the invasion of superior competitors or by evolution (speciation and character displacement) within the community, productivity should determine diversity when the opportunities for speciation, colonization, and extinction have been approximately equal. My observations on communities of desert rodents confirm the predictions of the theory. When opportunities for colonization of sandy soil habitats have been comparable, species diversity of granivorous rodents is determined primarily by the availability of seeds. However, this is one of the few empirical demonstrations of a positive relationship between productivity and species diversity (see Cody, 1974, and Terborgh, in preparation, for other examples). Most studies of species diversity have demonstrated correlations with habitat isolation (e.g., MacArthur and Wilson, 1967; Vuilleumier, 1970), habitat structure (e.g., MacArthur and MacArthur, 1961; Cody, 1968; Rosenzweig and Winakur, 1969), climatic stability (e.g., Sanders, 1968), or predation (e.g., Paine, 1966).

There seem to be two important reasons why comparative studies of communities rarely implicate productivity in explaining variations in species diversity. First, since species diversity in most habitats represents an equilibrium between opposing rates of colonization (or speciation) and extinction (MacArthur and Wilson, 1967; Terborgh, 1973; Rosenzweig, Chapter 5), the historical events and geographic features that affect these processes often have more influence than productivity. This is obvious in the case of communities inhabiting islands or insular habitats on continents (MacArthur and Wilson, 1967; Vuilleumier, 1970; Brown, 1971; Barbour and Brown, 1974; Diamond, Chapter 14), but it may be equally important, although less apparent, in other situations (see Terborgh, 1973; Cody, 1970). Second, some of the demonstrated relationships between species diversity and habitat structure, climatic stability (cf. MacArthur, Chapter 3), and predation may actually reflect underlying, unappreciated correlations between these parameters and productivity. It is extremely difficult to measure the production and availability of the resources used by most species.

May and MacArthur (1972) developed a theory of limiting similarity, which predicts that the maximum tolerable overlap in resource utilization among coexisting species is directly related to resource predictability in drastically fluctuating environments, but approaches a constant value in less variable environments. The transition toward a constant value is predicted to occur as the variance to mean ratio ($\sigma^2/\bar{x}$) for resource production decreases towards unity. Empirically I have demonstrated a strong correlation between species diversity and average overlap in resource utilization in relatively unproductive, variable habitats that contain fewer than four species; overlap values approach a constant in predictably productive habitats that support four or more species. May and MacArthur suggested that it was unlikely that there were environments where $\sigma^2/\bar{x}$ of productivity exceeded unity and species-packing was affected by environmental fluctuations, but the most arid deserts probably constitute an exception.

Community Structure And Resource Utilization

Rodent communities that occur in sandy habitats in different deserts demonstrate remarkable parallels in structure despite significant differences in their component species. Readily colonized habitats of comparable productivity characteristically support communities composed of species of similar sizes and foraging techniques, even though these species may belong to different genera or even different families. Similar cases of convergence among geographically isolated communities composed of different species have been reported for other taxa, especially birds (e.g., Cody, 1966, 1973, and Chapter 10; Karr and James, Chapter 11).

Communities of granivorous rodent species are conspicuously structured on the basis of body size. The highly predictable ratios in size between coexisting species suggest that competition has selected for the subdivision of available seeds on the basis of size. Similar patterns of character displacement in body size and

trophic structures among species that share requirements for particulate foods are extremely common (e.g., Hutchinson, 1959; Hespenheide, 1973). Selection resulting from competition appears to have produced regular distributions of body size in sympatric species that devote considerable time and energy to searching for, pursuing, or attracting similar kinds of particulate foods. Examples include predacious aquatic and terrestrial insects (Hutchinson, 1959; Evans, 1970), carnivorous fishes (Barbour, 1973), insectivorous lizards (Schoener, 1970), insectivorous, raptorial, frugivorous, and granivorous birds (Hespenheide, 1971; Storer, 1966; Lack, 1947; Diamond, Chapter 14, Figure 30), and carnivorous, frugivorous, and granivorous mammals (Rosenzweig, 1966; McNab, 1971a, 1971b). When the food habits of such sympatric species have been analyzed, it sometimes has been found that regular, nonoverlapping patterns of body size (or trophic structure size) conceal irregular, highly overlapping distributions of food size (e.g. present study, Figure 7; Pulliam and Enders, 1971). In some cases this may be owing to the fact that the data were collected during periods when competition was not intense. Zaret and Rand (1971) showed that the diets of sympatric fish species overlapped least during the tropical dry season, when food was relatively scarce and competition probably was most intense.

It seems likely that many of the generalities concerning patterns of resource utilization and the role of competition in structuring natural communities (MacArthur, 1972) may not apply to many plants or to those animals that harvest relatively abundant food by grazing or filter feeding. Such organisms normally acquire nutrients without expending large amounts of time and energy in locating and harvesting them. For these organisms the impact of predators seems to play a more conspicuous role than competition for food, both in the regulation of populations (Hairston, Smith, and Slobodkin, 1960; Slobodkin, Smith, and Hairston, 1967) and in the structuring of communities (Paine, 1966; Janzen, 1970; Connell, Chapter 16). For example, within communities of rodent granivores, frugivores, and carnivores, closely related species frequently occur in sympatry, interspecific competition has produced clear patterns of body-size displacement and food-particle utilization, and the impact of predators is not obvious. In sharp contrast, within communities of rodent grazers, closely related species rarely coexist in the same habitat (Baker, 1971), the effects of predation are often obvious (Pearson, 1966, 1971), and those of interspecific competition are not (Krebs, Keller, and Tamarin, 1969).

Summary

Clear and repeatable patterns characterize the distribution and community ecology of seed-eating rodents in the North American deserts. These have been demonstrated by independent studies in deserts that are inhabited by distinctive rodent faunas. The major geographic patterns and the mechanisms that I have proposed to account for them may be summarized as follows:

1. Species diversity in rodents is dependent in part on historical and biogeographic events. Isolated desert basins have lower species diversity than comparable habitats in the major deserts, because geographic barriers have prevented colonization by several species, which otherwise could occur in the isolated habitats.

2. Productivity is the major factor that determines how many species can coexist in habitats to which approximately equal numbers of species historically have had access. Species diversity and total rodent density within similar habitats vary directly with the amount and predictability of seed resources, which in turn depend on the amount and predictability of precipitation. Habitats with sandy soils normally support from one to six granivorous rodent species, whereas rocky habitats contain one or two species. Because of differences in habitat specificity between areas of high and low productivity, α and β species diversities tend to be inversely related, and diversity within habitats bears no direct relation to species diversity within large geographic areas.

3. Structurally similar habitats with comparable productivity and historical access are inhabited by similar rodent communities. Since the rodent faunas of the major deserts differ, much of the structural similarity between communities is the result of the convergence of different species (often of different genera or even families) to fill similar ecological roles. Convergent patterns of body size, foraging behavior, and seasonal activity (or inactivity) are especially apparent.

4. Competition is a major force in the structuring of rodent communities. Coexisting species subdivide food resources

by collecting seeds of different sizes and by foraging in different microhabitats. Evidence of competition is provided by: a) character displacement in body size related to seed size selection, b) utilization of exceptionally wide ranges of seed sizes by populations occurring in the absence of close competitors, and c) attainment of unusually high population densities by the few species that have managed to colonize geographically isolated habitats.

5. Overlaps in resource utilization among species vary with species diversity and productivity in the most variable and unproductive habitats, but approach a constant value as productivity increases and fluctuates less.

Acknowledgments

My intellectual debt to Robert MacArthur should be apparent, but his personal encouragement and interest in my work were equally important in stimulating me to do this research. Astrid K. Brown, Alan MacArthur, Michael L. Rosenzweig, Gerald A. Lieberman, and the students in two field ecology classes helped to gather the original data. Discussions with students and colleagues, particularly William M. Schaffer and Michael L. Rosenzweig, have been of great value. The work was supported by grants from the University of Utah and the National Science Foundation (GB 8765 and GB 39260).

Appendix

Methods: Sonoran Desert Study

My earlier work in the Mojave and Great Basin Deserts was done on vegetated sand dunes (Brown, 1973). Because of differences in geology, sand dunes are virtually nonexistent in the Sonoran Desert. In the present study I sampled two different kinds of habitats—sandy flatlands and rocky hillsides. The former were extensive, relatively flat areas usually near the bottom of desert valleys. They had soft, sandy soils and were vegetated with a mixture of shrub species. These habitats were chosen because they were quite similar to the sand dune habitats of the Mojave and Great Basin deserts. The rocky hillsides were steep slopes covered almost entirely with bare bedrock and loose boulders. The dominant vegetation consisted of a mixture of shrubs, succulents (cacti and agaves), and occasional small trees. These rocky hillsides were chosen to provide maximum contrast with the sandy soil habitats nearby and in the other deserts.

The study sites in the Sonoran Desert were selected on the basis of weather records. Data from the U.S. Weather Bureau were used to select four weather stations, which were well spaced along the gradient of rainfall across southern Arizona. Within a radius of 25 km of each weather station two sandy flatland habitats (separated by at least 5 km) and one rocky hillside were sampled. Every effort was made to select study sites so that habitats of the same type (i.e., either sandy flatland or rocky hillside) were as similar as possible in soil and vegetation structure. The exact location of the study areas and the time and amount of sampling are given in Table 1.

The habitats were sampled as follows: "Museum Special" dead traps (normally

200 per night) were placed in sets of four, centered around shrubs. The four traps in each set were distributed: 1) in the center of a shrub; 2) at the edge of the shrub; 3) in the open, 1 m from the shrub; 4) in the most open area available, at least 2 m from the nearest shrub. These sets of four traps were placed approximately 10 m apart in an irregular line running through the most homogeneous habitat. Traps were set in the evening and collected the following morning. The position of capture (relative to a shrub) of each victim was recorded, the animals were weighed, and the contents of their cheek pouches were collected in labeled vials. The latter were returned to the laboratory where the seeds were passed through a graded series of sieves to determine their sizes. Most habitats were sampled only once during the summer of 1973, but Rodeo A was sampled three times between May 1972 and May 1973 (Table 1). The techniques employed in the present study were virtually identical to those used in my earlier work (Brown, 1973; Brown and Lieberman, 1973), which can be consulted to obtain a more detailed description.

In the Sonoran Desert a total of 552 rodents representing 18 species were captured in 3,700 trap nights (Tables 1 and 4). Of these, 382 individuals belonged to the 12 species that I considered primarily granivorous. The criteria for selecting the seed-eating species from the Sonoran Desert fauna were those used in the previous study. Those species defined as granivores included all species of the family Heteromyidae (the genera *Perognathus* and *Dipodomys*) and two species of the family Cricetidae, *Reithrodontomys megalotis* and *Peromyscus maniculatus*. The last species is an opportunistic feeder that takes many insects in the warm months (as do several heteromyids) but in the winter, when food is most likely to be limited, it feeds primarily on seeds. The nongranivorous species included forms that are leaf-eating herbivores (*Neotoma* and *Spermophilus*), carnivores (*Onychomys*), and largely insectivorous omnivores (*Peromyscus eremicus*).

My measure of species diversity within a habitat, the number of common species, was defined and used in my earlier paper (Brown, 1973). The number of common species of seed-eating rodents included all granivorous species which comprised 5% or more of the individuals in a sample. It also counted *Dipodomys deserti* and *D. spectabilis* whenever they were present. These large kangaroo rats were usually less dense than the smaller species, and they also tended to be underrepresented in samples because many individuals were able to break out of the traps.

Other Methods

In addition to data collected during my recent field work in the Sonoran Desert, I have used several other sources of information for this paper. First, I utilized some unpublished data collected during my field work in the Mojave and Great Basin deserts. These include the results of sampling one additional sand dune habitat in the isolated eastern part of the Great Basin, which was done during the course of other work in 1972. Dunes 15 km N and 5.5 km E of Gandy, Juab Co., Utah were sampled on September 5–7, 1972; 29

Table 4. Composition of the rodent communities of the study sites in the Sonoran Desert

	Body weight (g)	Yuma			Gila Bend			Casa Grande			Rodeo			Total
		A	B	C	A	B	C	A	B	C	A	B	C	
Seed-eating species														
Perognathus flavus	7.2	—	—	—	—	—	—	—	—	—	—	44	—	44
P. longimembris	7.1	—	—	—	—	—	—	—	3	—	—	—	—	3
P. amplus	10.7	—	—	—	—	1	—	—	—	—	—	—	—	1
P. intermedius	13.2	—	—	8	—	—	6	—	—	7	—	—	6	27
P. penicillatus	17.1	—	—	—	—	1	—	9	1	—	6	33	—	50
Dipodomys merriami	45.3	—	5	—	3	1	—	4	2	—	5	96	—	116
D. ordi	52.2	—	—	—	—	—	—	—	—	—	—	8	—	8
D. deserti	104.7	2	11	—	1	1	—	—	1	—	—	—	—	16
D. spectabilis	120.1	—	—	—	—	—	—	—	—	—	1	6	—	7
Reithrodontomys megalotis	11.4	—	—	—	—	—	—	—	—	—	1	28	—	29
Peromyscus maniculatus	24.3	—	—	—	—	—	—	4	1	—	7	67	—	79
Peromyscus boylei	23.5	—	—	—	—	—	—	—	—	—	—	—	2	2
Total		2	16	8	4	4	6	17	8	7	20	282	8	382
Other Species														
Peromyscus eremicus	25.2	—	—	6	—	—	—	—	—	5	1	107	1	120
Onychomys torridus	28.9	—	—	—	—	—	—	—	—	—	2	14	—	16
O. leucogaster	38.6	—	—	—	—	—	—	—	—	—	2	26	—	28
Neotoma lepida	77.2	—	—	3	—	—	—	—	—	—	—	—	—	3
N. albigula	133.6	—	—	—	—	—	—	1	1	—	—	—	—	2
Spermophilus spilosoma	67.8	—	—	—	—	—	—	—	—	—	—	1	—	1
Total		0	0	9	0	0	0	1	1	5	5	148	1	170
Total Rodents		2	16	17	4	4	6	18	9	12	25	430	9	552

The entries in this table represent the rodents captured at the study sites described in Table 1.

Peromyscus maniculatus, 19 *Dipodomys ordi,* and 1 *Onychomys leucogaster* were captured in 300 trap nights. Second, collections of seeds from heteromyid cheek pouches were made available by J. S. Findley of the University of New Mexico. These were analyzed as described above and in Brown and Lieberman (1973). Third, the maps of species distributions in Hall and Kelson (1959) and Burt and Grossenheider (1964) were used to prepare Figure 3.

References

Baker, R. H. 1971. Nutritional strategies of myomorph rodents in North American grasslands. *J. Mammal.* 52:800–805.

Barbour, C. D. 1973. A biogeographical history of *Chirostoma* (Pisces: Atherinidae): a species flock from the Mexican Plateau. *Copeia* 1973:533–556.

Barbour, C. D., and J. H. Brown. 1974. Fish species diversity in lakes. *Amer. Natur.* 108:473–489.

Beatley, J. C. 1967. Survival of winter annuals

in the Mojave Desert. *Ecology* 48:745–750.

Beatley, J. C. 1969. Dependence of desert rodents on winter annuals and precipitation. *Ecology* 50:721–724.

Brown, J. H. 1971. Mammals on mountaintops: nonequilibrium insular biogeography. *Amer. Natur.* 105:467–478.

Brown, J. H. 1973. Species diversity of seed-eating desert rodents in sand dune habitats. *Ecology* 54:775–787.

Brown, J. H., and G. A. Lieberman. 1973. Resource utilization and coexistence of seed-eating desert rodents in sand dune habitats. *Ecology* 54:788–797.

Burt, W. H., and R. P. Grossenheider. 1964. *A Field Guide to the Mammals.* Houghton Mifflin, Boston.

Cody, M. L. 1966. The consistency of inter- and intra-continental grassland bird species counts. *Amer. Natur.* 100:371–376.

Cody, M. L. 1968. On the methods of resource division in grassland bird communities. *Amer. Natur.* 102:107–147.

Cody, M. L. 1970. Chilean bird distribution. *Ecology* 51:455–464.

Cody M. L. 1973. Parallel evolution and bird niches. *In* F. DiCastri and H. A. Mooney, eds., *Ecological Studies 7,* pp. 307–338. Springer-Verlag, Wien.

Cody, M. L. 1974. *Competition and the Structure of Bird Communities.* Princeton University Press, Princeton.

Connell, J. H. 1961. The influence of interspecific competition and other factors on the distribution of the barnacle *Chthamalus stellatus. Ecology* 42:710–723.

Diamond, J. M. 1970. Ecological consequences of island colonization by southwest Pacific birds. II. The effect of species diversity on total population density. *Proc. Nat. Acad. Sci. U.S.A.* 67:1715–1721.

Evans, H. E. 1970. Ecological-behavioral studies of the wasps of Jackson Hole, Wyoming *Bull. Mus. Comp. Zool.* 140:451–511.

Hairston, N. G., F. E. Smith, and L. B. Slobodkin. 1960. Community structure, population control, and competition. *Amer. Natur.* 94:421–425.

Hall, E. R., and K. R. Kelson. 1959. *The Mammals of North America.* Ronald Press, New York.

Hespenheide, H. A. 1971. Food preferences and the extent of overlap in some insectivorous birds, with special reference to the Tyrannidae. *Ibis* 113:59–72.

Hespenheide, H. A. 1973. Ecological inferences from morphological data. *Ann. Rev. Ecol. Syst.* 4:213–229.

Hillel, D., and N. Tadmor. 1962. Water regime and vegetation in the central Negev highlands of Israel. *Ecology* 43:33–41.

Hutchinson, G. E. 1959. Homage to Santa Rosalia, or Why are there so many kinds of animals? *Amer. Natur.* 93:145–159.

Janzen, D. H. 1970. Herbivores and the number of tree species in tropical forests. *Amer. Natur.* 104:501–529.

Krebs, C. J., B. J. Keller, and R. H. Tamarin. 1969. *Microtus* population biology: demographic changes in fluctuating populations of *M. ochrogaster* and *M. pennsylvanicus* in southern Indiana. *Ecology* 50:587–607.

Lack, D. 1947. *Darwin's Finches.* Cambridge University Press, Cambridge, England.

MacArthur, R. H. 1972. *Geographical Ecology.* Harper and Row, New York.

MacArthur, R. H., and J. M. MacArthur. 1961. On bird species diversity. *Ecology* 42:494–598.

MacArthur, R. H., J. MacArthur, D. MacArthur, and A. MacArthur. 1973. The effect of island area on population densities. *Ecology* 54:657–658.

MacArthur, R. H., and E. O. Wilson. 1967.

The Theory of Island Biogeography. Princeton University Press, Princeton.

May, R. M., and R. H. MacArthur. 1972. Niche overlap as a function of environmental variability. *Proc. Nat. Acad. Sci. U.S.A.* 69:1109–1113.

McNab, B. K. 1971a. The structure of tropical bat faunas. *Ecology* 52:352–358.

McNab, B. K. 1971b. On the ecological significance of Bergmann's rule. *Ecology* 52:845–854.

Paine, R. T. 1966. Food web complexity and species diversity. *Amer. Natur.* 100:65–76.

Pearson, O. P. 1966. The prey of carnivores during one cycle of mouse abundance. *J. Animal Ecol.* 35:217–233.

Pearson, O. P. 1971. Additional measurements of the impact of carnivores on California voles (*Microtus californicus*). *J. Mammal.* 52:41–49.

Pulliam, H. R., and F. Enders. 1971. The feeding ecology of five sympatric finch species. *Ecology* 52:557–566.

Rosenzweig, M. L. 1966. Community structure in sympatric carnivora. *J. Mammal.* 47:602–612.

Rosenzweig, M. L. 1968. Net primary productivity of terrestrial communities: prediction from climatological data. *Amer. Natur.* 102:67–74.

Rosenzweig, M. L. 1973. Habitat selection experiments with a pair of coexisting heteromyid rodent species. *Ecology* 54:111–117.

Rosenzweig, M. L., and P. Sterner. 1970. Population ecology of desert rodent communities: body size and seed husking as bases for heteromyid coexistence. *Ecology* 51:217–224.

Rosenzweig, M. L., and J. Winakur, 1969. Population ecology of desert rodent communities: habitats and environmental complexity. *Ecology* 50:558–572.

Sanders, H. L. 1968. Marine benthic diversity: a comparative study. *Amer. Natur.* 102:243–283.

Schoener, T. W. 1970. Size patterns in West Indian *Anolis* lizards. II. Correlations with the sizes of particular sympatric species—displacement and convergence. *Amer. Natur.* 104:155–174.

Simberloff, D. S., and E. O. Wilson. 1969. Experimental zoogeography of islands. The colonization of empty islands. *Ecology* 50:278–296.

Slobodkin, L. B., F. E. Smith, and N. G. Hairston. 1967. Regulation in terrestrial systems, and the implied balance of nature. *Amer. Natur.* 101:109–124.

Smigel, B. W., and M. L. Rosenzweig. 1974. Seed selection in *Dipodomys merriami* and *Perognathus penicillatus*. *Ecology* 55:329–339.

Storer, R. W. 1966. Sexual dimorphism and food habits in three North American accipiters. *Auk* 83:423–436.

Terborgh, J. 1973. On the notion of favorableness in plant ecology. *Amer. Natur.* 107:481–501.

Vuilleumier, F. 1970. Insular biogeography in continental regions. The northern Andes of South America. *Amer. Natur.* 104:373–388.

Went, F. W. 1948. Ecology of desert plants. I. Observations on germination in the Joshua Tree National Monument, California. *Ecology* 29:242–253.

Went, F. W. 1955. The ecology of desert plants. *Sci. Amer.* 192:68–75.

Went, F. W., and M. Westergaard. 1949. Ecology of desert plants. III. Development of plants in the Death Valley National Monument, California. *Ecology* 30:26–38.

Zaret, T. M., and A. S. Rand. 1971. Competition in tropical stream fishes: support for competitive exclusion principle. *Ecology* 52:336–342.

14 Assembly of Species Communities

Jared M. Diamond

Contents

Summary

This chapter explores the origin of differences in community structure, such as those between different islands of the same archipelago, between different localities on the same island, between different adjacent habitats, and between different biogeographical regions. The working hypothesis is that, through diffuse competition, the component species of a community are selected, and coadjusted in their niches and abundances, so as to fit with each other and to resist invaders. Observations are derived from bird communities of New Guinea and its satellite islands, of which some are at, some above, and some below equilibrium in species number (S).

From exploration of numerous islands with various values of S, so-called incidence functions are constructed for individual species. These relate J, the incidence of occurrence of a particular species on islands of a certain S-class, to S. Species are classified according to their incidence functions into six categories: high-S species, confined to the most species-rich islands; A-, B-, C-, and D-tramps, present on the most species-rich islands and also on increasing numbers of increasingly more species-poor islands; and supertramps, confined to species-poor islands and absent from species-rich islands. Since different species have incidence functions of different shapes, the fauna of any real island is a very nonrandom subset of the total species pool.

The high-S category consists partly of endemic species of forest on large islands, partly of non-endemic species of scarce habitats often unrepresented or barely represented on smaller islands. Tramps, especially C- and D-tramps, are mostly nonendemic species characteristic of habitats that occur on virtually any island.

The dependence of incidence on area involves several factors, which vary from species to species: whether the required habitat of a species occurs on small islands; minimum territory size for species in which each pair maintains an exclusive territory; minimum year-round support area for species dependent on patchy or seasonal food supplies; population size in relation to short-term and long-term population fluctuations; and the role of "hot spots" (areas of locally-high utilizable resource production) in colonization and in recovery from population crashes.

Dispersal ability of species in different incidence categories has been assessed from data sources such as recolonization of islands defaunated by volcanic explosion or tidal wave, long-term records of vagrants, and direct observations of overwater colonization. Especially in the tropics, many bird species capable of strong flight refuse to cross water barriers of even a few miles. Dispersal rates are highest for supertramps and D-tramps, followed by C-tramps, B-tramps, and nonendemic A-tramps of scarce habitats. For high-S species, such dispersal as there is may be associated with rare population "blooms."

There is no obvious correlation between clutch size and incidence category. However, supertramps and D- and C-tramps have longer breeding seasons and raise more broods per year than do other species.

Supertramps have extraordinarily catholic and unspecialized habitat preferences, high reproductive potential, and high dispersal ability. They are competitively excluded from species-rich islands by "K-selected" species. However, faunas dominated by supertramps maintain population densities up to nine times *higher* than those of *K*-selected faunas composed of the same number of species. Thus, the supertramp strategy may be contrasted with an inferred overexploitation ethic practised by high-*S* species, which are selected by competition to harvest early and overexploit. The high-*S* species thereby reduce resource levels below the point where other species can survive, even though this diminishes the rate of resource production and hence the population density of the harvesting species.

In a few instances, competition expresses itself in "simple" checkerboard distributions, by which species replace each other one-for-one. The frequent occurrence of "empty squares," however, shows that even these cases are complex. In the great majority of species groups or guilds, competitive exclusion involves so-called diffuse competition, i.e., the combined effects of several closely related species. Detailed examination of four guilds reveals the following types of assembly rules for species communities:

If one considers all the combinations that can be formed from a group of related species, only certain ones of these combinations exist in nature.

These permissible combinations resist invaders that would transform them into a forbidden combination.

A combination that is stable on a large or species-rich island may be unstable on a small or species-poor island.

On a small or species-poor island a combination may resist invaders that would be incorporated on a larger or more species-rich island.

Some pairs of species never coexist, either by themselves or as part of a larger combination.

Some pairs of species that form an unstable combination by themselves may form part of a stable larger combination.

Conversely, some combinations that are composed entirely of stable subcombinations are themselves unstable.

The forbidden combinations do not exist in nature because they would transgress one or more of three types of empirical rules: compatibility rules banning the coexistence of certain closely related species under any circumstances; incidence rules, implicit in incidence functions; and combination rules, which cannot be predicted from incidence functions.

Most of the evidence for these assembly rules is drawn from comparison of communities on different islands. However, examples are also drawn from communities at different localities, or in different habitats, or at different altitudes, or at different heights above the ground, on the same island. In some cases one can recognize simple effects of one-to-one competition. In other cases, one can recognize assembly rules describing more complex competitive effects and permitted combinations of several related species. In still more complex cases, competitive effects must be described by incidence functions relating the occurrence or niche limits of one species to diffuse competition from many other species. Thus, recognition of assembly rules may help us understand competitive effects on the spatial niche limits of a given species, and the puzzling tropical phenomenon of patchy distributions.

Much of the explanation for assembly rules has to do with competition for resources and with harvesting of resources by permitted combinations so as to minimize the unutilized resources available to support potential invaders. Communities are assembled through selection of colonists, adjustment of their abundances, and compression of their niches, in part so as to match the combined resource consumption curve of all the colonists to the resource production curve of the island. Members of permitted combinations must also be "companions in starvation"—i.e., must be similar in their tendencies to overexploit and in their tolerances for lowered resource levels, thereby starving less tolerant species off the island. Thus, consumer species form hierarchies with respect to exploitive strategy. The conditions under which overexploitation becomes a useful strategy for its practitioners are examined by loop analysis. Also relevant to the origin of assembly rules are two further factors: dispersal abilities, which permit only certain species to have a high incidence on small islands with high extinction rates; and transition probabilities, i.e., ease of assembling a species combination in one or a few steps from other permitted combinations.

Major unsolved problems include: the development of mathematical models for incidence functions; extensions to habitat communities and to locally patchy communities; the relative roles of chance and of predestination (i.e., detailed matches of different species combinations to slightly different local production curves) in the build-up of alternate communities; and applications to conservation problems.

Introduction

The understanding of alternate, stable, invasion-resistant communities of co-adjusted species poses a major current problem in ecology. Sets of such communities occur in similar habitats in different biogeographical regions, in similar habitats on different islands colonized from the same species pool, in similar habitats at different localities on the same large island or continent, and in different adjacent habitats. The theoretical basis for the existence of alternate stable communities was brilliantly explored by Robert MacArthur (1972) in *Geographical Ecology*. A conceptual framework is now available within which field observers can approach such unsolved problems as the following:

To what extent are the component species of a community mutually selected from a larger species pool so as to "fit" with each other?

Does the resulting community resist invasion? If so, how?

To what extent is the final species composition of a community uniquely specified by the properties of the physical environment, and to what extent does it depend on chance events (e.g., the question of which colonists arrive first, possibly also affecting which subsequent arrivals are compatible with the successful first colonists)?

The present chapter discusses such problems in the light of observations on bird communities of New Guinea satellite islands. It will be shown that (a) the probabilities or incidences of occurrence of particular species in a community bear neat empirical relations to the total species number in the community; (b) these so-called incidence functions can be interpreted in terms of island area plus a species' habitat requirements, dispersal ability, birth and death schedule, exploitation strategy, and competitive relations; (c) the various species in a guild can coexist only in certain combinations; (d) these permitted combinations resist invaders that would result in forbidden combinations; and (e) lowering of resource levels by coadjusted constellations of species, to below the point where invaders can survive, may be an important mechanism of competitive exclusion.

Statement of the Problem

The structure of a species community may be described in terms of its species composition, together with the resource utilization, and distribution and abundance in space and time, of each component species. Comparison of different communities at any one of four levels generally reveals some differences in structure:

1. Differing but adjacent habitats differ in community structure, even though there may be no physical barriers preventing species of one habitat from invading another habitat (cf. Cody, Chapter 10).

2. Differences in community structure may exist between similar habitats in different areas of the same continent or large island, or even between similar habitats in areas that are in immediate contact and constitute artificially defined sections of a continuum. This phenomenon is es-

pecially marked in the tropics. The result is often that tropical species are patchily distributed with respect to the available habitat. Figures 33–38 will present examples of these baffling distributional patterns.

3. Communities on similar islands colonized from the same species pool may differ. For example, the islands Sakar and Tolokiwa lie 29 miles apart in the Bismarck Sea near New Guinea, differ in area by only 13%, are geologically similar, support similar forest, have derived their birds from the same sources, and support similar numbers of lowland bird species (36 and 40, respectively). Yet Tolokiwa lacks three of the seven most abundant species of Sakar, Sakar lacks eight of the 15 most abundant species of Tolokiwa, and only 23 species are shared. In the Pearl Archipelago off Panama, MacArthur, Diamond, and Karr (1972) cite equally striking differences in bird species composition between Chitre and Contadora islands, which are only 1 mile apart. Furthermore, a species that is shared between similar islands may still occupy different habitats and have different abundances. For example, the fruit pigeon *Ptilinopus insolitus* is present both on Sakar and on Tolokiwa, but on Sakar it is widespread whereas on Tolokiwa it is confined to mid-montane forest. Its congener *Ptilinopus solomonensis* is present both on Sakar and on Tolokiwa and occupies similar habitats on the two islands, but is approximately six times more abundant on Tolokiwa than on Sakar.

4. The examples mentioned so far involve communities formed from the same species pool and lying within the same

biogeographic region or faunal province. Much larger differences are observed between more distant communities lying in different faunal provinces. For more than a century, from the time of Sclater and Wallace until the publication of *The Theory of Island Biogeography* by MacArthur and Wilson (1967), these differences formed the principal subject matter of biogeography. Although similar habitats in South America, Africa, and Australia may share few species in common, these communities may exhibit remarkably detailed convergent similarities in structure (Cody, Chapter 10; Karr and James, Chapter 11). The borders of the world's major faunal provinces are formed by present and past barriers to movement of organisms. These barriers have not served to eliminate colonization, but rather to reduce it to a level where great differences are maintained indefinitely between the communities on opposite sides of the barrier. If the communities did not possess some resistance to invasion, colonization across the barriers for millions of years would have smoothed many of the differences between even the major faunal provinces. Thus, the differences between the Australian Region and the Oriental Region present many of the same problems, albeit in more marked form, as the differences between Sakar and Tolokiwa islands in the Bismarck Sea.

These examples suggest (but do not prove) that the species in a community are somehow selected, and their niches and abundances somehow coadjusted, so that the community possesses some measure of "stability." Stability implies the existence of several different properties, some of which are easier to demonstrate than others. The most obvious thing we mean in describing a community as "stable" is that its present species composition is likely to persist with little change if there is no change in the physical environment. This property is easy to assess by comparing historical surveys with recent surveys. For instance, faunal surveys of a given New Guinea satellite island a century ago and today yield much more similar species compositions than do surveys of several different islands of similar size at the same time. The property of stable species composition suggests the existence of an additional property, namely, ability of a community to resist invasion by new species. This property is more difficult to document, because one needs much more than two faunal surveys at different times. A particular species may be absent from a particular island because the existing community prevents colonizing individuals of the new species from establishing themselves, or merely because colonizing individuals of the species may never reach the island at all. To document resistance to invasion requires sufficiently extensive observations so that arrivals of colonizing individuals, and their failures to establish stable populations, are detected. Finally, the property of resistance to invasion suggests a further property, which is still more difficult to document as well as to formulate, namely, that the existing community utilizes available resources in some optimal manner (MacArthur, 1970; MacArthur, 1972, pp. 231–234).

It seems likely that competition between species plays a key role in the integration of species communities. Real or potential utilization of some of the same resources could be an obvious explanation for why similar species do not occur in the same community, unless their resource utilizations are somehow coadjusted. Numerous recent studies have provided clear-cut distributional evidence for competition between members of a pair of related species. These examples are valuable in documenting the existence of competition, but by themselves they do not account for much of the real world. Far more often, the presence or absence of a given species, and intercommunity variation in its abundance or spatial distribution, cannot be understood predominantly in terms of a correlated distribution of any single other species. It is then a logical extension of simple two-species distributional checkerboards to invoke "diffuse competition"—i.e., the complex situations resulting from the sum of competitive effects from many other somewhat similar species (Diamond, 1970a, p. 530; 1970b, pp. 1716–1717; MacArthur, 1972, pp. 43–46 and 249; Pianka, Chapter 12). The power of this concept is that, in principle, it can explain anything. Its heuristic weakness is that, if it is important at all, its operation is likely to be so complicated that its existence becomes difficult to establish and impossible to refute. Such a concept deserves to be greeted with skepticism until its importance can be documented. A profitable biogeographic approach to documenting diffuse competition would seem to be, first, to seek evidence whether variation in the incidence, niche, or abundance of a given species is correlated with variation in total species number; then, to seek to trace out cases in which the distribution of a given species can be clearly related to the distribution of certain *combinations* of a few other species, yielding patterns that are analogous to two-species distributional checkerboards but more complex.

Such a test of the hypothesis of alternate, stable, invasion-resistant communities integrated by diffuse competition requires a field situation or experimental situation with the following properties: (a) a large number of communities that provide a similar physical environment and habitat structure; (b) a large species pool, varying fractions and combinations of which occur in the available communities; (c) availability of evidence that a species absent from a given community actually has had access, and that its absence is not simply due to a total lack of immigrants; (d) availability of evidence that the community does resist invasion, and that failure of attempted colonizations is not simply due to unsuitable habitat; (e) availability of cases in which a community has been displaced from equilibrium, so that relaxation towards equilibrium can be studied.

The avifauna of New Guinea and its satellite islands provides a favorable test situation. Considerable ecological and evolutionary information exists about the New Guinea species pool of 513 breeding nonmarine bird species. Surrounding New Guinea, and colonized by varying fractions of this species pool, are thousands of islands of varying sizes and at varying distances, providing numerous sets of replicate communities. Ornithological ex-

Species

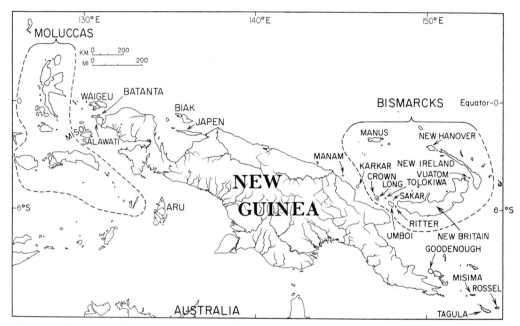

Figure 1 Map of the New Guinea region with names of some of the islands to be discussed.

ploration has been sufficiently intensive to provide not merely species lists but, for some islands, instances of successful and unsuccessful colonizations. Species numbers on some islands have been displaced above what would be their present value at equilibrium by Pleistocene episodes of lowered sea level, which joined some islands to New Guinea, joined other islands to each other, and expanded still other islands in area. Species numbers on other islands have been displaced below equilibrium by Krakatoa-like volcanic explosions or by tidal waves. Some species called supertramps are particularly useful in studying community integration, be-

cause of their high colonization rates and sensitivity to competition. We shall see that the distributions of most species can be neatly related to total species number in a community; and that, in a few cases, it is possible to relate species distributions to diffuse competitive effects from specific combinations of related species.

Background: The New Guinea Biogeographic Scene

New Guinea lies near the equator, north of Australia, at the eastern end of the Indonesian Archipelago, and at the faunal gateway to the islands of the southwest Pacific. Figure 1 gives the positions of some of the islands that will be mentioned. The predominant natural vegetation is rainforest, but there are also savannas in low-rainfall regions of south New Guinea, alpine grassland at high elevations on New Guinea and a few of the highest satellite islands, and glaciers above 16,000 feet on the highest peaks of New Guinea's mountain backbone (the so-called central cordillera).

On satellite islands near New Guinea the number of lowland bird species S increases with island area (A, in square miles), approximately according to the empirical relation (Diamond, 1972b)

$$S = 15.1A^{0.22} \qquad (1)$$

Within the satellite archipelagoes, such as the Bismarcks or the New Hebrides, the exponent of area has a lower value (Figures 2, 3). The deviation below a linear $\log S - \log A$ relation at very low A values (Figure 3) is in the direction predicted by May (section 5 of Chapter 4). On mountainous islands each 1000 feet of elevation L is associated with a number of montane species equal to 2.7% of the species number at sea level. S decreases exponentially with distance d from New Guinea, by a factor of 2 for each 1620 miles. Thus, the total number of species on an island is given by the relation

$$S = 15.1(1 + 0.027\ L/1000)$$
$$(e^{-0.693d/1620})A^{0.22} \qquad (2)$$

On islands within a few miles of a more species-rich island, S is higher than pre-

dicted from eq. 2. New Guinea has 513 species, of which 325 are in the lowlands. The most species-rich satellite island, Aru, has 160 lowland species, 49% of the number for New Guinea. At the other extreme, some small or remote islands have only one species.

Species numbers on three types of islands were displaced above the present equilibrium value (predicted from eq. 1) by injection of species during the Pleistocene, at times of lower sea level. One group of present-day islands (points +, Fig. 2) was then connected to New Guinea itself and must have received most of the New Guinea lowlands avifauna. Other islands (points ⊕, Fig. 2) were connected to larger satellite islands, though not to New Guinea itself, and must have received most of the avifauna of these islands. Still other islands (points ☐, Fig. 2) that lie on large shallow shelves were formerly much larger in area and must formerly have held an equilibrium species number larger than the value appropriate to their present shrunken area. When sea level rose about 10,000 years ago, all these islands must have found themselves supersaturated with bird species in relation to the equilibrium value for their new condition. As shown by Figure 2, species numbers on the smaller of these islands have already relaxed to the equilibrium value given by eq. 1, whereas the larger of these islands are still "supersaturated." Salawati, for example, has 134 lowland species, considerably less than the New Guinea total of 325, most of which it must have supported at times of lower sea-level up to about 10,000 years ago, but still more

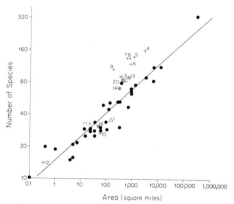

Figure 2 Number of resident land and fresh-water bird species on New Guinea satellite islands, plotted as a function of island area on a log-log scale. ●, islands on which species number is presumed to be at equilibrium. The avifaunas of the remaining, numbered islands are in various stages of "relaxation" after displacement of species number from an equilibrium value. △, exploded volcanoes: 1 = Long. □, contracted islands: 2 = Goodenough, 3 = Fergusson. ⊕, islands formerly connected by land bridges ("first-order bridges") to New Guinea itself: 4 = Aru, 5 = Waigeu, 6 = Japen, 7 = Salawati, 8 = Misol, 9 = Batanta, 10 = Pulu Adi, 11 = Ron, 12 = Schildpad. +, islands formerly connected by land bridges ("second-order bridges") to some larger island but not to New Guinea itself: 13 = Batjan, 14 = Amboina, 15 = New Hanover, 16 = Tidore. The straight line is eq. 1 ($S = 15.1A^{0.22}$), the least-mean-squares fit to all points except the first-order land-bridge islands. Note that large land-bridge islands and contracted islands deviate in the direction of an excess of species, that small islands do not, and that the deviation is more marked for the first-order land-bridge islands or contracted islands. (From Diamond, 1972b.)

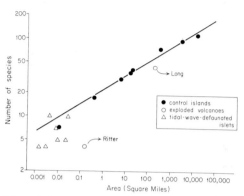

Figure 3 Number of resident, nonmarine, lowland bird species S on New Britain and neighboring islands in the Bismarck Archipelago, plotted as a function of island area on a double logarithmic scale. Symbols: ●, relatively undisturbed "control" islands (from left to right, Midi, Malai, Crown, Sakar. Tolokiwa, Umboi, New Ireland, New Britain); ○, exploded volcanoes; △, coral islets inundated by the Ritter tidal wave in 1888. The straight line $S = 22.4A^{0.18}$ was fitted by least-mean-squares through points for the seven larger control islands. Species number is still below equilibrium on the two exploded volcanoes, especially on Ritter, because of incomplete regeneration of vegetation. The scatter in S values for the smallest islets, and their general deviation below the line for larger control islands, is attributed to their very low S values and few individuals per species (cf. Table 4), and is in the direction predicted by May (section 5 of Chapter 4). (After Diamond, 1974.)

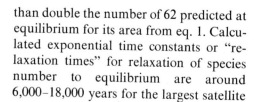

than double the number of 62 predicted at equilibrium for its area from eq. 1. Calculated exponential time constants or "relaxation times" for relaxation of species number to equilibrium are around 6,000–18,000 years for the largest satellite islands ($A = 300$–3,000 sq. mi.). That is, larger islands remain supersaturated for much longer than do smaller islands (Diamond, 1972b). Cody (Chapter 10, Figure 11) has discovered a similar relaxation effect in a continental avifauna. On the small island of Barro Colorado, which was part of the Panama mainland until about 1914, the relaxation time of the avifauna is short enough for decreases in species number to have been detectable each decade (Wilson and Willis, Chapter 18; Willis, 1974; see also Terborgh, 1974a, 1974b).

Species numbers on three other types of islands have been displaced below, rather than above, the equilibrium value. First, Long and Ritter were defaunated in historical times by cataclysmic, Krakatoa-like volcanic explosions (Diamond, 1974). The species number on Ritter, which exploded in 1888, is still 75% below the equilibrium value (Figure 3). Lowland species number on Long, which exploded about two centuries ago, is 25% below the equilibrium value and has remained "stuck" in this depressed quasi-steady state for at least the last 40 years. These bird species deficits on Ritter and Long are clearly attributable to the stunted, subclimax, lowland forest on both islands; lowland bird species number must have equilibrated rapidly with the available vegetation. In the mountains of Long, however, where the forest is already structurally mature, there is a large deficit of montane bird species that cannot be attributed to stunted vegetation and that indicates much slower dispersal by montane species than by lowland species, as already assumed by ornithologists for other reasons (Stresemann, 1939; Mayr, 1942). A second group of defaunated islands consists of seven coral islets defaunated by a tidal wave in 1888. All now support a climax forest and an equilib-

rium number of bird species. Finally, on one very small islet (3 acres) at a distance of 2 miles from New Guinea, I carried out experiments in removing existing bird populations and observing recolonization. Equilibration was apparently complete in less than two days and occurred at an initial rate of one colonizing species arriving per hour.

For three satellite islands much larger than this experimentally defaunated islet, estimates have been obtained of the rates at which bird species "turn over" at equilibrium, by natural immigration and extinction. On Karkar, for which surveys are available for the years 1914 and 1969, the estimated average turnover rate for the intervening years is 0.32%/year (Diamond, 1971). That is, during each decade a number of established species equal to 3.2% of the total bird species number disappeared, and an approximately equal number of immigrants successfully established breeding populations. For Vuatom, the turnover rate between 1906 and 1930 is 0.40%/year. The rate for Long, measured not at equilibrium but in the quasi-steady state that has persisted at least from 1933 to 1972, is 0.18%/year. These turnover rates are comparable to values calculated for the California Channel Islands (Diamond, 1969; Jones and Diamond, 1975) and for Mona in the Caribbean (Terborgh and Faaborg, 1973).

Incidence Functions

If one examines the detailed distributions of numerous individual species in the New Guinea region, several types of patterns emerge. Certain species are found only on the largest, most species-rich islands. At the other extreme, certain species occur only on the smallest or most remote, most species-poor islands. There is a whole spectrum of intermediate cases. These general impressions can be quantified and depicted for each species by the type of graph illustrated in Figures 4–15, which we shall term an incidence function of the species. Such a graph can serve as a "fingerprint" of the distributional strategy of a species, once the forms that the graph may assume have been interpreted.

Incidence functions are constructed by grouping well-surveyed islands together into classes, the islands in each class sharing a similar total number of species. One class, for example, may be defined as consisting of islands with 4–6 species, another of islands with 7–10 species, another of islands with 11–20 species, etc. Construction of such functions is practical only if there are a large number of well-surveyed islands, permitting arbitrary definition of classes, each spanning a narrow range of total species number. For a given species, one calculates the incidence of occurrence on islands of a given class—i.e., the fraction of the class's islands on which the species actually occurs. Incidence of occurrence of the given species, J, is then plotted against the average total number of species S in each island class. Sufficient information is available to construct sets of functions for the avifaunas of half-a-dozen archipelagoes of the New Guinea region and southwest Pacific.

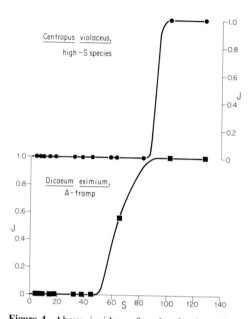

Figure 4 Above, incidence function for the high-S cuckoo *Centropus violaceus* in the Bismarck Archipelago, identical to the function for several other high-S species (the plover *Charadrius dubius*, the pigeon *Gymnophaps albertisii*, the parrot *Micropsitta bruijnii*, the wood swallow *Artamus insignis*). Below, incidence function for the A-tramp berrypecker *Dicaeum eximium*. To construct Figure 4–15, islands of the Bismarck Archipelago were divided into groups such that value of the total species number S of all islands in a given group fell within a narrow range ΔS. ΔS was generally ≤ 3 for $S < 20$, 2–10 for $20 < S < 65$, and 2–20 for $S > 65$. The ordinate J is the incidence of the given species (i.e., the fraction of the islands in the group on which the species occurs), and the abscissa S is the average species number for the islands of the group. Thus, $J = 1.0$ or $J = 0$ means that a species occurs on all islands or on no island, respectively, that has approximately the indicated species number. Each point is usually based on 3–13 islands, except that the two-right-most points usually represent one island each ($S = 101$ and 127).

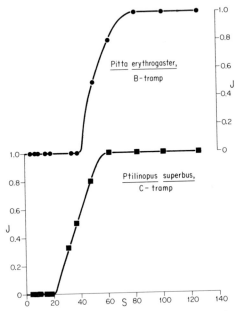

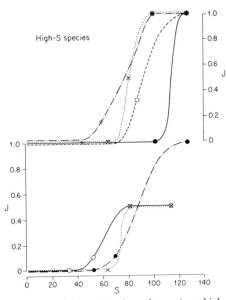

Figure 5 Incidence functions for the B-tramp pitta *Pitta erythrogaster* (above), and for the C-tramp pigeon *Ptilinopus superbus* (below), in the Bismarcks.

Figure 7 Incidence functions for various high-*S* species in the Bismarcks. Above, points ● and line —, the rail *Rallus insignis,* hawks *Accipiter brachyurus* and *Milvus migrans,* kingfisher *Halcyon albonotata,* parrot *Cacatua galerita,* owl *Tyto aurantia,* warbler *Ortygocichla rubiginosa,* flycatcher *Monachella muelleriana,* honeyeater *Melidectes whitemanensis,* finch *Lonchura melaena,* and several other species. Points ○ and line − − −, the ducks *Dendrocygna guttata* and *D. arcuata,* heron *Egretta intermedia,* and honeyeater *Myzomela cineracea.* Points X and line · · · ·, the pigeon *Ducula finschii,* parrot *Vini rubrigularis,* and hornbill *Aceros plicatus.* Points + and line − · − · −, the pigeon *Columba pallidiceps* and cuckoo-shrike *Coracina lineata.* Below, points ● and line − · − · −, the hawk *Henicopernis longicauda* and button-quail *Turnix maculosa.* Points ○ and line —, the rail *Rallina tricolor.* Points X and line · · · ·, the heron *Butorides striatus.* Most *J* = 0 points are omitted for clarity.

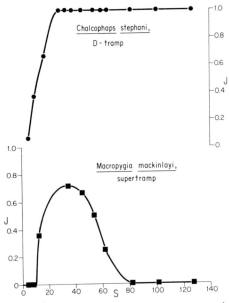

Figure 6 Incidence functions for the D-tramp pigeon *Chalcophaps stephani* (above), and for the supertramp pigeon *Macropygia mackinlayi* (below), in the Bismarcks.

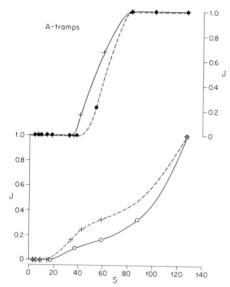

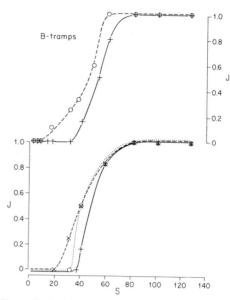

Figure 8 Incidence functions for various A-tramps in the Bismarcks. Above, points + and line —, the pigeon *Ptilinopus rivoli*, parrot *Geoffroyus heteroclitus,* and kingfisher *Ceyx lepidus*. Points ● and line − − −, the white-eye *Zosterops hypoxantha*. Below, points ○ and line —, the bee-eater *Merops philippinus*. Points + and line − − − −, the rail *Porphyrio porphyrio*.

Figure 9 Incidence functions for various B-tramps in the Bismarcks. Above, points + and line —, the grass warbler *Cisticola exilis*. Points ○ and line − − −, the cuckoo *Cacomantis variolosus*. Below, points X and line − − −, the rail *Amaurornis olivaceus*. Points ○ and line · · · ·, the pigeon *Ducula rubricera*. Points + and line —, the nightjar *Caprimulgus macrurus*.

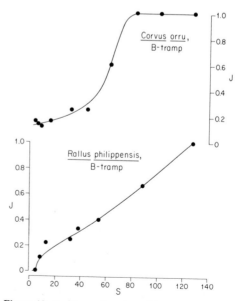

Figure 10 Incidence functions for two B-tramps in the Bismarcks. Above, the crow *Corvus orru*. Below, the rail *Rallus philippensis*.

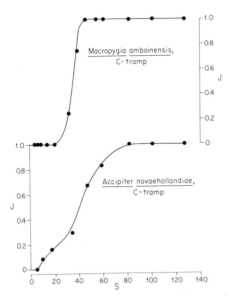

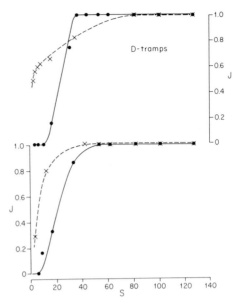

Figure 11 Incidence functions for two C-tramps in the Bismarcks. Above, the pigeon *Macropygia amboinensis*. Below, the hawk *Accipiter novaehollandiae*.

Figure 13 Incidence functions for various D-tramps in the Bismarcks. Above, points ● and line —, the hawk *Haliastur indus*. Points X and line — — —, the kingfisher *Halcyon saurophaga*. Below, points ● and line —, the starling *Aplonis metallica*. Points X and line — — —, the incubator-bird *Megapodius freycinet*.

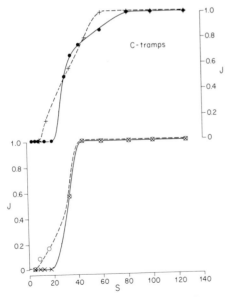

Figure 12 Incidence functions for various C-tramps in the Bismarcks. Above, points ● and line —, the cuckoo-shrike *Coracina tenuirostris*. Points + and line — — —, the parrot *Micropsitta pusio*. Below, points X and line —, the kingfisher *Alcedo atthis*. Points ○ and line — — —, the heron *Nycticorax caledonicus*.

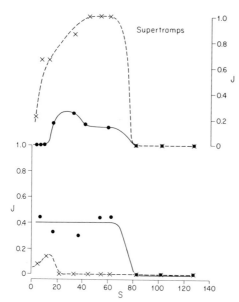

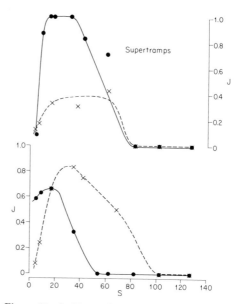

Figure 14 Incidence functions for various super-tramps in the Bismarcks. Above, points X and line − − −, the pigeon *Ducula pistrinaria*. Points ● and line −, the white-eye *Zosterops griseotincta*. Below, points ● and line −, the whistler *Pachycephala melanura dahli*. Points X and line − − −, the starling *Aplonis feadensis*.

Figure 15 Incidence functions for various super-tramps in the Bismarcks. Above, points ● and line −, the flycatcher *Monarcha cinerascens*. Points X and line − − −, the honeyeater *Myzomela sclateri*. Below, points ● and line −, the honeyeater *Myzomela pammelaena*. Points X and line − − −, the pigeon *Ptilinopus solomonensis*.

Figures 4–15 are examples of incidence functions based on 50 well-surveyed islands of the Bismarck Archipelago, the group of islands extending from New Britain to Manus and lying between New Guinea and the Solomon Islands (Figure 1). A detailed analysis of the Bismarck avifauna, including species lists for each island, is given by Mayr and Diamond (1975), and speciation in the Bismarcks and Solomons has been discussed by Mayr (1969). Within the Bismarck Archipelago most of the variation in S is accounted for by variation in area A, but some variation in S is also due to variation in isolation or to variation in elevation.

Thus, the abscissa scale of Figures 4–15 is approximately but not perfectly equivalent to an A^z scale, where z (or x in the terminology of May, Chapter 4, eq. 5.1) is the exponent of area in the species/area relation. In general, if one constructs incidence functions $J_i(S)$ for all i species of an archipelago, the relation

$$\sum_i J_i(S') = S' \qquad (3)$$

holds for any arbitrary value $S = S'$.

The most obvious and immediate conclusion of Figures 4–15 is that the incidence function $J(S)$ may assume very different forms for different species. That is, the fauna of any island of any size, large or small, is a highly nonrandom selection of the archipelago's species pool and is systematically biased by island size (and, to some extent, by island elevation, isolation, and historical effects such as land bridges). Figures 4–6 assemble, for comparison, examples of the main types of incidence functions, and Figures 7–15 give further examples of each type of function.

Species distributions define a spectrum of incidence functions, which may be divided into six arbitrary categories. Table 1 summarizes, for each category, the number of species in the category, the range of number of islands inhabited by each species, the habitat preferences of each species, and the level of endemism of each species (i.e., whether the Bismarck population represents a full species, semispecies, or subspecies endemic to the Bismarck Archipelago, or belongs to the same subspecies as populations outside the Bismarcks). Table 1 also summarizes upper and lower values of critical species number, or so-called S_{crit} values, for each category. The meaning of S_{crit} values is that, for many species (see Figures 4, 5, 6 below, 7–9, 11 above, and 12), the incidence $J(S)$ decreases with decreasing S and goes to zero at some value of S that we may term the lower critical value or S_{crit}. For most species $J(S)$ approaches 1.0 or at least some nonzero value with increasing S, the highest S on any single island of the archipelago being 127. However, some species (Figures 6 below, 14, 15) also have an upper S_{crit} value, such that $J(S) = 0$ for higher S.

The six categories may be briefly described in words as follows:

One category (Figures 6 below, 14, 15), which we shall term "supertramps," consists of species characterized (by definition) by an upper S_{crit} and entirely absent as residents on species-rich islands (see Table 6 for a complete list). Several of the supertramp species (e.g., starling *Aplonis feadensis*) have an upper S_{crit} of only 16, i.e., are confined to islands with 16 or fewer species. It will be shown that supertramps represent the extreme of r-selection, in MacArthur-Wilson terminology. Only two supertramp species (cf. the pigeon *Macropygia mackinlayi*, Figure 6) have a lower S_{crit}; all other supertramps have a nonzero incidence on even the smallest or most species-poor islets. The geographical limits of most supertramp species extend beyond the Bismarcks, but most have evolved endemic subspecies in the Bismarcks.

The opposite extreme consists of species confined to the most species-rich islands (Figures 4 above, 7). It will be shown that some, but not all, of these high-S specialists represent the extreme of what is considered to be K-selection in MacArthur-Wilson terminology. We shall arbitrarily delimit a category of high-S specialists as consisting of species confined to the four most species-rich Bismarck islands ($S = 81$–127) plus not more than one island with 43–80 species. The great majority of the endemic species (8 out of 10) and semispecies (21 out of 28) of the Bismarcks belong to this high-S category. However, it is not true that, conversely, a great majority of high-S species are endemics: only a slight majority (56%) are

Table 1. Distributional categories in the Bismarck Archipelago

Category	No. species in category	No. islands inhabited per species	Endemism level				Lower S_{crit}			Upper S_{crit}		
			None	Subspecies	Semispecies	Species	<10	15–31	>32	≤16	45–83	None
High-S	52	1–5	9	14	21	8	0	0	52	0	0	52
A-tramp	26	3–9	8	14	3	1	1	1	24	0	0	26
B-tramp	17	10–14	4	12	1	0	3	5	9	0	0	17
C-tramp	19	15–19	8	10	1	0	6	9	4	0	0	19
D-tramp	14	20–35	9	5	0	0	11	3	0	0	0	14
Supertramp	13	2–33	2	8	2	1	10	2	1	4	9	0

For each of the six distributional categories listed in column 1 and defined in the text, column 2 gives the number of species in the category. Column 3 gives the range of number of islands inhabited by the species in the category (e.g., the entry 1–5 for high-S species means that the most widely distributed high-S species occurs on five islands, whereas some high-S species occur on only one island). Columns 4–7 give the number of species in the category endemic to the Bismarcks at the indicated level (none = not differentiated even subspecifically, subspecies = endemic subspecies, semispecies = endemic semispecies or allospecies of a superspecies, species = endemic full species). Columns 8–10 and 11–13 give the number of species in the category with the indicated lower and upper S_{crit}, respectively (e.g., 4 C-tramps are confined to islands with 32 or more species, 9 C-tramps reach at least one island with 31 or fewer species but no island with fewer than 15 species, 6 C-tramps reach islands with 10 or fewer species, and all 19 C-tramps lack an upper S_{crit} and have nonzero incidence on even the largest islands). It happens that no species has a lower S_{crit} value of 10–14 or an upper S_{crit} value of 17–44. Note that as one proceeds from supertramps to high-S species, the species tend to become endemic at higher levels (supertramps fall somewhat out of line as regards endemism) and to be confined to increasingly species-rich islands. The analyses in this table, Tables 2–6, and the text are based on all Bismarck bird species except for approximately a dozen that are difficult to assign to distributional category. Most of the omitted species are ones that have a wide range outside the Bismarcks and are confined in the Bismarcks to one or two peripheral, medium-sized islands at the edge of the species' geographical range.

endemic to the Bismarcks at the semi-species or species level, and a large minority (44%) have evolved only endemic subspecies or have not differentiated at all. This suggests that high-S species are a heterogeneous group, as will be confirmed by other evidence.

The remaining species are "tramps" that occur on the most species-rich islands but also on varying numbers of the poorer (smaller, or more remote) islands. We shall divide tramps into four arbitrary categories according to the total number of Bismarck islands inhabited: A-tramps on 3–9 islands, B-tramps on 10–14 islands, C-tramps on 15–19 islands, and D-tramps on 20–35 islands. When defined in this way, A-tramps prove to be almost exclusively confined to islands with more than 32 species; half of the C-tramps reach some islands with 15–29 species (as well as richer islands); and most of the D-tramps reach some islands with only 4–10 species (as well as richer islands). The level of endemism decreases progressively as one passes from high-S species to A-, B-, C-, and finally D-tramps, and then increases slightly again with supertramps. There are no endemic species among the B-, C-, and D-tramps, and most D-tramps have not even differentiated subspecifically.

Is the distribution of a given species determined by island area alone, so that we should properly speak of large-island and small-island species rather than of high-S species and supertramps? Or, do individual species distributions depend on distance as well as on area? Or, is S itself the relevant variable, some species being able to fit only into diverse communities and others only into poor communities? Expressed in another way, is S the best choice of abscissa for the incidence functions, or would area be a better choice? Detailed answers to these questions cannot yet be given. The answers are certain to be complex and must vary with the particular species considered. However, several preliminary conclusions can be drawn. First, for some species with lower S_{crit} values that are low but nonzero (e.g., C- and D-tramps reaching islands with 10 or 15 species but not reaching poorer islands), the S_{crit} value seems to be determined by minimum territory or minimum area requirements. In this case area is the significant variable. Second, we shall see that some species are excluded by certain combinations of other species; and that large land-bridge islands slowly relaxing towards equilibrium, and with S values presently greatly in excess of equilibrium values for their area, have compositions of some species guilds appropriate to their S and inappropriate to their area. In both of these cases S rather than area may be the more directly relevant variable. Third, some insight can be gained by making use of the fact that the Bismarck island that is third in area (Manus) is only twelfth in S, because it lies some 200 miles from the four other large islands (New Britain, New Ireland, New Hanover, and Umboi). No species is confined to and shared by the three largest Bismarck islands (New Britain, New Ireland, Manus), but 31 species have distributions confined to and shared by islands some of which are much smaller than Manus but all of which are more species-rich than Manus. Thus, the distribution of these species does not depend on large area alone; either they are stopped by the distance to Manus, or else they require high-S communities, whatever the factors that produced the high S. Finally, the relative proportions of high-S species, A-, B-, C-, and D-tramps, and supertramps on Manus are *fairly* similar to the proportions on six islands with similar S values but progressively decreasing area (down to 1% of the area of Manus) and increasing proximity to the large Bismarck source islands. More detailed examination indicates that if one maintains S constant by increasing island area with increasing distance from the largest and richest island (New Britain), the proportion of high-S species and A-tramps is higher on large distant islands (like Manus) and on small close islands than on medium-size islands of intermediate distance. This suggests again that the categories of high-S species and A-tramps are heterogeneous and consist partly of species limited by dispersal, partly of species limited by other factors.

Relation Between Incidence Functions and Habitat Requirements

To what extent are restricted species distributions due to restricted availability of required habitat? Habitat diversity is likely to correlate with island size (hence with bird species number). Three habitats (lowland forest, lowland aerial, and seacoast) exist on every Bismarck island, but five habitats (fresh-water lakes or streams, mangrove, mountains, open country or grassland, and marshes or swamps) may be entirely absent from small islands and present only in limited amounts on large islands. Species confined to these five scarce habitats must have zero incidence on small islands devoid of the habitat, and low incidence on islands with small areas of the habitat. For example, 13 species are confined to high elevations in the Bismarcks. Only five Bismarck islands exceed 3500 feet in elevation; one of these is a small island (18 square miles) and another was recently defaunated by a volcanic explosion. Naturally, the 13 montane species are restricted to these five islands, and 12 of the species are restricted to the three larger or nondefaunated mountainous islands. Similar examples can be cited for water birds, grassland birds, marsh birds, and mangrove birds.

Table 2 summarizes the habitat requirements of the species in each of the six incidence categories. All of the supertramps, and almost all of the D- and C-tramps, can live in one of the three ubiquitous habitats that exist on any island. However, half of the high-S species, and one-third of the A- and B-tramps, are confined to the five scarce habitats that occur only on certain islands. This indicates that up to 43% of the species with restricted island distributions (high-S species, A- and B-tramps) could possibly be restricted by habitat availability, whereas 57% are not.

Table 3 summarizes the endemism level of species in each incidence category according to habitat requirement. For this purpose we assemble habitats into three groups: ubiquitous (lowland forest, low-

Table 2. Species habitat preferences, by distributional category

Habitats	Species					
	High-*S*	A-tramp	B-tramp	C-tramp	D-tramp	Supertramp
Ubiquitous habitats						
Lowland forest	25	16	12	13	8	13
Sea coast	0	0	0	2	3	0
Lowland aerial	0	1	0	3	1	0
Subtotal	25	17	12	18	12	13
Scarce habitats						
Fresh water	5	1	1	0	0	0
Mangrove	2	0	0	0	0	0
Montane	12	1	0	0	0	0
Open country	5	6	4	1	2	0
Marsh, swamp	3	1	0	0	0	0
Subtotal	27	9	5	1	2	0

For each Bismarck distributional category, the table lists how many species in each category live in each of the three "ubiquitous" habitats that occur on every island, and how many species are confined to each of the five "scarce" habitats that occur only on certain larger islands. A species is assigned to the category "lowland forest" if it lives in lowland forest on some islands; some tramps that live in lowland forest on species-poor islands are competitively excluded from forest and restricted to forest edge on species-rich islands. Note that almost all lowland aerial and sea-coast species are widespread (C- or D-tramps), that most species of scarce habitats have restricted distributions (high-*S* species, A- or B-tramps), that almost all montane birds have very restricted distributions (high-*S* species), and that almost all widespread species (most C- and D-tramps, all supertramps) are in ubiquitous habitats.

land aerial, sea-coast); montane; and scarce lowland (fresh-water, mangrove, grassland, and marsh). The scarce habitats have been separated into montane and lowland, because these two groups behave very differently as regards endemism. Table 3 and the detailed distributions on which it is based yield the following conclusions:

1. Endemic species are entirely (10 out of 10), and endemic semispecies are largely (24 out of 28), confined to lowland forest or mountains. Only 5 of these 38 endemics occur on more than 9 Bismarck islands, the remaining 33 endemics being high-*S* species or A-tramps.

2. Conversely, 12 out of 13 montane species have differentiated at least at the subspecies level, and 7 out of 13 have differentiated further. All montane species

are high-*S* species except for one A-tramp.

3. Most of the high-*S* species and A-tramps that are confined to scarce lowland habitats (19 out of 23) have failed to differentiate beyond the subspecies level. In contrast, of the high-*S* species and A-tramps in lowland forest, half (22 out of 42) have differentiated beyond the subspecies level.

4. Not a single one of the 9 B-, C-, and D-tramps confined to scarce lowland habitats has differentiated beyond the subspecies level. In addition, only a few (3 out of 13) of the supertramps, and almost none (just 2 out of 41) of the B-, C-, or D-tramps, in widespread lowland habitats have passed the subspecies level.

In most cases, low endemism can probably be interpreted to mean high dispersal rates between Bismarck islands, and be-

Table 3. Levels of endemism, by habitat preference and distributional category

Category	Endemism level			
	None	Subspecies	Semispecies	Species
High-S	1/1/7	5/4/5	12/6/3	7/1/0
A-tramp	4/0/4	10/1/3	2/0/1	1/0/0
B-tramp	2/0/2	9/0/3	1/0/0	0/0/0
C-tramp	8/0/0	9/0/1	1/0/0	0/0/0
D-tramp	6/0/3	5/0/0	0/0/0	0/0/0
Supertramp	2/0/0	8/0/0	2/0/0	1/0/0

In each entry, the first, second, and third numbers give the number of species, respectively, in ubiquitous habitats (lowland forest, sea-coast, lowland aerial), confined to mountains, or confined to scarce habitats other than montane habitats (fresh water, mangrove, open country, marshes and swamps). For each of these three habitat categories, the table gives how many Bismarck species in each distributional category are endemic at the indicated level. See text for discussion.

tween the Bismarcks and neighboring archipelagoes. This criterion would suggest the following conclusions:

1. Most species confined to scarce lowland habitats have high dispersal rates, whatever the form of their incidence function. It is adaptive for species in these habitats to have high dispersal rates so as to make up for the high extinction rates of their populations, associated with the small areas of the habitat patches and instability of the patches.

2. All lowland aerial and sea-coast species in the Bismarcks without exception have high dispersal rates. (This is not so strictly true, however, for the New Guinea region, where there are several swifts that have failed to colonize a single island.)

3. In lowland forest, almost all of the widespread species, and half of the high-S species and A-tramps, have high dispersal rates. The remaining half of the high-S species and A-tramps have low dispersal rates.

4. Almost all montane species have low dispersal rates.

Although conclusions about dispersal based on degree of endemism are not unassailable, these conclusions are confirmed by the direct observations of dispersal to be discussed in the section "Relation between incidence functions and dispersal." These relations between habitat preference, degree of endemism, and geographical range for Bismarck birds share some similarities with the corresponding relations for Melanesian ants, as formulated by Wilson (1961) and described as the "taxon cycle." Colonization of the Bismarcks by birds has originated almost exclusively from New Guinea. Evidently most of the colonists disperse initially into lowland forest and nonforest habitats, and many of them initially retain the high interisland dispersal rates that enabled them to reach the Bismarcks in the first place. With time, many of the colonists lose their dispersal ability, become extinct on most small or species-poor islands, adapt to forest or shift entirely into forest, and sometimes shift into montane forest. One of the differences between birds and ants on New Guinea is that most colonizing ants occupy lowland nonforest habitats and enter forest only on reaching a species-poor island, whereas many colonizing birds already occupy lowland forest. These relations are discussed in detail by Mayr and Diamond (1975).

For the present, we note that high-S species and A-tramps are actually a mixture of two types of species: birds whose absence from small islands is due to specialized habitat requirements; and birds that live in ubiquitous habitats and whose absence from small islands must be interpreted in other ways.

Relation between Incidence Functions and Island Area

Since the sum of all incidence functions equals total species number (eq. 3), and since area contributes most of the variation in species number, the interpretations of incidence functions and of the species/area curve must be closely related. There is an extensive literature on the interpretation of the species/area relation (Preston, 1962; MacArthur and Wilson, 1967; May, Chapter 4, Section 5). Despite this literature it is still neither simple nor generally agreed how to answer the simple question, "Why do smaller islands have fewer species?" Let us examine the ways in which area affects species distributions in the light of examples from the New Guinea region. We begin with trivial effects and proceed to more interesting ones.

Absence of suitable habitat.

Small islands may entirely lack certain habitats and hence species restricted to these habitats. For instance, ponds and permanent streams in the Bismarcks require a large watershed, and natural grassland requires mountains high enough to create a rainshadow; hence these habitats are lacking on most Bismarck islands smaller than 15 square miles. The lower S_{crit} values of species of these habitats, such as the duck *Anas superciliosa*, the montane river flycatcher *Monachella muelleriana*, the montane honeyeater *Melidectes whitemanensis*, the grass warbler *Cisticola exilis*, and the grass nightjar *Caprimulgus macrurus*, are due to this reason.

Island size less than minimum territory requirement.

Depending on the territory size of a single pair, there will be some minimum island size below which a single pair of even a highly vagile species cannot exist, even if the entire surface of the island is covered by suitable habitat. For example, the eagle *Haliaeetus leucogaster* forages over a territory of about 5–15 square miles. On islands smaller than this range, a resident pair of this eagle cannot maintain itself. The lower S_{crit} values of some other rapidly dispersing C- and D-tramps that maintain pair territories, such as the hawks *Haliastur indus* and *Accipiter novaehollandiae*, the herons *Nycticorax caledonicus* and *Ixobrychus flavicollis*, and the dollarbird *Eurystomus orientalis*, may also correspond to an island area comparable to the territory of one pair.

Seasonal or patchy food supply.

In the great majority of temperate-zone bird species, a single pair of birds defends an exclusive breeding territory against conspecific birds. This is possible because all resources that the pair needs are available within an area small enough to be feasibly defended. For many tropical species, however, the required resources are patchily distributed both in time and in space. Karr (1971) has shown that many such species in Panama lowland forest have large territories held jointly by a flock of individuals, or else large, extensively overlapping territories each held by a pair, rather than a smaller, exclusive and self-sufficient territory. All the individuals sharing these large territories may depend upon food in one part of the territory in one month, in another part in another month.

To an observer in the southwest Pacific, it becomes obvious that patchy and seasonal availability of food supply is important especially for fruit-eaters, seed-eaters, and species dependent on flowering trees. For example, a guild of pigeon species specializes on tall trees bearing soft fruits, whereas a guild of lorikeets and honeyeaters specializes on nectar and insects obtained from flowering trees. When a particular tree transiently comes into fruit or flower, dozens of birds congregate to feed. Yet suitable feeding trees may be spaced a mile or more apart, even in an area where fruiting or flowering is occurring (Terborgh and Diamond, 1970). Correlated with the cycle of wet and dry seasons, the occurrence of fruiting and flowering shifts from locality to locality throughout the year, and the associated bird guilds undertake local seasonal movements (Diamond, 1972a). These seasonal movements have been documented

not only for fruit-eaters (*Ptilinopus* and *Ducula* pigeons, *Aplonis* starlings, the cuckoo *Eudynamis scolopacea*, etc.) and flower-feeders (*Myzomela* honeyeaters, parrots of numerous genera), but also for seed-eaters such as *Lonchura* finches, and some insectivores such as *Collocalia* swiftlets, the tree swift *Hemiprocne mystacea*, the swallow *Hirundo tahitica*, and the flycatcher *Rhipidura leucophrys*.

The result of these seasonal shifts in food supply is that an island that seems to an ornithologist large and covered with fruits and flowers still may not be able to support the expected bird guilds, unless a food supply is available within the island's confines for all twelve months of the year. Availability of a year-round food supply may require sufficiently varied topography to cause local intra-island variation in fruiting or flowering season (due to intra-island variation in slope, exposure, rainfall, and altitude). Thus, either an island does provide a year-round food supply and supports an entire flock of a species, or else it supports no individuals of the species at all. Even on an island as large as Sakar (17 square miles), the small resident flock of the flower-feeding parrots *Vini placentis, Trichoglossus haematodus,* and *Lorius hypoinochrous* could be seen flying back and forth among a few flowering trees at opposite ends of the island. The requirement that an island be large enough to provide a year-round food supply is probably the reason for the lower S_{crit} values in some incidence functions of species dependent on shifting or patchy food, such as the D-tramp parrot *Trichoglossus haematodus* or the C-tramp parrot *Vini placentis*. The incidence function of the former parrot is very steep, falling from $J = 1.0$ at $S = 15$ to $J = 0$ at $S = 9$.

In the cases discussed above, resource seasonality or patchiness is due to seasonality or patchiness in resource production itself. For some other tropical species, resource *availability* may be patchy despite widespread resource *production,* because most of the resources are preferentially consumed by a diverse constellation of competitors. Suppose that a guild consists of many species with broadly overlapping resource-utilization functions, but that each species has a competitive advantage in a different region of the resource or habitat spectrum. Then a given species will find resources available to it only at those scattered points in time or space where it has a competitive advantage or where some of its competitors are absent (see Figure 44). This may be the reason why, among New Guinea birds as among the Panama birds Karr studied, individuals of some insectivorous species as well as frugivorous species associate in flocks or have overlapping territories or are patchily distributed.

Thus, seasonal or patchy food supplies may directly reflect patterns of resource production, but may also be created by the activities of a guild of overlapping consumers. In either case, a species can persist on an island only if the area is sufficient to provide a year-round food supply.

Population fluctuations.

We have discussed the three preceding effects of area as all-or-nothing effects—as if an island either were big enough to hold a given species, or else not big enough, with no intermediate states. Had this all-or-nothing interpretation been true, some incidence functions might have been step functions, which is never the case. In fact, of course, an important ingredient of the MacArthur-Wilson equilibrium theory, developed further by Leigh (Chapter 2), is that a population of any given size runs a finite risk of extinction due to population fluctuations, and that this risk increases with decreasing population size. Let us consider four examples illustrating this risk:

(a) Table 4 summarizes bird censuses on five small islets, ranging in area from 0.8 to 12.8 acres, and supporting from 3 to 7 species. Of the 23 populations on these islands, 12 consisted only of a single individual, 5 of two individuals each, 3 of three individuals each, 2 of five individuals each, and 1 of 18 individuals. Thus, most of these populations were so small that the death of a single individual would either eliminate the population or else remove its potential for reproduction. These populations must go extinct so often that only the most rapidly recolonizing species can exist on such small islets. In practice, 20 of the 23 populations were either supertramps or D-tramps. Four of the 23 populations (one supertramp and three D-tramps) belonged to species that certainly or possibly move from islet to islet in the course of their normal foraging, but the other 19 belonged to territorial species whose individuals probably spend their whole adult lives on the islet on which they settle. It

Table 4. Bird censuses on small islets

Islet	Area (acres)	Number of species	Population size (number of individuals)				
			18	5	3	2	1
Midi	6.3	7	—	—	—	3	4
Hein	12.8	5	—	—	2	1	2
Matenai	1.6	4	—	1	—	1	2
Araltamu	0.8	4	—	1	—	—	3
Little Pig	2.9	3	1	—	1	—	1

For each coral islet column 2 gives the area; column 3, the number of resident Bismarck species found on the island; and columns 4, 5, 6, 7, and 8, the number of species whose island population consisted of 18, 5, 3, 2, and 1 individuals, respectively. The conclusion is that populations on small islets consist of few individuals and must be very unstable.

is not the case that these were probably vagrant individuals that were just flying by and had alighted; instead, these were the characteristic resident species of small islets.

(b) Crown, an island of 5 square miles, supports 30 species. Of these, two were represented by a single individual, one by a pair, two by three individuals, four by about five individuals, two by about five pairs, and one by about ten pairs. Sakar, an island of 15 square miles, supports 36 species. Of these, two were represented by a single individual, two by a pair, one by three individuals, seven by about five individuals, and four by about five pairs. For many of the 28 marginal populations on these two islands, there were obvious reasons why the population was so small: four were species that could utilize the whole island area but required large pair territories (e.g., the eagle *Haliaeetus leucogaster,* the forest heron *Nycticorax caledonicus*); eight were species that could utilize only small fractions of the island area as habitat (e.g., the coastal heron

Egretta sacra, the grassland bee-eater *Merops philippinus*); and five were flower-feeding parrots that maintain a large flock territory. Two or three of the species on each island wander routinely among islands as adults, but the remaining species are ones that probably spend their adult lives on one island. Thus, even on islands as large as 5 or 15 square miles, about 40% of the populations are small enough to face a high risk of extinction. This point is worth mentioning in view of the erroneous assumption sometimes made that island populations are generally abundant and therefore immune to risk of extinction arising from population fluctuations.

(c) When rising sea-level severed late-Pleistocene land bridges, populations of New Guinea lowland species stranded on the resulting land-bridge islands began selectively to go extinct (Diamond, 1972b). Today there are 32 species that are widespread in the New Guinea lowlands but absent from every single land-bridge island, including islands as large as 3000 square miles. For most of these species the land-bridge islands must formerly have supported populations, which must therefore have disappeared after the land bridge was severed. In most cases it is clear that these initial populations must have been very small and extinction-prone. Three of the species live in forest but require very large territories (e.g., the New Guinea Harpy Eagle *Harpyopsis novaeguineae*); 13 species live in forest and do not have very large territories but live at low densities (e.g., the kingfisher *Clytoceyx rex,* the bird of paradise *Drepanornis bruijnii*); 13 species are nonforest birds with specialized habitat requirements (e.g., the swamp rail *Megacrex inepta,* the grass warbler *Malurus alboscapulatus*); and nine species, including some of the above ones, depend on seasonal and patchy supplies of fruit, flowers, and seeds. Only two widespread and common forest species disappeared from all land-bridge islands without obvious reason. Wilson and Willis (Chapter 18, Table 1) document similar selective extinctions of bird populations on Barro Colorado Island, and Brown (1971) did so for mammal populations on mountaintops of the Great Basin.

(d) Equilibrium turnover studies have been carried out on three Bismarck or New Guinea islands: Karkar, Vuatom, and Long. Of the 13 populations that disappeared, nine were small because of being confined to specialized habitats or because of a large territory requirement. The remaining four populations were fruit-eaters or flower-feeders dependent on a patchy, seasonal food supply. Turn-

over studies on the California Channel Islands have similarly shown that populations that for one reason or another (scarce habitat, large territory, recency of colonization, small island, presence of a close competitor) consist of few individuals are the ones most likely to go extinct (Diamond, 1969, 1971; Jones and Diamond, 1975).

As a final general point about area-dependent extinction rates due to fluctuations in small populations, the significance of rare events and long-term fluctuations should not be underestimated. For normally large populations, the risk of extinction due to stochastic variation in births and deaths will be negligible. Much more significant is the risk of extinction when the population has crashed temporarily to a low level following a rare event (i.e., the problem of resilience as opposed to steadiness, in the sense of Leigh, Chapter 2), such as the drought that occurred in the Madang area in 1972; forest fires, such as the one that swept the Madang lowlands 50 years ago; temporary local superabundance of other bird species, such as is associated with the seeding of some bamboo species once or twice a century; invasions of other species, such as the invasion of New Zealand around 1856 by *Zosterops lateralis* (now one of the most abundant New Zealand birds); cyclones, such as the one that blew down much of the forest on Ngela and Santa Ysabel islands in 1972; volcanic explosions and the resulting ash fall-out, such as occur often along the south edge of the Bismarck Sea; and tidal waves, such as one that inundated the islands of Dampier

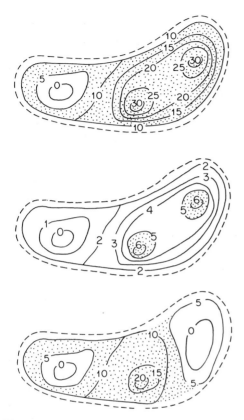

land, either because of an unproductive year, or because of an abundant competitor that utilizes the given resources in the same proportion. The species now can survive only on the two hot spots. Below: as in top sketch except that the island has now been invaded by a competitor with an overlapping but different resource utilization, such that the competitor cannot survive at the west end but becomes increasingly common towards the east end. The original species still occupies most of the island, but only one of the two spots is left. This illustrates that hot spots depend not only on local microhabitat variation in relation to a species' requirements, but also on microhabitat variation in relation to the requirements of competing species.

Strait in 1888. These events may be rare on the time-scale of an ornithologist's scientific career, but common on the time-scale of population lifetimes for species with low dispersal rates. If the recolonization frequency is comparable to or lower than the frequency of disaster-induced population crashes, the incidence of a species on otherwise suitable islands must be significantly less than 1.

"Hot spots."

Especially on large islands one must take account of the fact that even within suitable habitat the area inhabited by an established species, or else the area potentially available to an invading species, is not all equally suitable. The rate of production of those resources that can be utilized by a given species varies locally, both as a function of the physical environment and as a function of the distributions of competing species in the same guild. If one connects points with equal resource levels by contours analogous to isotherms, areas of especially high utilizable resource production stand out as "hot spots." The

Figure 16 Hypothetical example of the importance of "hot spots." The example refers to a forest bird species on a forested island (coast line indicated by dashed outline), on which the rate of production of resources utilizable by the species (R) varies locally with exposure, slope, and soil type. Contours connect points with equal values of R, in arbitrary units. The species can survive only in areas where $R > 5$, distinguished by stippling. Above: situation in the absence of a competitor and during a productive year. The species occupies most of the island except the unproductive "cold spot" near the west end. An especially productive area in the center and another at the east end ($R > 25$) function as "hot spots," where steady-state population density is highest, and where the species could increase most rapidly if it were initially colonizing the island or reexpanding after a population crash. Middle: same, except that R is uniformly reduced by 80% over the whole is-

presence of competing species, or a year of low production, may submerge resource levels over most of an island below minimum support levels, leaving the hot spots as islands of supraminimal resource production and hence as distributional islands (Figure 16). In general, the number or total extent of hot spots will increase with island area. For many species this may be the principal reason why, on large islands, rates of successful immigration continue to increase with increasing island area. It may also, in conjunction with population fluctuations, be for many species the main reason why extinction rates continue to decrease with increasing island area on large islands.

Consider a newly arrived species on an island sufficiently large that, if the species could equilibrate over the island, its population would be large enough to be virtually extinction-proof. Whether the species actually succeeds in reaching this saturating population level *K* depends on whether it escapes extinction in the initial phase of very small numbers. In this phase, the immigrant's chance of success depends upon the presence of hot spots where it can quickly start generating a population surplus. An immigrant that can adequately search an island would be expected to select the hot spots for its first settlement. In fact, even for organisms of high dispersal ability such as birds, recent immigrant populations on a large island are often very local. For example, in 1933 the sole fruit pigeon of genus *Ptilinopus* on Long Island was *P. solomonensis*. At some time between 1933 and 1972, the related *P. insolitus* invaded. In 1972, *P. insolitus* was well established at 500–1500

feet on the northeast ridge of Mount Reumur but was encountered nowhere else on Long. Since *P. insolitus* and *P. solomonensis* coexist on numerous islands smaller than Long, the chances of eventual success for an invasion by *P. insolitus* must be rated a priori as excellent. By comparison with other islands, *P. insolitus* may finally come to share all of Long from sea level to about 2000 feet with *P. solomonensis*. Thus, the northeast ridge of Mount Reumur may constitute a hot spot that permitted *P. insolitus* quickly to pass through the early critical phase.

Correspondingly, consider a species in risk of extinction on a large island. Hot spots are the most likely place for the species to make its last stand, or to be saved from extinction at the last moment. If a year of low production drastically reduces the population and the species contracts back to the hot spots, it may then quickly escape danger by producing population surpluses from the hot spots in better years and reexpanding over the island. If the species does finally go extinct, hot spots may have permitted it to escape the inevitable end for a long time. The fact that the larger land-bridge islands around New Guinea still have double their equilibrium species number, more than 10,000 years after the land bridges were severed, emphasizes for what long times many species can escape their inevitable doom on a large island. The distributions of numerous patchily distributed bird species on New Guinea itself are most plausibly interpreted in terms of gradual extinction and contraction onto hot spots. As illustrated in Figures 35–38 (see also Diamond, 1972a), many New

Guinea bird species that belong to distinct monotypic genera without close relatives (e.g., the bowerbird *Archboldia papuensis,* the cuckoo-shrike *Campochaera sloetii*), or else to large genera with many ecologically similar species (e.g., the honeyeater *Ptiloprora plumbea,* the berrypecker *Melanocharis arfakiana*), have fragmented, relict-like distributions with respect to the available habitat. Many of these species are probably slowly going extinct, either because they are the last survivors of unsuccessful evolutionary lines (monotypic genera), or because they are being outcompeted by ecologically similar congeners. For example, the Umboi population of the fruit pigeon *Ptilinopus solomonensis,* which was widespread in 1913 but is now probably doomed to extinction as a result of successful simultaneous invasions by two relatives, has contracted back to a small area at the base of Mount Birik. The flycatcher *Microeca leucophaea,* which was formerly widespread in the Port Moresby savanna of southeast New Guinea, has contracted since 1945 down to two local populations. Both of these cases probably exemplify species that are making their last stands on hot spots. Such species represent an appreciable fraction of the total species number on large islands like New Guinea.

Relation between Incidence Functions and Dispersal

Absence of a species from an island could be due to interesting competitive interactions, but it could also simply mean that no individuals of the species ever immigrated. Or, there might have been immigrants, but in too small numbers and at too rare intervals to yield much probability of success. Thus, before we can examine incidence functions for evidence of community integration and resistance to invasion, we need to know something about the relative rates at which species disperse. In the section "Relation between incidence functions and habitat requirements" we drew tentative conclusions based on degree of endemism. In this section we shall consider five, more direct, lines of evidence about dispersal ability.

Note that if a bird species is known never to cross a 10-mile water gap, we should not assume that it is physiologically and mechanically incapable of a 10-mile flight but that, much more likely, the species has an insuperable psychological barrier to crossing water. This reluctance of some strong-flying bird species, especially tropical species, to cross water gaps is insufficiently appreciated. For example, of the 325 bird species of the New Guinea lowlands, only 4 are flightless, yet only 191 have crossed water gaps more than 5 miles wide, as shown by their having been recorded from one or more islands lacking a recent land bridge. Figure 17 illustrates the distribution of the flycatcher *Monarcha telescophthalmus,* which occurs on every large island that had a Pleistocene land bridge to New Guinea but on no island that lacked a land bridge. The list of species absent from oceanic islands includes some soaring hawks, swifts, and itinerant parrots that fly 100 miles a day over the New Guinea mainland in the course of their normal foraging, and that could reach

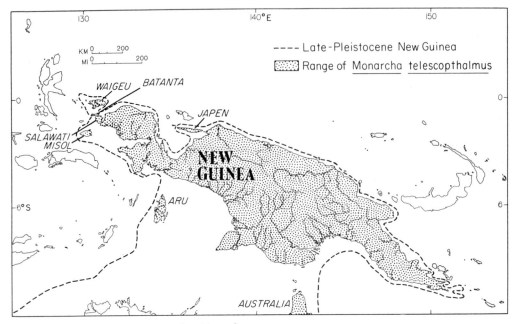

Figure 17 Distribution of the flycatcher *Monarcha telescophthalmus* (shaded islands). The dashed line is the contour of 200-m ocean depth, which is very close to the 100-m contour. The whole area within the dashed line formed a single land mass at times of lower sea level during the Pleistocene. Thus, present-day islands within this dashed line are "land-bridge islands." Note that this flycatcher now occurs on every large (>175 square miles) land-bridge island, named on the map, but on no oceanic island nor on any small land-bridge island, not even those within a few hundred yards of New Guinea. The cockatoo *Probosciger aterrimus* has the same distribution. Both species are capable of normal flight, and the cockatoo is a strong flier.

some large, close, readily visible offshore islands in a fraction of an hour, but never do. Similarly, hundreds of Asian bird species, including numerous whole families, have completely failed to cross Wallace's line separating the land-bridge islands of the Sunda Shelf from the deep-water islands of Indonesia. An even larger num-

ber of neotropical bird species, including the whole family Formicariidae with 221 species, have failed to reach any oceanic island off Central America or South America (MacArthur et al., 1972). Wilson and Willis (Chapter 18) note that many forest species have not crossed the water gap 500 meters wide separating Barro

Colorado Island from the Panama mainland. Even in the temperate zones, extensive records from the Channel Islands and Farallon Islands off California show that some strong overland fliers, such as the Common Crow *Corvus brachyrhynchos* and Red-shouldered Hawk *Buteo lineatus,* rarely or never cross water to reach California's offshore islands (Jones and Diamond, 1975). Thus, it would be a serious mistake to assume that all birds that can fly are potentially good colonists.

Let us now examine the direct evidence for dispersal abilities of species.

Recolonization of defaunated islands.

Two centuries ago, a volcanic explosion defaunated Long Island in the southwest Bismarcks, covering the island with nuées ardentes and ash deposits up to 100 feet thick or more. Long is 76 miles from the richest Bismarck island (New Britain) and 37 miles from the third-richest island (Umboi). By 1972, 54 bird species had recolonized Long, as detected by surveys in 1933 and 1972 (Diamond, 1974). These species are 10 of the 13 Bismarck supertramps, all 14 Bismarck D-tramps, 11 of the 19 C-tramps, 9 of the 17 B-tramps, 6 of the 26 A-tramps, 1 of the 52 high-*S* species, and 3 unclassified species. Of the A-tramps and high-*S* species, all but one (an endemic Bismarck subspecies) are species that have not differentiated even subspecifically in the Bismarcks, and all but one are species of scarce and unstable habitats (marshes, fresh water, grassland). Thus, all the colonists of Long are species that were suspected of dispersing rapidly on the basis of wide distribution and/or low en-

demism. Although species number in the lowlands of Long is already close to the equilibrium value predicted for Long's area, the number of Long's montane avifauna is far below equilibrium, in agreement with the postulated poor dispersal ability of montane birds. The sole obligate montane species that has recolonized Long is the sole montane A-tramp of the Bismarcks, the ubiquitous montane finch *Erythrura trichroa,* which opportunistically follows the seeding of bamboo.

Ritter Island in the southwest Bismarcks was defaunated by a volcanic explosion in 1888. In 1972, seven species were found on Ritter, four as residents and three as vagrants. These were three supertramps, one D-tramp, two C-tramps, and one nonendemic A-tramp (the Peregrine Falcon, which had also recolonized Long).

Seven small coral islets near Ritter were defaunated by the tidal wave caused by the Ritter explosion. In 1972 these islets supported a total of 45 populations: 17 supertramps, 24 D-tramps, 2 C-tramps, and 2 B-tramps.

Thus, the evidence from defaunated islands indicates that the highest dispersal rates are among supertramps and D-tramps, followed by C- and B-tramps and nonendemic A-tramps of scarce habitats. Dispersal of endemic forest A-tramps, high-*S* species, and montane species has been negligible.

Dispersal to Vuatom.

Vuatom Island is 4 miles offshore from New Britain, the largest and most species-rich Bismarck island. The area of Vuatom is 5 square miles, and its elevation is somewhat over 1000 feet. It supports forest, grassland, and permanent streams but lacks marshes and high mountains. The observations of Father Otto Meyer (1906, 1930), resident missionary on Vuatom for several decades, provide the best long-term record of vagrants in the Bismarck Archipelago.

In 1906, Father Meyer reported about 47 breeding species on Vuatom. By 1930, the populations of 5 of these breeding species had disappeared, 4 new breeding species had become established, and 24 other species that breed on New Britain and adjacent islands (plus 15 wintering species from Asia, Australia, and New Zealand) had appeared transiently but had not established breeding populations.

Table 5 summarizes some facts about the successful and unsuccessful colonists.

By 1930 every Bismarck D-tramp, all but two each of the C-tramps and B-tramps, half of the A-tramps, and only 13% of the high-S species had been recorded from Vuatom. A similar sequence is obtained if one calculates the 1906–1930 arrivals (whether successful or unsuccessful) as a percentage of the available species pool not already resident on Vuatom in 1906. The low percentage of super-tramp arrivals is initially surprising, but in fact the five species classified as super-tramps that were not recorded have no populations within 150–250 miles of Vuatom.

Table 5.   Colonization of Vuatom Island, 1906–1930

	Category					
	High-S	A-tramp	B-tramp	C-tramp	D-tramp	Supertramp
1. Total species for category	52	26	17	19	14	13
2. Resident species, 1906	1	11	11	11	8	5
3. Total colonists, 1906–1930	6	3	4	6	6	3
4. Successful colonists	0	0	1	0	2	1
5. Unsuccessful colonists	6	3	3	6	4	2
5a. On smaller island?	0	0	1	4	4	2
5b. On more species-poor island?	0	1	3	6	4	2
6. Available pool, 1906	51	15	6	8	6	8
7. 100 × total colonists/ available pool	12	20	67	75	100	37
8. 100 × total recorded/total in category	13	54	88	89	100	62

For each of the six Bismarck distributional categories, row 1 gives the total number of species (from Table 1, column 2); row 2, the number breeding on Vuatom in 1906; row 3, the number of species that were not breeding on Vuatom in 1906 but at least one individual of which was observed on Vuatom any time between 1906 and 1930; row 4, the number of colonists listed in row 3 that had breeding populations on Vuatom in 1930; row 5, the number of colonists listed in row 3 that did not have breeding populations in 1930; row 5a, the number of unsuccessful colonists in row 5 that breed on at least one island smaller than Vuatom; row 5b, the number of unsuccessful colonists in row 5 that breed on at least one island with fewer species than Vuatom; row 6, the species pool available for colonization in 1906 (row 1 minus row 2); row 7, the percentage of this pool that did colonize between 1906 and 1930 (row 3 divided by row 6); and row 8, the percentage of all the species in the category that had been recorded from Vuatom by 1930 (row 2 plus 3, divided by row 1). The data are extracted from Meyer (1906, 1930).

Of the 28 arrivals (successful or unsuccessful), 14 have not differentiated racially in the Bismarcks, 8 have formed endemic subspecies, and only 6 (including 5 of the 6 high-S species) are endemic semispecies.

Of the 24 unsuccessful colonists, the preferred habitat type of all but one (a montane high-S species) was available on Vuatom. Table 5 summarizes for each of the unsuccessful colonists whether it has a resident population on any Bismarck island smaller than Vuatom, or on any Bismarck island with a smaller resident species total than Vuatom. All of the unsuccessful supertramps, D-, C-, and B-tramps occur on islands more species-poor than Vuatom, and most occur on smaller islands. Thus, there is nothing intrinsic in the size, species richness, or habitat spectrum of Vuatom that doomed these colonists to failure. Their failure is presumed to be due either to chance (as calculated by MacArthur and Wilson, 1967, Chapter 4, from the numbers and from the expectations of births and deaths of the colonists), or else to exclusion by competitors or constellations of competitors. A role of competition is indicated by the fact that 8 of these 15 plausible but unsuccessful colonists found congeners or close relatives already established on Vuatom. In contrast, of the 9 unsuccessful high-S species and A-tramps, none has a resident population on an island smaller than Vuatom, and only one on a more species-poor island. Thus, incidence functions would already have forewarned us that these colonists were doomed to failure. A role of competition in these cases as well is indicated by the fact that 6 of

these 9 implausible colonists found congeners or close relatives already established on Vuatom (and would have found the same or similar competitors on almost any other Bismarck island of Vuatom's size and poverty). For instance, the sole vagrant individual of the high-S kingfisher *Ceyx websteri* that Father Meyer ever saw on Vuatom was being chased by the C-tramp kingfisher *Alcedo atthis*, which is established on Vuatom.

As striking as what Father Meyer saw is what he failed to see. Of the 127 species present on New Britain, only 4 miles over water from Vuatom, 55 were never observed by Father Meyer on Vuatom during his several decades there. These include 14 of New Britain's 15 montane species, 8 of the 9 fresh-water and marsh species, 32 of the 74 forest species, and 5 of the 18 species of grassland and open country, but all 10 of New Britain's aerial and sea-coast species were recorded. The lack of vagrants of montane and water species could be associated with the absence of these habitats on Vuatom. However, the 32 forest species (24 of them high-S species or A-tramps) had no such excuse for failing to appear as vagrants, and the conclusion is inescapable that attempted colonization even of suitable habitats across 4-mile water gaps must be a rare event for these species.

The patterns of relative dispersal ability derived from Long and from Vuatom are in general agreement, in indicating the high dispersal of supertramps and D-, C-, and B-tramps contrasted with the low dispersal of A-tramps and high-S species. However, there is a quantitative difference, in that dispersal of high-S species to Vuatom was not quite so negligible as it was to Long (7 vs. 1 species, respectively). Part of the explanation may be that Vuatom is much closer to New Britain than is Long (4 vs. 76 miles), and overwater dispersal of high-S species may fall off very steeply with distance. The other reason may be that the Vuatom high-S species records are mainly of vagrants observed over the course of three decades, whereas the Long records were obtained during two surveys lasting one month each and consisted largely of established populations. High-S species may have difficulty not only in dispersing but also in establishing new populations, because of low reproductive potential.

Supertramp dispersal.

In the case of the 13 supertramp species, dispersal rates are sufficiently high that instances may be observed even during a short residence in the Bismarcks. As summarized in Table 6, Umboi has only one resident supertramp species, but in 25 days on Umboi I observed vagrants of six others. Two additional supertramps breed on islets 2 miles from Umboi, and two more on an island 13 miles distant, so that frequent vagrants of these species are also expected. New Britain has no resident supertramps, but vagrants of six species have been recorded, and two more breed on islets a few hundred yards from New Britain and must also produce vagrants. These dispersing vagrants often prove to be juvenile birds.

Witnessed instances of over-water dispersal.

On five occasions in 1972, while surveying the islands to the west of New Britain, I observed birds dispersing over water from island to island while I was in a boat or on the coast. On July 14 two C-tramp swallows *Hirundo tahitica* left the coast of Sakar and flew at a height of 100–200 feet towards Umboi, 10 miles away, until they were lost to binocular view. On July 15, while I was on a boat one mile south of Sakar, a flock of 12 *Collocalia* swiftlets (either the D-tramp *vanikorensis* or the C-tramp *esculenta*) flew past me 50 feet over the water from the direction of New Britain towards Umboi, which is about 20 miles from New Britain at this point. On July 18 several flocks consisting of a dozen birds each of the supertramp pigeon *Ducula pistrinaria* were seen flying at a height of several thousand feet for at least 8 miles along the north coast of Umboi, heading in the direction of Tolokiwa 13 miles away. On August 6 a flock of seven *Ducula pistrinaria* took off from the coast of Crown, were joined half-a-mile offshore by an eighth bird, gained an altitude of about 1000 feet, and headed north-north-west, where there was no land in sight (and no land actually to be found until Manus, 200 miles distant). Within the first few miles the flock occasionally veered left or right by up to 90° for a fraction of a minute but never turned back towards Crown, and eventually disappeared from binocular view at a distance of 5–10 miles, still heading away from Crown. Finally, on August 12, while

Table 6.   Dispersal and characteristics of supertramps

Species	Family	New Britain	Umboi	Food	Body weight (g)
Ptilinopus solomonesis	Columbidae	vagrant	resident	soft fruit	91
Ducula pistrinaria	Columbidae	vagrant	vagrant	soft fruit	470
Ducula pacifica	Columbidae	—	—	soft fruit	435
Ducula spilorrhoa	Columbidae	vagrant	vagrant	soft fruit	560
Macropygia mackinlayi	Columbidae	vagrant	vagrant	soft fruit	87
Caloenas nicobarica	Columbidae	vagrant	vagrant	hard fruit	635
Turdus poliocephalus	Turdinae	—	(13 miles)	insects, fruit	60
Monarcha cinerascens	Muscicapinae	(<1 mile)	vagrant	insects	26
Pachycephala melanura dahli	Pachycephalinae	vagrant	vagrant	insects	26
Aplonis feadensis	Sturnidae	—	—	fruit, insects	85
Myzomela pammelaena	Meliphagidae	(18 miles)	(3 miles)	nectar, insects	16
Myzomela sclateri	Meliphagidae	(<1 mile)	(13 miles)	insects, nectar	10
Zosterops griseotincta	Zosteropidae	—	(13 miles)	insects, fruit, nectar	14

Column 1 lists the 13 Bismarck supertramp species; column 2 gives the family; columns 3 and 4 indicate whether each species is resident on the species-rich islands of New Britain and Umboi, respectively; if not, whether it has been recorded as a vagrant; if not, whether it is resident on an island < 1, 3, 13, or 18 miles from New Britain or Umboi (hence likely to appear as a vagrant but to have been overlooked so far). A dash means that there are no records of vagrants and no resident population on an island within 18 miles. All of these supertramp species except *Ducula pacifica, Turdus,* and *Aplonis* occur on Long and Crown, and all except *Ducula pacifica, Aplonis,* and *Pachycephala* are on Tolokiwa. For each species, columns 5 and 6 give, respectively, the main components of the diet and the average body weight in grams.

in a boat halfway between Long and Tolokiwa (23 miles apart), I saw a flock of four of the A-tramp grass finch *Lonchura spectabilis,* flying 5–10 feet above the waves, heading straight from Long towards Tolokiwa in moderate wind, with rapid short wing-beats and an undulating, seemingly weak flight, progressing at about 14 miles/hour. I watched the flock from our boat for the next 10 minutes, at the end of which only three birds were left, still heading for Tolokiwa.

It is noteworthy that all five of these instances involved a flock of a normally gregarious species: if the target were reached, the chances of colonizing success would be higher for a flock arriving simultaneously than for separately dispers-ing individuals. It is also interesting that in four of the five instances the island towards which the colonists were heading was clearly visible at the height at which the colonists flew, and the colonists were heading straight for the target. Finally, unbeknownst to all five groups of colonists, a population of their species was already established on the target island (unless the swiftlets heading for Umboi were *Collocalia esculenta* rather than *C. vanikorensis*).

Rare dispersal events: gales and blooms.

The preceding five witnessed instances involve species that probably send out colonists frequently. However, the dispersal of other species depends on rare events, either gales or else a rare "bloom" of the species.

Between March 3 and 7, 1952, a gale blew pelicans *Pelecanus conspicillatus* from Queensland or New Guinea for several thousand miles over much of Melanesia (Bradley and Wolff, 1956). A flock of 20 reached Rennell; others reached Bougainville, Guadalcanal, San Cristobal, Santa Anna, Nissan, and many other islands in the Solomons, Feni in the Bismarcks, and various islands in the New Hebrides. Some of these colonists bred, and a population may still exist on Nissan. In June 1972 Cyclone Ida blew flocks of the parrot *Eos cardinalis* from the main chain of the Solomons 150 miles to the outlying atoll Ongtong Java, where the birds are now flourishing (Bayliss-Smith, 1973). The same cyclone brought a flock of the hornbill *Aceros plicatus* to Savo Island in the Solomons, and brought back the parrot *Geoffroyus simplex,* which had formerly been present on Savo but had disappeared. In the early 1960s a pair of the ibis *Threskiornis spinicollis,* moving north in Australia during a drought, were caught in strong winds and blown 1000 miles to New Ireland in the Bismarcks (Coates, 1973).

Occasionally an otherwise sedentary species experiences a "bloom" that disperses large numbers of individuals over a wide area. Father Meyer (1934) recorded blooms of the New Britain lowland forest pigeon *Ducula finschii* and montane forest pigeon *D. melanochroa* reaching Vuatom once in a period of three decades. He also recorded *D. melanochroa,* another montane forest pigeon *Gymnophaps albertisii,* and the lowland forest pigeon *Columba pallidiceps,* which are normally confined to the interior of New Britain, "blooming" at certain times on the coast. All of these pigeons are high-S species that occur on only a few Bismarck islands and probably colonize rarely. In the summer of 1875 a bloom of the Fijian Mountain Lorikeet *Vini amabilis* spread it widely over three Fijian islands (Mayr, 1945). In 1962, coincident with the seeding of mountain bamboo, two seed-eating birds, the A-tramp finch *Erythrura trichroa* and C-tramp pigeon *Gallicolumba beccarii,* became superabundant for a year in the Okapa Subdistrict of New Guinea. The last such bloom of these bird species and seeding of bamboo in this district had occurred several decades previously. In general, these blooms may reflect either colonists generated by a transient local superabundance of food, as in the last example, or else colonists expelled by a transient local failure of the normal food supply, as in the blooms of *Gymnophaps albertisii.* In either case, these blooms, like the rare dispersal events caused by gales, mean that even an ornithologist resident in an area for much of his lifetime may fail to observe directly the dispersal that is responsible for the long-term colonization potential of some species.

Relation Between Incidence Functions and Reproductive Rates

All other things being equal, good colonist species should have a high reproductive potential, so that propagules can fill up an island or recover from a population crash as quickly as possible. Otherwise, fluctuations in population size may wipe out the population at an early stage, when population size is still small. MacArthur and Wilson (1967, Chapter 4) showed that, for quick success in achieving carrying capacity, colonists should maximize the ratio $(\lambda - \mu)/\lambda = r/\lambda$, where λ and μ are the *per capita* birth rate and death rate, respectively. Crowell (1973) confirmed this prediction for rodents colonizing islands off the coast of Maine. One therefore expects this ratio to be maximal in the Bismarcks for supertramps, D- and C-tramps, and possibly for some other species of scarce lowland habitats. In the complete absence of knowledge of λ or μ for any Bismarck species, this expectation cannot be tested. However, we can at least examine whether colonizing ability is related to clutch size, number of broods per year, and length of breeding season, as tabulated for many Vuatom and New Britain species by Meyer (1930, 1933).

Most Bismarck species lay clutches of 1, 2, or 3, occasionally 4 or 5, eggs. No correlation is apparent between clutch size and type of incidence function, because a given clutch size is so characteristic of a given taxonomic group. For instance, clutch size is 1 in all except one Bismarck pigeon, 2 in myzomelid honeyeaters, 2 in *Monarcha* flycatchers, 3–4 in rails, and 3–5 in *Lonchura* finches, even though each group contains supertramps or tramps as well as high-S species. Of the seven supertramp species whose clutch size is known, four lay a single egg, while three lay 2

eggs. Clearly, superior colonists do not depend upon large clutches.

The Bismarcks lie near the equator, so that seasonal variation in daily hours of sunlight is negligible. However, each locality exhibits an idiosyncratic seasonal rainfall pattern. For example, the southern watershed of New Britain receives 80% of its rain in the months May–October, whereas the northern watershed receives 70% of its rain in the months December–April. Certain species concentrate their breeding activities in specific two-to-six-month periods of the seasonal cycle (e.g., most hawks in the dry season, the crow *Corvus orru* in the wet season, the kingfisher *Alcedo atthis* in the transition from wet season to dry season). Other species breed twice a year, each time at a specific stage in the rain cycle, generally at the transition from wet to dry season and again at the transition from dry to wet season. In the cases of still other species, breeding individuals can be found at any month of the year, and the same pair may rear several broods per year. For example, a pair of the D-tramp flycatcher *Rhipidura leucophrys* has been observed to raise six broods within six months (Mackay, 1970). The 22 Vuatom species known to breed throughout the year, or else to rear two or more broods per year, consist of 14 supertramps, D-tramps, or C-tramps; 3 A- or B-tramps of grassland or open country; and 5 A- or B-tramps of forest. The 16 Vuatom species known to have a single concentrated breeding season, and not known to rear multiple broods during this period, consist of 6 D- or C-tramps, 4 A- or B-tramps of grassland or open country,

and 6 A- or B-tramps of forest. Supertramps and D- and C-tramps are definitely good colonists, A- and B-tramps of
forest are definitely poorer colonists, and
A- and B-tramps of grassland and open
country may be either good colonists
whose distribution is limited by habitat
availability, or poorer colonists. All five
supertramps studied breed year-round,
and two of them are known to rear two
or more broods. Thus, the ratio of year-
round breeders and/or multiple-brooded
species to single-brooded seasonal breeders is higher among definitely good colonists ($^{14}/_6 = 2.3$) than among definitely
poor colonists ($^5/_6 = 0.8$).

In other words, many but not all good
colonists have higher-than-average per
capita birth rates λ. A perfect correlation
is not expected, since colonist success depends on the ratio $(\lambda - \mu)/\lambda$, not just on
λ, and it may be easier to maximize this
ratio by decreasing μ than by increasing
λ. For example, the three D-tramp large
hawks of the Bismarcks (the eagle *Haliaeetus leucogaster*, osprey *Pandion haliaetus*,
kite *Haliastur indus*) have low birth rates
(single-brooded seasonal breeders, one or
two eggs per clutch) but disperse over
water very readily and are probably
long-lived (low μ).

The Supertramp Strategy Versus the Overexploitation Ethic

Ornithologists who have observed
mainly on New Britain are accustomed to
thinking of the supertramps as habitat
specialists, confined to small coral islets off
the New Britain coast. A visit to Long, a
large (173 square miles) and high (4300
feet) island defaunated by an eruption two
centuries ago, shows how misleading this
impression is. Eight of the ten supertramp
species of Long are abundant in literally
every land habitat with low or high trees:
in strand forest, open savanna, coconut
groves, gardens, tall forest of the hill
slopes, a closed-canopy forest without
understory or middle-story foliage on the
crater rim, montane cloud forest, and
dense elfin forest or subalpine shrubbery
at the highest elevations; in bright sunlight, in forest shade, or in virtually permanent cloud cover; in the dryness of the
coast, where for four months there is no
rain and only a few fresh-water springs,
or in the perpetual moisture and mists of
the cloud forest; in the heat at sea level,
or in the cold of the summits. Similarly,
seven of the ten supertramps of Tolokiwa,
an extinct volcano near Long, are abundant from sea level to the cloud forest on
the summit at 4650 feet. On Karkar two
of the five supertramps occur from sea-
level up to cloud forest around 5000 feet,
and one occurs through subalpine shrubbery to the summit at 6000 feet, where it
is the sole bird species. Thus, if supertramps occur on an island at all, they are
extraordinarily catholic in their habitat
tolerance.

Supertramps are virtually absent as residents on the species-rich, large, central
islands of the Bismarcks (New Britain,
New Ireland, New Hanover, Umboi).
With one set of exceptions, the islands
that have more than four resident supertramp species are either small (less than
6 square miles), or isolated by a water gap

of at least 50 miles from the large central islands, or both. The set of exceptions is particularly striking in that it constitutes the most supertramp-rich islands of the Bismarcks: Long, Tolokiwa, and Crown, with ten species of supertramps each. Long is large and close (37 miles) to the species-rich central island Umboi, and Tolokiwa and Crown are the two islands nearest Long. The obvious reason for the presence of so many supertramps on Long is that, by virtue of their rapid dispersal and breeding, they were the first species to arrive and become reestablished after the recent volcanic holocaust. Tolokiwa and Crown, as the nearest neighbors of Long, and with respective areas only 10% and 3% of Long's, have been the most immediate and visible targets for birds dispersing from the supertramp breeding vat on Long and must have been inundated by supertramp colonists. Most of the other species of Tolokiwa, Crown, and Long are D- and C-tramps.

The strategy of the supertramps may be summarized as "Breed, disperse, tolerate anything, specialize in nothing." They represent the extreme of r-selected species (cf. discussions by Shapiro, Chapter 8, and Patrick, Chapter 15). In this respect supertramps resemble the species that have been variously called "fugitive species," "opportunistic species," "weeds," or "tramps." The difference is that supertramps are virtually confined to small islands, where they occur in practically any habitat, whereas fugitive species also occupy or are characteristic of unstable habitats on large islands. The investment of supertramps in reproduction and in dispersal powers presumably comes at the expense of competitive ability and harvesting efficiency. On species-rich islands the supertramps are outcompeted and permanently excluded by guilds of K-selected high-S specialists, each of which is better adapted to some fraction of the supertramps' potential niche. For example, on New Britain or New Ireland the supertramp fruit dove *Ptilinopus solomonensis* is replaced by three congeners, one confined to mountains, another to lowland forest, a third to open lowland habitats; the medium-sized supertramp fruit pigeon *Ducula pistrinaria* is excluded by three congeners, one large species and one small species in the lowlands, a third in the mountains; the supertramp cuckoo-dove *Macropygia mackinlayi* is replaced by three cuckoo-doves that form a graded size series; and the supertramp flycatcher *Monarcha cinerascens* is replaced by four related flycatchers, one in the mountains, a second in lowland second-growth, a third in the canopy of lowland forest, and a fourth in the interior of lowland forest. On small islands, where extinction rates are high, only the supertramps can recolonize frequently enough to have a high incidence. On larger islands defaunated by volcanic activity or tidal waves, the supertramps are first to arrive and rapidly fill the island. By the time the more slowly dispersing specialists follow and crowd out the populations of supertramps, the latter have already found another defaunated island and thereby ensured their survival as a species. With the explosion of a large island like Long, the supertramps "struck it rich" as first arrivals. For some supertramp species the colonization of Long quadrupled the population of the whole species.

To an observer who comes to the super-tramp-rich islands Long, Tolokiwa, or Crown from a "normal" island, the feature of these islands that becomes most obvious within the first hour of field observations is the extraordinary abundance of birds. As quantitated by mist-net techniques (Figure 18), population densities of all species combined are up to six times higher in lowland forest, and up to nine times higher in montane cloud forest, on the supertramp-rich islands than on "normal" control islands with the same number of species. These excess densities are largely due to the abundance of the supertramps themselves, which account for 40–70% of all bird individuals in lowland forest of these three islands, and 73–90% of all bird individuals in montane cloud forest.

Several possible explanations that immediately come to mind for these startlingly high abundances do not withstand scrutiny. Could bird populations on "normal" islands be kept below carrying capacity by predators (as is true for many of the species discussed by Connell in Chapter 16), so that supertramp abundance represents a release from predation pressure? In fact, predators of birds on Long (two species of falcon, an eagle, and the nest-robbing marsupial *Phalanger orientalis*) are at least as diverse as on normal islands, and far more numerous, mainly because of the abundance of the marsupial that has struck every visitor to Long. Could supertramps be taking over the

niches of lizards, mammals, or some other nonavian competitors that are present on normal islands but unable to reach Long? Long in fact has small and large lizards, frogs, rats, the phalanger, and insectivorous, nectar-feeding, and frugivorous bats; and the supertramps include omnivores, insectivores, nectar-feeders, and specialists on soft or hard fruit, and cover a wide range of sizes (Table 6). Could the supertramps be drawn disproportionately from small species, so that the excess abundance of individuals does not reflect an excess biomass? In fact, there is little difference between supertramp-rich islands and normal islands in the average weight of a netted bird. Could mist-netting techniques greatly overestimate the abundance of supertramps? The netting results are at least qualitatively confirmed by field impressions of obvious abundance. Furthermore, more individuals were netted of the territorial than the nonterritorial supertramp species (the latter happen to be species that forage largely above the height of the nets), so that wanderings of nonterritorial supertramps are unlikely to have significantly inflated the mist-net yields. Could excess abundances on Long, Crown, and Tolokiwa reflect high productivity of rich volcanic soil resulting from ash fall-out during the Long explosion? But Tolokiwa received little ash from the Long explosion.

Rather, the high densities of birds on Long, Crown, and Tolokiwa must be a property of the supertramp-rich communities themselves, since physically similar volcanic islands that are more distant

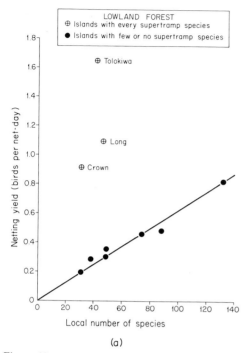

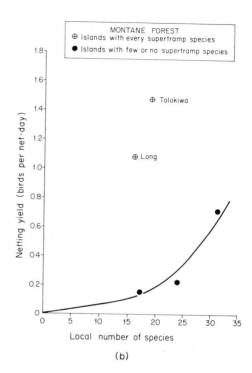

(a)

(b)

Figure 18 Ordinate: rate of catching land birds under standardized conditions in mist nets on a given island (expressed in birds caught per net-day), in lowland forest (Figure 18a, left) or montane cloud forest (Figure 18b, right). Abscissa: local number of species present on the island. If one compares the same type of habitat on different islands (but not if one compares different habitats), this catch rate is proportional to the combined population densities of all species present (Diamond, 1970b). Symbols: ●, control islands with not more than three supertramp species; ⊕, the exploded volcano of Long and two control islands, which support ten supertramp species each. Note that, on islands with few supertramps, netting yields (and total population densities) increase with the local number of species (as interpreted by Diamond, 1970b); and that the islands with many supertramp species have much higher yields than islands with the same total number of species but few supertramps. (From Diamond, Copyright 1974 by the American Association for the Advancement of Science.)

from Long and were not inundated by supertramps do not have the excess population densities. This interpretation is confirmed by the situation on the mountainous island Karkar, where spatially separate supertramp-rich and supertramp-poor communities with very different population densities contrast within the same island. On Karkar and the adjacent lower island of Bagabag, supertramps constitute a considerable minority of species in all major bird guilds except at high elevations on Karkar, where the supertramp honeyeater *Myzomela sclateri* is the sole species in the insectivorous gleaner guild. Below 4000 feet *Myzomela sclateri* is excluded by two sunbird species, and in subalpine shrubbery above 5000 feet there is no other bird species at all besides this supertramp. In lowland forest on Karkar (third-from-left-most solid point, Figure 18a) and Bagabag (left-most solid point, Figure 18a), and in montane forest below 4000 feet on Karkar (left-most solid point, Figure 18b), total population densities measured by mist-netting fall on the curve for "normal" islands. At the highest elevations on Karkar, however, *Myzomela sclateri* was obviously much more abundant than all bird species combined in similar habitats on supertramp-poor islands I have studied, although no mist-netting was done to quantify this finding.

At first, it seems paradoxical that supertramp guilds, where they are present, maintain much higher population densities than the high-*S* competitors that exclude them from most islands. If one thinks of the high-*S* species as *K*-selected, then *K*-selected guilds have much *lower*

equilibrium abundances than do supertramp guilds, and the term *K*-selection becomes misleading and contradictory. However, as MacArthur predicted (1972, p. 31), "self-renewing resources can be overexploited to the detriment of the predator's population, and this over-exploitation will be a natural consequence of competition among the predator species." That is, competition in a high-*S* fauna selects for those species that can reduce resource levels below the point where other species can survive, even though this diminishes the rate of resource production and hence the population density of harvesting species (see MacArthur, 1972, pp. 56–57 for a simple model). For example, high-*S* guilds of insectivorous birds, by catching insects more efficiently than do supertramps, may be depressing sustainable insect yields far below the level that exists on supertramp-rich islands. In the most species-rich community on earth, the neotropical rainforest, Elton (1972) has already emphasized the low total *densities* (despite high diversity) of insects, and the relation of these low prey densities to sustained predator pressure. The mechanism of fruit over-exploitation by high-*S* frugivores must differ from the mechanism among insectivores, since the number of fruits produced is independent of the harvest rate so long as the reseeding of new fruit trees is not suppressed. What may be happening is that high-*S* frugivores are selected by competition to harvest the fruit at an earlier stage of ripeness than supertramps can tolerate, thereby eliminating the supertramps but also reducing the energy yield available to the remaining species from eating less ripe or smaller fruit.

Thus, the supertramp strategy may be contrasted with the overexploitation ethic forced on high-S faunas by competition: "Harvest early, overexploit, and starve out the next species, lest the next species harvest earlier and closer." The overexploitation ethic adopted by high-S species under the pressure of interspecific competition, and to the detriment of their own population densities, strikingly exemplifies the conclusion reached by Levins in Chapter 1, that the course of natural selection in a particular species may not be comprehensible without reference to the whole matrix of community interactions. Lowering of resource levels, whether by a single species or by a mutually adjusted guild of species, may be the principal mechanism by which invaders are competitively excluded from integrated communities. If these interpretations are correct, they imply that predators in high-S communities have a major impact on the densities of their prey. Later in this chapter we shall apply Levins's method of loop analysis to determining the conditions under which overexploitation is a useful strategy for its practitioners.

In the remainder of this chapter we shall search for evidence that local species guilds are in fact composed of coadjusted species and do resist invaders. A lesson to draw from our discussion of incidence functions is that the clearest evidence is likely to come from exclusion of supertramps by other species. The catholic habitat tolerance of most supertramps means that they are capable of thriving on virtu-ally any island of the southwest Pacific. Because of the demonstrated high dispersal of at least some of the Bismarck supertramps, we can assume that every island of the Bismarcks is continuously bombarded by their colonists, and we can reasonably conclude that absences of supertramps from islands are due to competitive exclusion.

The Meaning of Assembly Rules

In the section "Relations between incidence functions and island area" we considered reasons why the incidence J of a given species should in general increase with island area, or why the total number of species S should increase with island area. For most species J has to be and is an increasing function of S. There are also obvious reasons why the function should not be identical for all such species. The form of the function depends, for instance, on each species' preferred habitat type, minimum territory requirement, patchiness of food supply, magnitude of population fluctuations, dispersal ability, and birth and death rates. In discussing incidence functions so far, we have hardly mentioned interspecific competition or coadjustment of species. Does competition fail to contribute to the form of $J(S)$? Do the properties of each species considered individually tell us most of what we need to know in order to predict how communities are assembled from the total species pool? Note, for instance, that only if all species had the same incidence functions would one expect communities to contain random samples of the species pool. Since properties of individual species already lead us to expect incidence functions of different shapes (though usually increasing functions of S), we expect even communities of noncompeting species to appear as if they are assembled in nonrandom ways. Are all the rules for community assembly contained in the incidence functions?

There are three clear indications that the answer to these questions is "No." We have already seen one type of evidence that competition affects the form of $J(S)$, and that there must exist some assembly rules besides "incidence rules" (i.e., besides constraints that can be deduced from incidence functions alone): evidence derived from supertramps. For these species J does not approach 1.0 with increasing S but instead goes to zero at some upper S_{crit}. Naturally, this does not mean that a dispersing juvenile supertramp flies into an island, performs a quick avifaunal census, and flies off if he sees that some preprogrammed upper S_{crit} is exceeded. Naturally too, we should not suppose that all species have equally detrimental effects on a supertramp, and that when a certain species number has been exceeded, the combined effect of all the species outcompetes or starves out the supertramp. Instead, the supertramp is outcompeted or starved out by the combined effects of those species that harvest similar resources, the other species in the supertramp's guild. Since the number of these guild species increases with S just as does the number of species in any arbitrary subset of the species pool not biased towards supertramps, the incidence of the supertramp is seen to decrease at high S.

Evidently the supertramp can fit into only certain small subsets of his guild. The problem thus becomes one of determining exactly which of the species that are being recruited into communities with increasing S are responsible for excluding the supertramp, and to what extent and in what combinations.

A second clear indication of competitive effects on $J(S)$ is provided by changes in $J(S)$ as a function of the species pool. Although each of the three island archipelagoes of Melanesia (Bismarcks, Solomons, and New Hebrides) supports a somewhat different species pool, many bird species of the Bismarcks occur in the other two archipelagoes. In an archipelago where a species finds fewer competing guild members in the species pool, the incidence category of that species shifts from supertramp towards D- or C-tramp, or from high-S species towards D-tramp. For example, in the Bismarcks the cuckoo-dove *Macropygia mackinlayi* belongs to a guild of four species and is a supertramp, absent from the largest islands (Figure 6 below). The total species pool of the Solomons is approximately as large as that of the Bismarcks, but only one other cuckoo-dove species occurs in the Solomons besides *Macropygia mackinlayi,* which becomes a D-tramp (Figure 19), present on the largest islands as well as on smaller islands. *Macropygia mackinlayi* is also a D-tramp in the New Hebrides, where there is no other cuckoo-dove. Similarly, *Ptilinopus solomonensis,* a supertramp in the Bismarcks, is present on large islands of the Solomons; and the Bismarck supertramps *Ducula pacifica* and *Turdus poliocephalus* become C-

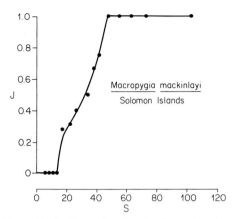

Figure 19 Incidence function for the cuckoo-dove *Macropygia mackinlayi* in the Solomon Islands, where only one other species of cuckoo-dove is present and *M. mackinlayi* is a D-tramp. Compare Figure 6 below, the incidence function for this species in the Bismarck Archipelago, where three other cuckoo-dove species occur and *M. mackinlayi* is a supertramp, excluded from the largest islands by competing congeners.

tramps in the poorer species pool of the New Hebrides, where they occur on large as well as small islands. Thus, the exclusion of supertramps from large islands is due to the presence of competing species in the same archipelago. Conversely, the Bismarck A-tramp rail *Porphyrio porphyrio* and B-tramp flycatcher *Pachycephala pectoralis* become C-tramps in the New Hebrides, where there are fewer competing species of rails and flycatchers. Thus, the absences of these species from some small Bismarck islands, like the absences of the supertramps from all large Bismarck islands, involve effects of competitors.

The other type of unequivocal evidence for additional assembly rules besides incidence rules comes from competitive exclusion patterns. Occasionally one finds

two closely related species that occur in a checkerboard distribution pattern without ever coexisting. This would not be predicted from the incidence functions of the two species unless the ranges of S values over which J was nonzero were nonoverlapping, and such nonoverlap of $J(S)$ is rarely the case for species with checkerboard distributions (cf. *Pachycephala pectoralis* and *P. melanura dahli* in Figure 28, *Macropygia mackinlayi* and *M. nigrirostris* in Figure 25, and *Myzomela sclateri* and *M. cruentata* in Figure 29). There are also more complex cases in which combinations of more than two species, the feasibility of whose existence is compatible with incidence functions, never exist. This evidence also indicates that there are constraints on how species fit together besides the constraints implicit in incidence functions.

To understand the role of competition in the assembly of communities, we shall begin with the most clearly illustrative but least important example, the distributional checkerboard. We shall then go through more complicated but still decipherable examples from four guilds that illustrate *diffuse competition,* the phenomenon of a species that is unable to fit into a community because it is excluded by specific *combinations* of other species.

"Simple" Checkerboard Distributions: the Dilemma of the Empty Squares

If we suppose that stable communities consist of coadjusted species that exclude invaders by competition, the simplest distributional pattern that might be sought as possible evidence for competitive exclusion is a checkerboard distribution. In such a pattern, two or more ecologically similar species have mutually exclusive but interdigitating distributions in an archipelago, each island supporting only one species. An increasing number of instances of such checkerboard distributions have been reported in the recent literature (cf. Stresemann, 1939, for the islands of Indonesia; Mayr, 1942, for islands of the southwest Pacific; Lack, 1971, for islands of the West Indies; MacArthur, Diamond, and Karr, 1972, for Panamanian islands; Diamond, 1972a, for habitat islands on New Guinea). Figures 20–24 illustrate five examples from the southwest Pacific. Figure 20 involves two species of *Macropygia* cuckoo-doves in the Bismarcks; Figure 21, two species of *Pachycephala* whistlers in the Bismarcks; Figure 22, two species of *Ptilinopus* fruit doves in the Bismarcks; Figure 23, five species of small *Myzomela*

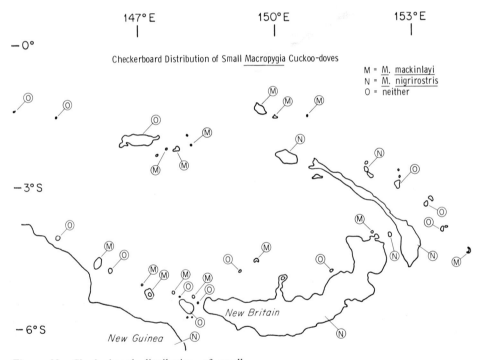

Figure 20 Checkerboard distribution of small *Macropygia* cuckoo-dove species in the Bismarck region. Islands whose pigeon faunas are known are designated by *M* (*Macropygia mackinlayi* resident), *N* (*Macropygia nigrirostris* resident), or 0 (neither of these two species resident). Note that most islands have one of these two species, no island has both, and some islands have neither.

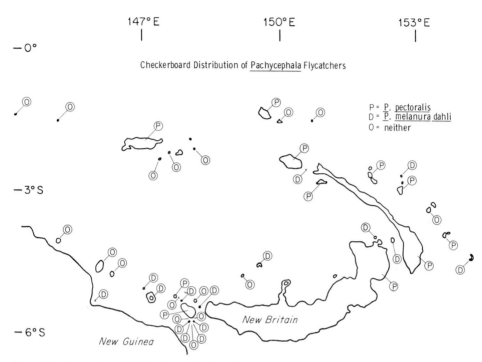

147°E   150°E   153°E

−0°

Checkerboard Distribution of <u>Pachycephala</u> Flycatchers

P = <u>P. pectoralis</u>
D = <u>P. melanura dahli</u>
0 = neither

−3°S

−6°S

New Britain

New Guinea

Figure 21 Checkerboard distribution of *Pachycephala* flycatcher species in the Bismarck region. Islands whose flycatcher faunas are known are designated by *P* (*Pachycephala pectoralis* resident), *D* (*Pachycephala melanura dahli* resident), or 0 (neither of these two species resident). Note that most islands have one of these two species, no island has both, and some islands have neither.

honeyeaters in the Bismarcks; and Figure 24, 12 species of *Zosterops* white-eyes in the Moluccas, New Guinea region, and Bismarcks. In each instance, an island supports only one species of the pair or group, but the distributions of the locally successful colonists replace each other in a checkerboard-like, irregular, geographical array. On each island the identity of the locally successful colonist may have been determined either on a first-

come-first-served basis, or on the basis of slight competitive advantages related to slight ecological differences among islands (see section on "Chance or Predestination?"). Once established, one species would be able to exclude its relatives indefinitely or at least for a long time.

Figures 20–24 also illustrate a complication that has generally been glossed over in discussions of checkerboard distributions. In every figure, it is not sim-

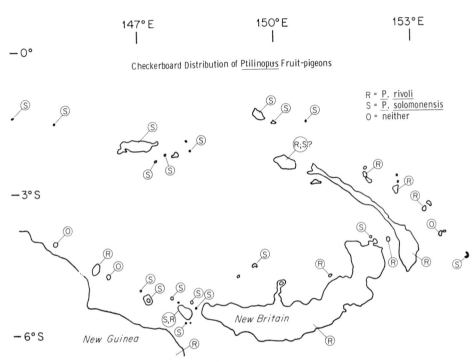

Figure 22 Checkerboard distribution of two closely related fruit pigeons in the Bismarcks. Islands whose pigeon faunas are known are designated by *R* (*Ptilinopus rivoli* resident), *S* (*Ptilinopus solomonensis* resident), or 0 (neither of these two species resident). Note that most islands have one of these two species, only one or perhaps two islands have both, and only a few islands have neither.

ply that each island supports one member of a group, no more and no less: some islands support no member at all. For instance, 11 Bismarck islands support *Pachycephala pectoralis,* 18 support *Pachycephala melanura dahli,* and 21 support neither. One might be tempted to explain the vacant islands by postulating that they are unsuitable for either species, or that by chance neither species has arrived, or that some empty islands are in-

evitably expected in the low-*S* portion of incidence curves, or that some other related species has been overlooked and fills the vacant islands. However, none of these explanations is plausible for the cases illustrated in Figures 20–24. In each of these five examples, at least one of the vicariant species is a supertramp of catholic habitat tolerance and high dispersal ability, so that it undoubtedly has sent colonists to the vacant islands and does exist elsewhere in physical environments similar to those left vacant. The vacant islands do have other species not depicted in Figures 20–24, which are ecologically similar to the supertramp, although not as similar as the strictly vicariant species. However, in each case some islands can be found on which the supertramp co-exists with each of the other related species. For instance, in addition to the similar-sized small cuckoo-doves *Macropygia nigrirostris* and *M. mackinlayi* of Figure 20, there are two larger species, *M. amboinensis* and the *Reinwardtoena* su-

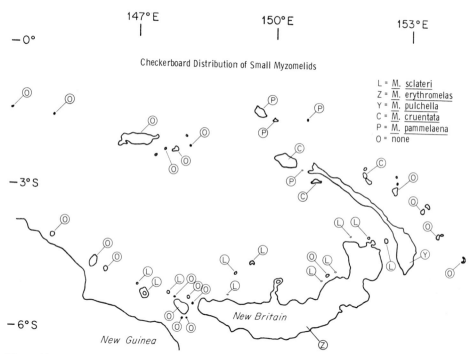

Figure 23 Checkerboard distribution of small myzomelid honeyeaters in the lowlands of Bismarck islands. Islands are designated by *L* (*Myzomela sclateri* resident), *Z* (*Myzomela erythromelas*), *Y* (*Myzomela pulchella*), *C* (*Myzomela cruentata*), *P* (small-sized races of *Myzomela pammelaena;* there are additional large-sized races in the northwest Bismarcks and on Long and its neighbors), or 0 (none of these species resident). Note that half of the islands have one lowland myzomelid species, no island has two species, and half of the islands have none. The two largest islands have an additional species in the mountains.

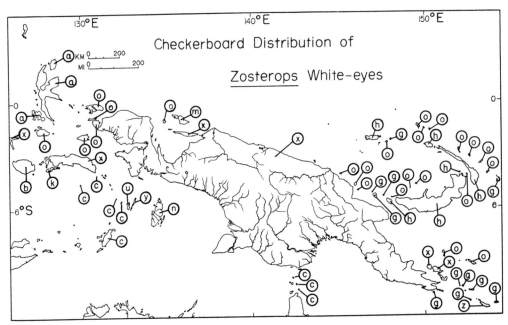

Figure 24 Checkerboard distribution of twelve white-eyes in lowland and/or hill forest of islands of the New Guinea region. Islands are designated by *a* (*Zosterops atriceps*), *b* (*Z. buruensis*), *c* (*Z. chloris*), *g* (*Z. griseotincta*), *h* (*Z. hypoxantha*), *k* (*Z. kuehni*), *m* (*Z. mysorensis*), *n* (*Z. novaeguineae*), *u* (*Z. uropygialis*), *x* (*Z. atrifrons*), *y* (*Z. grayi*), *z* (*Z. meeki*), or 0 (no species). Five of the largest islands have an additional species in the mountains. Species *a, h, m, x,* and *z* belong to a superspecies.

perspecies. Each island left vacant both by *M. nigrirostris* and *M. mackinlayi* does have either *M. amboinensis* or *Reinwardtoena* or both. However, numerous islands can be found as illustrations of the ability of *M. nigrirostris* to coexist with either of the two larger species, and similarly for *M. mackinlayi*. Thus, no other species can be found to fill the vacant islands of the Figure 20 checkerboard on a one-species-per-island basis. In Figures 21–24 as well,

one-to-one competitive exclusion fails to explain why the checkerboards have numerous empty squares.

Checkerboard distributions are of great interest in demonstrating the existence of competitive exclusion. Nevertheless, the complication of the empty squares shows that there is no distributional pattern in the Bismarcks that can be understood completely in terms of one-to-one competition. In addition, there is a far more serious difficulty in trying to understand community resistance to invasion in terms of one-to-one competition: the distributions of only a few species can be understood even partly on this basis. For most Bismarck species one cannot point to any close relative with which it is unable to coexist: most pairs of close relatives are demonstrably compatible on one or another island. Thus, the innumerable instances of vagile species excluded from physically suitable islands—instances that are implicit in the $J < 1$ region of many incidence functions—must have a more complex explanation than one-to-one competitive exclusion.

Assembly Rules for the Cuckoo-Dove Guild

Let us consider the cuckoo-dove guild in detail, because it is the only Bismarck guild for which the assembly rules can at present be deduced completely, and because there is considerable historical evidence that permissible combinations of species do resist forbidden invaders.

Cuckoo-doves are a group of arboreal, long-tailed, fruit-eating pigeons that live in the middle story of shaded forest from India east to Melanesia. In the eastern portion of this region, from the Moluccas to the New Hebrides, occur four species or superspecies: the *Reinwardtoena* superspecies, which is large (weight, average 297 g, range 279–315); *Macropygia amboinensis,* which is medium-sized (149 g, range 128–178); and *M. mackinlayi* (87 g, range 73–110) and *M. nigrirostris* (86 g, range 73–97), which are smaller. As with

other guilds in which successive members differ by a factor of 2 in weight (Diamond, 1972a, pp. 43–44; Figures 30 and 31 of this chapter; see also Brown, Chapter 13, Figures 5 and 6, and Hespenheide, Chapter 7, Figure 4), ecological segregation among sympatric cuckoo-doves depends on the fact that larger species can eat larger fruits, but smaller species can perch on lighter twigs and perches. All four species coexist in the Bismarcks. To the east, only *Reinwardtoena* and *mackinlayi* occur in the Solomons, and only *mackinlayi* in the New Hebrides. To the west, New Guinea has three species (*mackinlayi* absent), and the Moluccas have *Reinwardtoena* and *amboinensis.* The two most similar species in phylogeny and in size are *nigrirostris* and *mackinlayi,* which are products of a recent speciation, largely allopatric, and overlap only in the Bismarcks.

Figure 25 compares the incidence functions of the four species in the Bismarcks. *M. nigrirostris* is an A-tramp, *Reinwardtoena* a B-tramp, *amboinensis* a C-tramp, and *mackinlayi* a supertramp.

Figure 26 plots the number of cuckoo-dove species on each Bismarck island as a function of total species number on the island. Only some of the poorest islands ($S \leq 18$) lack cuckoo-doves, and all islands with $S \geq 20$ have one or more species. Islands with $S = 9$–47, 20–83, and 62–127 have 1, 2, and 3 cuckoo-dove species, respectively.

Cuckoo-dove species are able to coexist on a single island only in certain combinations. Table 7 summarizes "permitted combinations" and "forbidden combina-

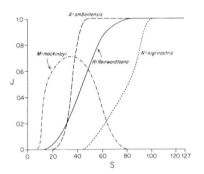

Figure 25 Incidence functions for the cuckoo-doves *Macropygia mackinlayi, M. amboinensis, M. nigrirostris,* and *Reinwardtoena browni* in the Bismarcks.

tions," i.e., combinations that occur on many islands, as opposed to combinations that occur on no island. Two of the four possible one-species combinations, three of the six possible two-species combinations, three of the four possible three-species combinations, and the sole possible four-species combination are forbidden. In the cuckoo-dove guild, as in other guilds, it turns out empirically that combinations may be forbidden on the basis of any one of three types of rules, which we shall term incidence rules, com-

patibility rules, and combination rules. In the following discussion we symbolize the four cuckoo-doves as $A = amboinensis$, $M = mackinlayi$, $N = nigrirostris$, and $R = Reinwardtoena$.

Incidence rules.

Examination of incidence functions alone tells us, without need for any further information, that certain combinations cannot occur. Thus, Figure 25 shows that J_A equals 1.0 (where J_A symbolizes the incidence of *amboinensis*) for all S exceeding the lower S_{crit} for *nigrirostris*, the S value below which $J_N = 0$. That is, all islands that have *nigrirostris* also have *amboinensis*. We therefore know that any combinations containing N but not A—i.e., the combinations N, NR, MN, and MNR—are forbidden. We shall use the term "incidence rules" to refer to constraints on species assembly that follow immediately from incidence functions.

Compatibility rules.

Distributional information combined with knowledge about species ecologies may suggest to us that a given pair of species is incapable of coexistence on an island under any circumstances, either alone as a pair, or as part of a larger combination. Absolute incompatibility is likely to be encountered only between two species that are products of a recent speciation and so similar ecologically that they cannot occupy the same space. An example is provided by the cuckoo-doves *M. mackinlayi* and *M. nigrirostris*, which must have constituted a superspecies until a relatively recent reinvasion of the Bismarcks brought them into contact. These

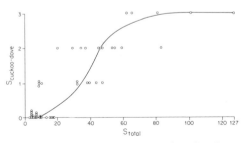

Figure 26 Number of resident species of cuckoo-doves on each Bismarck island (ordinate), as a function of the total number of resident bird species on the island (abscissa). The solid curve is drawn through points at the average value of S_{total} for each $S_{cuckoo-dove}$ value.

Table 7. Assembly rules for the cuckoo-dove guild

Number of species	Assembly observed	Permitted combination	Forbidden combination	Collapse observed	Reason why forbidden
1		A (3,13)			
	$-, + M \rightarrow$	M (8,41)			
			N		incidence
			R		combination
2	$M, + A \rightarrow$	AM (5,0)			
		AR (4,15)			
		MR (2,12)			
			AN		combination
			MN		compatibility and incidence
			NR		incidence
3		ANR (5,4)			
			AMN	$AM + N \rightarrow |$	compatibility
			AMR	$AM + R \rightarrow |,$ $AR + M \rightarrow |$	combination
			MNR	$MR + N \rightarrow |$	compatibility and incidence
4			$AMNR$	$ANR + M \rightarrow |$	compatibility

For all possible 1-species, 2-species, 3-species, and 4-species combinations of the four cuckoo-dove species (A = *Macropygia amboinensis*, M = *Macropygia mackinlayi*, N = *Macropygia nigrirostris*, R = *Reinwardtoena* superspecies), the table indicates whether the combination does occur ("permitted combination") or never occurs ("forbidden combination") on islands. The two numbers after each permitted combination give, respectively, the number of Bismarck islands, and the number of islands outside the Bismarcks (Moluccas, Papuan islands, Solomons, New Hebrides), on which the combination is resident. "Assembly observed" indicates cases in which assembly of a permitted combination has been observed historically, when the second-named species (named after the plus sign) has successfully invaded a community of the first-named species (named before the plus sign; the dash in the second row indicates that no cuckoo-dove was present before species M invaded). "Collapse observed" indicates cases in which collapse of a forbidden combination has been observed historically, when the second-named species has unsuccessfully invaded a community of the first-named combination (e.g., in the last row, the entry means that invasions of ANR by M have been observed and have been unsuccessful). The last column gives the type of assembly rule that disallows each forbidden combination.

species are of the same size, coexist on no island, are similar ecologically, and differ mainly in distributional strategy (supertramp vs. A-tramp). Thus, the combinations MN as well as MNR, AMN, and $AMNR$ are forbidden by "compatibility rules." The first two of these combinations are also forbidden by incidence rules.

If *mackinlayi* and *nigrirostris* were not incompatible, their highest probability of coexistence would be on islands with $53 \leq S \leq 66$, where the product $J_M J_N$ exceeds 0.04, reaching a maximal value of 0.06 at $S = 58$. However, there are only six Bismarck islands in this S range whose resident cuckoo-doves are well known. In general, the probability $p_{i,ab}(S)$ of *not* observing two species a and b together, whether as ab or as part of a larger combination, on an island i of species number S_i is given by

$$p_{i,ab}(S_i) = 1 - J_a(S_i)J_b(S_i) \qquad (4)$$

Then the no-coexistence probability Z_{ab} of not finding a and b coexisting on any of a set of n islands, which need not have the same species numbers, is given by the product

$$Z_{ab} = \prod_{i=1}^{i=n} [1 - J_a(S_i)J_b(S_i)] \quad (5)$$

where the product is taken over all n islands. Evaluation of eq. 5 for the species *M. mackinlayi* and *M. nigrirostris*, substituting J values from Figure 25 corresponding to the S values for all islands for which $J_M \neq 0$, $J_N \neq 0$, yields a no-coexistence probability Z_{ab} of 0.74. That is, incidence functions alone predict the probability of not finding M and N together on any of these six islands, and we do not have statistical evidence for invoking a compatibility rule. By analogy with other similar cases, however (cf. following section, "Assembly of the gleaning flycatcher guild"), where statistical evidence for incompatibility does exist, and because of the close similarity of *Macropygia mackinlayi* and *M. nigrirostris*, my guess is that availability or surveys of more islands would show their coexistence to be incompatible and not just improbable. It should also be realized that incompatibility of these two species is an important *reason* for the form of their incidence functions: viz., that the supertramp *M. mackinlayi* is absent from large islands where occurrence of the A-tramp *M. nigrirostris* is likely, and vice versa for small islands. This is demonstrated by Figure 19, which shows that removal of *M. nigrirostris* and *M. amboinensis* from the species pool converts *M. mackinlayi* from a supertramp (in the Bismarcks) to a D-tramp (in the Solomons).

Combination rules.

The incidence functions of Figure 25 suggest that the combination *AMR* should occur frequently on islands with a medium species number. This combination does not violate the presumed compatibility rule forbidding *MN*. Neither does it violate incidence rules, since Figure 25 shows J_A, J_M, and J_R all to be high in an S range where J_N is zero or at least far below 1.0. The value of the product $J_A J_M J_R$ suggests that, if there were no restrictions on cuckoo-dove combinations other than those implicit in compatibility rules and incidence rules, the combination *AMR* should occur on about one-third of all islands supporting 43 to 55 species, and on a somewhat lower proportion of islands with 25–43 or 55–70 species. Nevertheless, there is no known island with the combination *AMR*. Since most islands of sufficient size or S potentially to support this combination have been ornithologically explored, it appears likely that *AMR* does not exist in nature.

The statistical test of this conclusion is similar to the one just considered for the compatibility of M and N. Whether we should attach significance to the actual nonexistence of *AMR* depends on whether there are enough real islands with sufficiently high $J_A J_M J_R$ products in an appropriate range of the incidence functions. If there were only a small number of islands in the range of high J_A, J_M, and J_R and low J_N, it could easily arise by chance that *AMR* would be found on none of these islands despite its being a permissible combination. To evaluate this possibility, we define $P_{i,X}$ as the probability that a certain combination X will occur on a certain island i, where $P_{i,X}$ takes account only of incidence functions.[1] For an arbitrary guild consisting of species a, b, c, d, $e, \ldots, j - 1, j$, the probability that the combination abc will occur on an island i of species number S_i is given by

$$P_{i,abc}(S_i) = [J_a(S_i)][J_b(S_i)]$$
$$[J_c(S_i)] \prod_{x=d}^{x=j} [1 - J_x(S_i)] \quad (6)$$

$[1 - P_{i,abc}(S_i)]$ is then the probability that the island will not support abc. The product

$$Y(X) = \prod_{i=1}^{i=n} [1 - P_{i,X}(S_i)] \quad (7)$$

evaluated for the combination $X = AMR$ with the product taken over all n islands of the Bismarck Archipelago, then has the following meaning: $Y(AMR)$ is the combination probability that AMR would not be observed on any actual Bismarck island if there were no rules about combinations other than those implicit in incidence functions.[1]

The evaluation of $Y(AMR)$ is simplified by the fact that $P_{i,AMR}(S) = 0$, $[1 - P_{i,AMR}(S_i)] = 1$ for all islands with

[1] Equations 6 and 7 take account of incidence functions but not of compatibility rules. For a more exact evaluation of $Y(X)$, the probabilities of combinations forbidden by compatibility rules must somehow be redistributed among compatible combinations. The forms of eqs. 4 and 6 differ, because eq. 4 calculates the probability that species a and b will coexist, either by themselves or as part of a larger combination, whereas eq. 6 calculates the probability that a, b, and c will coexist by themselves, not as part of a larger combination.

$S > 77$, since J_M then is zero, and for all islands with $S < 22$, since J_R then is zero. Table 8 lists J_A, J_M, J_N, and J_R values read off from Figure 25, and $P_{i,AMR}$ calculated from eq. 6, for all Bismarck islands that have $22 < S < 77$. As evaluated from the product of the $(1 - P_{i,AMR})$ terms in the last column of Table 8, $Y(AMR)$ is 0.027. That is, given the actual number of well-explored Bismarck islands and actual incidence functions, the probability of failing to find AMR on any island would be only 0.027 if all constraints were implicit in incidence functions. We conclude that there must be additional *combination rules* whose existence could not have been suspected from Figure 25, and one of which forbids the coexistence of *amboinensis*, *mackinlayi*, and *Reinwardtoena*. Note that the subcombinations AM, MR, and AR all occur on real islands, yet AMR is forbidden.

There are two further combinations whose existence is not forbidden by incidence rules or compatibility rules but which nevertheless do not exist on real Bismarck islands: R and AN. The AN combination is also absent from the New Guinea region, where *amboinensis* and *nigrirostris* are both present; and the R combination is also absent from the New Guinea region, Moluccas, and Solomons, where *Reinwardtoena* is present along with either one or two other species of cuckoo-doves. Thus, there may be additional combination rules forbidding R and AN. In these cases, however, there are not enough islands with appropriate incidence values to give us as much confidence as in the AMR case that the combination

Table 8.   Combination probability of the cuckoo-dove species AMR for
16 Bismarck Islands

S	J_A	J_M	J_R	J_N	$P_{i,AMR}$	$1 - P_{i,AMR}$
29	0.13	0.70	0.16	0	0.01	0.99
32	0.25	0.71	0.21	0	0.04	0.96
32	0.25	0.71	0.21	0	0.04	0.96
34	0.34	0.72	0.26	0	0.06	0.94
37	0.59	0.72	0.32	0	0.14	0.86
38	0.69	0.71	0.34	0	0.17	0.83
39	0.75	0.71	0.36	0	0.19	0.81
43	0.97	0.68	0.46	0	0.30	0.70
45	1.0	0.67	0.50	0	0.34	0.66
46	1.0	0.66	0.53	0	0.35	0.65
53	1.0	0.52	0.72	0.08	0.34	0.66
55	1.0	0.48	0.77	0.10	0.33	0.67
59	1.0	0.35	0.85	0.17	0.25	0.75
60	1.0	0.31	0.86	0.18	0.22	0.78
62	1.0	0.25	0.88	0.20	0.18	0.82
65	1.0	0.17	0.92	0.24	0.13	0.87

The cuckoo-dove combination *AMR* occurs on no known island. The table calculates whether this finding is expected by chance, given only the constraints of incidence functions. Only in the range $22 < S < 77$ are the incidences J_A, J_M, and J_R all nonzero. Column 1 gives the species number on all 16 Bismarck islands in this S range whose resident cuckoo-doves are known. Columns 2–5 are the incidences corresponding to the S of each island for each cuckoo-dove species, read from the incidence functions of Figure 25. Column 6 gives the probability $P_{i,AMR}$ [calculated from these incidences as $J_A J_M J_R (1 - J_N)$, eq. 6] that the combination *AMR* will occur on an island with this S; column 7 gives $1 - P_{i,AMR}$, the calculated probability that *AMR* will not occur on the island. The product $Y(AMR)$ of all the terms in the last column (eq. 7) is 0.027, the probability that *AMR* would occur on none of the 16 islands if the only constraints were those of incidence functions. Since this probability is very low but *AMR* in fact occurs on none of the 16 islands, the combination is presumed to be forbidden for some reason not implicit in incidence functions.

would have been observed if it were not explicitly forbidden.

Figure 27 plots the observed combination incidence $C_X(S)$ as a function of island species number S, for the six permitted cuckoo-dove combinations of the Bismarcks. For any $S = S_i$, the sum $\Sigma\, C_X(S_i)$, taken over all X, is 1.0. For any given S_i, there are generally only two or three combinations that have any significant probability of occurring. By considering separately the dependence of S on island area and isolation, we may eventually be able to predict with higher proba-

bility what combination a particular island will support.

Historical proof of resistance to invasion, or of feasibility of assembly.

The derivation of the assembly rules summarized in Table 7 was based entirely on observation of existing distributions. These rules imply that certain, but not all, combinations of cuckoo-doves form stable subcommunities that resist the invasion of other cuckoo-dove species. Invasion of such stable communities by a species

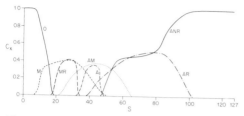

Figure 27 Ordinate, the observed combination incidence C_X for each of the six permitted cuckoo-dove combinations of the Bismarcks, as a function of the total number of bird species on an island (abscissa). This graph is constructed in the same way as the incidence functions of Figure 4–15, except that what is plotted is the incidence of the indicated combination, not the incidence of a species regardless of what combination it occurs in. Symbols for combinations have the same meaning as in Table 7, except that the additional symbol 0 means no resident species of cuckoo-dove.

whose addition would result in a forbidden combination is doomed to failure, unless the invader can eliminate one of the original members of the community. In five cases it has been possible to document directly, by historical evidence, that permitted combinations of species do resist forbidden invaders:

(a) *Reinwardtoena browni*, *M. nigrirostris*, and *M. amboinensis* have been the resident cuckoo-doves of New Britain throughout this century (combination *ANR*). There is no resident population of *M. mackinlayi*, whose addition would create a combination forbidden by compatibility rules (*AMNR*). However, vagrant individuals of *M. mackinlayi* have been encountered by O. Meyer and A. Eichhorn. Since *M. mackinlayi* is an abundant breeder on small islands within one mile of New Britain, it must be a frequent invader.

(b,c) *M. amboinensis* and *M. mackinlayi* are the resident cuckoo-doves of Vuatom. O. Meyer recorded repeated unsuccessful invasions of *Reinwardtoena browni*, which would have yielded *AMR* (forbidden by combination rules), and of *M. nigrirostris*, which would have yielded *AMN* (forbidden by compatibility rules).

(d) *Reinwardtoena reinwardtsi* and *Macropygia mackinlayi* have been resident on Karkar for at least the last 60 years. Several vagrants of *M. nigrirostris* were collected in 1969. *MNR* is forbidden by compatibility rules.

(e) *Macropygia amboinensis* and *Reinwardtoena browni* have coexisted on Umboi since at least 1913. There are four records of vagrants of *M. mackinlayi* (*AMR* forbidden by combination rules).

On the other hand, two successful invasions of cuckoo-dove subcommunities have been documented, both resulting in the permitted combination *AM*. *M. mackinlayi* was the sole species breeding on Long in 1933, but by 1972 *M. amboinensis* as well had become a widely distributed breeder, though it was still less common than its congener. On Vuatom, *M. amboinensis* was the sole breeding species in 1906, but *M. mackinlayi* had also become an abundant breeder by 1930.

In all five cases of unsuccessful invasions cited, populations of the forbidden species breed on islands within a short distance of the island from which they are excluded (*M. mackinlayi* 1 mile from New Britain, *Reinwardtoena browni* and *M. nigrirostris* 4 miles from Vuatom, *M. nigrirostris* 10 miles from Karkar, *M. mackinlayi* 10 miles from Umboi). Only for

Vuatom are records of a long-term resident observer available. However, the conclusion seems inescapable for the other cases as well that the documented invasions of forbidden species are not isolated occurrences.

We can now reconsider the problem of the "empty squares" on the *Macropygia* checkerboard (Figure 20). The dilemma was that most Bismarck islands are occupied by either *M. nigrirostris* or *M. mackinlayi* in checkerboard fashion. However, six islands support neither species, and there is no single additional species whose distribution would complete the checkerboard. Of these six islands, the four most species-rich ($S = 40$–59) prove to support both *M. amboinensis* and *Reinwardtoena*. Although *M. mackinlayi* can coexist separately with either of these two species, the combination *AMR* is forbidden, as shown by the analysis of Table 8 as well as by *M. mackinlayi's* unsuccessful invasion of Umboi's *AR* community. While the combination *ANR* is not forbidden, its probability of occurrence exceeds 0.5 only on islands with $S > 80$ (Figure 27), indicating that an established *AR* combination can frequently exclude *M. nigrirostris* on more species-poor islands. The two most species-poor islands ($S = 38, 39$) that correspond to empty squares prove to support *M. amboinensis*. The combination *AN* is forbidden by incidence rules (lower $S_{crit} = 45$ for *M. nigrirostris*). The combination *AM* is not forbidden, but its probability of occurrence is only 0.35 on islands with $S = 38$ or 39, indicating that an established *M. amboinensis* population can often exclude *M. mackinlayi* on islands

with this *S*. Thus, the empty squares are actually filled by a species or a combination of species that make invasion difficult or impossible both for *M. nigrirostris* and for *M. mackinlayi.*

Assembly of the Gleaning Flycatcher Guild

Whereas some flycatchers sally to catch insects in mid-air, seven Bismarck species glean insects from leaves and branches: the B-tramp *Pachycephala pectoralis* (abbreviated *P*), supertramps *Pachycephala melanura dahli* (*D*) and *Monarcha cinerascens* (*C*), A-tramps *Monarcha verticalis* superspecies (*V*), *Monarcha chrysomela* (*R*), and *Myiagra hebetior* (*H*), and C-tramp *Myiagra alecto* (*A*). One or more of these flycatcher species is present on every Bismarck island larger than 10 acres, and on most smaller islets as well. No single island, however, supports all seven species. The smallest or most species-poor islands have only one or two species, whereas the richest islands have four or five species. Figure 28 gives incidence functions of all seven species.

Because the gleaning flycatcher guild contains more species and more possible combinations than does the cuckoo-dove guild, a complete analysis of assembly rules would require more islands than are available or have been explored in the Bismarcks. However, the available information does make certain regularities in assembly obvious.

Pachycephala pectoralis and *P. m. dahli* never coexist on an island, either by themselves or as part of a larger combina-

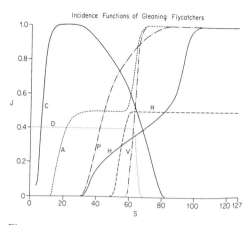

Figure 28 Incidence functions for the gleaning fly-catchers *Monarcha cinerascens* (*C*), *Pachycephala melanura dahli* (*D*), *Myiagra alecto* (*A*), *Pachycephala pectoralis* (*P*), *Myiagra hebetior* (*H*), *Monarcha verticalis* (*V*), and *Monarcha chrysomela* (*R*) in the Bismarcks.

tion. These species are products of a recent speciation and are the two most similar gleaning-flycatchers, both ecologically and morphologically. J_P is 0 for $S < 32$, and J_D is 0 for $S > 69$, but there are 14 well-explored islands in the range $32 < S < 69$ in which coexistence is not prohibited by incidence functions. The no-coexistence probability Z_{PD} that P and D by chance would not be found together on any of these 14 islands is calculated from eqs. 4 and 5 in Table 9, as 0.036. That is, the failure of P and D to coexist is not expected from their incidence functions, and we must assume P and D to be incompatible, as we might have guessed on biological grounds.

Pachycephala m. dahli and *Monarcha chrysomela* also coexist on no island, either by themselves or as part of a larger combination. In this case there is less

overlap of incidence functions than for *P. pectoralis* and *P. m. dahli:* only in the range $50 < S < 69$, where there are seven well-explored islands, are J_D and J_R both nonzero. Evaluation of the noncoexistence probability Z_{DR} from eq. 5 yields 0.37. That is, incompatibility is not established at a statistically significant level, and non-coexistence might perhaps have been ex-

Table 9. No-coexistence probability of the flycatcher species P and D

Species number	Observed incidences		Calculated probabilities	
S	J_P	J_D	$J_P J_D$	$1 - J_P J_D$
34	0.05	0.40	0.02	0.98
37	0.19	0.40	0.08	0.92
38	0.23	0.40	0.09	0.91
39	0.27	0.40	0.10	0.90
43	0.44	0.40	0.18	0.82
45	0.51	0.40	0.20	0.80
46	0.55	0.40	0.22	0.78
53	0.72	0.40	0.29	0.71
55	0.75	0.40	0.30	0.70
59	0.81	0.40	0.32	0.68
60	0.82	0.40	0.33	0.67
62	0.84	0.40	0.34	0.66
65	0.89	0.22	0.20	0.80
65	0.89	0.22	0.20	0.80

The flycatcher species P and D coexist on no known island. The table calculates whether this finding is expected by chance, given only the constraints of incidence functions. Only in the range $32 < S < 69$ are the incidences J_P and J_D both nonzero. Column 1 gives the species number of all 14 Bismarck islands in this S range whose resident flycatchers are known. Columns 2 and 3 are the incidences J_P and J_D corresponding to the S of each island, read from the incidence functions of Figure 28. Column 4 gives the probability (calculated from these incidences as $J_P J_D$) that species P and D will coexist on an island with this S; column 5 gives $p_{i,PD} = 1 - J_P J_D$ (eq. 4), the calculated probability that P and D will not coexist on the island. The product Z_{PD} of all the terms in the last column (eq. 5) is 0.036, the probability that P and D would coexist on none of the 14 islands if the only constraints were those of incidence functions. Since this probability is very low but P and D in fact coexist on none of the islands, their coexistence is presumed to be forbidden for some reason not implicit in incidence functions: i.e., they are presumed incompatible.

pected from incidence functions alone. *P. m. dahli* and *M. chrysomela* are sufficiently different (the latter smaller, moving more rapidly, avoiding dense vegetation) that incompatibility does not seem likely.

If an island has a single flycatcher species, that species is either species *C* (12 islands), *D* (8 islands), or rarely *A* (1 island). *H, P, R,* and *V* never occur alone.

If an island has two flycatcher species, only four combinations are permissible: *AC* (3 islands), *CD* (6 islands), *CP* (1 island), *AD* (1 island). As discussed above, the combination *DP* and conceivably the combination *DR* are incompatible, but all other forbidden two-species combinations occur as parts of larger combinations on islands with more flycatcher species.

If an island has three flycatcher species, the observed combinations are *APV* (3 islands), *CPR* (1 island), *ACP* (1 island), and *ACV* (1 island).

If an island has four flycatcher species, the observed combinations are (1 island each) *AHPV, CHPV, ACDV, ACDH, ACPR,* and *ACRV.* Note that the combination *CD* is favored over *AC* on two-flycatcher islands, while the reverse is true on four-flycatcher islands.

The only permissible five-species combination is *AHPRV* (3 islands).

In order to assess the extent to which these assembly rules are predictable from incidence functions alone, the simple case of the two-flycatcher islands was analysed approximately, as follows (Table 10). From the total species number S_i on each of the known two-flycatcher islands, $J(S_i)$ was read off of the incidence functions (Figure 28) for each species, the product

$J_a J_b$ was calculated for each possible fly-catcher pair *a* and *b* on each island, except for the incompatible pair *DP,* and the relative values of the products $J_a J_b$ for each pair summed over all islands were compared with the actual incidence of occurrence of each flycatcher pair. As shown in Table 10, this calculation predicts approximately the permitted two-flycatcher combinations and their relative frequency, except that the supertramp combination *CD* seems more frequent than expected. That is, these species may "fit together" particularly well, and this fit is even better than expected simply from their supertramp incidence functions. Observations on more real islands are desirable to be certain that this conclusion is statistically significant.

The favored combinations are resistant to invasion. Vagrants of the supertramp *Monarcha cinerascens* (*C*) have been recorded, but have apparently not established resident populations, on Umboi (resident combination *APV*), New Ireland (*AHPRV*), and Dyaul (*AHPRV*). Vagrants of the supertramp *Pachycephala melanura dahli* (*D*) have appeared on Tolokiwa (resident combination *CP*), Umboi (*APV*) and New Britain (*AHPV*). The regular occurrence of these supertramps on small islets close to larger islands from which they are absent suggests continuous bombardment of islands with supertramp-excluding combinations by supertramp vagrants.

Reconsideration of the "empty squares" of the *Pachycephala pectoralis–P. m. dahli* checkerboard (Figure 21) shows that each island without a *Pachycephala* species is

Table 10. Two-species combinations of gleaning flycatchers

S	Actual	J_A	J_C	J_D	J_P	J_H	J_R	J_V	J_AJ_C	J_AJ_D	J_CJ_D	J_CJ_P	J_AJ_P	J_HJ_A	J_HJ_C	J_HJ_D	J_HJ_P	J_RJ_P	J_RJ_C	J_RJ_A
10	CD	0	0.90	0.40	0	0	0	0	0	0	0.36	0	0	0	0	0	0	0	0	0
13	CD	0.08	0.98	0.40	0	0	0	0	0.08	0.03	0.39	0	0	0	0	0	0	0	0	0
17	CD	0.24	1.00	0.40	0	0	0	0	0.24	0.10	0.40	0	0	0	0	0	0	0	0	0
18	AC	0.28	1.00	0.40	0	0	0	0	0.28	0.11	0.40	0	0	0	0	0	0	0	0	0
32	CD	0.50	1.00	0.40	0	0	0	0	0.50	0.20	0.40	0	0	0	0	0	0	0	0	0
34	CD	0.50	0.97	0.40	0.06	0.03	0	0	0.49	0.20	0.39	0.06	0.03	0.02	0.03	0.01	0	0	0	0
37	AC	0.50	0.95	0.40	0.19	0.10	0	0	0.48	0.20	0.38	0.18	0.10	0.05	0.10	0.04	0.02	0	0	0
39	AC	0.50	0.93	0.40	0.28	0.14	0	0	0.47	0.20	0.37	0.26	0.14	0.07	0.13	0.06	0.04	0	0	0
45	CP	0.50	0.86	0.40	0.51	0.21	0	0	0.43	0.20	0.34	0.44	0.26	0.12	0.18	0.08	0.11	0	0	0
53	CD	0.50	0.75	0.40	0.72	0.28	0	0.14	0.38	0.20	0.30	0.54	0.36	0.14	0.21	0.11	0.20	0.10	0.11	0.07
62	AD	0.69	0.58	0.40	0.85	0.35	0.50	0.37	0.40	0.28	0.23	0.49	0.59	0.24	0.20	0.14	0.30	0.43	0.29	0.35
ΣJ_XJ_Y, all islands									3.75	1.72	3.96	1.97	1.45	0.64	0.85	0.44	0.67	0.53	0.40	0.42
Number of islands, predicted									2.2	1.0	2.3	1.2	0.9	0.4	0.5	0.3	0.4	0.3	0.2	0.2
Number of islands, actual									3	1	6	1	0	0	0	0	0	0	0	0

This table attempts to predict the relative frequency of each two-species combination of gleaning flycatchers on Bismarck islands, from the incidence functions of Figure 28. The first column lists the total species number on each Bismarck island supporting two (and only two) flycatcher species; the second column gives the actual two-species combination on each island. Columns 3–9 give J values for each species corresponding to the S of each island, read off from Figure 28. Columns 10–21 give the pair products of these J values, omitting the incompatible pair PD (cf. Table 9) and all pairs for which the sum of the product J_XJ_Y (summed over all 11 islands) is less than 0.4 (i.e., DR, HR, and all pairs that include species V). This sum of the products J_XJ_Y is given for all remaining pairs in row 12. The sum of these J_XJ_Y sums for all pairs except the incompatible pair PD is 18.53. Row 13 gives for each pair listed in the table the value of $(\Sigma J_XJ_Y)(11/18.53)$, the predicted number of islands on which the pair should occur if the only constraints were those of incidence functions plus incompatibility of PD. The last row gives the actual number of islands of occurrence. The actual numbers and predicted numbers agree fairly well, except that CD apparently occurs more often than expected.

occupied by one or more other species of the gleaning flycatcher guild. One "empty" island supports the combination *ACV,* 5 support *AC,* and 13 support *C* alone. Each of these combinations can coexist with one or either of the *Pachycephala* species on a larger (more species-rich) island but can exclude *Pachycephala* on a smaller (more species-poor) island.

Assembly of the Myzomelid-Sunbird Guild

In the New Guinea region two superficially similar groups of birds serve as ecological counterparts of the New World's hummingbirds (cf. Karr and James, Chapter 11, Figure 1): sunbirds (family Nectariniidae), represented in the Bismarcks by the D-tramp *Nectarinia jugularis* and the C-tramp *Nectarina sericea* (abbreviated *J* and *S,* respectively), and honeyeaters (family Meliphagidae) of genus *Myzomela,* represented in the Bismarcks by the supertramps *M. sclateri* (*L*) and *M. pammelaena* (*P*), the A-tramp *M. cruentata* (*C*), and the high-*S* species *M. cineracea* (*X*), *M. pulchella* (*Y*), and *M. erythromelas* (*Z*). Each of the latter three species is present only on one or two of the largest islands, but the other species are more widespread. Figure 29 presents incidence functions. All known Bismarck islands except one tiny islet and the small, recently defaunated volcano Ritter support at least one member of this guild. *M. cineracea* and *M. pammelaena* are larger than the other myzomelids, which are similar in size to the two sunbirds. In the Bismarcks, as elsewhere, the small myzo-

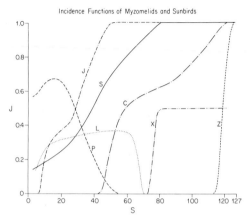

Figure 29 Incidence functions for the myzomelid-sunbird guild in the Bismarcks: *Myzomela pammelaena* (*P*), *M. sclateri* (*L*), *M. cineracea* (*X*), *M. erythromelas* (*Z*), *M. cruentata* (*C*), *Nectarinia jugularis* (*J*), and *N. sericea* (*S*). The remaining species, *M. pulchella,* is confined to an island with $S = 101$.

melids all maintain average spatial segregation. Low islands have only one small myzomelid, whereas the two highest Bismarck islands have two species segregating by altitude. The two sunbirds overlap spatially with each other but tend to segregate spatially from the similar-sized small myzomelids (cf. Figure 41). Large myzomelids such as *X* and *P* overlap spatially with small myzomelids or sunbirds. For instance, *X* and *S, X* and *Z, X* and *C, P* and *J,* and *P* and *L* overlap spatially in the Bismarcks. All eight species are similar in taking nectar and insects from flowers, and in gleaning insects off leaves and twigs. The guild is of interest in consisting of two distinct families whose species are nevertheless exceptionally similar in ecology and exhibit obvious interfamilial aggressive behavior (Ripley,

Table 11. Two-species combinations of myzomelids and sunbirds

S	Actual	J_P	J_L	J_J	J_S	J_C	J_LJ_P	J_JJ_P	J_PJ_S	J_JJ_L	J_LJ_S	J_JJ_S	J_CJ_L	J_CJ_J	J_CJ_S	J_CJ_P
9	LS	0.63	0.24	0.15	0.18	0	0.15	0.09	0.11	0.04	0.04	0.03	0	0	0	0
9	LS	0.63	0.24	0.15	0.18	0	0.15	0.09	0.11	0.04	0.04	0.03	0	0	0	0
15	JP	0.67	0.30	0.32	0.22	0	0.20	0.21	0.15	0.09	0.07	0.07	0	0	0	0
16	JP	0.68	0.30	0.33	0.22	0	0.20	0.22	0.15	0.10	0.07	0.07	0	0	0	0
17	JP	0.68	0.30	0.34	0.23	0	0.20	0.23	0.16	0.10	0.07	0.08	0	0	0	0
32	JP	0.40	0.35	0.61	0.36	0	0.14	0.24	0.14	0.21	0.13	0.22	0	0	0	0
37	JS	0.27	0.35	0.74	0.44	0	0.09	0.20	0.12	0.26	0.15	0.33	0	0	0	0
38	JS	0.24	0.35	0.76	0.46	0	0.08	0.18	0.11	0.27	0.16	0.35	0	0	0	0
43	JP	0.14	0.36	0.88	0.58	0	0.05	0.12	0.08	0.32	0.21	0.51	0	0	0	0
45	PL	0.11	0.36	0.92	0.61	0.02	0.04	0.10	0.07	0.33	0.22	0.56	0.01	0.02	0.01	0
55	JS	0	0.37	1.00	0.75	0.44	0	0	0	0.37	0.28	0.75	0.16	0.44	0.33	0
59	JS	0	0.36	1.00	0.80	0.50	0	0	0	0.36	0.29	0.80	0.18	0.50	0.40	0
ΣJ_XJ_Y, all islands							1.30	1.68	1.20	2.49	1.73	3.80	0.35	0.96	0.74	0
Number of islands, predicted							1.1	1.4	1.0	2.1	1.5	3.2	0.3	0.8	0.6	0
Number of islands, actual							1	5	0	0	2	4	0	0	0	0

This table attempts to predict the relative frequency of each two-species combination of myzomelids and sunbirds on Bismarck islands in the same manner as Table 10, from the incidence functions of Figure 29. The first column lists the total species number on each Bismarck island supporting two (and only two) myzomelid-sunbird species; the second column gives the actual two-species combination on each island. Columns 3–7 give J values for each species corresponding to the S of each island, read off from Figure 29 and omitting species X, Y, and Z, for which J is 0 throughout the range $9 \leq S \leq 59$. Columns 8–16 give the pair products of these J values, and row 13 gives the sum of the product J_XJ_Y (summed over all 12 islands) for each pair. The sum of all these J_XJ_Y sums for all pairs is 14.25. Row 14 gives for each pair the value of $(\Sigma J_XJ_Y)(12/14.25)$, the predicted number of islands on which the pair should occur if the only constraints were those of incidence functions. The last row gives the actual number of islands of occurrence. The actual numbers and predicted numbers agree fairly well, except that JP occurs more often, and PS and JL less often, than expected.

1959) and competitive effects on distributions (Diamond, 1970a).

As with the gleaning flycatcher guild, the number of species in the myzomelid-sunbird guild precludes a complete analysis of assembly rules based on the available number of real islands. However, some regularities in assembly are obvious:

If a Bismarck island contains a single guild member, it is either *P* (14 islands), *L* (5 islands), *S* (3 islands), or *J* (2 islands), never *C, X, Y,* or *Z*.

If an island contains two members, the permissible combinations are *JP* (5 islands), *JS* (4 islands), *LS* (3 islands), or *LP* (1 island). The four permuted species are the same ones as are permitted on one-species islands. However, two of the six possible pairs drawn from these four species, *PS* and *JL,* never occur on two-species islands, although *J* and *L* do coexist on three-species islands. The species *C, X, Y,* and *Z* are absent from two-species islands as well as from one-species islands.

On three-species islands the permissible combinations are *JSL* (3 islands), *JSX* (1 island), *JSC* (3 islands), and *JPL* (2 islands). Only 2 islands have higher combinations (*JSCY* and *JSCXZ,* each on 1 island). In the Bismarcks as elsewhere, the combination *JS* appears as part of most three-species combinations and all four- and five-species combinations.

Table 11 attempts to predict the permitted two-species combinations and their relative frequencies on the basis of incidence functions alone. The calculation is carried out in the same manner as for the gleaning flycatchers in Table 10. It appears that the incidence functions fail to predict the nonoccurrence of the combinations *PS* and *JL,* or the high frequency of *JP.* These combinations must have special difficulties or special advantages of fit. Frequencies of other combinations are approximately as predicted from incidence functions.

Reexamination of the numerous empty squares in the checkerboard formed by the distributions of the four small myzomelids *C, L, Y,* and *Z* shows that 29 of the 31 empty squares are occupied by another guild member (*J, P,* or *S*), or else by a pair or trio of guild members (*JP, JS,* or *JSX*).

Assembly of the Fruit-Pigeon Guild

On New Guinea live 18 species of fruit pigeons, 12 smaller species in genus *Ptilinopus* and 6 larger species in genus *Ducula.* All are ecologically similar in being arboreal, living in the crowns rather than in the middle story, being exclusively frugivorous, not taking stones into the gizzard, and hence restricted to eating soft fruits that can be crushed by the gizzard wall. At a single locality in New Guinea lowland rainforest one encounters no more than 8 species, which form a graded size series, each species weighing approximately 1.5 times the next smaller species (average weights 49, 76, 123, 163, 245, 414, 592, and 802 g: see Figure 30). Size differences permit resource partitioning in two ways. First, larger pigeons can eat larger fruits, and may not find it energetically worthwhile to harvest smaller fruits (Figure 31; see also Hespenheide, Chapter 7). Second, smaller pigeons can perch on

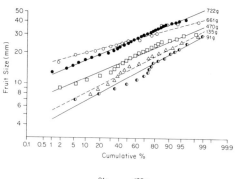

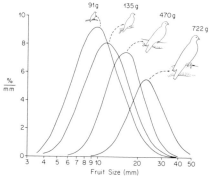

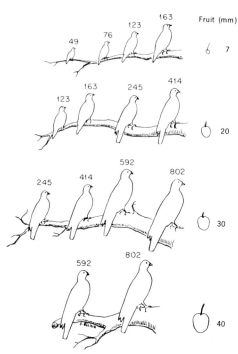

Figure 30 Schematic representation of niche relations among the eight species of *Ptilinopus* and *Ducula* fruit pigeons in New Guinea lowland rain forest. On the right is a fruit of a certain diameter (in millimeters), and on the left are pigeons of different weights (in grams) arranged along a branch. Each pigeon weighs approximately 1.5 times the next pigeon. Each fruit tree attracts up to four consecutive members of this size-sequence. Trees with increasingly large fruits attract increasingly large pigeons. In a given tree the smaller pigeons are preferentially distributed on the smaller, more peripheral branches. The pigeons having the weights indicated are: 49 g, *Ptilinopus nanus;* 76 g, *P. pulchellus;* 123 g, *P. superbus;* 163 g, *P. ornatus;* 245 g, *P. perlatus;* 414 g. *Ducula rufigaster;* 592 g, *D. zoeae;* 802 g, *D. pinon* (from Diamond, Copyright 1973 by the American Association for the Advancement of Science.)

Figure 31 Proportions of fruit of different sizes eaten by fruit-pigeon species of different sizes. Above: cumulative proportion plotted on probability paper (cf. Cody, Chapter 10, Figure 7). Diameters of fruits in the stomachs of *Ducula rubricera* (symbols ●, body weight 722 g), *D. melanochroa* (○, 661 g), *D. pistrinaria* (□, 470 g), *Ptilinopus rivoli* (△, 135 g), and *P. solomonensis* (○, 91 g) were measured. The abscissa is the "cumulative" number of fruits up to a given diameter (ordinate, logarithmic scale), expressed as percentage of total number of fruits. Of the fruits measured as having a particular diameter, half were considered to have been accumulated in the interval preceding that diameter and half in the subsequent interval. The abscissa scale is such that if fruit sizes followed a log-normal distribution, the experimental points would define straight lines. This is approximately but not perfectly true; the deviations may be due to only a limited number of tree species in the size range suitable for each pigeon species being in fruit at the sampling time. All data were obtained in June–August 1972 on the Vitiaz-Dampier Islands. In the figure below, the straight lines fitted through the experimental points of the upper figure have been replotted as conventional log-normal curves (abscissa, log fruit size; ordinate, frequency distribution of sizes, expressed as percentage of all fruit lying within a size span of 1-mm width around the size value of the abscissa). Because the abscissa is logarithmic but the width-span linear, the areas under the curves of pigeons eating larger fruits are smaller. Note that larger pigeons eat larger fruits.

thinner twigs that would not support the weight of a heavier bird, and they can thereby reach some fruits inaccesible to a heavy bird. The remaining species besides the eight lowland rainforest birds are ten habitat vicariants that replace these rainforest species at higher altitudes, in savanna, in dry forest, or coastal forest. For instance, *Ptilinopus pulchellus* (76 g) in high-rainfall areas is replaced by *P. coronulatus* (75 g) in lower-rainfall areas; *P. superbus* (123 g) in forest is replaced by *P. iozonus* (112 g) in open country; and *Ducula zoeae* (592 g) at low elevations is replaced by *D. chalconota* (613 g) at high elevations. The eight New Guinea lowland rainforest species and their habitat vicariants can be considered as filling eight size levels, all of which except levels 1 (~49 g) and 5 (~245 g), corresponding to the smallest and largest species of genus *Ptilinopus*, are degenerate, i.e., occupied by several species. On a satellite island colonized by a vicariant but not by the corresponding lowland rainforest species, the vicariant generally expands into lowland rainforest. Assembly of fruit-pigeon guilds in lowland rainforest of satellite islands exhibits regular patterns with respect to the sequence in which levels are emptied as the guild becomes impoverished (Figure 32). (Figures 30, 31, and 32 resemble Figures 5, 7, and 4, respectively, in Chapter 13 by Brown, demonstrating similar size levels and vicariant species in desert rodent communities.) The fact that a given level may be occupied by any one of up to six species makes these regularities of community structure all the more striking, in their independence of species composition:

Five islands, exemplified by Salawati, Waigeu, and Japen in Figure 32, support 10–13 guild species. On two of these five islands (Salawati and Waigeu) all 8 levels are full, on two islands (e.g., Japen) level 1 (~49 g) is empty, and on one island level 5 (~245 g) is empty. Only level 8 (~800 g) is filled by the same species on all five islands and on New Guinea. For example, level 2 (~75 g) is occupied by *P. pulchellus* on three islands and on New Guinea, but by *P. coronulatus* on the remaining two islands.

Four islands, exemplified by Batanta, Manam, and Goodenough in Figure 32, support seven or eight guild species. Each island has either 5 or 6 levels occupied. Level 1 (~49 g) is empty on all four islands, and is also empty on all islands with

Figure 32 This figure shows how size levels of the fruit-pigeon guild are filled on islands with progressively poorer faunas. For each island named on the right, the rainforest fruit-pigeon species or habitat vicariant at each of eight size levels is indicated by a mark whose abscissa position indicates the bird's average weight (abscissa in grams, logarithmic scale), together with a letter abbreviation identifying the species. The numbers at the left are the number of fruit-pigeon species on each island. Most weights are average weights for the range of the species, but if geographical variation in size is well established, the weight for the local population is given (notice geographical variation in species *q*, *m*, and *p*). Abbreviations: *p* = *Ducula pinon*, *u* = *D. chalconota*, *z* = *D. zoeae*, *t* = *D. spilorrhoa*, *j* = *D. myristicivora*, *k* = *D. pistrinaria*, *r* = *D. rufigaster*, *y* = *Ptilinopus perlatus*, *m* = *Megaloprepia* (= *Ptilinopus*) *magnifica*, *o* = *P. ornatus*, *q* = *P. rivoli*, *s* = *P. superbus*, *l* = *P. solomonensis*, *c* = *P. coronulatus*, *h* = *P. pulchellus*, *n* = *P. nanus*. Notice that on progressively more impoverished islands (i.e., going from top to bottom in the figure) the first level to empty is level 1, followed by levels 2 and 5, followed by level 8.

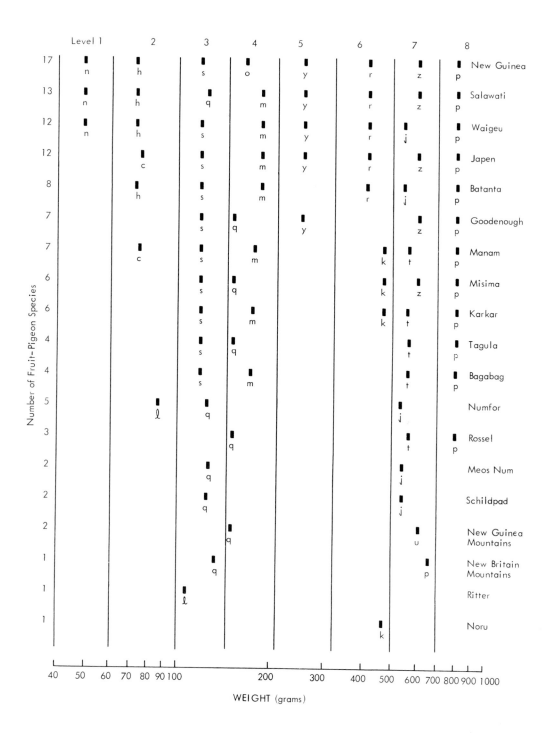

fewer than seven guild species. Both level 2 (~75 g) and level 5 (~245 g) are empty on two islands, and levels 3 (~123 g), 4 (~163 g), 7 (~592 g), and 8 (~802 g) are occupied on all four islands. For instance, level 7 (~592 g) is occupied by *Ducula myristicivora* (~535 g) on Batanta, by *D. spilorrhoa* (~563 g) on Manam, and by *D. zoeae* (~592 g) on Goodenough and Fergusson.

Seven islands, exemplified by Misima, Karkar, Tagula, Bagabag, and Numfor in Figure 32, support four to six guild species. Again, level 1 (~49 g) is empty on all islands. Level 5 (~245 g) is also empty on all islands, and levels 2 (~75 g) and 6 (~414 g) on most. Levels 7 or 8 or both, and 3 or 2 or both (usually just 3), are occupied on all islands.

Six islands, exemplified by Rossel and Meos Num in Figure 32, support three guild species. Levels 1, 2, and 5 are empty on all islands, and levels 4 and 8 are empty on all but one. Each island has one *Ptilinopus* species in level 3 (usually) or 4 (one island), and one *Ducula* species in level 7 (usually) or 8 (one island). The remaining species is always either another *Ptilinopus* vicariant in level 3 or another *Ducula* vicariant in level 7.

Four islands, exemplified by Schildpad in Figure 32, support two guild species. Three further "habitat islands" supporting two guild species are the mountains of New Guinea, the mountains of New Britain, and the mountains of New Ireland. One species is always a *Ptilinopus* in level 3 (usually) or level 4 (once), and the other is always a *Ducula* in level 6 or 7.

Eleven islands, exemplified by Ritter and Noru in Figure 32, support one guild

species. This is always either a *Ptilinopus* in level 3 or a *Ducula* in level 6.

Thus, as the guild is progressively simplified on smaller or more remote islands, the first level to become empty is level 1, followed by levels 2 and 5, followed by level 8. These can be considered levels for high-S species only. Not only are these particular levels precarious, they are also the levels that on New Guinea either lack a habitat vicariant (levels 1 and 5) or have just one vicariant (levels 2 and 8). The precarious levels are the largest or smallest species of each genus, while the simplified communities generally consist of medium-size *Ptilinopus* and *Ducula*.

Of particular interest is a comparison of fruit-pigeon guilds on incompletely equilibrated islands with guilds on fully equilibrated islands. As discussed in the section "Background: the New Guinea biogeographic scene," species numbers on large islands that were connected to New Guinea by land bridges until the end of the Pleistocene are still far above equilibrium values expected from the species/area relation. For example, the land-bridge islands Japen and Waigeu each has about the same area as the oceanic island Biak, but each of the former has about twice as many total species, and three times as many species of fruit pigeons, as Biak. Some "contracted" islands that lacked land bridges to New Guinea but whose late-Pleistocene areas were much larger than their present areas are also still "supersaturated" in species number. Yet guild composition on these supersaturated islands is very similar to that on larger islands whose species number is at equilibrium. For instance, the same levels are occupied on the supersaturated land-bridge island Batanta as on the equilibrated island Manam (levels 2, 3, 4, 6, 7, 8); the supersaturated land-bridge islands Salawati and Waigeu have all 8 levels occupied, as on New Guinea, which is several hundred times larger; and the level occupancy on the supersaturated contracted islands Goodenough and Fergusson (levels 3, 4, 5, 7, 8) is nearly the same as that on the equilibrated islands Misima and Karkar (levels 3, 4, 6, 7, and 8). Misima actually shares five of its six species in common with Goodenough. Similarly, the exploded volcanoes Long and Ritter, which are still "underfilled" as to species number, have the same sets of fruit-pigeon species as those on islands one-tenth and one-hundredth, respectively, of their area.

Thus, a community not at equilibrium with respect to species number is similar, in the structure of its fruit-pigeon guilds, to an equilibrium community on a larger or smaller island. This implies that the time required for an island's species to become coadjusted to each other, and to sort into stable combinations by expulsion of excess species or by incorporation of suitable invaders, is much shorter than the relaxation time for equilibration of species number on the island. As a fauna relaxes towards equilibrium species number, community structure traverses a series of internally coadjusted states. Consideration of transition probabilities among permitted combinations suggests that relaxation may sometimes occur in steps (see section "Transition probabilities" for further discussion).

Finally, the fruit-pigeon guild is of in-terest in illustrating how an established species that could resist a single invader may be eliminated by two simultaneous invasions. Three *Ptilinopus* species are widespread in the lowlands of Bismarck islands: *P. superbus, P. insolitus,* and *P. solomonensis.* Each of the three possible two-species combinations (*superbus-insolitus, superbus-solomonensis, insolitus-solomonensis*) occurs on at least five islands, but there is no well-established instance of stable coexistence of all three species. The *solomonensis-insolitus* combination has successfully resisted *superbus* invaders on Vuatom and possibly on Tolokiwa, and the *superbus-insolitus* combination has successfully resisted *solomonensis* invaders on New Britain. In 1913 *P. solomonensis* was the sole *Ptilinopus* in the lowlands of Umboi, and was apparently abundant. Since Umboi is a large and species-rich island that could easily support two *Ptilinopus* species, an invasion either by *P. insolitus* or by *P. superbus* might have been predicted to have excellent prospects of success leading to coexistence with *P. solomonensis.* In fact, some time between 1913 and 1933 *both* of these two other species invaded, and by 1933 had become established though still uncommon. By 1972 both *P. insolitus* and *P. superbus* had become widespread, and *P. solomonensis* had contracted into a small area and narrow altitudinal range, where its chances of survival seem precarious. MacArthur (1972, pp. 43–45) has pointed out theoretical reasons why a species sandwiched between two close competitors is in a much more precarious situation than if it coexisted with just one competitor.

Replicate Communities
on the Same Island

So far, we have been comparing the species composition of communities in physically similar environments on different islands. Let us now compare communities in physically similar environments at different localities within the same island.

Figure 33 presents a simple example. Natural grassland in New Guinea is largely confined to lowland marsh and savanna areas, mountaintops above timberline (ca. 11,000 feet), and landslide areas and strips along lakes and streams. Within the last few millenia human agriculture at 3000–8000 feet has created midmontane grassland "islands," which have been colonized by eight different species of grass finch of genus *Lonchura* (Diamond, 1972a, pp. 409–410). Four of these species are widespread in the lowlands, one is confined to the lowlands of southeast New Guinea, one is widespread in alpine grassland, and two are localized species of a midmontane lake or river system; three other localized lowland species failed to colonize midmontane grassland. In any given area a single finch species is ubiquitous in midmontane grassland over a considerable range of grass types and heights, altitudes, and rainfall conditions. However, the identity of the locally successful colonist varies in a geographically irregular checkerboard (Figure 33). This distribution pattern is very similar to the simple island checkerboards of Figures 20–24.

Figure 34 is a more complex checkerboard. Three similar honeyeaters of genus

Melidectes (*M. ochromelas, M. belfordi,* and the *M. rufocrissalis* superspecies) live in midmontane forest. When considered individually, each has a peculiarly disjunct range and is absent from several portions of the New Guinea cordillera. Each species can coexist with each other species, so that the individual ranges do not form a checkerboard. When the ranges of the three species are considered together, however, it is clear that (a) each mountainous area supports two species that exclude the third; (b) the identity of the locally successful combination varies in irregular checkerboard fashion; and (c) each of the three possible combinations occurs in several areas (Diamond, 1972a, pp. 387–389). Among the three alpine species of *Melidectes* there is a similar checkerboard also involving all three possible two-species combinations, and there is evidence from one area that the locally missing third species has unsuccessfully invaded and is excluded. This pattern is similar to the two-species island checkerboard of lowland *Ptilinopus* pigeons in the Bismarcks (see discussion of *Ptilinopus superbus, P. insolitus,* and *P. solomonensis* in the section just preceding).

Figures 33 and 34 present in effect the ranges of patchily distributed species whose closest competitors are few in number and easy to guess biologically, and whose distributions combined with the distributions of the competing species form neat single-species (Figure 33) or two-species (Figure 34) checkerboards. Many more species are patchily distributed without any neat pattern involving competitors being apparent (Figures 35–38). For example, the tree creeper

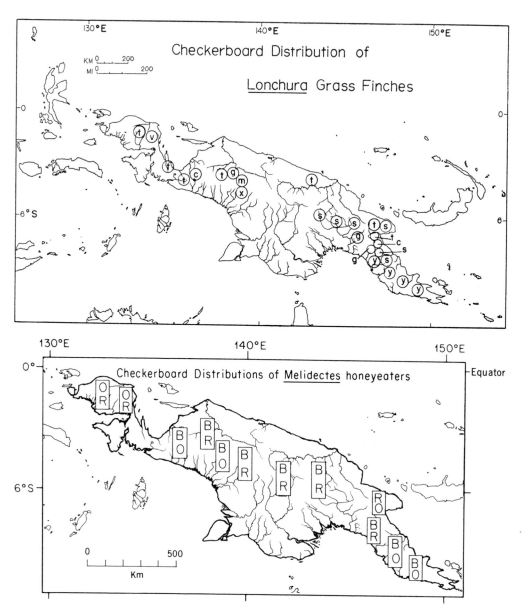

Figure 33 Checkerboard distribution of *Lonchura* grass finches in midmontane grasslands of New Guinea. These grasslands, which are largely a recent by-product of human agriculture, have been colonized by eight finch species of lowland grassland, alpine grassland, or midmontane lake or river systems. Most localities support only one species, but the identity of the locally successful colonist shows irregular geographical variation. *c = Lonchura castaneothorax, g = L. grandis, m = L. montana, s = L. spectabilis, t = L. tristissima, v = L. vana, x = L. teerinki, y = L. caniceps.*

Figure 34 Distributions of three *Melidectes* honeyeaters in the mountains of New Guinea (*O = M. ochromelas, B = M. belfordi, R = M. rufocrissalis* superspecies). Most mountainous areas of New Guinea support two species with mutually exclusive altitudinal ranges. At each locality depicted on the map, the letters above and below indicate the species present at higher and lower altitudes, respectively. All three possible two-species combinations occur in several disjunct areas, and the identity of the locally successful combination varies in irregular checkerboard fashion. (After Diamond, 1973.)

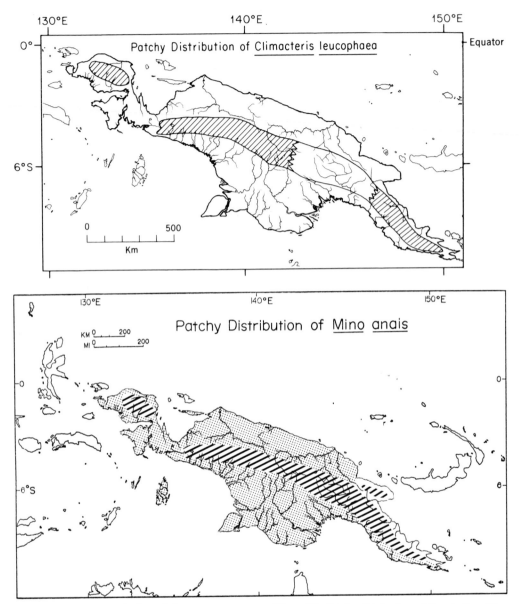

Figure 35 Patchy distribution of the tree creeper *Climacteris leucophaea* in the mountains of New Guinea. Although mountains and forest with similar tree bark extend uninterrupted for 1000 miles, and although there is no other New Guinea bird in the same family, this bark-feeding species has a distributional gap (unshaded area) of 250 miles in the middle of its range (hatched area). (From Diamond, 1972a.)

Figure 36 Patchy distribution of the starling *Mino anais* in the lowlands of New Guinea. Areas of occurrence are shaded, and unsuitable mountainous areas are crosshatched. Notice that the species is absent from the lowlands on the northern watershed of southeast New Guinea (unshaded area), despite the presence of suitable habitat, lack of any geographical representative in this area, and continuity of habitat with areas populated by the species.

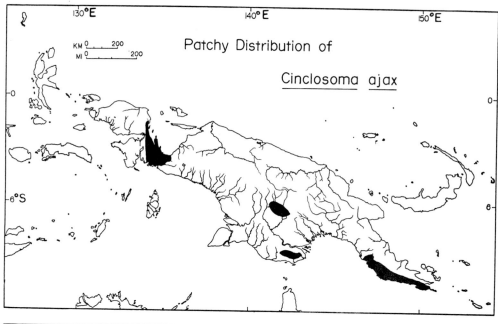

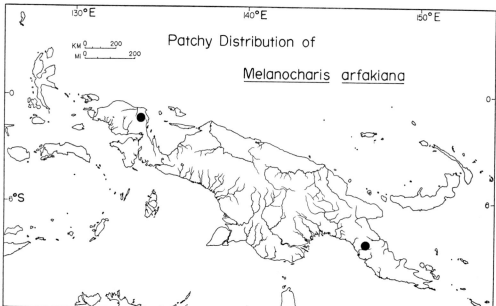

Figure 37 Patchy distribution of the logrunner *Cinclosoma ajax* on New Guinea. The species occurs only in four far-flung areas.

Figure 38 Distribution of the berrypecker *Melanocharis arfakiana* on New Guinea, the ultimate in patchiness. This species is known only from two localities at opposite ends of New Guinea, 1000 miles apart.

Climacteris leucophaea, the sole New Guinea bird in its family, is present in the mountains of eastern and western New Guinea but absent for a distance of 250 miles in the center of New Guinea (Figure 35). The starling *Mino anais* is widespread in the New Guinea lowlands but absent for a distance of 450 miles in northeast New Guinea (Figure 36). The logrunner *Cinclosoma ajax* occurs in four scattered areas of the lowlands (Figure 37). The berrypecker *Melanocharis arfakiana* is known from two localities 1000 miles apart, at opposite ends of the central cordillera (Figure 38). Dozens of similar cases can be cited of conspicuous New Guinea bird species that are present in certain disjunct areas but are absent from hundreds of ecologically similar, ornithologically well-explored localities elsewhere in New Guinea. A skeptic can always dismiss these cases by suggesting that the distributional gaps are deficient in some unspecified but ecologically essential factor. However, many of these cases involve species whose ecological requirements are sufficiently well understood to permit reasonable confidence that their distributional gaps are not due to any feature of the physical environment. This phenomenon of "inexplicably" patchy distributions, i.e., patchy distributions not explicable in terms of habitat conditions, is a distinctive feature of tropical communities compared with temperate communities (MacArthur, 1972, pp. 207–210 and 231–235; Diamond, 1972a, 1973). Many of these distributional patterns may prove to involve species excluded by constellations of competitors, or species that are gradu-

ally disappearing and are being outcompeted except on patchy "hot spots". It is suggestive, for example, that the patchily distributed flycatcher *Poecilodryas placens* occurs mainly at localities where its two closest relatives are either both absent, both rare, or only one present.

The problems posed by different species communities in physically similar environments at different localities within the same island, and in physically similar environments on different islands, are formally identical. Both situations involve the question how replicate communities formed from the same species pool can differ. The differences between the inter-island and intra-island situations are twofold. First, immigration is obviously over water in the former case, over land in the latter. It should not be assumed, however, that dispersal between nearby regions of the New Guinea mainland is much more efficient for New Guinea species than is inter-island dispersal for Bismarck species, since many New Guinea mainland species are highly sedentary. Second, there are more than three times as many species on New Guinea as in the Bismarcks, and a tendency towards patchiness is expected to increase with increasing number of species in the species pool (MacArthur, 1972, pp. 233–234).

Communities at the Habitat Level

So far, we have considered a community as consisting of all the species on a single island, or all the species in a certain geographical area of a large island like

New Guinea. However, within each geographical area, species segregate to varying degrees by habitat. A geographical area consists of a group of habitat communities, close enough to each other so that physical barriers to dispersal are negligible, but each community is still co-adjusted in its species composition and is continuously resisting invasion by species of adjacent habitat communities. The assembly of such habitat communities is discussed in detail in Chapter 10 by Cody. Numerous papers in the recent ecological literature have studied "niche shifts" in which the range of habitat communities to which a given species belongs is constricted by competition. Although interpretations in terms of predation or other environmental variables must also be considered in such cases (cf. Connell, Chapter 16), especially for species at lower trophic levels, effects of competition are easiest to recognize when range of habitat occupancy can be related to the presence or absence of a single competitor.

Figure 39 presents a simple example of competitive effects on altitudinal range among three very similar parrots of genus *Vini. V. rubronotata* is confined to the New Guinea region, and *V. rubrigularis* to the Bismarcks, while *V. placentis* occurs in both regions. On islands or mountain ranges that *V. placentis* shares with a congener (New Britain shared with *V. rubrigularis,* the North Coastal Range of New Guinea shared with *V. rubronotata*), *V. placentis* is confined to low elevations, the congener to high elevations, and there is little or no altitudinal overlap. On islands or mountain ranges where *V. placentis*

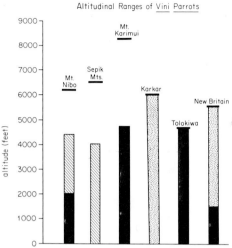

Figure 39 Altitudinal ranges of three similar, congeneric parrots on six mountains of the New Guinea region. Solid shading, *Vini placentis;* diagonal bars, *V. rubronotata;* dots, *V. rubrigularis.* The elevation of the summit of each mountain is indicated by a heavy horizontal bar just under the mountain's name. The left-most three mountains are in different mountain ranges of New Guinea, and the right-most three are separate islands. On mountain ranges or islands where *V. placentis* plus either *V. rubronotata* or *V. rubrigularis* is present, the former is confined to low elevations, the latter to high elevations, with little altitudinal overlap. Where *V. placentis* occurs alone, it ascends to high elevations. Where *V. rubronotata* or *V. rubrigularis* occur alone, they descend to sea level.

occurs alone (Adelbert Range, Mount Karimui, Tolokiwa Island), it spreads upwards to elevations of 4000 feet or higher. On islands or mountain ranges where *V. rubronotata* (Biak Island, Sepik Mountains) or *V. rubrigularis* (Karkar Island) occurs alone, it descends to sea level.

Figure 40 is another simple example, demonstrating competitive effects on type of habitat occupied. On New Guinea, as one moves progressively inland from the

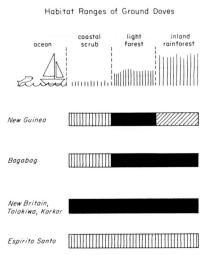

Figure 40 Habitats occupied by three species of ground doves on various islands. As one proceeds inland from the coast on New Guinea, one encounters each species in turn: *Chalcophaps indica* (vertical bars) in coastal scrub, *Chalcophaps stephani* (solid shading) in light forest, and *Gallicolumba rufigula* (diagonal bars) in inland rainforest. On Bagabag, where *G. rufigula* is absent, *C. stephani* expands into inland rainforest. On New Britain, Tolokiwa, and Karkar, where *C. indica* is also absent, *C. stephani* also expands into coastal scrub to occupy the whole habitat spectrum. On Espiritu Santo, where only *C. indica* occurs, it occupies the whole habitat spectrum.

coast, from coastal scrub to light forest (or second-growth) to rainforest, one encounters in sequence three similar species of ground doves: *Chalcophaps indica, C. stephani,* and *Gallicolumba rufigula.* On Bagabag, where *G. rufigula* is absent, *C. stephani* expands inland into rainforest. On Karkar, Tolokiwa, New Britain, and numberous other islands where *C. indica* is also absent, *C. stephani* expands coastally to occupy the whole habitat gradient. In the New Hebrides, where *C. indica* is

the sole species, it occupies the whole habitat gradient.

A far larger number of cases differ from the neat examples of Figures 39 and 40, in that the geographically varying habitat spectrum occupied by a given species cannot be correlated with the presence or absence of a single competitor. The variation may instead be correlated with changes in a constellation of competitors. Figure 41 illustrates a case that is complex, but out of which some sense, though not yet complete sense, can be made. The figure shows that within the myzomelid-sunbird guild, which we previously discussed at the gross level of whether or not a species occurs on an island, each species may strikingly alter its habitat occupancy from island to island, as a function of what other guild members are present. *Myzomela sclateri* may be ubiquitous (Long), confined to cloud forest (Karkar), or confined to small offshore islets (New Britain); *Nectarinia sericea* may be ubiquitous (Bagabag), only in lowland and midmontane forest (Karkar), only in lowland forest (Umboi), only in coastal vegetation and forest edge (New Britain), or only in forest edge (New Guinea); *Myzomela cruentata* may be confined to montane forest (New Britain), or in lowland forest (Tabar); and *Myzomela cineracea* may be at all elevations in forest (Umboi), only in lowland and mid-montane forest (New Britain, only in lowland forest (widespread New Guinea allospecies), or only in lowland savanna (south New Guinea allospecies). When information about enough islands becomes available, it seems likely that it will be possible to

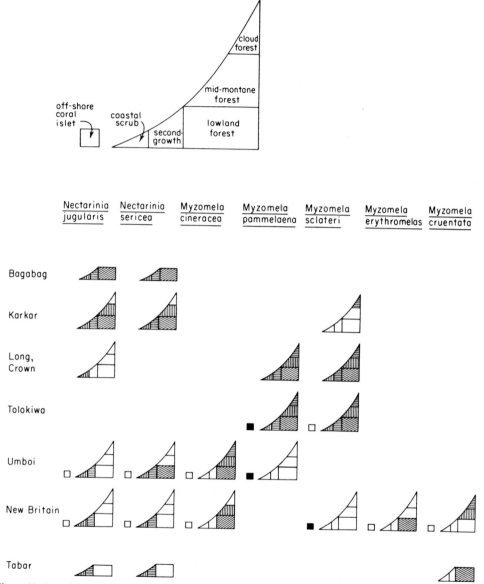

Figure 41 Inter-island variation in habitat partitioning among species of the myzomelid and sunbird guild. The sketch in the upper left shows how the major habitats are pictured. For each island, the habitats occupied by each guild species are shown as variously shaded. Note that the range of habitats occupied by each species varies greatly from island to island, depending on what other species are present. Bagabag and Tabar Islands are not high enough to support midmontane forest or cloud forest.

formulate assembly rules for the subsets of this guild within a habitat. At present the following rules may be tentatively suggested: 1. Two small *Myzomela* species cannot extensively share a habitat. 2. Two large *Myzomela* species cannot extensively share a habitat. 3. The two *Nectarinia* species can extensively share a habitat. 4. One small *Myzomela* species and one large *Myzomela* species can share forest, but cannot share the scrub of a small islet. 5. A small *Myzomela* species can coexist with one but not with both *Nectarinia* species. 6. One *Nectarinia* species can coexist with one small *Myzomela,* but not with one small and one large *Myzomela,* except that *N. jugularis* can sometimes do so precariously in coastal vegetation. Similar assembly rules for habitat subsets can be tentatively formulated for the gleaning flycatcher guild.

Figure 42 illustrates a still more complex case, in which the competition that one species faces is spread even more diffusely over more species. The thrush *Turdus poliocephalus* has a very variable niche. On New Guinea it is confined to the edge of alpine forest and alpine grassland, above 9000 feet. Its habitat on Ceram and probably on Goodenough is similar. On some other islands, such as Bougainville, Guadalcanal, Karkar, and Tolokiwa, it is in montane forest, descending to a lower level that varies between 2460 and 4000 feet. On still other islands, such as Espiritu Santo and Viti Levu, it occurs in forest at all altitudes down to sea level. It also lives on Rennell, a raised coral atoll whose elevation nowhere exceeds 360 feet. Although some of the finer differences may be due to differences in forest physiognomy at a given elevation on islands of different sizes and elevations (the so-called Massenerhebung effect), some other explanation is required for the much grosser differences, given the extraordinary ability of this thrush variously to adapt to alpine grassland, cloud forest, midmontane forest, lowland forest, and coral atolls. The species is equally catholic in its diet (both insects and fruit), so that it will be difficult to delimit its guild and identify a critical constellation of competitors. However, a suggestive clue is that on any Pacific island the number of species decreases with increasing elevation, and that a forest community with a given number of species may therefore be encountered at sea level on a species-poor island but only at much higher elevations on a species-rich island. As indicated on Figure 42, the maximum number of species in the habitat communities occupied by *Turdus poliocephalus* on each island shows much less inter-island variation (23 to 36 species) than does the number of total species on each island (42 to 513). By analogy, then, with incidence functions such as Figures 6 below, 14, and 15, one can describe *Turdus poliocephalus* as a supertramp with an upper S_{crit} in the range 23–36, capable of invading only those habitat communities with this number of species or fewer.

Competitive effects similar to the effects on habitat range illustrated in Figures 40 and 41, and similar to the effects on altitudinal range illustrated in Figures 39 and 42, also exist for vertical foraging range. An example of simple effects involving

Altitudinal Range of <u>Turdus</u> poliocephalus

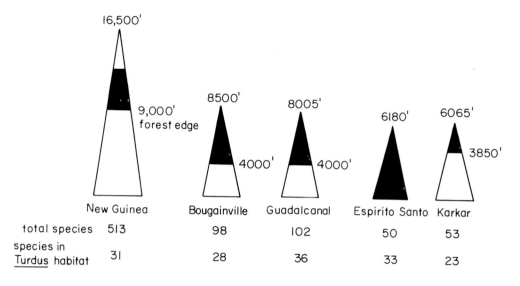

	New Guinea	Bougainville	Guadalcanal	Espirito Santo	Karkar
total species	513	98	102	50	53
species in <u>Turdus</u> habitat	31	28	36	33	23

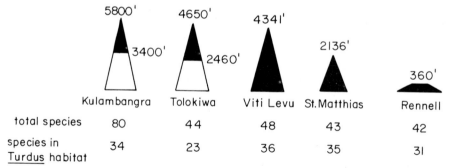

	Kulambangra	Tolokiwa	Viti Levu	St. Matthias	Rennell
total species	80	44	48	43	42
species in <u>Turdus</u> habitat	34	23	36	35	31

Figure 42 Interisland variation in altitudinal range of the thrush *Turdus poliocephalus.* For each island the numbers represent the total number of bird species on the island, the number of bird species in the habitat occupied by *Turdus poliocephalus,* the highest elevation on the island (written above the island pyramid), and the lower altitudinal limit of *Turdus poliocephalus* (if this species does not descend to sea level); the altitudinal range inhabited by *Turdus poliocephalus* is shaded. *Turdus* lives in forest and/or scrub on all islands except New Guinea, where it lives mainly at the edge between forest and alpine grassland. On each island it adjusts its altitudinal range or habitat preference so as to encoun- ter a fauna of about the same richness (23–36 species). In more detail, it can encounter more species (31–36) on islands without the thrush-like insectivorous ground birds called pittas (Guadalcanal, Kulambangra, Espiritu Santo, Viti Levu, St. Matthias, Rennell) than on islands with pittas (23–31 species: New Guinea, Bougainville, Karkar, Tolokiwa), because pittas may be disproportionately important competitors of *Turdus poliocephalus.* This may underlie the different altitudinal ranges on Espiritu Santo vs. Karkar, or on Tolokiwa vs. Viti Levu, the two islands of each pair being otherwise similar in total species number and maximum elevation.

one or two readily identifiable competitors, analogous to Figures 39 and 40, is provided by the flycatchers *Pachycephala hyperythra* and *P. soror*. Where these species coexist on the central cordillera of New Guinea, *P. soror* spends much time in the understory and is often caught in mist-nets, whereas *P. hyperythra* remains 10 or more feet above the ground and is seldom or never netted. In the North Coastal Range of New Guinea, where *P. soror* is absent, *P. hyperythra* often utilizes the understory and is regularly netted. A more complex case analogous to Figure 42, in that numerous, more distantly related competitors are involved, is provided by the omnivorous cuckoo-shrike *Lalage leucomela*. On New Guinea (total number of bird species, 513) it is confined to the tree crowns; on New Britain (127 species) it occasionally descends to the lower story; and on Espiritu Santo (50 species) and Samoa (28 species) the related species *L. leucopyga* and *L. maculosa*, respectively, descend to the ground (see Figure 43).

In general, the ease with which we were able to construct assembly rules for guilds on islands was inversely proportional to the number of species in the guild, and directly proportional to the ecological isolation of the guild from other guilds. The same factors determine the ease with which we can understand patchy distributions (Figures 35–38) or competitive effects on spatial niche parameters (Figures 39–43). If most of the competition faced by a particular species comes from just one or two other species, we stand a good chance of being able to construct assembly rules for island communities

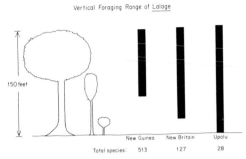

Figure 43 Inter-island variation in vertical foraging range of *Lalage* cuckoo-shrikes. On New Guinea, with a total of 513 bird species, *Lalage leucomela* is confined to the crowns; on New Britain, with 127 species, it occasionally descends to shrub level; and on Upolu (Samoa), with 28 species, the related *L. maculosa* regularly descends to the ground. Evidently the vertical foraging range is compressed on species-rich islands by diffuse competition.

(e.g., Table 7), to recognize what fills the missing patches (e.g., Figures 33 and 34), or to construct assembly rules for habitat communities (e.g., Figure 41). If instead there is diffuse competition from many species and no single competitor is overwhelmingly important, we remain baffled by patchiness (Figures 35–38), and we can make only statistical predictions of the probability that a species will occur on a particular island, from total species number and incidence functions. The cases of *Lalage* (previous paragraph) and *Turdus poliocephalus* (Figure 42) suggest the possibility that statistical predictions of diffuse competitive effects on spatial niche parameters could similarly be obtained, by constructing graphs analogous to incidence functions. These graphs would have total species number on an island as the abscissa, and the spatial limit of a particular species' range (highest or lowest altitude, vertical foraging height, or position on habitat gradient) as the ordinate.

The Origin of Assembly Rules

Community assembly involves the following patterns:

> If one considers all the combinations that can be formed from a group of related species, only certain ones of these combinations exist in nature.

> Permissible combinations resist invaders that would transform them into forbidden combinations.

> A combination that is stable on a large or species-rich island may be unstable on a small or species-poor island.

> On a small or species-poor island, a combination may resist invaders that would be incorporated on a larger or more species-rich island.

> Some pairs of species never coexist, either by themselves or as part of a larger combination.

> Some pairs of species that form an unstable combination by themselves may form part of a stable larger combination.

> Conversely, some combinations that are composed entirely of stable subcombinations are themselves unstable.

Let us now attempt to understand assembly rules in terms of resource utilization, overexploitation strategies, dispersal strategies, and transition probabilities between species combinations.

Resource Utilization

It seems likely that competition for resources is a major factor underlying assembly rules. Permissible combinations of species may often be those combinations (out of all the combinations containing the same total number of species) that leave the fewest resources unutilized, that are therefore relatively immune to displacement of one component species by an invading species, and that still provide each component species with enough resources to maintain a stable population. A full analysis of how the species of a guild are coupled to each other through utilization of resources requires knowledge of the so-called utilization function U of each species, taking account of the multidimensionality of U's, the differences in compressibility of U's along different resource axes, and the relative ability of each species to displace each other species from each portion of resource space. We shall discuss the principles involved in such an analysis and then apply them to the fruit-pigeon guild.

We begin by considering one-dimensional incompressible U's, for illustrative purposes. Suppose that the biologically important differences in the resources harvested by different species of a guild—i.e., the differences that enable the species to coexist—can be described in terms of one resource variable, x. (An example would be the diameter of fruit eaten by *Ptilinopus-Ducula* fruit pigeons in the same habitat, but these pigeons segregate along a tapering tree branch as well as along a fruit-diameter axis (Figure 30). A better example would be altitude in the case of the trios and quartets of New Guinea bird species that strictly segregate by altitude (cf. Diamond, 1972a, pp. 27–35 and p. 217; 1973)). Then the frequency distribution $R(x)$ of resource production available to the guild can be plotted along one axis. The frequency distribution of resources consumed by each individual of each species, the utilization function $U(x)$ of the species, can be plotted along the same axis (Figure 44). The total resource production available to the guild is $\int_x R(x)dx$; the resources consumed by the n_i individuals of each species i present are $n_i \int_x U_i(x)dx$; the frequency distribution of unutilized resource production $V(x)$ is given by

$$V(x) = R(x) - \sum_i n_i U_i(x) \qquad (8)$$

and $\int_x V(x)dx$ is the total production of unutilized resources. Depending upon the forms of all the functions $U_i(x)$, $\int_x V(x)dx$ will in general be lower for certain combinations of species than for others. Combinations with high $\Sigma n_i \int U_i dx$ and low $\int V(x)dx$ will maintain higher populations, and will leave fewer resources that could support an invader, than combinations with the reverse properties. Assuming that $\int R(x)dx$ is proportional to island area, and that (for a given combination of species) the n_i's are also proportional to area, then $\int V(x)dx$ will also be proportional to area. Thus, the resources left unconsumed by a certain combination of species may be too small to support a stable population of an additional species if the island is small, but may be sufficient to support an additional species if the island is larger.

For example, consider the four-species guild depicted in Figure 44a. The only species that could maintain a stable population on the small island whose production curve is depicted on the left of Figure 44b would be species 3. On a somewhat larger island (Figures 44c and 44d) species 2 and 4 could both persist together, would be preferred over species 3 alone by virtue of leaving fewer unutilized resources, and would leave too few resources for either species 1 or 3 to coexist in addition. On a still larger island (Figure 44e) species 1 could maintain itself along with species 2 and 4. Thus, Figure 44 suggests explanations for many of the findings listed at the beginning of the section "The origin of assembly rules": that certain combinations of n species are preferred over others (e.g., in Figure 44e, species $1 + 2 + 4$ preferred over $2 + 3 + 4$); that permissible combinations resist certain invaders (e.g., species $2 + 4$ resist species 3 in Figure 44d); that a combination stable on a small island may be unstable on a large island (e.g., species 3 stable in Figure 44b, unstable in Figure 44c–44d); that, conversely, a combination stable on a large island may be unstable on a small island (e.g., species $1 + 2 + 4$ are stable in Figure 44e but not in 44d); that, correspondingly, invaders resisted by a combination on a small island may be incorporated on a large island (e.g., species $2 + 4$ incorporate species 1 in Figure 44e but not in 44d); and that combinations composed entirely of stable subcombinations may be unstable (e.g., species $2 + 4$ stable, species 3 stable, species $2 + 3 + 4$ unstable).

Figure 44 Illustration of how the match of utilization functions $U(x)$ to resource production curves $R(x)$ could help explain why certain combinations of species occur in nature and others do not. Resource production is assumed to be distributed along a single dimension (e.g., fruit size or altitude) according to curves such as the solid curves on the left of Figures 44b–44e. Each of four species i has a characteristic utilization function or frequency distribution of resources consumed, illustrated by the four curves of Figure 44a. The areas under these curves, $\int U_i(x)dx$, represent the resource production rate required for maintaining the smallest population of each species that could survive for a reasonable length of time. Thus, multiplying the ordinate values of each $U_i(x)$ by M_i yields a population M_i times the minimum size. The left sides of Figures 44b–44e all depict curves $M_i U_i(x)$ that fit entirely under the illustrated $R(x)$ but are still greater than 1.0 times $U_i(x)$, meaning that the illustrated $R(x)$ is sufficient for species i to maintain a stable population on the island. The right sides of Figures 44b–44e give as solid curves $R(x) - \Sigma M_i U_i(x)$, i.e., the unutilized resource distribution $V(x)$; the dashed curves represent 1.0 times $U_j(x)$ for species j that do not fit under the $V(x)$ curve and hence cannot maintain stable populations on the island in coexistence with species i. In Figure 44b (left) only 1.0 $U_3(x)$ [i.e., 1.0 times $U_3(x)$] fits under $R_b(x)$, and 1.0 $U_1(x)$, 1.0 $U_2(x)$, or 1.0 $U_4(x)$ would not. $R_b(x) - 1.0\ U_3(x)$ equals $V_b(x)$, the solid curve on the right; $U_1(x)$, $U_2(x)$, and $U_4(x)$ all exceed $V_b(x)$. Thus, none of these three species could maintain itself alone on this island nor share the island with species 3. Figure 44c represents an island whose area is double that of the island of Figure 44b, so that $R_c(x)$ is twice as high as $R_b(x)$, and twice as large a population of species 3 can be maintained (dashed curve 2.0 $U_3(x)$ on the left). On the right, $R_c(x) - 2.0\ U_3(x) = V_c(x)$; this exceeds 1.0 $U_2(x)$ or 1.0 $U_4(x)$ and barely exceeds $U_1(x)$, so that species 1 might be marginally capable of sharing this island with species 3, but species 2 or 4 could not. Figure 44d depicts the same island as Figure 44c, so that $R_d(x) = R_c(x)$; 1.9 $U_2(x)$ or 1.9 $U_4(x)$ fit under $R_d(x)$, alone or summed, so that these two species can coexist on this island. On the right $R_d(x) - 1.9\ U_2(x) - 1.9\ U_4(x) = V_d(x)$; this exceeds 1.0 $U_3(x)$ and barely exceeds $U_1(x)$. Since the area under the curve $V_d(x)$ is less than the area under $V_c(x)$, species 2 and species 4 together leave fewer unutilized resources on this island than species 1 and are more likely to occur. Figure 44e represents an island four times as large as that of Figure 44b, with $R_e(x)$ four times as high as $R_b(x)$. The maximum value of M such that $R_e(x) - MU_2(x) - MU_4(x) \geq 0$ for all x is now 3.8, so that this island can simultaneously support populations of species 2 and 4 that are 3.8 times the minimum size (dashed curves on left). On the right $R_e(x) - 3.8\ U_2(x) - 3.8\ U_4(x) = V_e(x)$; this exceeds 1.0 $U_1(x)$ but is less than 1.0 $U_3(x)$, so that species 1 but not species 3 could share this island with species 2 and 4. Thus, in this guild one would usually find species 3 on a small island, species 2 + 4 or occasionally 1 + 3 on a larger island, and species 1 + 2 + 4 on a still larger island.

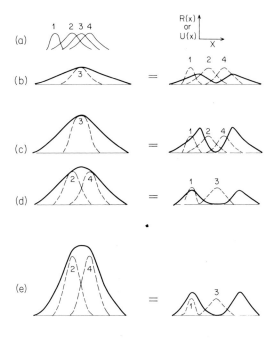

More realistically, U's and R's are multi-dimensional and involve spatial variables (habitat type, altitude, and vertical height above the ground) and variables related to foraging behavior as well as food variables. For example, among New Guinea frugivorous pigeons (including not only the fruit pigeons of the genera *Ptilinopus* and *Ducula* but also cuckoo-doves, ground doves, and some other groups) the most significant food variables are size and hardness of fruit. Among flycatchers, variables related to foraging behavior include gross foraging technique (gleaning, sallying, hovering, etc.) and quantitative foraging characteristics (ratio between stationary and moving time, duration of stationary or moving bouts, and rate of travel). Other examples of significant resource axes include calendar time (for butterflies: Shapiro, Chapter 8), seed size and microhabitat (for desert rodents: Brown, Chapter 13), clock time, microhabitat, and food type (for desert lizards: Pianka, Chapter 12), prey size, prey type, foraging tactic, and foraging zone (for oak-woodland insectivorous birds: Hespenheide, Chapter 7), and position on habitat gradient (for Mediterranean birds: Cody, Chapter 10).

A further important property of real U's is that they are compressible under competition, and are much more compressible along certain axes than along other axes. When a species shifts from a species-poor fauna to a species-rich fauna, its U is often found to become compressed along spatial axes but rarely along axes related to foraging tactics (Crowell, 1962; Diamond, 1970a, 1970b, 1973). This spatial compressibility has been illustrated in Figures 39–43. For this reason measurements of so-called α's or niche overlaps (cf. Chapter 13, Brown; Chapter 7, Hespenheide; and Chapter 12, Pianka) may provide good estimates of competition coefficients for dimensions along which U's are relatively incompressible, but may give not even the crudest approximation to competition coefficients along spatial axes. For example, strict spatial segregation between two species means that along a spatial axis the niche overlap is 0, but it may also mean that this is precisely because the competition coefficient for the fundamental niches is close to 1.0. If one measures the fundamental niches of several related species in each other's absence and finds the niches to overlap in some portion of niche space, then prediction of assembly rules may require knowing two further properties of each species, both as a function of position in niche space: the relative resource-harvesting rate as a function of resource density (see section "Companions in starvation"), and the ability to displace other species by aggressive behavior.

The simplest application of U's to understanding assembly rules is to the members of incompatible pairs or sets, the closely related species that are of similar size and foraging technique and never coexist on an island (cf. *Macropygia mackinlayi* and *M. nigrirostris*, Figure 20, Table 9, and the small lowland myzomelids, Figure 23). These are species whose U's of the fundamental niche must overlap so broadly that there is no way for the two species to divide an island spatially such that each species has a significantly higher fundamental U than the other in some habitat. Within these incompatible pairs or sets, the high-S or tramp species has the advantage on large islands because of its ability to overexploit and tolerate low resource levels, while the supertramp has the advantage on small or remote islands because of its high dispersal rates and reproductive potential (see section "Companions in starvation"). Much more often, however, species whose U's are very similar along nonspatial axes differ sufficiently along some spatial axis that they can coexist on an island by segregating spatially. For instance, the fruit size distributions of the similar-sized pigeons *Ducula melanochroa* and *D. rubricera* are very similar (curves labeled 661 g and 722 g, upper half of Figure 31), but all islands on which *D. melanochroa* lives are nevertheless shared with *D. rubricera,* the former species being concentrated in the mountains, the latter in the lowlands. Other habitat vicariants of similar size presumably also have similar U's along axes other than some spatial axis.

In the remainder of this section let us consider in detail the Bismarck fruit-pigeon guild, the case in which we can come the closest to reconstructing assembly rules from inter-island competitive shifts in resource utilization along more than one dimension. We can compact this system to two dimensions, because all the species live in the canopy, eat soft fruit, and pick fruit in the same manner; the habitat dimension and altitude dimension can be combined into one axis; and the remaining axis is taken as fruit size, with which the microhabitat dimension (diameter of branch used as perch) is directly correlated, since both are related to pigeon body size. For simplicity of graphical representation in preference to utilization contours in this two-dimensional space, Figure 45 displays more coarsely the utilization function of each of eight common pigeon species on each of four islands: each axis is quantized into four or five values, and shading on the resultant checkerboard (quantized in three values) indicates intensity of utilization. The left-most column of checkerboards approximates the fundamental niche by depicting the utilization function of each species on the most species-poor Bismarck island where it occurs, but even this underestimates the fundamental niche in most cases; *Ducula finschii* and *D. melanochroa,* for instance, occur only on islands with at least six species of fruit pigeons. Figure 45 yields the following conclusions:

(a) The fundamental niches of the eight species overlap massively (compare corresponding squares in the first column of eight checkerboards entitled "fundamental"). One square of the niche-space checkerboard may be utilized by up to six species, resulting in up to 537 units of utilization (see Figure 45 legend for definition).

(b) Rarely does a species achieve its fundamental niche on an island. Instead, the realized utilization functions are considerably compressed, as may be seen by comparing the left-most checkerboard in any row with the other four checkerboards. For instance, *Ptilinopus insolitus* (third row of checkerboards) underutilizes the lowland forest of Umboi (compare

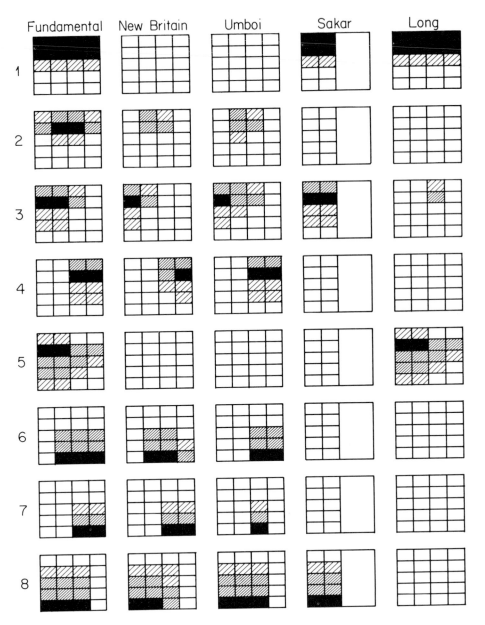

(a)

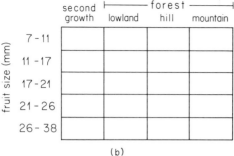

Figure 45 Assembly of the fruit-pigeon guild on some Bismarck islands. Each of the 40 checkerboards (a) represents the resource utilization of one fruit-pigeon species on one island, in a two-dimensional niche space. As shown in the enlarged checkerboard (b) giving the key to the 20 squares of each checkerboard, the ordinate of each checkerboard is a fruit-size axis divided into five intervals, and the abscissa is a habitat-plus-altitude axis divided into four intervals. The shading of each of the 20 squares of each checkerboard indicates how heavily the given species utilizes the given portion of niche space on the given island: solid shading = heavy use, close diagonal bars = moderate use, sparse diagonal bars = light use. Utilization units were calculated by rating abundance of the species in the habitat as 0 (absent) 1, 2, or 3 (abundant), and by multiplying this abundance value times the percentage of fruit in the diet that fell within the given size range, obtained from graphs similar to Figure 31 above. A product of ≥ 100, 40–99, and 20–39 was equated with heavy, moderate, and light use, respectively. For instance, in hill forest on New Britain *Ptilinopus rivoli* was fairly common, abundance rating 2; 44% of the fruit it consumed was in the size range 11–17 mm; hence the utilization product in this square of niche space (second row and third column of the checkerboard in the fourth row (species "4") and second column ("New Britain") of checkerboards) is 88, construed as moderate use and depicted by close diagonal bars. The species, coded by the eight numbers to the left of each row of checkerboards, are: 1 = *Ptilinopus solomonensis* (91 g), 2 = *P. superbus* (123 g), 3 = *P. insolitus* (144 g), 4 = *P. rivoli* (135 g), 5 = *Ducula pistrinaria* (470 g), 6 = *D. finschii* (383 g), 7 = *D. melanochroa* (661 g), 8 = *D. rubricera* (722 g). The second through fifth columns of checkerboards represent the single island named at the top of the column; the left-most column depicts the niche of the species on that Bismarck island (not necessarily one of the four depicted) where the species encounters the fewest other fruit-pigeon species; this is taken as an approximation of the fundamental niche. Utilization on Tolokiwa and Crown is virtually identical to utilization on Long. Hill forest and mountain forest were not studied adequately on Sakar. See text for discussion.

shading in second column of its Umboi and fundamental checkerboards), under-utilizes lowland forest and is absent from hill forest on New Britain, and is absent from second-growth and lowland forest and confined to hill forest on Long, Tolokiwa, and Crown. Since we have underestimated fundamental niches because we could not study each species in the absence of the other species, we have also underestimated the degree of compression.

(c) The compressions of various species on the same island are correlated, in a way such that no square of the niche-space checkerboard on an island is actually utilized by more than four species or receives more than 353 (usually not more than 240) units of utilization. Conversely, except on Long (plus Tolokiwa and Sakar), no square receives less than 75 units of utilization. [Utilization drops to 9–42 units for large fruit in the hills and mountains of Long, as reflected in the fact that the largest pigeon of Long (*Ducula pistrinaria*) weighs only 470 g and it becomes uncommon at higher elevations.] Some qualitative examples of how this distribution of utilization is achieved through selection and compression of colonists deserve notice before we consider a quantitative analysis. The medium-large supertramp *Ducula pistrinaria* (species 5 of Figure 45), which has sent colonists to all the islands of Figure 45 (and undoubtedly to all other Bismarck islands), is established on Long, but on the three other islands of Figure 45 it is "squeezed out" between some combination of the large species 7 and 8, the similar-sized species 6, and the medium-small species 2 and 3. The small supertramp *Ptilinopus solomonensis* (species 1), which is ubiquitous on Long as a consumer of 7–17-mm fruits (first two rows of each checkerboard), is replaced in this capacity on New Britain and Umboi by three habitat specialists: *P. insolitus* (species 3) in second-growth and lowland forest, *P. superbus* (species 2) in lowland forest and hill forest, and *P. rivoli* (species 4) in hill forest and mountain forest. The medium-small *P. insolitus* (species 3), which mainly eats fruit in the range 11–17 mm, shares both Sakar and Long with the small *P. solomonensis* (species 1, mainly 9–14-mm fruit), but differs strikingly from island to island in habitat preference: ubiquitous on Sakar, confined to hill forest on Long (plus Tolokiwa and Crown), and squeezed out of hill forest by the combination of the similar-sized *P. superbus* and *P. rivoli* on New Britain. To an ornithologist who has learned *P. insolitus* as the second-growth small fruit pigeon of New Britain, this inversion of its habitat preference on Long is astonishing. The reason for the difference between Sakar and Long is that the remaining species of Sakar besides *P. insolitus* and *P. solomonensis,* the very large *D. rubricera* (species 8), eats mainly 22–30-mm fruit, leaving much medium-sized fruit to support a medium-small pigeon; but the remaining species of Long, the medium-large *D. pistrinaria* (species 5), eats mainly 15–21-mm fruit, squeezing out *P. insolitus* between *D. pistrinaria* and *P. solomonensis* except in hill forest, where *D. pistrinaria* becomes less common.

Thus, Figure 45 probably illustrates *how communities are assembled through selection of colonists, adjustment of their abundances, and compression of their niches, in part so as to match the resource consumption curve of the colonists to the resource production curve of the island.* For a quantitative test of this hypothesis, we need both the consumption curve and the production curve. The form of the consumption curve is available from analyses of stomach contents as expressed in Figure 31, combined with the relative abundances and habitat distributions of each species. The production curve is unfortunately difficult to measure and is unknown; we assume, however, that it will be roughly bell-shaped, and observation suggests that there is relatively more large fruit in the lowlands than in the mountains. We can therefore proceed by (a) constructing consumption curves of actual pigeon communities, (b) comparing the shapes of these curves with the expected shapes of production curves, and (c) comparing the actual consumption curves with those calculated for imaginary communities of forbidden combinations of species.

Figure 46 depicts the actual consumption curves for the fruit-pigeon combinations in the lowlands of Sakar, in the lowlands of Long, and in the mountains of Long. These curves were constructed by summing the products of single-species consumption curves such as Figure 31 times the relative abundance of each species in the given habitat. All three curves are approximately bell-shaped. The curves for the Long and Sakar lowlands are quite similar, especially in the range of small fruit; slightly more large fruit and

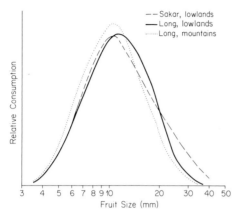

Figure 46 Actual consumption curves for the fruit-pigeon guild in the lowlands of Long Island and Sakar Island and at an elevation of about 1500 feet in the mountains of Long. The ordinate is the relative consumption (arbitrary units), by all species of the guild present, of fruit of the diameter given by the abscissa (logarithmic scale). To obtain these local-guild consumption curves, the frequency distribution of fruit sizes in the stomach of each species present was measured as in Figure 31 and was weighted by the relative abundance of the species. The three curves were then scaled to give approximately the same area. Relative abundances are: Long lowlands, *Ptilinopus solomonensis: Ducula pistrinaria*, 60:40; Long mountains, *P. solomonensis: P. insolitus: D. pistrinaria*, 77: 9: 15; Sakar lowlands, *P. solomonensis: P. insolitus: D. rubricera*, 43: 29: 29. Curves for Tolokiwa and Crown Islands are very similar to those for Long. Note that all three curves are bell-shaped (as one would expect for the distribution of fruit production), that the two lowland curves are quite similar, and that the curve for the mountains of Long is slightly shifted towards smaller fruit.

less medium-sized fruit are consumed on Sakar than on Long. The curve for the Long mountains is shifted somewhat towards small fruit (more consumption of small fruit and less of large fruit). Thus, it is at least possible that the actual consumption curves provide good matches to production curves.

We can now fit the actual consumption curves of Figure 46 to curves calculated for other combinations of species. Figure 47 compares the actual lowland curves to the curve of *P. solomonensis* (91 g) + *D. rubricera* (722 g), the smallest and the largest species of Figure 45; the ratio of their abundances (2.0:1.0) is chosen so as to give the best fit to the actual curves. This combination of species exists on no Bismarck island. Compared with the actual curve for Sakar [*P. solomonensis* (91 g) + *P. insolitus* (144 g) + *D. rubricera* (722 g) in the proportions 1.5:1.0:1.0], the imaginary combination fits well in the fruit-size range above 12 mm but consumes too much small fruit. In effect, the actual presence of an additional medium-small species on Sakar lets the otherwise excessive consumption of small fruit be eliminated by reducing the *P. solomonensis/D. rubricera* ratio without at the same time developing a deficit in consumption of medium-small (11–17-mm) fruit. Compared to the actual curve for Long [*P. solomonensis* (91 g) + *D. pistrinaria* (470 g), proportions 1.5:1.0], the fitted combination consumes too much large fruit (because of the replacement of *D. pistrinaria* with the larger *D. rubricera*) and too little medium-small fruit but too much small fruit (because the small *P. solomonensis* in the imaginary combination is in effect trying to eat alone what it actually eats together with the medium-sized *D. pistrinaria* on Long). Figure 48 compares the actual lowland curves to the best-fit combination of *P. insolitus* (144 g) + *D. rubricera* (722 g) (proportions 2.0:1.0). This

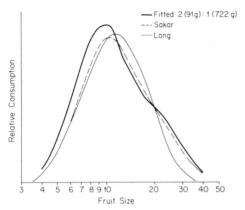

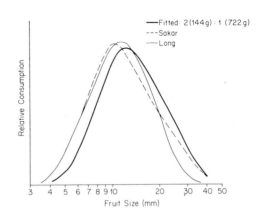

Figure 47 Comparison of actual and fitted consumption curves by combinations of fruit pigeons. The curves labeled "Sakar" and "Long" are the actual curves for the guilds present in the lowlands of these islands, copied from Figure 46. The "fitted" curve was calculated for the imaginary combination of *P. solomonensis* (91 g) and *D. rubricera* (722 g) from the single-species consumption curves of Figure 31 below, with the abundance ratio adjusted to 2:1 to give the best fit to the two actual curves. Note that the fitted curve for the imaginary combination utilizes more small fruit of <11 mm diameter than the actual guilds on either island, and utilizes less medium-sized fruit and more large fruit than the actual guild on Long. The combination *P. solomonensis* and *D. rubricera* actually exists on no island, perhaps because it gives a poor match to actual production curves, as suggested by this figure if the actual consumption curves are similar to the production curve.

Figure 48 Like Figure 47, except that the fitted curve is for the imaginary combination *P. insolitus* (144 g) and *D. rubricera* (722 g), with abundances adjusted to the ratio 2:1. The imaginary combination consumes less small fruit and more medium-sized or large fruit than the actual guilds. This combination does not exist in nature.

combination also exists on no Bismarck island. On both Sakar and Long the absence of a small pigeon in the imaginary combination would result in an unutilized production of small fruit and an excess consumption of medium or large fruit. Some other forbidden combinations, such as *P. solomonensis* + *P. insolitus* or *D. pistrinaria* + *D. rubricera,* yield much worse best-fit curves.

Thus, if the actual consumption curves

for Sakar and Long are similar to the production curves, the reason for the non-existence of so-called forbidden combinations could be that they give too poor a fit to the production curves of real islands. Without knowledge of the actual production curves, we cannot decide whether the actual species combination on Long could equally well be on Sakar and vice versa, and whether these two islands support different combinations for fortuitous reasons; or whether there are slight differences in the production curves that selected for different successful combinations on the two islands (see discussion of "Chance or predestination?" in the final section of this chapter).

Figures 47 and 48 obviously represent only a preliminary attempt to understand assembly rules in terms of matches between consumption curves and production curves. Some of the technical problems that will have to be overcome in a more detailed analysis include the following: 1. The shape of the production curve must be measured directly. 2. Absolute values must be obtained both for production and for consumption. The consumption curves of Figures 47 and 48 are relative; I estimated ratios of species abundances, not absolute densities in birds per acre. 3. My measurements so far show that, for a given species, there is far less variation between islands or habitats in the fruit consumption curve (vertical cross-sections through the checkerboards of Figure 45, or curves as in Figure 31 below) than in the habitat distribution (horizontal cross-sections through the checkerboards of Figure 45, or figures such as Figure 41) or than in abundance. However, careful comparisons of single-species consumption curves in different habitats or islands are necessary, as a consumption curve could shift with a shift in the available resource curve caused by a change either in the competing species pool or in the production curve. 4. The contribution of species segregation by perch position (see Figure 30 legend) to segregation based on body size requires quantitation. I have quantitated only the contribution of segregation by fruit size, so far. 5. An analysis needs to be done at different seasons. It is possible that the shape of the production curve varies seasonally, and that a species combination that fits poorly at one season may nevertheless be preferred because of a superior fit for the rest of the year.

Companions in starvation:
a mechanism of coadjustment?

The preceding discussion of utilization functions fails to explain why certain species or combinations of species are preferred over others with similar U's in the same habitat. For instance, distributions of fruit sizes eaten by supertramp fruit pigeons seem similar· to those of other pigeons of comparable body size. How are supertramps normally excluded by other species?

The high population densities on supertramp-rich island or habitats (Figure 18) suggest that lowering of resource levels to mutually acceptable levels by members of successful combinations could be an important mechanism of coadjustment and exclusion. Among the incidence categories, supertramps may be most prone to starve in times of scarcity, and high-S species least. Figure 49 illustrates how this hypothetical coadjustment by resource level might operate. In simplified form, one hypothesizes an equilibrium on a medium-sized island, such that at a given time some high-S species are absent because of low K's, too few hot spots, too low reproductive potential and recolonization frequency, etc., and some super-

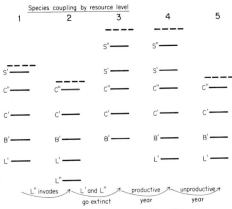

Species coupling by resource level

1   2   3   4   5

L″ invades → L′ and L″ → productive → unproductive →
go extinct   year   year

Figure 49 Hypothetical example of how occurrences of species in a community might be coupled by resource level. Members of the total species pool differ in their ability to survive at low resource levels, and also in their tendency to depress resource levels by overexploitation: the worst overexploiters can tolerate the lowest levels. A guild of species exploiting similar resources consists of supertramps S'' and S', C-tramps C'' and C', B-tramp B', and high-S species L' and L'', whose tolerance of low resource levels and tendency to overexploit increases in that sequence. For the same community in each of five states, solid bars give, for each guild member present, the lowest resource level it can tolerate (assumed invariant with community state), and the dashed bars give the actual resource level. Any species whose solid bar lies above the dashed bar is eliminated. Initially (state 1, left), species S', C'', C', B', and L' are present. A successful invasion by the worst overexploiter, L'', lowers the resource level, eliminating the species present that requires the highest resource level, S' (state 2). The overexploiters L'' and L' both go extinct owing to population fluctuations, relieving overexploitation, letting the resource level rise above the initial value, and permitting both S' and S'' to invade (state 3). An unusually productive year permits L' to invade again without lowering the resource level or eliminating species (state 4). A drought or unproductive year then lowers the resource level, eliminating S'' and S' (state 5). Among the respects in which the figure is an oversimplification is that the dashed bar lying above the solid bar of a species permits but does not guarantee its presence (e.g., the drought might also have eliminated L', and the productive year might have permitted invasion by some species B'' of intermediate tolerance between B' and C'). If this were not true, all permitted species combinations could be formulated as groups of consecutive species in a single list, which is not the case (cf. permitted combinations of cuckoo-doves, gleaning flycatchers, etc.). The figure may apply either to a whole island, or to individual habitats on an island.

tramps are starved out because the remaining species overharvest the resources to a point that these remaining species but not the supertramps can tolerate. Fluctuations in resource levels, and corresponding fluctuations among permitted species combinations, occur around this equilibrium state. Occasionally, another high-S species temporarily establishes itself or there is an unproductive year, in either case lowering the resource level. Then another supertramp is starved completely off the island, and some tramp is starved out of forest into a less stable habitat. Occasionally, there is an especially productive year, or a high-S species temporarily disappears, letting the resource level rise again. Then another supertramp is able temporarily to colonize, and an established tramp temporarily shifts back into the forest.

The plausibility and internal consistency of this proposed mechanism of resource coupling may be assessed by means of loop analysis, the technique used by Levins in Chapter 1 for studying evolution in communities near equilibrium. The treatment in this and the next paragraphs is based on Levins's formulation of loop analysis, which may be consulted for an explanation of loops, feedback, stability conditions, and methods for determining effects of selection for a particular trait on the equilibrium abundance of the species in a community. Figure 50 represents the simplest possible system that could exhibit resource coupling and competitive exclusion by overexploitation. The system consists of two species (X_1 and X_2) that harvest the same nutrient (X_3). The coeffi-

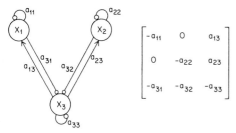

Figure 50 Left: loop diagram of a three-variable community consisting of two species X_1 and X_2 harvesting the same nutrient X_3. See Levins, Chapter 1, for explanation of loop diagrams. The coefficients a_{ij} represent the effect of species j on the growth of species i in the equation $dX_i/dt = f_i(X_i,X_j)$, where $a_{ij} \equiv \partial f_i/\partial X_j$. On the right is the community matrix for this system, taking all the a_{ij} as positive numbers and representing the directions of their effects by the sign in front. See text for discussion.

cients a_{ij} in the growth equation (Levins, Chapter 1, eq. 1)

$$dX_i/dt = f_i(X_1, X_2, X_3)$$

are defined by

$$a_{ij} \equiv \partial f_i/\partial X_j$$

For example,

$$dX_3/dt = f_3(X_1, X_2, X_3)$$
$$a_{31} \equiv \partial f_3/\partial X_1$$

Thus, the coefficients a_{31} and a_{32}, which represent the effects of X_1 and X_2 on the growth of X_3 in the growth equation for dX_3/dt, are negative; and the coefficients a_{13} and a_{23}, which represent the effect of X_3 on the growth of X_1 and X_2, respectively, are positive. The growth of X_3 will in general either be self-damped or will incorporate the damping effect of non-self-reproducing resources at a lower trophic level that are not otherwise represented in Figure 50. Thus a_{33}, the effect

of X_3 on its own growth, is negative. The growths of X_1 and/or X_2 may or may not be self-damped (a_{11}, a_{22} negative?). We ask the question: under what conditions is an overexploitation strategy profitable for its practitioner? That is, suppose one of the consumer species, X_1, evolves a trait C that tends to let the species grow faster but tends to deplete nutrient ($\partial f_1/\partial C > 0$, $\partial f_3/\partial C < 0$, $\partial f_2/\partial C = 0$). Under what conditions can this species thereby eliminate the other consumer species without also eliminating itself?

First, we write the expressions for feedback F_1, F_2, and F_3 at all three levels in the system, to determine stability conditions. From eqs. 8–10 of Levins, Chapter 1, these expressions become:

$$F_1 = (-a_{11}) + (-a_{22}) + (-a_{33}) \qquad (9)$$
$$\begin{aligned} F_2 = {} & (a_{13})(-a_{31}) + (a_{23})(-a_{32}) \\ & - (-a_{11})(-a_{22}) - (-a_{11})(-a_{33}) \\ & \hspace{3em} - (-a_{22})(-a_{33}) \quad (10) \end{aligned}$$
$$\begin{aligned} F_3 = {} & (-a_{11})(-a_{22})(-a_{33}) \\ & - (-a_{11})(a_{23})(-a_{32}) \\ & \hspace{2em} - (-a_{22})(a_{13})(-a_{31}) \quad (11) \end{aligned}$$

(In eqs. 9–11, in the following analysis, and in the right side of Figure 50 we take all a_{ij} as positive numbers and indicate their effect on dX_i/dt by the sign). One stability condition is that feedback at all levels be negative. This will be true of F_1 and F_2 whether or not a_{11} and/or a_{22} are zero, but if both a_{11} and a_{22} are zero, F_3 is zero and the system is unstable. Thus, one or both of the consumer species must be self-damped for the two species to coexist. The other stability condition (eq. 27 of Levins, Chapter 1) is that $F_1F_2 + F_3$ must be positive. Expressing $F_1F_2 + F_3$ in terms of eqs. 9–11 yields 12 positive terms if a_{11} and a_{22} are both negative; two of these terms remain if $a_{11} = a_{22} = 0$. Thus, this condition imposes no new restraints.

The effect of natural selection for the trait C on the abundance of the species X_i is obtained by replacing column i of the community matrix (Figure 50) with the vector

$$\begin{bmatrix} -\partial f_1/\partial C \\ 0 \\ +\partial f_3/\partial C \end{bmatrix}$$

and dividing by F_3, $\partial f_1/\partial C$ and $\partial f_3/\partial C$ both being positive quantities. This yields:

$$\frac{\partial X_1}{\partial C} = \frac{\begin{vmatrix} -\partial f_1/\partial C & 0 & a_{13} \\ 0 & -a_{22} & a_{23} \\ \partial f_3/\partial C & -a_{32} & -a_{33} \end{vmatrix}}{F_3}$$

$$= \frac{\begin{matrix} [(\partial f_1/\partial C)a_{22}a_{33} \\ +(\partial f_1/\partial C)a_{23}a_{32} \\ -(\partial f_3/\partial C)a_{13}a_{22}] \end{matrix}}{a_{11}a_{22}a_{33} + a_{11}a_{23}a_{32} + a_{22}a_{13}a_{31}} \quad (12)$$

$$\frac{\partial X_2}{\partial C} = \frac{\begin{vmatrix} -a_{11} & -\partial f_1/\partial C & a_{13} \\ 0 & 0 & a_{23} \\ -a_{31} & \partial f_3/\partial C & -a_{33} \end{vmatrix}}{F_3}$$

$$= \frac{\begin{matrix} [-(\partial f_3/\partial C)a_{11}a_{23} \\ -(\partial f_1/\partial C)a_{23}a_{31}] \end{matrix}}{a_{11}a_{22}a_{33} + a_{11}a_{23}a_{32} + a_{22}a_{13}a_{31}} \quad (13)$$

$$\frac{\partial X_3}{\partial C} = \frac{\begin{vmatrix} -a_{11} & 0 & -\partial f_1/\partial C \\ 0 & -a_{22} & 0 \\ -a_{31} & -a_{32} & \partial f_3/\partial C \end{vmatrix}}{F_3}$$

$$= \frac{\begin{matrix} [-(\partial f_3/\partial C)a_{11}a_{22} \\ -(\partial f_1/\partial C)a_{22}a_{31}] \end{matrix}}{a_{11}a_{22}a_{33} + a_{11}a_{23}a_{32} + a_{22}a_{13}a_{31}} \quad (14)$$

Since $\partial f_1/\partial C$, $\partial f_3/\partial C$, and all the a_{ij} in eqs. 12–14 are positive quantities, $\partial X_2/\partial C$ and $\partial X_3/\partial C$ are always negative, whereas $\partial X_1/\partial C$ may be either positive or negative (two of the terms make it positive, one makes it negative). Thus, selection in X_1 for increase at the expense of nutrients always decreases the abundance of the competitor X_2 and the nutrient X_3, but the direction of the effect on the abundance of X_1 itself depends on the values of the system parameters.

Examining the effects of individual parameters in eqs. 12–14, we draw five further conclusions:

If $a_{22} = 0$, then $\partial X_1/\partial C$ consists only of one positive term, and $\partial X_3/\partial C = 0$. Furthermore, the denominator of eqs. 12–14 loses two of its three terms, increasing the damage to X_2's abundance and the stimulation to X_1's abundance. That is, if the competitor does not self-regulate his growth at all, the evolving consumer is guaranteed to increase his own abundance and harm his competitor without even lowering the nutrient level. In effect, the species undergoing selection simply diverts to his own use some resources that would previously have gone to his unregulated rival.

Conversely, large a_{11} (X_1 highly self-damped) hurts X_2 more than X_1, since a_{11} terms appear in the denominator for both $\partial X_1/\partial C$ and $\partial X_2/\partial C$ but only in the numerator, as a negative term, in $\partial X_2/\partial C$.

If a_{23} is large, the rival is hurt greatly, since a_{23} appears in both terms of $\partial X_2/\partial C$; and the evolving species is more likely to increase his own abundance, since a_{23} appears in a positive term of $\partial X_1/\partial C$. If a_{13} is small, the evolving species is less likely to decrease in abundance, since a_{13} appears in a negative term of $\partial X_1/\partial C$. The greater the inequality $a_{13} < a_{23}$, the more valuable does the trait C become. To interpret the meaning of this inequality, recall that a_{i3} gives the magnitude of the effect of changes in nutrient level on the growth of the species X_i. Thus, a species that can increase very rapidly under conditions of abundance but that survives very poorly under conditions of scarcity will be particularly hard hit if a competitor evolves to overexploit. Conversely, a species that at best increases slowly but that also weathers bad times well can afford to overexploit: depletion of nutrient will differentially affect the rival's population.

A large a_{32}, by contributing to a positive term of $\partial X_1/\partial C$, tends to make X_1 increase rather than decrease in abundance. Large a_{32} would mean that the rival initially harvests much of the resources, so that the evolving species has much to gain by preempting the rival's share.

Finally, large $(\partial f_1/\partial C)$ contributes to two positive terms in $\partial X_1/\partial C$ and one negative term in $\partial X_2/\partial C$. Even if large $(\partial f_1/\partial C)$ inevitably means large $(\partial f_3/\partial C)$ for biological reasons (growth of X_1

achieved at the expense of nutrient depletion), the $(\partial f_1/\partial C)$ and $(\partial f_3/\partial C)$ terms both act to depress X_2 but tend to cancel in their effects on X_1 because of opposite signs.

We can summarize by saying that overexploitation is a viable strategy for a species faced with competitors; and that it is an especially profitable strategy for a species that regulates its own population closely and has low reproductive potential but good ability to survive scarcity, when the species is confronted with a rival that has high reproductive potential, breeds without limit, squanders resources, and crashes in times of scarcity. By overexploiting, the former species can decrease its rival's abundance more than its own, and may even increase its own abundance. But this description of the rival matches exactly with the properties of supertramps and r-selected species, whereas the successful overexploiter matches the description of a K-selected species. Thus, as one proceeds through the incidence categories from high-S species to supertramps, one passes down a hierarchy from plausible practitioners to plausible victims of overexploitation. The striking abundance of bird individuals in supertramp-dominated islands or habitats (Figure 18) is consistent with this interpretation of resource coupling derived from loop analysis and depicted in Figure 49.

Dispersal and Assembly

We have so far discussed assembly rules as if they were determined solely by how species utilize resources. This is obviously not the full explanation, else supertramps would be starved out everywhere. In fact, on a small island no population survives long. Among two competing species of mutually exclusive requirements, an equilibrium is established on a small island, such that the fraction of time each species is present depends not only on competitive ability (or resource utilization) but also on population size and interisland dispersal rate. As discussed by Levins and Culver (1971) and Horn and MacArthur (1972), a species with a sufficiently high dispersal rate may thereby occupy the island for more of the time than does the species with superior competitive ability. Overexploitation, by reducing population size, becomes a dangerous strategy on a small island. Thus the supertramps, with their twin advantages of high dispersal rates and large populations, are increasingly favored on islands of decreasing size. The species of small islands will be to some extent the species that share these advantages.

Transition Probabilities

It is conceivable that certain species combinations that would be superior in utilization of resources may not exist in nature because it is too difficult to assemble them from other permitted combinations. For instance, suppose that a certain four-species combination could be assembled from permitted three-species combinations only by eliminating two of the species and adding three others; and could be assembled from permitted five-species combinations only by eliminating three species and adding two others. Since such changes are much less likely to occur

simultaneously than is the addition or subtraction of a single species, and since the intermediate combinations would be forbidden, such a four-species combination might never be assembled because of low transition probabilities.

To assess the significance of this consideration, let us compare the ease of assembly of the permitted and forbidden combinations in the guilds we have discussed:

Among the cuckoo-doves (Table 7), all permitted combinations of n species, $p.c._n$, can be assembled from below (i.e., from $p.c._{n-1}$) by addition of a single species (e.g., $AR + N \rightarrow ANR$). One of the $p.c._2$ and both of the $p.c._1$ can be assembled from above (from $p.c._{n+1}$) by loss of a single species (e.g., $AR - R \rightarrow A$), but two of the $p.c._2$ (AM and MR) cannot. Of the forbidden combinations of n species ($f.c._n$), five (R, AN, AMN, AMR, MNR) can be assembled in a single step from below or above, three ($N, MN, AMNR$) from below but not from above, and one (NR) from above but not from below.

Among the gleaning flycatchers, nine $p.c.$'s can be assembled in one step from above or below, five just from below, one just from above (APV on Umboi, probably derived by loss of H from $AHPV$ on New Britain, formerly in near-contact with or joined by a land-bridge to Umboi), and two ($CHPV$ and $ACDH$, on one island each) in neither direction. These last two combinations could have been derived either by addition of two species to a $p.c._2$, by addition of two species and loss of one species in a $p.c._3$, by replacement of one species in a $p.c._4$, or by two-for-one ($CHPV$) or three-for-two ($ACDH$) substitution in the sole $p.c._5$, $AHPRV$. $ACDV$ on the land-bridge island Duke of York was probably in fact assembled by loss of HP from the parent New Britain combination $AHPV$ and immediate colonization by the supertramps C and D. On St. Matthias $CHPV$, which includes the supertramp C, could have been assembled in several plausible ways (e.g., $AHPV - A + C$, $APV + H - A + C$). Among the $f.c._2$ and $f.c._3$, 13 can be assembled in one step from above or below, 11 from below but not from above, 9 from above but not from below, and 15 from neither direction. None of these forbidden combinations would require more steps to assemble than the permitted combinations $CHPV$ and $ACDH$. The myzomelid-sunbird guild yields a qualitatively similar picture.

Thus, most permitted combinations can be assembled in one step, although a few require more numerous but still plausible transition steps. Simultaneous invasions by several species in a guild may occur if the species are all stimulated by the same environmental change to send out immigrants. For instance, Meyer (1906) recorded an invasion of Vuatom, then inhabited by the fruit pigeons *Ducula rubicera* and *D. pistrinaria*, by mixed flocks of *Ducula melanochroa*, *D. finschii*, and *D. spilorrhoa;* the invaders disappeared, leaving the original two species, one of which then disappeared during the next few decades. On Umboi invasions of both *Ptilinopus superbus* and *P. solomonensis* within a few decades virtually supplanted *P. solomonensis*. Near-simultane-

ous extinctions of several species may occur if the extinction of one species of a $p.c._n$ yields a $f.c._{n-1}$ and another species is then rapidly expelled to reach a $p.c._{n-2}$. Thus, relaxation of guilds on land-bridge islands could proceed in steps. (cf. p. 411).

Conversely, most but not all forbidden combinations can be assembled in one step. Whether low transition probabilities are the explanation for a few forbidden combinations remains uncertain.

Summarizing our discussion of the origin of assembly rules, we conclude that resource utilization must be a major factor. Permitted combinations may leave fewer resources unutilized than forbidden combinations with the same total number of species. In addition, the various species of a permitted combination must be "companions in starvation," i.e., must have roughly comparable abilities to tolerate lowered resource levels. On increasingly small or remote islands, the ability to arrive frequently and maintain a large population becomes an increasingly significant consideration in determining what species will co-occur. Finally, the possibility exists that some combinations are just difficult to assemble, especially if some of their component species colonize infrequently.

Unsolved Problems

The finding that community assembly obeys the rules discussed above raises numerous unsolved problems of interpretation, some of which will be briefly summarized.

Reconstructing incidence functions.

The incidence function of a species is expected to depend on the species' territory size, population density, birth and death rates, and dispersal rates; on similar properties of competing species; and on island area and isolation. It should be feasible to develop mathematical models of incidence functions, initially for guilds of a few species.

Applications to habitat communities,
and to locally patchy communities.

The incidence functions of Figures 4–15, 19, 25, 28, and 29 were all constructed for communities each of which corresponds to a single island. Such functions were feasible to construct because islands are the easiest communities to delimit, hence species lists (the starting point for construction of incidence functions) were available for many Bismarck islands. Correspondingly, the discussion of assembly rules on pp. 393–411 was based on island communities. However, the discussions of pp. 412–416 and of pp. 416–423 suggest that assembly rules exist, and incidence functions can be profitably constructed, for communities corresponding to local geographical areas, and for communities corresponding to single habitats. Applied to local geographical areas, this approach may help rationalize the phenomenon of tropical patchiness resulting from diffuse competition. Applied to habitat communities, the approach may permit prediction of expansion and compression in habitat preference associated with changes in diffuse competition. The main practical problem may be to obtain enough good local lists for a tropical area with patchy distributions, or enough islands with good information about variation in species habitat preference. In addition, the abundance of a species often varies with competition from other species (see discussions of "density compensation" by Crowell, 1962; Diamond, 1970a, 1970b; MacArthur et al., 1972; Cody, Chapter 10; Brown, Chapter 13). Where competition is diffuse, it may prove profitable to plot abundance as a function of the number of competing species (cf. Yeaton and Cody, 1974, Fig. 1); such curves may have the form of incidence functions inverted.

Chance or predestination?

At the one extreme, the species composition of an island fauna might be uniquely determined by an island's physical properties. Combinations of colonists might be reshuffled through invasion and extinction until the best-suited groups of colonists had been assembled, and these would then persist. (This view is surely valid if applied to supertramp exclusion from large islands.) Differences between the faunas of apparently similar islands might really be due to slight physical differences in the islands selecting those combinations of colonists that had a slight selective advantage. For instance, the consumption curve for the fruit-pigeon guild shows a slight proportionate excess of large fruits on Sakar compared with Long (Figure 46). If this same finding applied to the fruit production curves, it might help explain why Sakar supports the pigeon combination *Ptilinopus solo-*

monensis (91 g), *P. insolitus* (144 g) and *Ducula rubricera* (722 g), whereas the Long lowlands support the combination *Ptilinopus solomonensis* (91 g) and *Ducula pistrinaria* (470 g): the greater abundance of large fruits on Sakar would favor the very large *D. rubricera* over the medium-large *D. pistrinaria,* leaving a gap between the *Ducula* and the very small *P. solomonensis* into which the medium-small *P. insolitus* could fit. At present such reasoning would be totally circular, since the production curves are unknown. Such detailed differences in production curves would be hard to detect or measure but could be very important.

At the other extreme, chance in the form of random historical events might play a large role in building up nonidentical communities that represent alternative stable equilibria. Of several closely related species, whichever happens to arrive first may become so numerous by the time a competitor arrives that it may be impossibly difficult for the later arrival to establish itself. Or, selection of the first successful colonist within a group of related species may then prejudice the chances of success among the remaining species, some of which will "fit" better than others with the first arrival. Any given community would thus represent one out of many possible, alternative, stable communities, drawn in turn from a much larger number of less stable communities that could be constructed on paper from the same species pool.

Numerous findings suggest at least some role of chance. After all, there does occur turnover in species compositions of islands that are at equilibrium with respect to species number, so that species composition wanders among adjacent stability maxima. Often, a habitat that suddenly becomes available may be colonized by whatever species is experiencing a "bloom" in the vicinity at that time (Connell, Chapter 16). For instance, the incidence of the supertramp *Zosterops griseotincta* (known from no Bismarck volcano except Long and its two neighbors) is so low that its colonization of Long after the explosion seems much more likely to have been the product of a bloom at the right time than of a uniquely good fit to local conditions. Yet this supertramp is now the most abundant and ubiquitous bird of Long, is omnivorous, must have severely restricted the selection of what other small bird species could colonize Long, and thereby played a key role in the assembly of the present Long community. Similarly, each of the eight *Lonchura* grass finches that have distributed themselves checkerboard-fashion in the New Guinea midmontane grasslands (Figure 33) occupies a wide range of altitudes, rainfall conditions, and grassland types. This suggests that the locally successful colonist was selected on a first-come first-served basis rather than because of locally superior attributes.

The problem of chance vs. predestination is still more acute for plants. Among plants that seed at long intervals, happening to seed at the right time may be even more important to success in colonizing suddenly vacant habitats, than among birds. If the first-arriving plant grows a foot tall before the seeds of an ultimately superior competitor arrive, and if in addition the first arrival is profusely self-replacing, the competitor may be kept out for a long time (cf. Horn, Chapter 9). Plant communities may also be much more closely integrated and resistant to invaders than bird communities, as suggested by such biogeographic chimaeras as the old island of New Caledonia, which supports an old endemic flora but an avifauna composed mostly of recent arrivals.

Applications to conservation problems.

Recently, there has been increasing recognition of the potential practical contributions that the theory of island biogeography may make to a rational design of nature preserves (Willis, 1974; Diamond, 1972b, 1975; Terborgh, 1974a, 1974b; Wilson and Willis, Chapter 18) A piece of threatened habitat, set aside for conservation purposes, becomes a distributional island for species tied to that habitat. The number of species in a refuge is expected to be a function of refuge area. In addition, the incidence functions of Figures 4–15 show that species composition will be a function of refuge area. A refuge of a size capable of holding a certain species number S_i at equilibrium will eventually lose all species whose lower S_{crit} values exceed S_i, and will lose half of the species for which $J(S_i) = 0.5$. This could be a serious problem, because high-S species include a disproportionate fraction of an island's endemic forms and may be especially in need of protection.

An example is provided by the fourth-largest Bismarck island, New Hanover, which was joined by a land bridge to the second-largest Bismarck island, New Ireland, during the Pleistocene. At that time New Hanover must have shared most of New Ireland's species. Today, New Hanover has lost 23 New Ireland species but still has a total of 81 species and is still somewhat supersaturated for its area. A loss of 23 species, or about 22% of the original number, does not seem serious. However, among these lost species, New Hanover lost 19 of the 26 high-S species of New Ireland, including every one of the 5 New Ireland high-S species that are endemic to the Bismarcks at the species level. As a faunal preserve, New Hanover would rate as a disaster. Yet its area of 458 square miles is not small by the standards of many of the tropical rainforest parks that one can realistically hope for today.

The incidence function of a species that becomes threatened may be useful in suggesting the minimum area of a refuge that would be adequate to ensure its survival (i.e., to have a high J for that species). In addition, assembly rules may suggest what combinations of guild relatives can share a refuge with a species requiring protection, and not threaten its survival.

Acknowledgments

This paper owes much to stimulating discussions with Robert MacArthur and with Ernst Mayr. It is a pleasure further to record my debt to Martin Cody, for valuable suggestions on the manuscript; to numerous residents of the southwest Pacific, for cooperation in the field work;

to numerous workers on southwest Pacific birds, for information; and to the National Geographic Society, American Museum of Natural History, Explorers Club, National Science Foundation through the Alpha Helix-New Guinea program, and American Philosophical Society, for support.

References

Bayliss-Smith, T. P. 1973. A recent immigrant to Ongtong Java atoll, Solomon Islands. *Bull. Brit. Ornith. Club* 93:52–53.

Bradley, D., and T. Wolff. 1956. The birds of Rennell Island. *In* T. Wolff, ed., *The Natural History of Rennell Island,* vol. 1, pp. 85–120. Danish Science Press, Copenhagen.

Brown, J. 1971. Mammals on mountaintops: nonequilibrium insular biogeography. *Amer. Natur.* 105:467–478.

Coates, B. J. 1973. Straw-necked Ibis *Threskiornis spinicollis. New Guinea Bird Society Newsletter* No. 84, p. 2.

Crowell, K. L. 1962. Reduced interspecific competition among the birds of Bermuda. *Ecology* 43:75–88.

Crowell, K. L. 1973. Experimental zoogeography: introductions of mice to small islands. *Amer. Natur.* 107:535–558.

Diamond, J. M. 1969. Avifaunal equilibria and species turnover rates on the Channel Islands of California. *Proc. Nat. Acad. Sci. U.S.A.* 64:57–63.

Diamond, J. M. 1970a. Ecological consequences of island colonization by southwest Pacific birds. I. Types of niche shifts. *Proc. Nat. Acad. Sci. U.S.A.* 67:529–536.

Diamond, J. M. 1970b. Ecological consequences of island colonization by southwest Pacific birds. II. The effect of species

diversity on total population density. *Proc. Nat. Acad. Sci. U.S.A.* 67:1715–1721.

Diamond, J. M. 1971. Comparison of faunal equilibrium turnover rates on a tropical island and a temperate island. *Proc. Nat. Acad. Sci. U.S.A.* 68:2742–2745.

Diamond, J. M. 1972a. *Avifauna of the Eastern Highlands of New Guinea.* Nuttall Ornithological Club, Cambridge, Mass.

Diamond, J. M. 1972b. Biogeographic kinetics: estimation of relaxation times for avifaunas of southwest Pacific islands. *Proc. Nat. Acad. Sci. U.S.A.* 69:3199–3203.

Diamond, J. M. 1973. Distributional ecology of New Guinea birds. *Science* 179:759–769.

Diamond, J. M. 1974. Colonization of exploded volcanic islands by birds: the supertramp strategy. *Science* 184:803–806.

Diamond, J. M. 1975. The island dilemma: lessons of modern biogeographic studies for the design of natural preserves. *Biological Conservation,* in press.

Elton, C. S. 1973. The structure of invertebrate populations inside neotropical rain forest. *J. Animal Ecol.* 42:55–104.

Horn, H. S., and R. H. MacArthur. 1972. Competition among fugitive species in a Harlequin environment. *Ecology* 53:749–752.

Jones, H. L., and J. M. Diamond. 1975. Species Turnover at Equilibrium. In preparation.

Karr, J. R. 1971. Structure of avian communities in selected Panama and Illinois habitats. *Ecol. Monographs* 41:207–233.

Lack, D. 1971. *Ecological Isolation in Birds.* Harvard University Press, Cambridge, Mass.

Levins, R., and D. Culver. 1971. Regional coexistence of species, and competition between rare species. *Proc. Nat. Acad. Sci. U.S.A.* 68:1246–1248.

MacArthur, R. H. 1970. Species packing and

competitive equilibrium for many species. *Theoret. Pop. Biol.* 1:1–11.

MacArthur, R. H. 1972. *Geographical Ecology.* Harper and Row, New York.

MacArthur, R. H., J. M. Diamond, and J. R. Karr. 1972. Density compensation in island faunas. *Ecology* 53:330–342.

MacArthur, R. H., and E. O. Wilson. 1967. *The Theory of Island Biogeography.* Princeton University Press, Princeton.

Mackay, R. D. 1970. *The Birds of Port Moresby and District.* Thomas Nelson, Melbourne.

Mayr, E. 1942. *Systematics and the Origin of Species.* Columbia University Press, New York.

Mayr, E. 1945. The correct name of the Fijian Mountain Lorikeet. *Auk* 62:139.

Mayr, E. 1969. Bird speciation in the tropics. *Biol. J. Linn. Soc.* 1:1–17.

Mayr, E., and J. M. Diamond. 1975. Speciation in the birds of Northern Melanesia. In preparation.

Meyer, O. 1906. Die Vögel der Insel Vuatom. *Natur und Offenbarung* 52:513–657.

Meyer, O. 1930. Uebersicht über die Brutzeiten der Vögel auf der Insel Vuatom (New Britain). *J.f. Ornithologie* 78:19–38.

Meyer, O. 1933. Vogeleier und Nester aus Neubritannien, Südsee. *Beiträge z. Fortpflanzungsbiol. der Vögel* 9:122–185.

Meyer, O. 1934. Seltene Vögel auf Neubritannien. *J.f. Ornithologie* 82:568–578.

Preston, F. W. 1962. The canonical distribution of commonness and rarity. *Ecology* 43:185–215, 410–432.

Ripley, S. D. 1959. Competition between sunbird and honeyeater species in the Moluccan islands. *Amer. Natur.* 93:127–132.

Stresemann, E. 1939. Die Vögel von Celebes. *J.f. Ornithologie* 87:299–425.

Terborgh, J. 1974a. Faunal equilibria and the design of wildlife preserves. *In* F. Golley and E. Medina, eds., *Tropical Ecological Systems: Trends in Terrestrial and Aquatic Research.* Springer, New York.

Terborgh, J. 1974b. Preservation of natural diversity: the problem of extinction-prone species. *BioScience* 24:715–722.

Terborgh, J., and J. M. Diamond. 1970. Niche overlap in feeding assemblages of New Guinea birds. *Wilson Bull.* 82:29–52.

Terborgh, J., and J. Faaborg. 1973. Turnover and ecological release in the avifauna of Mona Island, Puerto Rico. *Auk* 90:759–779.

Willis, E. O. 1974. Populations and local extinctions of birds on Barro Colorado Island, Panama. *Ecol. Monographs* 44:153–169.

Wilson, E. O. 1961. The nature of the taxon cycle in the Melanesian ant fauna. *Amer. Natur.* 95:169–193.

Yeaton, R. I., and M. L. Cody. 1974. Competitive release in island song sparrow populations. *Theoret. Pop. Biol.* 5:42–58.

15 Stream Communities   *Ruth Patrick*

This chapter will give a broad review of the species communities in fresh-water streams, and of the factors affecting the structure of these communities. As study objects, stream communities have a major advantage over vertebrate communities (such as those discussed in Chapters 10–14) in that they lend themselves to experimental manipulation. For example, because of the small sizes and high population densities (in individuals per square meter) of stream diatoms, it is feasible to create realistic "islands" on glass slides less than a few inches in diameter. The rate of colonization of these islands can be varied simply by changing the flow rate of stream water over the slides. Because of the short life cycles of diatoms, one can relatively rapidly perform detailed analyses of how environmental variables such as light, temperature, trace metal concentrations, and quality of the substrate affect the growth rates and relative competitive abilities of the species. Similar experiments on vertebrates usually require modifying many acres or square miles of habitat and waiting months or years for the results, because vertebrates are relatively large, live at low densities, and have long life cycles.

I shall begin by summarizing some distinctive characteristics of the natural history of stream communities, and the patterns of species diversity in these communities. Next I shall describe what factors bound the niches and affect the relative competitive abilities of stream species, and how predation affects community structure. Finally, species succession in stream communities will be briefly discussed.

Natural History of Stream Communities

In this section I point out three distinctive features of stream communities: the sizes, life cycles, and trophic relations of their species.

Size

Fresh-water stream communities are composed of a wide variety of both microscopic and macroscopic species. The important herbivores and carnivores not only exhibit a wide range of size but also belong to many different phyla, classes, and orders of animals (e.g., Protozoa, Annelida, Mollusca, Crustacea, Insecta, and Vertebrata). The primary producers are mainly microscopic algae, although macrophytes or large plants, such as Spermatophytes, Pteridophytes, or Bryophytes, may be common in some areas.

Most of the species in stream ecosystems are small compared with their terrestrial counterparts, and many are small compared with their marine counterparts.

445

For example, macroscopic plants are the primary producers in virtually all terrestrial communities, and also in the intertidal and rocky-shore zones of estuaries and oceans (e.g., the macroscopic algae *Fucus* and *Ascophyllum*). Even those river organisms that are relatively large, such as the porpoises of the Amazon River, are still generally smaller than their marine counterparts. Since stream communities are evolutionarily ancient, why are their species skewed towards small size? The answer may be partly related to the total available area of habitat, because fish, for example, tend to be bigger in larger rivers than in small rivers. Perhaps the number of individuals of a large species that could be maintained in a stream system is sufficiently low that such populations and species would have short survival times to extinction. A further contributing factor may be the mechanical constraints of narrow channels, and the expenditure of energy necessary to hold a fixed position in swift currents.

Life Cycles

Compared with the species of terrestrial communities and some deep-lake and marine communities, stream organisms have short life cycles. In many species the life cycle only lasts one day, in others a few weeks or months. In only a few groups of stream organisms, especially fish and molluscs, do we find species whose cycles last several years. Yet even the longest such cycles are short compared with those of terrestrial communities, where many predators live for decades and many trees for centuries.

These short generation times mean that stream organisms typically have very high reproductive rates. For example, asexual reproduction of diatoms may occur daily, the blackfly *Simulium vitatum* often has three generations per year, and many other aquatic insects and several invertebrates of other classes have more than one generation per year. The offspring of each reproductive bout are numerous, relatively small in size, and rarely receive parental care (cf. Connell, Chapter 16). Clearly, stream species, like multivoltine butterflies (Shapiro, Chapter 7) and annual weeds (Schaffer and Gadgil, Chapter 6), have developed *r*- rather than *K*-strategies.

Food Webs

We routinely find four or five stages of energy transfer in stream communities. At any given place and time the species at a given trophic level belong to many taxonomic groups and have very diverse ecological requirements. For instance, at the base of food webs are primary producers and detritivores. The former are mainly algae of many major groups (Chlorophyta, Euglenophyta, Myxophyta, Bacillariophyta, and Chrysophyta), and the latter consist of bacteria, fungi, and sometimes invertebrates. The bacteria and fungi degrade detritus metabolically into much simpler chemical radicals that are utilized by algae and some invertebrates as a direct food source. In contrast, the invertebrate detritivores, such as certain worms and insects, usually break detritus particles into sizes acceptable to other detritivore species. R. Vannote (personal communication), for example, found that leaf particles in the feces of tipulid larvae are in the size range of food particles selected by mayflies (*Ephemerella*). The general importance of particle size as a criterion of food preference has been stressed for plankton communities by Brooks and Dodson (1965) and duplicates similar findings for vertebrate communities (Hespenheide, Chapter 7, Figures 1, 2, and 4; Brown, Chapter 13, Figures 6 and 7; Diamond, Chapter 14, Figures 30 and 31). As a further instance of digestion by one species generating food for other species, algae that pass through the guts of a crustacean may reproduce more effectively than uneaten algae and may thereby increase in food value to their consumers (Porter, 1973).

While algae constitute the main plant food of stream ecosystems, they also serve other functions. Although macroscopic algae and other plants are eaten by many stream organisms, it is often unclear whether this is for the food value of the plants themselves or of their epiphytes. The main significance of rooted aquatic plants (Spermatophyta, Pteridophyta, and Bryophyta) seems to be in furnishing shelter and protection from predators to various animals (cf. Connell, Chapter 16). Rooted or floating plants also serve as substrates on which other sessile organisms live, and thus increase the diversity of available habitats.

The approximate number of species performing a given trophic function remains fairly constant over time. For example, in the Savannah River communities we studied, carnivores comprise 40–50% of all species of fish, 10–30% of insect species, and 3% of invertebrate species other than insects. These figures are only approximate, since little definitive information on diets of invertebrates and fish is available. An important factor making it difficult to classify species by food habits is that different developmental stages of certain species have different food preferences. Thus, caddisfly larvae are detritus feeders for several months, then switch to feeding primarily on diatoms. The reason for the switch seems to be that detritus and diatoms contain very different nutrients and that caddisfly larvae need different amounts of nutrients at different stages of development (R. Vannote, personal communication). Without gut analysis and feeding experiments one cannot always be confident in deciding whether a certain consumer species at a particular stage of its life cycle is a carnivore, a herbivore feeding exclusively on plants, an omnivore, or a detritivore.

Species Diversity

Patterns of Species Diversity

Natural stream communities often support a very large number of species, as first noted by Thienemann (1920, 1939) and subsequently confirmed for stream communities by Patrick (1949, 1953, 1972), Tarzwell and Gaufin (1953), and Patrick, Cairns, and Roback (1967). The species belong to many different systematic groups, and each genus or family present may be represented by several to many species.

When the number of species in any

given aquatic community is carefully examined (Patrick, 1949), this number is found to remain constant with time as long as no exogenous perturbations occur. This is illustrated by Table 1, which shows that the number of species in each of four different groups (algae, protozoa, insects, and fish) surveyed in the Savannah River was relatively similar from year to year. We also find little difference among species numbers in different sites or streams with structurally similar habitats. This similarity is exemplified by comparison of the four stations in the Savannah River in a given year (Table 1), or by comparison of the 12 stations in nine streams listed in Table 2. These nine streams were selected for their wide geographic range: the Savannah River, North Anna, Rock Creek, and Potomac drain to the Atlantic; the Escambia River and North Fork of the Holston drain via the Mississippi River to the Gulf of Mexico; and the Ottawa drains into the St. Lawrence. Despite this geographic diversity, the species numbers in each of four systematic groups rarely de

viate by more than 33% from the mean value for all rivers studied. When we separate hard-water rivers from soft-water rivers, the deviations from the means for each type of river are even less. Since these nine streams were censused by different groups of scientists, Table 2 provides compelling evidence that well-collected, similar-sized areas of different streams support similar numbers of species. This constancy of α-diversity or species packing level among similar stream communities parallels Cody's (Chapter 10) findings for terrestrial bird communities.

Although different streams share similar *total numbers* of species, the number of *species shared* between stations in different streams is low. Table 3, which is based on surveys of the same stations in these nine different streams, shows that more than half of the species in each of the four broad systematic groups were found at only one station, and that vanishingly few species were found at seven or more stations. This finding does not

Table 1. Numbers of species at four stations in the Savannah River.

Organisms	Stations															
	1	3	5	6	1	3	5	6	1	3	5	6	1	3	5	6
	Year: 1955				Year: 1956				Year: 1960a				Year: 1960b			
Algae	98	89	103	120	98	97	97	84	96	77	90	72	75	90	103	99
Protozoa	42	52	48	55	41	38	37	51	53	54	67	58	55	60	62	67
Insects	44	41	54	58	46	47	54	46	33	35	37	26	26	34	35	28
Fish	35	23	30	25	24	30	31	29	32	30	36	33	40	33	37	40

The table gives the number of species in each of four broad groups of organisms, at each of four collecting stations in the Savannah River at each of four times. Low flow prevailed during the 1955 and second 1960 (1960b) collection, high flow during the 1956 and first 1960 (1960a) collection. Silt loads were high in 1960 because of dredging. Note that the species number in each category is relatively constant in space and time.

Table 2. Total number of taxa in a local area in each of nine streams

Organisms	Soft-water Rivers						Hard-water Rivers								
	Escambia	Savannah 54	Savannah 55	North Anna	White Clay	Flint	N. Fork Holston	Rock Creek	Ottawa 55	Ottawa 56	Potomac 56	Potomac 57	Mean All Rivers	Mean Soft Rivers	Mean Hard Rivers
Algae	77	105	101	98	73	79	63	65	76	58	105	103	84	89	78
Protozoa	38	61	40	58	56	51		86		48	85	68	59	51	72
Insects	29	58	51	61	57		83	48	59	61	89	99	63	51	73
Fish	39	19	35	21	20	13	21	24	18	28	18	29	24	25	23

The number of taxa or species in each of four broad groups of organisms is given for a local collection at each of 12 stations in nine streams. The streams belong to three different drainage systems (Atlantic, Gulf of Mexico, St. Lawrence). Nevertheless, the number of species in a given category is relatively constant, especially if comparisons are confined to different hard-water or different soft-water streams.

imply that equally marked differences would be found if entire stream systems were collected and compared; the meaning is, instead, that if one collects small areas of different streams at particular times, the chance of finding the same species in such collections is low. In our Savannah River study (Table 1) there were few shared species between collections at the same station at different times but more shared species among different stations collected at the same time.

Table 3. Distribution of taxa in rivers cited in Table 2

Organisms	Total no. of taxa	Number of taxa and number of rivers in which they occur								
		1	2	3	4	5	6	7	8	9
Algae	354	197 (55.7%)	61 (17.2%)	38 (10.7%)	25 (7.1%)	16 (4.5%)	8 (2.3%)	6 (1.7%)		3 (0.9%)
Protozoa	299	188 (62.9%)	40 (13.4%)	32 (10.7%)	23 (7.7%)	8 (2.7%)	5 (1.7%)	1 (0.1%)	2 (0.7%)	
Insects	283	209 (73.9%)	31 (10.9%)	24 (8.5%)	9 (3.2%)	6 (2.1%)	3 (1.1%)	1 (0.1%)		
Fish	132	75 (56.8%)	22 (16.7%)	22 (16.7%)	9 (6.8%)		3 (2.3%)			

For collections from the twelve stations belonging to nine rivers listed in Table 2, and for each of four broad groups of organisms, the table gives the number of taxa or species found in 1, 2, 3, 4, 5, 6, 7, 8, or all 9 rivers. Column 2 gives the total number of taxa in each category; numbers in parentheses refer to percentages of total number of taxa in each category. Protozoa and insecta were studied in only eight rivers. Most species are confined to one or a few collections, indicating high species-turnover in space and time.

Thus, the relative constancy of α-diversity with geography or time conceals a high species turnover with geography or time, analogous to high β-diversity. Some of the reaons why different species occur at different places or times will be explored later. In the next section we consider how these high species diversities are established.

Experiments on
Stream Community "Islands"

Support for the MacArthur and Wilson (1967) equilibrium theory of island species diversities has come both from experimental tests (e.g., Simberloff & Wilson, 1969) and observational tests. Stream communities lend themselves much more readily to experimental study than do the mangrove-tree islands studied so fruitfully by Simberloff and Wilson. To create artificial islands in streams, I mounted small bits of glass on pedicels of glass slides, placed the slides in plastic containers through which water could flow, placed the containers in streams, and counted the numbers of individuals and species of diatoms that developed populations on the slides (Patrick, 1967, 1968). As long as one counts more than 8000 individuals, 95% or more of the individuals on each of a set of replicate slides belong to species encountered on the other slides of the set. Thus, one has very similar communities with which to experiment on the various slides. This system can be used to study the effects of island area, invasion rate, and species pool size on the species diversity of diatom communities.

To test the effect of island area on species diversity, I compared slides with areas of 9, 36, 144, or 625 square millimeters. Larger slides were found to support more species: e.g., 28–32 and 44–47 diatom species on slides with areas of 144 and 625 mm², respectively.

To test the effect of invasion rate, I let water from White Clay Creek flow over slides of constant area at either 1.5 or 650 liters per hour. As illustrated in Figure 1, reduction in flow rate caused a large decrease in the numbers of species forming the community, and also caused an increased variance in the population sizes of the species. The same two findings emerge from comparison of two very different flow rates in Darby Creek (Table 4, first four lines).

The effect of species pool size was studied in two series of experiments. In the first series I compared slides of various areas suspended in two Pennsylvania streams with diatom communities of very different sizes: Ridley Creek and Roxborough Spring Creek, which support about 250 diatom species and fewer than 100 diatom species, respectively, at any point in time. More than 160 species grew on the 36-mm² slides suspended in Ridley Creek, whereas only 14–29 species grew on slides of the same area in Roxborough Spring Creek. In the second series, I compared two species-rich continental streams (Hunting Creek in Maryland and Darby Creek in Pennsylvania on the North American mainland) with three species-poor insular streams (Canaries River on St. Lucia Island, and Layou River and Check Hall River on Dominica Island, in the West Indies). As summar-

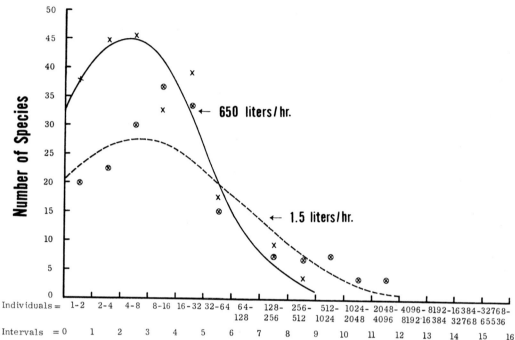

Figure 1 Glass slides were suspended in water from White Clay Creek, and the number of individuals of each diatom species that attached was counted. The ordinate is the number of species represented by the number of individuals given by the abscissa. The abscissa is a logarithmic scale, divided into intervals spanning a two-fold range of number of individuals. Stream water flowed over the slides either at 650 liters/hr (points x; the natural rate of stream flow), or at 1.5 liters/hr of new water recycled to give a total flow rate of 650 liters/hr (points ⊗). The curves are lognormal distributions fitted through the experimental points. Note that reduction in flow rate lowers the height of the mode and the total number of species represented, but increases the variance of abundance and the number of abscissa intervals covered by experimental points.

ized in Table 4, slides suspended in the continental streams supported 79–129 species, whereas slides of the same area in the insular streams supported only 46–61 species. The slides from insular streams also exhibited greater variation in population sizes than did the continental slides. Thus, lower species number, whether it results from lower invasion rate (Figure 1 and first four lines of Table 4)

Table 4. Species-abundance relations of diatom communities

Streams	(1) Height of mode	(2) σ^2	(3) Observed number of species	(4) Theoretical number of species	(5) Intervals covered by curve
Continental streams					
Darby Creek (Pa.), Sept.–Oct. 1964					
550–600 liters/hr.	22.5	6.9	129	148	9
1.5 liters/hr.	13.9	12.6	100	124	12
Darby Creek, Oct.–Nov. 1964					
550–600 liters/hr.	22.4	6.2	123	140	9
1.5 liters/hr.	15.3	12.0	97	133	15
Hunting Creek (Maryland)	12.0	9.1	79	92	14
Insular streams					
Canaries River (St. Lucia Island)	8.4	9.3	61	64	10
Layou River (Dominica Island)	5.3	26.0	49	67	14
Check Hall River (Dominica Island)	5.2	21.6	46	60	14

Glass slides of constant area were suspended in water from two continental (North American) and three insular (West Indian) streams, and the number of individuals of each diatom species that grew on the slides was counted. The species-abundance relation was then plotted as in Figure 1. The first column gives the number of species with modal abundance (i.e., ordinate values of the maxima of curves such as Figure 1); the second column, a measure of the width or variance of the distribution; the third column, the observed number of species; the fourth column, the theoretical total number of species calculated from a lognormal distribution curve fitted to the observed species-abundance values; and the last column, the number of abscissa intervals (each representing a two-fold range of number of individuals) covered by the observed species-abundance values. Two successive experiments were carried out in Darby Creek; each experiment used two different flow rates of stream water over the slides. Note that more species grow on slides in species-rich continental streams than in species-poor insular streams (compare rows 1–5 with rows 6–8 in columns 3 and 4); that number of attached species decreases with decreasing flow rate (compare rows 2 and 4 with rows 1 and 3 in columns 3 and 4); and that decrease in species number resulting either from reduced species pool or reduced flow rate decreases the height of the mode (column 1) but increases the variance (column 2) of the species-abundance relation.

or from lower species pool (insular vs. continental streams of Table 4), is associated with greater variation in population size.

These experiments show that species diversity of island communities increases with "island area," and is maintained by invasion. The invasion rate is proportional to the product of water flow rate times species pool size. In nature, invasion results from a combination of passive downstream drift and active upstream movement of stream organisms. As pointed out by Waters (1962), downstream drift of aquatic insects is a common phenomenon and is maximal at dusk and at dawn. In Darby Creek and White Clay Creek we noted that downstream drift of diatoms is much greater after sunset and before sunrise than during the daytime. The reason for the temporal pattern may be that at low light intensity aquatic insects forage on algae on the tops of rocks and dislodge the algae, which then drift. Active upstream movement of aquatic organisms has been noted by many authors. For example, many species of fish migrate upstream to spawn. To deposit eggs, aquatic insects may fly either upstream or downstream but usually fly upstream (Hynes, 1970).

Niche Limits and Competition

We have seen that brief, local sampling efforts show stream communities to be rich in species, but that there is much turnover of species in space and time. The spatial and temporal niche limits of each species are determined by physical factors, competition, and predation. In this section we shall discuss physical factors (especially current, substrate texture, light, temperature, nutrients, and trace metals) and competition, while predation is considered in the following section.

Effects of physical factors and of competition on niche limits are often linked. That is, a species may be confined to a region in time or space where a certain physical factor falls within a certain range of values, not because a value outside this range is directly lethal to the species but because at such a value the species is outcompeted by other species. For instance, trace metal concentrations and temperature greatly influence the relative abilities of algal species to outcompete each other. The species of stream communities have generally developed different strategies and different preferred physical conditions, in order to coexist and to reduce niche overlap and competition for resources. Thus, the same rock in a stream may support several species of caddisflies eating detritus and algae. However, these species eat different

amounts and types of food at different stages of development, and at any moment only one species is eating large amounts of algae. As a result, similar amounts of algae are being cropped from the rock throughout the year, but different species are the principal grazers at different times (R. Vannote, personal communication). This competitive spacing of species is often reinforced by interspecific aggression. For example, species of Trichoptera occupy areas with certain preferred current rates and attack other species that try to occupy the areas (Scott, 1958; Edington, 1965). Similarly, crayfish may divide stream habitats into riffle and pool territories when two species are present, instead of one species occupying the whole area (Bovbjerg, personal communication).

Among the factors determining where various species of invertebrates and fish live, current is one of the most important (Hustedt, 1938; Ruttner, 1953; Hynes, 1970). The various organisms coexisting in stream communities prefer different kinds of current structure. Species that are known as rheophils and prefer rapidly flowing water include stoneflies, mayflies, and some filter feeders such as blackfly larvae and caddisflies. In tropical streams many species of fish have special adaptations, such as suckers, to permit them to live in rapid currents. Other species, such as tubificid worms, chironomid larvae, and many Crustacea, prefer pools or slower currents.

Substrate structure and roughness provide further habitat axes along which species segregate. Soft-textured habitats are preferred by certain burrowing species such as burrowing mayflies, oligochaete worms (especially the Tubificidae), and some chironomid larvae such as *Chironomus plumosus.* Certain diatom species such as *Nitzschia palea* and *Navicula cryptocephala* prefer to grow on soft sediments. In contrast, other species (including many mayflies, stoneflies, caddisflies, and diatoms such as *Synedra rumpens* and the stalked diatoms *Gomphonema parvulum, G. intricatum,* and *Cymbella lanceolata*) prefer very hard-textured substrates, regardless of the current speed. When habitat roughness is reduced by siltation, as often happens in channelizations, species diversity is greatly reduced (Patrick and Vannote, in U.S. Government, 1973). This homogenization of the habitat by siltation reduces not only substrate diversity but also diversity of light patterns and of current patterns.

It is well known that different species of algae grow better at different light intensities. Thus, algal species can coexist even when one is shaded by the other. Species also differ in their temperature tolerances and preferences. For instance, a community that is dominated by blue-green algae above 34°C is dominated by diatoms at lower temperatures.

Differential effects of nutrients on species may also occur. For example, certain algae, including some blue-green algae and dinoflagellates, can live on much lower levels of nutrients such as nitrogen and silica than can *Fragilaria crotonensis* and *Gomphonema parvulum.* Thus, the latter two species are outcompeted by the former types of algae when concentrations

of these nutrients are low.[1] In bird communities this finding is paralleled by the suspected ability of so-called high-*S* bird species to survive at lower nutrient levels than so-called supertramps. Such a gradient of nutrient requirement lends itself to displacement of the species with high requirements by species with low requirements, through overexploitation (Diamond, Chapter 14, especially Figures 18, 49, and 50). However, competitive effects due to nutrient limitation or to heterotoxins (toxic substances that are produced by one species and that affect populations of other species) are far less pronounced in stream communities than in lakes and other aquatic communities (cf. Levins, Chapter 1, Figures 3 and 4), and are probably less pronounced than in terrestrial communities (cf. examples among terrestrial plants discussed by J. Connell, Chapter 16). The reason is that nutrients are always being supplied to streams from the watershed, and that the flow of water continually removes autotoxins from the species producing them and dilutes autotoxins and heterotoxins. Little local recycling occurs in streams, and the effects of upstream-to-downstream recycling are masked by nutrient renewal from watershed runoff or from groundwater.

In recent experiments on algal communities we have found that variations in concentrations of trace metals may have

---

[1] This need not mean that the outcompeted species is completely absent from the community. More often, it either becomes so rare as to escape collection, or passes into a different stage such as an egg, spore, or (in the case of insects) pupa.

profound effects on the kinds of species present and on their population sizes. For example, blue-green algae such as *Schizothrix calcicola* and *Microcoleus vaginatus* can live on much lower concentrations of manganese (Mn^{++}) than are needed to support diatom communities (Patrick, Crum, and Coles, 1969). When Mn^{++} concentrations exceed 40 micrograms per liter ($\mu g/l$), diatoms outcompete blue-green algae and maintain much larger populations. At Mn^{++} concentrations below about $15 \mu g/l$, blue-green algae outcompete diatoms. The reason may be related to lipid metabolism, since Mn^{++} is known to be required for lipid metabolism in the alga *Euglena gracilis* (Constantopoulos, 1970). Because diatoms metabolize and store lipids, it is reasonable that other species that do not store such large amounts of lipid should be able to outcompete diatoms when Mn^{++} is present in very low concentrations.

The effects of vanadium are opposite to those of manganese. At $3-4 \mu g/l$ vanadium, diatoms increase in biomass and dominate the community, whereas at $4000 \mu g/l$ blue-green algae replace diatoms as dominant species and attain high biomass. Similarly, blue-green algae are significantly stimulated by addition of $6 \mu g/l$ nickel and become dominant at $36-40 \mu g/l$ nickel (R. Patrick, unpublished observation). In some cases the effect of an element depends on its form. Thus, selenium in the form of selenite at $1 \, mg/l$ or $10 \, mg/l$ stimulates diatom growth, whereas selenate at $1 \, mg/l$ inhibits diatom growth but promotes blue-green algae and some unicellular green algae.

Effects of Predation

Although some predator species are relatively indiscriminate in their choice of food, other predators select for certain prey and against other prey. For instance, among ciliate Protozoa some species eat only diatoms, some prefer bacteria, and others that are generalists feed on a combination of diatoms and bacteria (Gizella and Gellert, 1958). Certain insect predators not only prefer diatoms as food but select particular diatom species, such as *Rhoicosphenia curvata*. Roop showed that the snail *Physa heterostropha* discriminates against the diatoms *Cocconeis placentula* and *Achnanthes lanceolata* in favor of other species (Patrick, 1970). Thus, if these two diatoms are allowed to continue to grow while snail predation reduces populations of other diatom species, selection by the predator results in a lowering of diatom species diversity. Similarly, it is well known that aquatic insects and other invertebrate predators greatly prefer diatoms to blue-green algae as prey. Thus, predation pressure allows populations of blue-green to multiply.

Depending on the degree of predator pressure and whether or not it is selective, predation can affect not only the biomass (standing crop) and species diversity but also the whole structure of a community. If herbivore grazing pressure is intense enough to control an aggressive species of primary producer that would otherwise become dominant, the result is increased plant diversity. This in turn supports a greater variety of herbivores, which in turn increases the diversity of carnivores

(Hairston et al., 1968). For example, the selective predation of large Crustacea by fish not only determines what species of Crustacea dominate the zooplankton, but also affects the composition of the phytoplankton, by permitting growth of species that had formerly been cropped by the larger zooplankton (Brooks and Dodson, 1965).

Connell (Chapter 16) has emphasized that the relative sizes of predator and prey often determine the intensity of predation, so that prey may be unable to escape predators until the prey somehow succeed in growing to a certain size. The prey population structure may then come to consist of widely spaced size classes, each reflecting a transient escape from predation. In stream communities, however, many species exhibit little variance in size in different stages of the life cycle. Because of the short life spans of most stream species, predators may eliminate many life stages of a prey species, including adults, and quickly reduce a large population of a prey species. Since the prey has no uneaten size classes left, the prey must either reinvade the community or else depend on a few surviving individuals to restore the population. Predation may have less marked effects on species in which the various age classes are of different sizes, since some size classes survive predation and continue to reproduce.

Connell also points out that prey species that can live in precarious or harsh environments are often able to develop larger populations than would be possible in benign areas where predators are com-

mon. In stream communities this phenomenon is exemplified by blackfly larvae, a prey species highly sought by predators. Except for a short time after the larvae hatch, their populations are small in most areas of natural streams. They become numerous only in rock crevices that are inaccessible to predators or in very fast water where the predators cannot survive.

Although competition affects the relative population sizes of species with similar ecological requirements and may reduce the number of species that can coexist, competition does not affect the structure of a whole food web or community as acutely as does predation. The surprising and far-reaching ways in which changes at one trophic level may affect other levels have been discussed at length by Levins (Chapter 1). This transmission of effects through the community is illustrated by the two cases discussed above, involving herbivore grazing pressure and fish predation on Crustacea. Effects of trace metals on stream communities provide another clear example. If changes in trace metal concentrations cause a shift among primary producers towards species less desirable as prey to herbivores (e.g., a shift from diatoms to blue-green algae, as caused by high vanadium and nickel concentrations or low manganese concentrations), the metals greatly affect the kinds and abundances of herbivores, hence also species at higher trophic levels. In the opposite direction along the trophic pyramid, the metals also shift the balance between primary producers and detritivores and change the levels of nutrients. Thus, the concentration of trace metals may alter the entire structure of a stream community.

Predictability and Succession of Communities

The values of some of the important parameters discussed in the section on Niche Limits and Competition, which control the existence and diversity of species in streams, vary unpredictably in space and time. For example, the identities and amounts of available nutrients vary as a function of the watershed use and runoff pattern at any moment. Storms and the resulting variable flows that scour the substrate of a stream often greatly reduce the abundances of previously dominant species and thereby open the community to invasion from the large available species pool. In analogy to observations by Simberloff and Wilson (1969) on recolonization of fumigated mangrove islands by arthropods, the species that invade a scoured stream often differ from the species reduced by scouring and are more diverse (Patrick, 1963). The new immigrants often correspond to what Diamond (Chapter 14) calls "tramp" or "supertramp" species.

As pointed out by Blum (1956) and observed by us at the Academy of Natural Sciences, temporal changes in stream communities and in terrestrial plant communities are very different. On land there is a predictable succession of communities, because shade, detritus, and other factors set by the community present at any instant determine the species composition of the subsequent community. Thus, species changes in succession can be predicted from a matrix of transition probabilities among species, as developed by Horn in Chapter 9. In contrast, the flushing of nutrients through streams, and the flushing-out of waste products, prevent the species of the moment from determining their successors. Streams do exhibit an annual cycle of seasonal succession, however, owing to seasonal changes in temperature, light, and nutrient levels. Superimposed on this seasonal regularity is a large element of unpredictability, as discussed in the preceding paragraph. Since the species of stream communities are relatively small in size and have short life cycles and high reproductive rates, the communities are always very young.

Many of the perturbations that man causes in stream communities arise from minute changes in the concentrations of substances such as trace metals, often by a few micrograms per liter. By affecting the physiology, behavior, and resistance to disease of individual species as well as competitive and trophic relations between species, such changes lead to large changes in ecosystem structure and to the dominance of certain species. Therefore, in examining the factors that develop and maintain diversity in stream ecosystems, one must consider not only obvious factors such as predator pressures competition, but also effects of substances present in minute amounts.

The species composition of a stream community at any place and time is determined by many independent variables, the unpredictable and ever-changing parameters of a randomly fluctuating environment. As discussed by May (Chapter 4), the Central Limit Theorem leads to the expectation that the species-abundance relation of such a community is lognormal (Figure 1). Although the fluctuations in species composition are largely unpredictable, the combination of high invasion rates, diverse physical structure of streams, and continual supply of nutrients and other factors serves to perpetuate stream communities.

References

Blum, J. L. 1956. Application of the climax concept to algal communities of streams. *Ecology* 37:603-604.

Brooks, J. L., and S. I. Dodson. 1965. Predation, body size, and composition of plankton. *Science* 150:28-35.

Constantopoulos, G. 1970. Lipid metabolism of manganese-deficient algae. *Plant Physiol.* 45:76-80.

Edington, J. M. 1965. The effect of water flow on populations of net-spinning Trichoptera. *Mitt. Internat. Verein. Limnol.* 13:40-48.

Gizella, T., and J. Gellert. 1958. Über Diatomeen und Ciliaten aus dem Aufwuchs der Ufersteine am Ostufer der Halbinsel Tihany. *Ann. Inst. Biol. Hungar. Acad. Sci.* (Tihany) 25:240-250.

Hairston, N. G., J. D. Allen, R. K. Colwell, D. J. Futuyma, J. Howell, M. D. Lubin, J. Mathias, and J. H. Vandermeer. 1968. The relationship between species diversity and stability: an experimental approach with protozoa and bacteria. *Ecology* 49:1091-1101.

Hustedt, F. 1938. Systematische and ökologische Untersuchungen über die Diatomeen-Flora von Java, Bali and Sumatra. *Arch. Hydrobiol.* Vol. 15 (Suppl.), pp. 131-177, 187-295, 393-506.

Hynes, H. B. N. 1970. *The Ecology of Running Water.* University of Toronto Press, Toronto.

MacArthur, R., and E. O. Wilson. 1967. *The Theory of Island Biogeography.* Princeton University Press, Princeton.

Patrick, R. 1949. A proposed biological measure of stream conditions based on a survey of Conestoga Basin, Lancaster County, Pennsylvania. *Proc. Acad. Nat. Sci. Philadelphia* 101:277-341.

Patrick, R. 1953. Biological phases of stream pollution. *Proc. Pennsylvania Acad. Sci.* 27:33-36.

Patrick, R. 1963. The structure of diatom communities under varying ecological conditions. *In: Conference on the Problems of Environmental Control of the Morphology of Fossil and Recent Protobionta. Trans. New York Acad. Sci.* 108:359-365.

Patrick, R. 1967. The effect of invasion rate, species pool, and size of area on the structure of the diatom community. *Proc. Nat. Acad. Sci. U.S.A.* 58:1335-1342.

Patrick, R. 1968. The structure of diatom communities in similar ecological conditions. *Amer. Natur.* 102:173-183.

Patrick, R. 1970. Benthic stream communities. *Amer. Scientist* 58:546-549.

Patrick, R. 1972. Benthic communities in streams. *Trans. Conn. Acad. Arts and Sci.* 44:272-284.

Patrick, R., J. Cairns, Jr., and S. S. Roback. 1967. An ecosystematic study of the fauna and flora of the Savannah River. *Proc. Acad. Nat. Sci. Philadelphia* 118:109-407.

Patrick, R., B. Crum, and J. Coles. 1969. Temperature and manganese as determining factors in the presence of diatom or blue-green algal floras in streams. *Proc. Nat. Acad. Sci. U.S.A.* 64:472–478.

Porter, K. G. 1973. Selective grazing and differential digestion of algae by zooplankton. *Nature* 244:179–180.

Ruttner, F. 1953. *Fundamentals of Limnology.* University of Toronto Press, Toronto.

Scott, D. 1958. Ecological studies on the Trichoptera of the River Dean, Cheshire. *Arch. Hydrobiol.* 54:340–392.

Simberloff, D. S., and E. O. Wilson, 1969. Experimental zoogeography of islands: the colonization of empty islands. *Ecology* 50:278–295.

Tarzwell, C. M., and A. R. Gaufin. 1953. Some important biological effects of pollution often disregarded in stream surveys. Proc. 8th Industr. Waste Conf., *Purdue University Engineering Bulletin* 295–316.

Thienemann, A. 1920. Untersuchungen über die Beziehungen zwischen dem Sauerstoffgehalt des Wassers und der Zusammensetzung der Fauna in norddeutschen Seen. *Arch. Hydrobiol.* 12:1–65.

Thienemann, A. 1939. Grundzüge einer allgemeinen Okologie. *Arch. Hydrobiol.* 35:267–285.

U. S. Government. 1973. Report on channel modifications submitted to Council on Environmental Quality. No. 4111-00015. *U. S. Government Printing Office,* Washington, D.C.

Waters, T. F. 1962. Diurnal periodicity in the drift of stream invertebrates. *Ecology* 43:316–320.

16 Some Mechanisms Producing Structure in Natural Communities: a Model and Evidence from Field Experiments

Joseph H. Connell

Introduction

Community Structure and
Niche Theory

If a local assemblage of organisms is to be regarded as a community with some degree of organization or structure, then it is in the interactions between the organisms that we must look to provide this structure. Two different interactions provide most of the organization: competition and predation. I will use competition in the sense of Birch's (1957) first meaning: "Competition occurs when a number of animals (of the same or of different species) utilize common resources, the supply of which is short; or if the resources are not in short supply, competition occurs when the animals seeking that resource nevertheless harm one or another in the process." I will use predation in the broad sense of an animal's eating another organism for its main source of food, including herbivores that eat plants as well as parasites or pathogens that eat their host.

Predation represents the interaction between trophic levels, whereas competition mainly represents interactions within trophic levels (although if space is the resource, it is possible for plants to compete with sessile animals in aquatic habitats). These two interactions may themselves interact: the numbers of two competing species may be reduced by their predators to such an extent that competition is prevented. In addition, the physical environment affects the intensity of both sorts of interactions, and the organisms affect the physical environment.

In studying community structure, one aspect of current interest is the theory of the ecological niche. One speaks of how niches are packed together, how much niches can overlap, what determines the breadth or shape of niches, etc. In these discussions the emphasis has been on only one of the two major interactions, competition, because of the way in which Hutchinson (1958) formally defined the niche. In his definition, the boundary of the "fundamental" niche is determined by the limiting states of all possible ecological variables, both physical and biological, which permit a species to exist indefinitely. The "realized" niche is defined as that portion of the fundamental niche within which a species is constrained by interactions with its competitors. "Interaction of any of the considered species is regarded as competitive in sense 2 of Birch (1957), negative competition being permissible, though not considered here. All species other than those under consideration are regarded as part of the coordinate system" (Hutchinson, 1958). Competition under sense 2 of Birch (1957) includes, in addition to the item in his

460

meaning quoted earlier, "an additional item, namely the interference with one species by another (with consequent change in birth rate or death rate) even when there is no demand for a common resource of space or food. But they exclude predation in which one animal eats another for its main source of food."

Vandermeer (1972) has pointed out that the idea of the niche presented by Grinnell (1917, 1928) is close to the idea of the fundamental niche, since he thought of it as "the concept of the ultimate distributional unit, within which each species is held by its structural and instinctive limitations." This concept of the niche excludes interactions with other species, both competitors and predators. In contrast, Elton's (1927) definition ("the 'niche' of an animal means its place in the abiotic environment, its relation to food and enemies") is closer to the realized niche, since it includes interactions with other species.

The distinction between fundamental and realized niches is a very useful one: if communities are organized by interactions, then the manner and degree of that organization will be reflected in the differences between the sizes and shapes of realized and fundamental niches.

The modern theory of the niche, as proposed by Levins (1968) and MacArthur (1968) and recently summarized and extended by Vandermeer (1972), is based upon the original definitions of Hutchinson (1958).

Vandermeer categorizes Hutchinson's distinction between fundamental and realized niche as a distinction between a preinteractive and a postinteractive niche (Vandermeer, 1972, p. 109), thus restricting the meaning of interaction to competition. This is also shown by his statement that the species ("operational taxonomic units") have only density-dependent feedback effects on each other and by his equating of "operational habitat" with "resource."

Thus, both in Hutchinson's original definition and in Vandermeer's recent summary and extension of the theory of the niche, only competitive interactions are considered. Other interactions, notably predation, are included in the factors bounding the fundamental niche, in the "conception of the niche as being preinteractive—the potential area within which a species can live as opposed to the area in which one actually finds it" (Vandermeer, 1972). Here, as elsewhere in discussions of niche theory, "preinteractive" means all aspects except competition, and "postinteractive" means "after competition occurs."

In this chapter I would like to challenge the idea that competition is the sole or even the principal mechanism determining the area in which one finds a species, as opposed to the potential area within which it can live. If this is not the case, then the formal structure of "niche theory" needs to be extended so that interactions other than competition are regarded as also constraining the realized niche.

Of course, it has long been recognized that predators can keep potential competitors so rare that they do not compete. Darwin's (1859, p. 67–68) experiment is probably the first demonstration of this: "If turf which has long been mown, and the case would be the same with turf closely browsed by quadrupeds, be let to grow, the more vigorous plants gradually kill the less vigorous, though fully grown plants; thus out of twenty species growing on a little plot of mown turf (three feet by four) nine species perished, from the other species being allowed to grow up freely." Hutchinson (1961), Paine (1966), and Connell (1971) have suggested the same thing, and MacArthur (1972) states: "If abundant predators prevent any species from becoming common, the entire picture changes. Resources are no longer of any concern and our Eqs. (1) and (2) are irrelevant. More correctly, resources are still a concern, but their manner of subdivision is irrelevant." The question I would like to try to answer is, when does predation, or any factor other than competition, limit the distribution or abundance of a species and so affect its role in community structure? Under which conditions does competition determine the shape and size of the realized niche and which not?

A Note on Methods

I will first consider the problem of how to detect and measure the extent of biological interactions such as competition or predation. In my opinion this has been the main stumbling block in testing the validity of models of community structure. There seem to have been at least three general methods used to detect or measure biological interactions under natural conditions. The first is to describe the pattern that exists at a point in time to see whether it does or does not conform to the predictions of the model. Some examples of this are gradient analysis (Whittaker, 1967, 1970; Terborgh, 1971), fitting relative abundances to various mathematical models (MacArthur, 1960; Whittaker, 1965; Levins, 1968), and examining the distribution patterns, food habits, etc., of closely related species to see how much their niches overlap. This method is very useful in detecting patterns and suggesting hypotheses for testing.

The second approach is an attempt to apply the experimental method but without using controls. It consists of searching for a "natural experiment." For example, if two similar species occupy the upper and lower halves of a mountain, respectively, with little or no overlap, one hypothesis is that they exclude each other by competition for resources. To test this hypothesis, one looks for a nearby mountain where one or the other is missing (e.g., Diamond, Chapter 14, Figure 39). If competition determines the boundary, the species present should extend beyond the boundary on this second mountain. The trouble with such "natural experiments" is that an essential part of all true experiments is missing: a control. There is no certainty that the only difference between the two mountains is the absence of one of the species. A predator may have been absent, or an essential food organism, soil nutrient, etc., may have been present beyond the boundary on the second mountain.

The third and, in my opinion the best, method for revealing the extent of biological interactions is the controlled field experiment, whose essential aspect is that everything varies in the same way between treatment and control except for the factor being tested (see review by Connell, 1974). In the example above, instead of looking for a mountain with one species missing, one finds another with both species in the same arrangement and then removes one species. The most efficient thing would be to remove it near the boundary; if the other species extends its range into the vacant territory, competition is the most likely mechanism producing the original boundary. If not, the hypothesis can be rejected forthwith.

Such field experiments need careful design. The trick is to make sure that all environmental factors, except the one being tested, vary in the same way and to the same degree on both experimental and control sites. Since there are bound to be some differences, replication is essential. Often it is more difficult to arrange adequate control sites than to perform the treatments.

The effect of the experimental treatment itself must be taken into account. The simplest treatment, removal of one or more species, is probably the best. Alternatively, animals are sometimes introduced into pens or cages (Jaeger, 1971; Wilbur, 1972; Grant, 1972). This may cause problems. First, if emigration is the usual result of aggressive encounters, this will be prevented and unusual outcomes may ensue (Krebs, Keller, and Tamarin,

1969). Secondly, other species of competitors or predators, which would normally have influenced the outcome, may be excluded. If enclosures must be used, these and other such effects need to be taken into account in the final interpretation of the experimental results.

The superiority of such controlled experiments to "natural" ones seems obvious, yet field experiments are rare in all ecological literature. One reason is that it is very difficult to change the abundance of large or highly mobile animals such as birds or large mammals. Also, tradition dictates that experiments are done in laboratories, whereas ecology is done outdoors. Field trials are relegated to agriculture or forestry, and ecology students seldom take courses in these areas. But applied scientists need answers, and necessity stimulated the development both of statistical techniques and of field experiments. Most ecologists have adopted the former but not the latter, to their cost.

In the following section I present a review of the evidence of where competition occurs in natural communities and where it is prevented by predation or physical conditions. The aim of this review is to see whether there is any pattern in the occurrence of competition. I have used evidence from controlled field experiments whenever possible, both to illustrate their value and to suggest where they could be used in the future. Although the number of instances where field experiments have been used are few, they differentiate between the alternatives clearly in a way that correlations between abundances, gradient analyses, maps of distributions, etc., cannot do.

The Pattern of Occurrence of Competition

Where Is Competition Common Under Natural Conditions?

On land. Competition between the roots of canopy trees and those in the understory was demonstrated long ago by controlled field experiments in natural forests in temperate latitudes. Trenches were dug around plots, cutting all roots and so eliminating competition for water or nutrients with adult trees (Fricke, 1904; Toumey and Kienholz, 1931; Korstian and Coile, 1938; Lutz, 1945; Shirley, 1945). In every case the smaller trees inside the trenched plots survived and grew much better than in nearby control plots. In the one published instance of trenching in a tropical forest, the plants in the experimental plot did not do better (Connell, 1971). This may have been due to the greater competition for light in this very dense forest and/or to a greater intensity of grazing.

Interference between bushes and herbs in the chaparral vegetation of California has been demonstrated in several studies. In laboratory experiments the bushes have been shown to contain volatile or waterborne chemicals that inhibit germination and early growth of herbs (Muller, 1966). New plants become established mainly after fires, which destroy the "allelopathic" effects of the original vegetation. However, as will be discussed later, graz-

ing may also affect seedling establishment in chaparral.

Direct evidence of interspecific competition between animals comes from field experiments on vertebrates and social insects. When DeLong (1966) reduced the numbers of the meadow mouse *Microtus,* the population of house mice (*Mus*) increased as compared to that on a control area. This increase was not due to greater growth or survival of adults, but rather to a greater production of young. This was probably due to lack of disturbance by *Microtus* of nests of *Mus,* since, in the laboratory, female *Mus* either deserted their nests or ate their young in the presence of *Microtus.* In another field experiment Koplin and Hoffman (1968) found that competition probably determined the spatial distribution of two species of *Microtus.* During a summer's trapping *M. montanus* was never caught in the wetter parts of the habitat where *M. pennsylvanicus* lives. Over the subsequent autumn and spring *M. pennsylvanicus* was systematically removed, greatly reducing its density. Then *M. montanus* was caught in the wetter places for the first time. Unfortunately the population densities of both species on the control area were much less than on the experimental area. Therefore, the fact that *M. montanus* were never caught in the wetter habitats on the control plot may have been the result either of the very small sample size or of the lack of intraspecific pressure to explore new areas, rather than from competitive aggression by *M. pennsylvanicus.* Apparently these two studies are the only field experiments on interspecific competition in rodents in which any sort of control area was established. Grant's (1972) review cites other field experiments that did not use control areas.

To my knowledge, the only field experiment testing for the existence of interspecific competition in birds is that of Davis (1973). By trapping over a period of three years in a field bordered by trees and bushes, he found that most of the golden-crowned sparrows were feeding both on seeds and young plants along one particular portion of the edge. Juncos, which feed almost entirely on seeds, occurred more or less equally around the edge of the field. In the fourth year he removed almost all the sparrows within two months. For the next two months, in the absence of the sparrows, a high proportion of the junco population shifted into the portion of the field where most of the sparrows had been feeding. Then half of the captured sparrows were returned to the field and the other half released 0.64 km away; about half of these reached the field within a week. Again a high number were found along the same portion of the edge where they had been commonest before. Concurrently the proportion of the total junco population in that area fell back nearly to the size it had been before the sparrows were removed. Although no separate control area was established, the fact that the junco population rose when the sparrows were removed and fell quickly when they were returned is strong evidence that the sparrows were excluding the juncos from that part of the habitat.

Other experimental evidence of competition between vertebrates comes from populations confined in pens. Jaeger (1971) found two species of salamanders occupying adjacent habitats. The first species, which normally lived only in shallow soil, survived well in pens in areas with deep soil, if the second species from the deep soil was excluded. However, if the species were penned together, the first survived poorly. No aggressive behavior was ever seen in the laboratory and the mechanism for this apparent competition is unknown. The number of animals used was small and the effects of the pen are unknown; manipulation of large numbers without using pens would be a welcome extension of these interesting experiments.

A series of controlled experiments with rodents in large pens has been done by Grant (see his review, 1972). He showed that each species was restricted to one type of habitat by competition with another species. These were mice which mainly used the surface of the ground rather than burrowing or climbing, so that they were likely to be competing for space.

Field experiments on competition between ant colonies have been done in at least two instances. Brian (1952) set up artificial nest sites by placing slabs of stone or slate on the ground or building turf banks. Initial colonization by different species was "near random," but then certain species were observed to drive out other species by direct aggression or to move in when the other species had left during a drought or cold spell. Pontin (1969) moved whole colonies into new positions. The colonies left behind, being less crowded, often produced more alate queens in the next two years. In the colonies that were more crowded after new colonies had been moved adjacent to them, fewer queens were sometimes produced. These effects were apparently more drastic on other colonies of the same species than on those of different species.

In the absence of experimental manipulation, observations of aggressive displacement of one species by another are convincing evidence of competition. For example, Way (1953) observed one species of ant gradually driving another species back through a coconut plantation in Tanzania. Levins, Pressick, and Heatwole (1973) observed aggressive interactions between different species of ants when bait was placed on the ground on small Caribbean islands. Aggressive displacement of one species by another has also been observed in birds (Pitelka, 1951; Orians and Collier, 1963; Orians and Willson, 1964) and mammals (Brown, 1971; Heller, 1971; Sheppard, 1971).

Indirect evidence of interspecific competition comes from observations of nonoverlapping adjacent ranges (Diamond, 1973), inverse correlations of abundance (Brown, Chapter 13), niche shifts on islands where competitors are absent (Crowell, 1962; Diamond, 1970 and Chapter 14), character displacement, etc. Such indirect evidence is open to explanations other than competition.

In aquatic habitats. Benthic invertebrates in fresh-water lakes sometimes occur at such high densities that they probably compete for space or food (Jonasson, 1971). Reynoldson and Bellamy (1971) have performed a field experiment which indicated that one species of flatworm was able to displace another in a small lake in Wales.

Wilbur (1972) enclosed populations of various combinations of six species of amphibians in pens along the edge of a pond. Competition occurred among the three local species of salamanders, affecting survivorship, length of larval period, and body weight. A fourth species not occurring locally was introduced; it survived well alone, but competed with the local species if grown with them. Another species of salamander was also a competitor unless it was able to grow quickly to a larger size, when it apparently became a predator on the other salamanders. Having frog tadpoles as prey may have given it this advantage of faster growth. To carry out these complex experiments it was necessary to enclose the populations in pens. These excluded other species of predators, both invertebrate and vertebrate, which probably would have affected the outcome of the interactions under natural conditions. However, this study is one of the most complete and rigorous analyses of the complex interactions that produce structure in communities.

Several controlled field experiments have demonstrated the existence of interspecific competition in marine organisms. The usual procedure consists in changing the abundance of one species and observing the survivorship or growth of another, with adjacent controls. All of the experiments have been done on rocky seashores. Connell (1961b) found that one species of barnacle was eliminated from lower shore zones by another that grew faster and either smothered, crushed, or undercut the first. Haven (1966, 1973) changed the population density of two species of grazing limpets and found that they affected each other's growth rates. A similar instance of competition for food between two predatory starfish was demonstrated in a controlled field experiment by Menge (1972). Competition for space producing a nonoverlapping mosaic between two species has been demonstrated with limpets by Stimson (1970, 1973) and with barnacles and anemones by Dayton (1971). Observations also suggest that competition may occur between barnacles and attached algae (Dayton, 1971).

Where is Competition Prevented
by Predation?

On land. Much of the experimental evidence for predation as an important factor in community structure deals with predation on plants by mammals. Grazing by wild rabbits kept grasses from displacing dicotyledonous herbs in southern England. When rabbits were excluded by fences, or after they had been eliminated by an epidemic of myxomatosis, grasses invaded, forbs disappeared and woody vegetation began to invade (Tansley and Adamson, 1925; Hope-Simpson, 1940; Watt, 1957, 1960; Thomas, 1960). When mice and voles were excluded with fences for seven years, the same thing happened in a conifer plantation in Wales (Summerhayes, 1941). Experimental exclusion of voles from grassland in California for two years showed that their grazing changed the relative abundance of species, reducing the palatable species (Batzli and

Pitelka, 1970). Thus small mammals, by selective predation, can determine the community structure of terrestrial vegetation.

Another example is the effect of insects on a species of hemiparasitic herb living in the understory of deciduous forests in Michigan. When these grazers were removed, the plant population increased greatly (Cantlon, 1970).

The regeneration of trees may be almost completely prevented by grazing mammals. In three species of trees in the New Forest in England, the populations consist of three "generations," established in the intervals 1648 to 1763, 1858 to 1915, and since 1938 (Peterkin and Tubbs, 1965). The two latter generations coincide with periods when grazers such as deer, cattle, and ponies were removed from the forest. Regeneration is limited to occasional escapes from predation. Once trees get tall enough, they are vulnerable to attack only by insects or pathogens. Instances of complete defoliation by insects, sometimes with mass mortalities of adult trees, have occurred, e.g., balsam fir in Canada (Morris, 1963) and *Eucalyptus delegatensis* in Australia (Readshaw and Mazanec, 1969). As pointed out by Murdoch (1971), balsam fir is vulnerable to this heavy attack only when mature and in continuous even-aged stands, where, presumably, the dispersal of insects on to other vulnerable trees is facilitated.

In the chaparral vegetation in California the competitive interference through release of chemicals (allelopathy) between bushes and herbs is sometimes modified by grazing. The periodic fires which remove the allelopathic effects also destroy or drive out the grazing animals (Halligan, 1972; Christiansen, 1973). Either experimental removal of the tops of allelopathic plants or exclusion of grazing vertebrates by cages or fences, or both, has been done in several instances. When vertebrate grazers alone were excluded, herbs have sometimes become established in abundance near allelopathic shrubs (Bartholomew, 1970; Halligan, 1972), sometimes much less abundantly or not at all (McPherson and Muller, 1969; Muller and del Moral, 1971; Chou and Muller, 1972; Christiansen, 1973). The biggest change has been reported when both grazing and allelopathy have been experimentally eliminated together in these studies. The conclusion from these studies is that both allelopathic interference between the plants and grazing by herbivores influence the community structure of chaparral vegetation.

Some field experiments testing the interaction between grazing and competition between grasses and forbs have been done on sheep pastures (Sagar and Harper, 1961; Putwain and Harper, 1970). They indicate that grazing may reduce competition, but without knowledge of whether wild herbivores graze natural vegetation at the same intensity as these sheep did, it is difficult to generalize from these results.

The effect of predators on terrestrial animals has seldom been estimated from field experiments under natural conditions. The only controlled field experiment I know of concerns the effect of woodpeckers on Englemann spruce beetles (Knight, 1958). The birds concentrate their attacks both in the groves and on the individual trees where the beetles are commonest. By using exclusion cages on 250 trees in the Rocky Mountains, Knight was able to show that the predators caused proportionately greater mortality where the prey were denser. Murdoch (1966) has shown that carabid beetles will survive well through the summer only when protected from predation.

Predators of terrestrial vertebrates do not seem to be so effective in reducing their prey, for reasons which will be discussed later. Predators of ruffed grouse were removed almost completely in two different places; nesting losses were reduced but the adult populations did not increase (Edminster, 1939; Crissey and Darrow, 1949). This supports the suggestion of Levins (Chapter 1) that a change in production of one life stage may not affect the abundance of another life stage limited by some other factor. When about half of the predators of the vole (*Microtus californicus*) were removed during a peak in vole numbers, the population declined to the same degree as it had from a previous peak (Pearson, 1966). The often-cited instance of an irruption of the deer population following the reduction in their predators on the Kaibab plateau of Arizona has been shown to be incorrect by Caughley (1970). There is one instance in which predators have been effective in reducing mammalian populations, i.e., wolves and moose on Isle Royale in Lake Superior. Although no controlled experiments have been done, the evidence strongly suggests that before the wolves arrived the moose were so dense that their feeding was injuring the vegetation. This is not happening now that wolves are feeding on the moose (Mech, 1966; Jordan Shelton, and Allen, 1967).

In freshwater. The zooplankton of open water in lakes without planktivorous fish usually consists of several species of relatively large crustaceans, together with smaller species of various groups. After planktivorous fish have been introduced and increased substantially, the larger species of zooplankton are absent from the open water and the smaller species are much commoner. Several studies have compared the same pond or lake before and after fish were introduced, and all have found that the large species disappeared and the smaller species increased, suggesting that the fish selectively ate the larger zooplankters (Hrabacek *et al.,* 1961; Brooks and Dodson, 1965; Macan, 1965; Reif and Tappa, 1966; Galbraith, 1967; Wells, 1970; Hall, Cooper and Werner, 1970; Warshaw, 1972). Brooks (1968) has confirmed that planktivorous fish selectively eat the larger zooplankton in laboratory experiments, as have Galbraith (1967) and Green (1967) by comparing stomach contents with plankton samples.

Where large aquatic vegetation is dense, the herbivorous zooplankton is much more abundant, even though phytoplankton productivity is less. A likely reason for this is that fish predation is less effective in dense vegetation. Hall *et al.* (1970), found that in ponds with dense vegetation (caused by addition of large amounts of mineral nutrients) the biomass of zooplankton was not reduced significantly by fish. But in replicate ponds with less vegetation, fish reduced the biomass of zooplankton significantly below that in control ponds.

Other predators besides fish may change the biomass and/or relative abundance of zooplankton. In a series of alpine ponds in Colorado where salamanders (*Ambystoma*) were common, the herbivorous zooplankton species were small. Where salamanders were rarer or absent, larger herbivores occurred (Dodson, 1970; Sprules, 1972). Like fish, salamanders also tend to select the larger individuals as prey, judging by stomach contents and laboratory experiments.

Field experiments on invertebrate predators have been conducted in only one study, by Hall *et al.* (1970). The biomass of zooplankton was consistently lower in treatments with increased invertebrate predation. The predators removed the larger herbivores first, thereby changing the species composition of the community in the same fashion as the vertebrate predators did.

It is difficult to decide whether fish or invertebrate predators have a greater effect on the relative abundance of species in the zooplankton. The field experiments of Hall *et al.* (1970) underestimated the effects of invertebrate predators since a small predator, *Chaoborus,* was left behind in the "reduced" treatment. Dodson (1970) calculated that the predation rate on *Daphnia* by the population of predatory midge larvae was about 10 times that of the salamander population. However, in the experimental ponds of Hall *et al.*

(1970), even with augmented numbers the invertebrate predators did not completely eliminate the larger herbivores as did the fish.

The benthic flora and fauna of freshwater lakes are also affected by predation. Field experiments excluding fish (Black, 1946; Threinen and Helm, 1954) and tadpoles (Dickman, 1968) resulted in rapid growth of larger plants. However, some of the fish were introduced asian carp, which may not have had such devastating effects in their native waters. Experimental removal and addition of fish (*Lepomis* sp.) in ponds or lakes have resulted in a greater standing crop of benthic invertebrates without fish (Ball and Hayne, 1952; Hayne and Ball, 1956), or no change (Hall *et al.*, 1970). This difference may have been the result of the much higher density of fish used in the two former studies (97 and 179 kg/ha, respectively) than the latter (50 kg/ha). Hall *et al.* (1970) found that the fish reduced the larger animals (insect larvae and amphipods), and the smaller species increased, maintaining the total biomass. The fish selected the larger pupae, thereby greatly reducing the numbers of emerging insects.

When invertebrate predators were reduced, the biomass of benthic invertebrates rose, but when these predators were increased, biomass rose in the first summer and fell in the second. Increased predation had predictable effects on the two principal species, the midge *Chironomus* being reduced and the ephemeropteran *Caenis* increasing.

Some experiments and observations have shown that fish may also be affected by their predators. In lakes and rivers young salmon are heavily preyed upon by predatory fish and birds. Removal of these predators increased the survival to both the smolt and adult stages (White, 1939; Foerster and Ricker, 1941; Foerster, 1954; Elson, 1962). Jackson (1961) observed lakes with and without the large predatory tiger fish (*Hydrocyon vittatus*), which eats all fish small enough to swallow. Only adults of species larger than this occur outside the shelter of aquatic plants in lakes with the tiger fish, but in lakes without it, small fish swim in open water.

In most of these studies, the predators selectively removed the larger invertebrates, regardless of species, and the smaller herbivores usually increased. This is interpreted by most of the authors to indicate that without the larger competitors, the less efficient smaller species could secure more of the limited resources and so increase. However, an alternative hypothesis is that after the predatory fish were added, they removed not only the larger competitors but also the predators of the smaller species. Fish removed the predatory insects in the experiments of Hall *et al.* (1970), and presumably also in some of the other lakes studied. However, when invertebrate predators were added by Hall *et al.* (1970), it is less likely that they ate the predators of the smaller herbivores. Thus *Caenis* increased, almost certainly because its competitor *Chironomus* had been reduced by invertebrate predators. Likewise the larger cladocerans in the zooplankton decreased progressively through the summer, and the rotifers increased. Invertebrate predators, by reducing the larger herbivores, which are probably more efficient collectors of suspended food (Brooks and Dodson, 1965), probably allowed the less efficient smaller species to increase. However, to elucidate the role of competition properly, field experiments are essential. Sprules (1972) has done a small-scale pilot experiment that indicated that the large *Daphnia pulex* has a deleterious effect on the smaller *Daphnia rosea*. Repetition of this interesting experiment on a larger scale would be welcome.

The effects of grazing by zooplankton on phytoplankton has recently been estimated by Porter (1973). She suspended clear plastic bags holding about 0.5 m³ of water a short distance below the surface of a lake. The numbers of zooplankton grazers were either reduced, increased, or kept the same in the bags. Counts of the phytoplankton after four days in the bags and examination of gut contents of the grazers showed that some groups of plants were reduced, others increased, and some were unaffected. The growth of some gelatinous green algae may have been increased by passage through the gut of zooplankters (as discussed by Patrick, Chapter 15, for grazing on algae by crustacea), whereas small edible species are reduced. Thus grazing may completely change the relative abundance of freshwater phytoplankton.

Where physical conditions become very harsh in fresh water, predation is reduced. For example, in the alpine ponds studied by Dodson (1970) and Sprules (1972), shallower ones froze to the bottom each winter and sometimes dried up in late summer, whereas deeper ones never did.

The predators in the deeper pools were the axolotl salamander and *Chaoborus*, neither of which apparently tolerates freezing or drying. Both were absent from all shallow pools, where the only predators were a large copepod, which apparently tolerates the harsh conditions, and/or a salamander, which could leave the pond and hibernate elsewhere; this salamander was never as abundant in these pools as the axolotl was in the deeper pools. Thus, predation was probably less intense in harsher conditions.

A second example involves large lakes. The deeper hypolimnion of temperate lakes often becomes anoxic in summer, reducing the activity of fish. Therefore, fish feed less of the year in the deeper than in the shallower depths, which do not become anoxic. Jonasson (1971) found that the midge *Chironomus anthracinus* reached much higher population densities in the hypolimnion than at shallower depths of Lake Esrom, Denmark. Since almost all of the mortality occurred during the short period when the predatory eel (*Anguilla*) could feed, between the autumn overturn and the onset of winter, Jonasson ascribed most of the mortality in deep water to predation. At shallower depths *Chironomus* never attain such high densities, possibly because the predators are able to feed on them for the entire warm season.

Because growth is also inhibited when oxygen is lacking in the hypolimnion, not all the *Chironomus* larvae complete development in one year. However, some do and emerge in May, leaving others behind. But in most years there are so many left that they still search the bottom completely with their feeding activities and eat all the eggs of their own species that are laid in early summer, thus completely suppressing the next generation. Only after the two-year-olds emerge can a new generation of eggs start to develop in the hypolimnion. In shallow water, growth and development are faster, all emerge at the end of one year, and space is left for a new generation each year. Thus predation thins out the population in shallow depths, but in deeper waters harsh physical conditions (anoxia) exclude predators for a greater proportion of the year. The prey apparently survive these harsh periods so that, with less predation, population density is greater, as is competition for feeding space.

The studies described above have demonstrated the dominant role of predation in determining the relative abundance and distribution of species in fresh-water lakes. The community structure is completely different, depending upon whether fish happen to be present or not. If not, the presence or absence of amphibian or invertebrate predators produces a different community structure. Competition may occur if the intensity of predation is so low that high population densities of herbivores develop.

In the sea. On rocky shores in temperate zones, grazers and predators often keep their prey populations so low that competition is prevented. The evidence for this comes from experiments in which grazers or predators were removed. The classic experiment of Jones (1948), who removed 15,000 limpets from a strip of rocky shore, demonstrated the great effect

of these herbivores on abundance of algae. The algae quickly colonized and covered the surface for several years. Others have repeated the experiment elsewhere with similar results (Southward, 1953, 1964; Castenholz, 1961; Haven, 1966, 1973; Dayton, 1970). The same effect of grazing by sea urchins on algae in tide pools or rocks below the intertidal region has been demonstrated by experimental removal (Kitching and Ebling, 1967; Jones and Kain, 1967; Paine and Vadas, 1969; Dayton, 1970). Lastly, by excluding herbivorous fish from intertidal and subtidal areas on coral reefs, Stephenson and Searles (1960) and Randall (1961) have found that they too keep algae and sea grasses grazed down. Unplanned experiments on a larger scale have shown the same effect. Grazers were killed by a spill of fuel oil from a tanker wreck (North, Neushal, and Clendenning, 1964), and by detergents used to clean shores of crude oil (Smith, 1968). Algae colonized and grew profusely after the grazers were gone.

On the middle and lower shore levels experimental removal of predators has shown that they also often eliminate sessile animals before these reach maturity. All barnacles were eaten by predatory snails within 18 months in Scotland (Connell, 1961a) or 12 to 15 months in Washington (Connell, 1970; Dayton, 1971) and New Zealand (Luckens, 1970), except where predators were excluded by cages. Mussels survived only in cages (Connell, unpublished; Dayton, 1971) or when the predatory starfish were picked off by hand (Paine, 1966, 1971).

Yet despite this very heavy grazing and

predation, some plants and sessile or sedentary animals survive to maturity and live a long time at low shore levels. The clue to how this happens is in the age structure of these populations: they often consist of widely spaced older year-classes. I have studied in detail the population dynamics of one such species, the barnacle *Balanus cariosus* on San Juan Island, Washington (Connell, in preparation). At the start the population consisted of classes aged 2, 4, and at least 10 years. Over the next 13 years, snail predators ate all the young that arrived every year at two different study sites, with three exceptions: once at one site and once at each of two different shore levels in different years at the other site. In these instances some individuals survived to the age of 2 years, at which time they were invulnerable to all the common predators except the very large starfish *Pisaster ochraceus*. I tested the invulnerability to smaller predators by protecting the barnacles in cages for varying lengths of time and then removing the cages or allowing predators to enter. All *B. cariosus* younger than 2 years were quickly eaten, whereas only seldom were older ones eaten. Dayton (1971) has since confirmed these findings.

As described earlier I have found that, under natural conditions, *B. cariosus* survives to this invulnerable age only occasionally, so that the populations consist of "dominant year classes." This age structure is produced by occasional escapes from the intense predation to which they are usually subjected.

Other organisms may also reach a size at which they are invulnerable to predators. Kitching, Sloane, and Ebling (1959)

found that the mussel *Mytilus edulis* became invulnerable to attack by crabs. Only one species of very large crab, *Cancer pagurus*, could break open the larger mussels available. This case is similar to that of *B. cariosus;* the prey can reach a size invulnerable to the smaller species of predators, but there exists a very large species of predator to which it never is invulnerable. Dayton (1971) suggests that *Mytilus californianus* may reach a size at which it becomes invulnerable to *Thais;* although this is probable, no evidence exists, since the predators ate all sizes offered in the experiments.

Some species of intertidal algae may become invulnerable to grazers. Southward (1964, Table 2) found that the alga *Fucus vesiculosus*, which colonized a shore after limpets, *Patella vulgata*, were removed, maintained a complete cover for the next 2 years. During this period many small limpets colonized the rock beneath the canopy of large algae. After $4\frac{1}{2}$ years the limpets were much larger and the algal cover had been reduced to 22%. The smaller limpets evidently did not attack the large algae, feeding instead on the smaller plants that colonized beneath the canopy. When they grew larger, they ate the large algae. It seems reasonable to draw the conclusion that large algae are invulnerable to small grazers but not to large ones.

Grazing and predation are often extremely intense under the more "benign" conditions of the lower shore, and "escapes" to invulnerable size occur only rarely. The only data as to the frequency of such escapes in natural conditions are those I have given for *Balanus cariosus*.

About every 5 years they escape from smaller predators; they may then survive for 15 years or so before being eaten by larger predators.

Where is competition prevented
by physical conditions?

On land. In certain very harsh deserts, the germination of seeds and the establishment of seedlings of perennial plants may be prevented by the absence of rainfall for many years. For example, in central Australia, *Acacia aneura* produces viable seeds in most years, but no seedlings get established. In the occasional years when rainfall is higher, seedlings may survive longer, but are usually destroyed by insects. Only if at least three successive years of higher than normal rainfall occurs will a crop of seedlings become established; this happens about every 40 to 50 years, judged by the age structure of the trees. Then competition for water between the seedlings ensues, and some may die (Slatyer, 1975). The population of *Acacia aneura* thus consists of widely spaced year-classes, produced by occasional escapes during rare periods of mild weather.

In aquatic habitats. Near the upper margin of their distribution in the intertidal zone, marine organisms are exposed for long periods to extreme and variable weather; young colonists are usually killed by the harsh physical conditions. Evidence for this comes from various field experiments in which the conditions were improved at high levels. For example, Hatton (1938), Frank (1965), and Dayton (1971) arranged streams of sea water above the intertidal zone; algae and bar-

nacles survived much higher than usual in these streams. Barnacles survived better under shades set up by Hatton (1938) on the upper shore. Conversely, barnacles transplanted to higher levels quickly died (Hatton, 1938; Foster, 1971); the smaller barnacles died before the larger ones. In general, younger or smaller individuals are more vulnerable to harsh physical conditions. This is probably a consequence of the greater surface-to-volume ratio of smaller individuals; they are relatively more exposed than larger ones to such hazards of the external environment as desiccation, increased radiation, extreme temperatures, fresh water, etc. (Lewis, 1964).

Weather being variable, there will occur periods when the harsh conditions are temporarily ameliorated. If favorable conditions last long enough, the young colonists may reach a size where they can survive the usual harsh weather. Many marine species at higher latitudes have short seasons of breeding and settlement each year, so that such "escapes" may happen only once every few years. For example, out of four year-classes of the barnacle *Balanus balanoides* that I observed at high shore levels, survival was high in only one, so that the population was composed mainly of the survivors of that year-class (Connell, 1961a, Figure 12). Other examples of populations dominated by older year-classes at high shore levels are three species of barnacles studied by Foster (1971) and a limpet, *Acmaea scabra,* studied by Sutherland (1970).

An interesting example for marine algae is given by Kain (1963). The physical conditions in the sublittoral zone de-

teriorate as the light intensity diminishes
with depth. At the lower limits of distribution of two populations of the large sea-
weed *Laminaria hyperborea,* the growth
rate and population density diminished,
and the age distribution consisted of
dominant year-groups. In contrast, the
shallower populations showed little evidence of such dominant age groups. Recruitment evidently happens inter-
mittently in the populations near the
lower edge of the range. Kain (1963) felt
that recruitment occurred only occasionally when the conditions become temporarily favorable, but there is no direct
evidence in support of this suggestion.

What Determines When
Competition Will Occur?

The evidence from controlled field experiments on the occurrence of competition and on the instances when it is prevented by predation or harsh weather has
now been reviewed. It seems clear that
competition is often prevented by predation, less often by harsh physical conditions. In fact, so many instances have been
demonstrated by controlled field experiments, in contrast to being simply suggested by correlations, that I suggest the
following priority be followed in adopting
simplifying assumptions to use in models
of community structure.

Predation should be regarded as being
of primary importance, either directly
determining the species composition or in
preventing competitive exclusion, except
where the effect of predation is reduced
for some reason. There seem to be two

principal situations in which predation is
reduced, both the result of evolution of
defensive adaptations by the prey.

1. Some prey species have evolved the
ability to live in refuges that the predator
cannot invade, either because the conditions are too harsh for the predator or the
habitat structure too difficult to search.
Outside the refuges the prey are eaten,
e.g., *Balanus glandula* on the middle and
lower seashore or larger zooplankton in
open waters. The highest levels on the
shore provide a refuge where the predators cannot drill and consume a barnacle
during the short period at high tide. In the
dense vegetation of ponds the fish cannot
effectively search for zooplankton.

There is a great difference in the relative abundance and species composition
of the zooplankton or benthic invertebrates in lakes with or without large predators such as fish or amphibians. Large
species of zooplankters or benthic invertebrates are eliminated where these large
predators are present. These larger invertebrates must have evolved in lakes without fish, and depend for their existence on
the fact that these larger vertebrate predators have low rates of movement between
lakes and so have not reached many small
lakes. Only in recent years have they done
so, thanks to the stocking by government
fishery departments.

This category of species protected by
refuges includes many prey species that
are smaller than their predators. They
may exist as "fugitive" species, invading
isolated patches of habitat such as lakes
or islands before their predators, which
have lower powers of dispersal. Alternatively, they live in the same habitat but

have absolute refuges from their predators.

Where the prey species are free of predators, they may increase in numbers until they compete for resources with other species. This is the usual explanation for the low abundance of smaller zooplankters in the lakes without large predators: that the larger zooplankters are more efficient competitors. However, the small zooplankters may be more common when fish are present because the fish also remove the predatory invertebrate species, which in lakes without fish reduce the smaller zooplankters.

A cautionary note is relevant here. The number of species is usually lower on smaller or more isolated islands. The fact that species there often undergo "niche shifts" on islands is usually interpreted as being due to the absence of one or more competitors (Crowell, 1962; Diamond, 1973, and Chapter 14, Figures 39–43). But an alternative hypothesis is that a significant predator may also be absent. Clearly, controlled field experiments are necessary to decide between the two alternatives in such instances.

2. In a completely different category are those species that defend themselves against predation by evolving adaptations allowing them to coexist with predators without having to live in refuges. The defenses may be morphological (spines, bark, stinging cells), chemical (tannins, alkaloids), behavioral (aggressive nature, social groups, parental care), or simply growing too large to be attacked successfully.

The problem with most of these defenses, most obvious with the last, is that they are less effective in younger than in older individuals. Parental care is an excellent adaptation to bridge the vulnerable young stage. Parental care is particularly well developed in groups such as birds and mammals. Having also evolved adaptations such as homeothermy and large size, these animals have escaped many of the hazards of the physical environment and of predation. Thus, it would be expected that their populations might be limited by competition for resources, as is assumed in many of the chapters in this book. A similar conclusion may be reached from consideration of survivorship curves: intraspecific competition among adults is more likely in species in which many young survive to adulthood (e.g., birds and mammals) than in species in which most young perish before adulthood.

In species with little parental care, the young survive until they reach a less vulnerable larger size only when predation is occasionally reduced, as in the case of *Balanus cariosus* described earlier. Once an individual or group survives to a size at which attack by a predator is much less probable, it will continue to grow and hold more resources, suppressing its neighbors or smaller individuals beneath it. Thus, competition involving the larger individuals is to be expected, once they escape their predators. Another advantage of large size is that it renders the individual less vulnerable to extremes of physical conditions. This is attested by dominant year-classes, representing escapes during occasional mild spells, in populations in normally harsh regimes.

An interesting aspect of physical harshness is that if prey and predator have similar physiological requirements, the prey species can sometimes tolerate physical extremes in which the predator cannot attack it. For example, sessile animals such as barnacles or mussels can survive on the seashore at such high levels that their predators cannot attack them, as described earlier. This relationship obviously does not apply if the two species have different physiological makeups, e.g., a land predator such as a seabird attacking a marine invertebrate.

If we consider only species of prey and predator from the same habitat, it follows that if physical conditions become more harsh, i.e., extreme, variable, unpredictable, or any combination of these, predation would be expected to be reduced.

This category includes the larger plants and animals on land and in aquatic habitats. The large plants and sessile aquatic animals provide much of the physical structure of ecosystems, modifying the climate or water movements and providing the vertical structure inhabited by many smaller species. Therefore a more detailed analysis of the mechanisms determining the relative abundance, distribution, and diversity of these species is in order here.

A Model of the Community Dynamics of Large Sessile Species of Animals and Plants

Communities dominated by large sessile species are the rule in most terrestrial habitats, and in shallow aquatic ones. Terrestrial plant communities, coral reefs, beds of turtle grass or kelp, oyster reefs, rocky shores covered with large barnacles,

mussels or algae, and the macrophytic vegetations in the littoral of lakes are common examples. All are mosaics of patches of "dominant" species, and gaps are continually appearing as individuals or groups are killed by predators, storms, floods, fires, etc.

The ecological events and evidence described above for marine communities can be summarized in a general scheme, which may serve as a testable model of use in predicting which species of "dominant" succeeds in occupying a gap.

Benthic plants and animals are arranged in a mosaic of patches, some held by long-lived dominants, others inhabited by a mixture of opportunists and young individuals of the dominants. The latter are usually killed within the first year after settlement. Many other species live only in the sheltered conditions created by the dominants. This is the situation shown in Figure 1 as step 1.

Let us now suppose that a patch of dominants is removed, by unpredictable variations in weather, damage by floating objects such as logs, increases in predation, or simply because the large older dominants die as they reach old age. Then the vacant patch is quickly colonized by opportunists, which characteristically have long breeding seasons and numerous motile spores or larvae. In addition, young stages of dominant species that happen to be available may be among the colonists (steps 2 and 3, Figure 1). The events that follow are different in harsh and in benign conditions.

In very harsh conditions, such as at the upper margin of the intertidal zone or in places with much abrasion, most newly

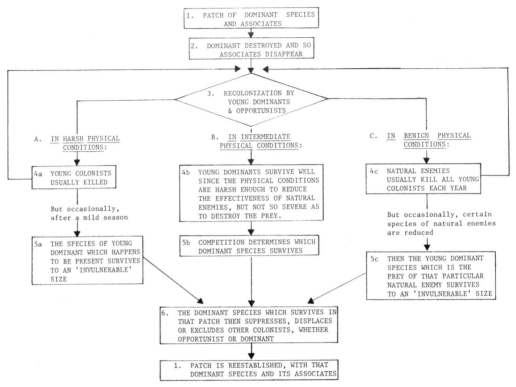

Figure 1 A model of the mechanisms determining which "dominants" (large plants or sessile aquatic animals) will fill a gap. See text for further discussion.

settled or young individuals are killed by harsh weather each year so that the patch is continually being colonized and vacated; this cycle is shown between steps 3 and 4a. They survive only in particularly favorable years, and then may live a long time. Populations consist of dominant year-groups, the survivors from these occasional favorable years. Because they occupy more and more space as they continue to grow, they exclude and may displace other organisms (steps 5a to 6).

In other words, once two or more individuals are past the vulnerable young stage, they may compete with one another. However, in many cases these populations never get dense enough to compete for resources.

In benign environments, natural enemies tend to be much more effective (Connell, 1971) and most colonists are eaten. This tends to reduce competition and also to prevent the growth to an invulnerable size of the young dominants.

If the populations of grazers or predators are not reduced by some external event, they are capable of keeping the patch in this state of continual recolonization (the cycle between steps 3 and 4c).

However, if some unpredictable event, such as a short spell of bad weather or an attack by their own predators or pathogens, reduces the population of these natural enemies, the prey will have a period of good survival and growth. This then allows whichever dominant species happens to be present to grow to an invulnerable size (step 5c). Which, if any, do so depends upon which natural enemy is reduced and whether the reduction lasts for a long enough period. Once the prey reaches invulnerable size, it excludes or displaces other organisms by further growth (steps 5c to 6). The situation then returns to the original state (steps 6 to 1) as "sheltering" species colonize the area beneath the dominant.

In intermediate environments there is less mortality from harsh weather and natural enemies are less effective, so that the young of dominant species more readily escape being eaten. Thus these young dominants may reach high population densities and begin to compete with one another. The eventual winner, i.e., the dominant that eventually fills the space, is likely to be the one able to displace the others during a period of competition (steps 4b and 5b).

Which species of dominant will succeed in filling the vacated patch depends upon several things. For example, in benign regions such as the lower seashore there are more species of predators, and some, such as starfish, are quite large and able to eat grazers and smaller predators as well as the dominant sedentary animals (Paine, 1966; Menge, 1972). Thus starfish tend to reduce both animal dominants and grazers. Plants should therefore be favored, since both their competitors and their natural enemies are being reduced. This may be the reason why, in temperate latitudes over the world, plants rather than animals are the dominant organisms covering the lower seashore. They form the "sublittoral fringe" in Stephenson and Stephenson's (1949) universal scheme of zonation. In contrast, at middle shore levels, large predators such as starfish are less common; the main predators are muricid snails. They feed much more heavily on sedentary animals such as mussels and barnacles than on grazing molluscs, so that grazers are less likely to be reduced than they were at lower shore levels. This means that both plant and animal dominants are under attack, so that no a priori prediction can be made as to which will succeed in filling the vacated patches of rock surface. Barnacles, mussels, tube worms, oysters, and various algal species—all may occur as dominants in middle shore zones.

What Determines Realized Niches and Community Structure?

The distribution and abundance in which one finds a species, as opposed to its potential area and population size, is obviously not determined solely by competition. Therefore the variables that constrain the realized niche of a species within its fundamental niche need not, as defined originally by Hutchinson (1958) and more recently by Vandermeer (1972), be limited to competitive interactions.

A major distinction was made earlier between species that reached a size much larger than their predators and those that did not. However, many species occur in both categories. A young member of a large species must spend some time in the latter category when predators are larger than it is. Even birds and mammals are vulnerable for a short time during infancy, and other groups may spend a much greater proportion of their lives in a vulnerable state. In addition, small prey are usually unattractive to much larger predators. Thus at any one time there is a restricted range of sizes of predators willing and able to attack an individual. As an individual grows, it may be attacked by a series of ever larger predators, or by predators operating in larger social groups.

Another important assumption underlying the present model is that predators attack with less intensity as physical conditions become harsher, i.e., more severe, variable, or unpredictable. Evidence supporting this generalization comes from gradients across the edges of lakes or the seashore, or from comparisons such as those between deep and shallow alpine ponds or between tropical and temperate latitudes, e.g., that nest predation of juvenile birds is greater in tropical than in temperate latitudes (Ricklefs, 1969; also data of G. Orians quoted in MacArthur, 1972, p. 218). Although further evidence is clearly desirable, this generalization seems to apply to a wide range of species.

Thus we have found two relationships to be of prime importance: 1) the change in vulnerability of prey as its size varies in relation to that of the predator, and 2) the change in intensity of predation as physical conditions vary.

The first of these applies if prey and predator are to live intermingled together. The prey must be either so much smaller than the predator as to be economically unattractive, or so much larger as to be impossible to attack successfully. The size of either predator or prey refers to the operational unit, whether it be an individual or a pack.

The second relationship is important for prey that do not grow large enough to escape their predators. Then prey must have a refuge from predation. One way this may occur is when predation diminishes as physical harshness increases; the prey may then evolve tolerance to physical regimes too harsh for the predator to attack effectively.

These points are combined in a diagrammatic way in Figures 2 to 5. Each of these represents a surface of varying probability of mortality per unit time, showing "contours" of equal probability. The shading indicates a region of very low probability of mortality. A gradient from left to right in all the figures represents increasing harshness of the physical regime, such as would obtain for a marine species from low to high levels in the intertidal zone. A gradient from bottom to top on the figures represents increasing body size of the prey.

Figure 2 indicates the probability of being killed by a predator under different physical regimes and body sizes. Let us

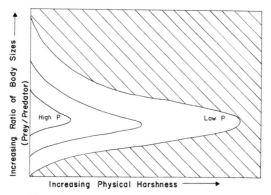

Figure 2 Mortality due to predation under different physical regimes and body sizes. A point on the surface represents a particular probability *P* that a prey individual will be killed by a predator per unit time. The curves are contours of equal probability, high toward the left, the shaded region being one of very low probability of mortality from predation. A gradient from left to right represents increasing physical harshness. A gradient from bottom to top represents increasing relative body size of prey compared with predator.

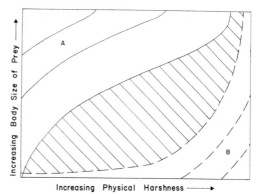

Figure 4 Mortality due to both predation and physical conditions, for prey that are small in relation to the size of their predators. The solid curves in region A represent probabilities of mortality from predation, as in the lower portion of Figure 2 with the scale expanded. The dashed curves in region B represent probabilities of mortality from physical factors, as in Figure 3. The shaded region represents a region of very low probability of mortality. See Figure 2 for more explanation.

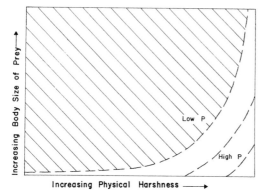

Figure 3 Mortality due to the direct effects of harsh physical conditions on prey of different body sizes. The gradients and curves are explained in Figure 2. A point on the surface now represents the probability P that a prey individual will die due to harsh physical conditions per unit time.

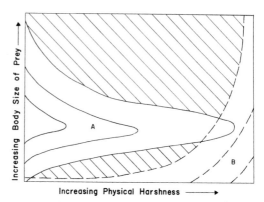

Figure 5 Mortality due to both predation and physical conditions, for prey that grow to a large size in relation to the size of their predators. The curves for mortality from predation (solid lines, region A) are taken from Figure 2, those for physical effects (region B) from Figure 3. See Figure 2 for more explanation.

examine the situation in which the prey is relatively much smaller than the predator, i.e., as in the lower portion of the figure. Along any horizontal line, with a constant ratio of body sizes of prey and predator, the probability of mortality from predation decreases toward the right as the physical conditions become more harsh, because of a progressive reduction in the period of activity of the predator. Along a vertical line, representing a particular degree of physical severity, the probability of mortality increases upwards, as the prey become relatively larger. The reason is that very small prey are usually not eaten by a large predator; they are ignored, or not caught by the meshes of its filter, etc. Presumably there is a size of prey below which there is too low an energy return for the effort expended. These two tendencies cause the contour lines of equal probability of death to slant upwards to the right in the lower portion of Figure 2.

In the upper part of Figure 2, the prey are relatively larger. As before, along any horizontal line the probability decreases as the physical regime becomes steeper. However, along a vertical line at a particular physical regime, the probability of death decreases upward. As a prey individual grows, it becomes too large for a particular predator to attack. This also tends to reduce the number of species of predators that can attack it. These tendencies cause the contour lines to slant downwards to the right.

The shape of the contour lines would vary with different combinations of prey, predator, and physical regime. I have drawn them as they probably apply to the species on rocky seashores. The situation in Figure 2 applies to predators and prey that have similar physiological characteristics, for example, marine organisms. In other instances, they may have quite different physiologies. For example, along shores of lakes or oceans, land animals feed on aquatic species. In this case, a mirror image of Figure 2 would be appropriate, since as the physical regime becomes less severe for the aquatic prey (toward the left), it would become more severe for the terrestrial predator.

The effect of the direct action of physical conditions on organisms of different sizes is shown in Figure 3. Along a horizontal line the probability of death per unit time increases toward the right. Upward along a vertical line the probability decreases since, as an individual grows, it becomes less vulnerable to the physical environment. These trends cause the contour lines of equal probability to slant upwards to the left. Again, the shape of the curves will depend upon the characteristics of the species and physical regime.

In Figures 4 and 5 I have combined the first two diagrams to illustrate the effect both of predators and of different physical regimes on the mortality of the prey. Figure 4 represents small prey, so uses only the lower portion of Figure 2. When prey are very small, they tend to be ignored by larger predators so that mortality is due mainly to physical harshness. As prey grow, they escape this hazard except where conditions are very severe. In very benign conditions, larger prey soon attract the notice of predators that tend to be active much of the time. In conditions of intermediate severity, they escape the attentions of the less active predators for a longer time, and having passed the very young stages at which they are vulnerable to variations in physical conditions, they may be safe for a while. If local population densities are high enough, competition may take place during this interval before predation begins to reduce the populations.

Figure 5, which combines most of Figure 2 with Figure 3, represents the situation in which the prey grow to a relatively large body size. Once they have grown large enough to be less vulnerable to physical factors, they may or may not be safe from predators for a short while, but soon are attacked. However, if they survive long enough, they may reach an invulnerable size. (These diagrams do not take into account the fact that prey eventually become vulnerable again in old age.)

The situation in Figure 4 is exemplified by the two species of barnacles that competed only at intermediate intertidal levels in Scotland (Connell, 1961b). After eliminating *Chthamalus* in competition, *Balanus* was eaten by predatory snails, except in a refuge high on the shore. *Balanus cariosus* fits both Figures 4 and 5, depending upon the relative size of its predators. With small predators it is able to grow too large to be attacked (Figure 5), but with the large starfish it fits Figure 4. Many prey species probably pass through a similar series of ever larger predators.

Species of prey that never grow very large compared with their predators, such as the aquatic plankton or small land plants or animals, are represented by Figure 4. Large plants and animals would be more closely represented by Figure 5.

Summary

The distribution and abundance of a species are ultimately determined by tolerances to extremes of physical conditions, but a species is usually limited to a smaller range of habitats and population size by interactions with other organisms. The evidence reviewed in this paper, taken mainly from controlled field experiments on invertebrates and plants, suggests that many species seldom reach population densities great enough to compete for resources, because either physical extremes or predation eliminates or suppresses them in their young stages.

Species may sometimes escape these hazards to reach high densities in various ways. One way is through large size, which reduces vulnerability to harsh weather and predation, but the problem for a young individual is to reach a large enough size before it is killed. Strong parental care helps as in many vertebrates, but offspring without care will escape only during occasional reductions in predation or harsh weather. Such escapes may produce widely spaced "dominant year-classes." Once they reach large size, they may compete for resources and suppress smaller individuals. Examples are forest trees, large sessile aquatic animals, etc. A model of this situation is presented.

Second, some species can never grow so large as to escape predation. They may escape for a while if they are too small to be attractive to the predator, and they may compete then. However, as they grow larger, they will be eaten unless they have a refuge where they are safe. If their habitat is patchy, e.g., small lakes or islands, they may colonize some patches before the predators reach them. Most smaller species have a refuge, permanent or temporary.

Since predation seems to be more intense in more benign physical conditions, competition should be prevented more regularly in more benign regimes. Escapes from predation and subsequent competition would be more likely in harsher regimes.

Acknowledgments

I would like to thank the participants in the symposium for helpful comments on this paper. In addition, the often pungent criticisms by S. J. Arnold, S. Avery, J. J. Childress, J. D. Dixon, E. P. Ebsworth, M. H. Fawcett, S. G. Fuzessery, R. P. Howmiller, P. S. McNulty, W. W. Murdoch, A. Oaten, C. P. Onuf, J. C. Roth, S. J. Rothstein, S. C. Schroeter, R. O. Slatyer, W. P. Sousa, and T. C. Tutschulte contributed substantially to clarifying the manuscript.

My ideas about some of the problems discussed here benefited immensely from arguments I had with Robert MacArthur.

References

Ball, R. C., and D. W. Hayne. 1952. Effects of the removal of the fish population on the fish-food organisms of a lake. *Ecology* 33:41–48.

Bartholomew, B. 1970. Bare zone between California shrub and grassland communities: the role of animals. *Science* 170:1210–1212.

Batzli, G. O., and F. A. Pitelka. 1970. Influence of meadow mouse populations on California grassland. *Ecology* 51:1027–1039.

Birch, L. C. 1957. The meanings of competition. *Amer. Natur.* 91:5–18.

Black, J. D. 1946. Nature's own weed killer, the German carp. *Wisconsin Conservation Bull.* 11:3–7.

Brian, M. V. 1952. Structure of a dense natural ant population. *J. Animal Ecol.* 21:12–24.

Brooks, J. L. 1968. The effects of prey size selection by lake planktivores. *Syst. Zool.* 17:273–291.

Brooks, J. L. and S. I. Dodson. 1965. Predation, body size, and the composition of plankton. *Science* 150:28–35.

Brown, J. H. 1971. Mechanisms of competitive exclusion between two species of chipmunks. *Ecology* 52:304–311.

Cantlon, J. E. 1970. The stability of natural populations and their sensitivity to technology. *In* G. M. Woodwell and H. H. Smith, eds., *Diversity and Stability in Ecological Systems,* Brookhaven Symposium in Biology No. 22, pp. 197–205. U.S. Department of Commerce, Springfield, Va.

Castenholz, R. W. 1961. The effect of grazing on marine littoral diatom populations. *Ecology* 42: 783–794.

Caughley, G. 1970. Eruption of ungulate populations with emphasis on Himalayan thar in New Zealand. *Ecology* 51:53–72.

Chou, C., and C. H. Muller. 1972. Allelopathic mechanisms of *Arctostaphylos glandulosa var. zacaensis. Amer. Midl. Natur.* 88:324–347.

Christiansen, N. L. 1973. Effects of fire on factors controlling plant growth in *Adenostoma* chaparral. Ph.D. thesis, University of California, Santa Barbara.

Connell, J. H. 1961a. Effects of competition, predation by *Thais lapillus,* and other factors on natural populations of the barnacle *Balanus balanoides. Ecol. Monogr.* 31:61–104.

Connell, J. H. 1961b. The influence of interspecific competition and other factors on the distribution of the barnacle *Chthamalus stellatus. Ecology* 42:710–723.

Connell, J. H. 1970. A predator-prey system in the marine intertidal region. I. *Balanus glandula* and several predatory species of *Thais. Ecol. Monogr.* 40:49–78.

Connell, J. H. 1971. On the role of natural enemies in preventing competitive exclusion in some marine animals and in rain forest trees. *In* P. J. den Boer and G. Gradwell, eds., *Dynamics of Populations* (Proc. Advan. Study Inst., "Dynamics of Numbers in Populations," Oosterbeek, 1970). pp. 298–312. Centre for Agricultural Publishing and Documentation. Wageningen, The Netherlands.

Connell, J. H. 1974. Field experiments in marine ecology. *In* R. Mariscal, ed., *Experimental Marine Biology.* Academic Press, New York.

Crissey, W. F., and R. W. Darrow. 1949. A study of predator control on Valcour Island. New York State Conservation Dept., Division of Fish and Game, Res. Ser. No. 1.

Crowell, K. L. 1962. Reduced interspecific competition among the birds of Bermuda. *Ecology* 43:75–88.

Darwin, C. 1859. *The Origin of Species by Means of Natural Selection.* Reprinted by The Modern Library, Random House, New York. John Murray, London. Facsimile of 1st. Edit., Harvard University Press, 1964.

Davis, J. 1973. Habitat preferences and competition of wintering juncos and golden-crowned sparrows. *Ecology* 54:174–180.

Dayton, P. K. 1970. Competition, predation,

and community structure: the allocation and subsequent utilization of space in a rocky intertidal community. Ph.D. thesis, University of Washington, Seattle.

Dayton, P. K. 1971. Competition, disturbance, and community organization: the provision and subsequent utilization of space in a rocky intertidal community. *Ecol. Monogr.* 41:351–389.

DeLong, K. T. 1966. Population ecology of feral house mice: interference by *Microtus. Ecology* 47:481–484.

Diamond, J. M. 1970. Ecological consequences of island colonization by Southwest Pacific birds. I. Types of niche shifts. *Proc. Nat. Acad. Sci. U.S.A.* 67:529–536.

Diamond, J. M. 1973. Distributional ecology of New Guinea Birds. *Science* 179:759–769.

Dickman, M. 1968. The effect of grazing by tadpoles on the structure of a periphyton community. *Ecology* 49:1188–1190.

Dodson, S. I. 1970. Complementary feeding niches sustained by size-selective predation. *Limnol. Oceanogr.* 15:131–137.

Edminster, F. C. 1939. The effect of predator control on ruffed grouse populations in New York. *J. Wildlife Mgt.* 3:345–352.

Elson, P. F. 1962. Predator-prey relationships between fish-eating birds and Atlantic salmon. *Bull. Fish. Res. Bd. Canada* No. 133.

Elton, C. 1927. *Animal Ecology.* Sidgwick and Jackson, London.

Foerster, R. E. 1954. On the relation of adult sockeye salmon (*Oncorhynchus nerka*) returns to known smolt seaward migrations. *J. Fish. Res. Bd. Canada* 11:339–350.

Foerster, R. E., and W. E. Ricker. 1941. The effect of reduction of predaceous fish on survival of young sockeye salmon at Cultus Lake. *J. Fish. Res. Bd. Canada* 5:315–336.

Foster, B. A. 1971. On the determinants of the upper limit of intertidal distribution of barnacles (Crustacea: Cirripedia). *J. Animal Ecol.* 40:33–48.

Frank, P. W. 1965. The biodemography of an intertidal snail population. *Ecology* 46:831–844.

Fricke, K. 1904. "Licht und Schattenholzarten": ein wissenschaftlich nicht begründetes Dogma. *Centralbl. f.d. gesamte Fortwesen* 30:315–325.

Galbraith, M. G. 1967. Size-selective predation on *Daphnia* by rainbow trout and yellow perch. *Amer. Fish. Soc., Trans.* 96:1–10.

Grant, P. R. 1972. Interspecific competition between rodents. *Ann. Rev. Ecol. Syst.* 3:79–106.

Green, J. 1967. The distribution and variation of *Daphnia lumholtzi* (Crustacea: Caldocera) in relation to fish predation in Lake Albert, East Africa. *J. Zool.* 151:181–197.

Grinnell, J. 1917. Field tests of theories concerning distributional control. *Amer. Natur.* 51:115–128.

Grinnell, J. 1928. Presence and absence of animals. *Univ. Calif. Chronicle* 30:429–450.

Hall, D. J., W. E. Cooper, and E. E. Werner. 1970. An experimental approach to the production dynamics and structure of freshwater animal communities. *Limnol. Oceanogr.* 15:839–928.

Halligan, J. P. 1972. The herb pattern associated with *Artemesia californica.* Ph.D. thesis. University of California, Santa Barbara.

Hatton, H. 1938. Essais de bionomie explicative sur quelques espèces intercotidales d'algues et d'animaux. *Ann. Inst. Monaco* 17:241–348.

Haven, S. B. 1966. Ecological studies on coexisting limpet species (Gastropoda) in the high intertidal of Central California. Ph.D. thesis. University of California, Berkeley.

Haven, S. B. 1973. Competition for food between the intertidal gastropods *Acmaea scabra* and *Acmaea digitalis. Ecology* 54:143–151.

Hayne, D. W., and R. C. Ball. 1956. Benthic productivity as influenced by fish predation. *Limnol. Oceanogr.* 1:162–175.

Heller, H. C. 1971. Altitudinal zonation of chipmunks (*Eutamias*): interspecific aggression. *Ecology* 52:312–319.

Hope-Simpson, J. F. 1940. Studies of the vegetation of the English Chalk. VI. Late stages in succession leading to chalk grassland. *J. Ecol.* 28:386–402.

Hrabacek, J. M., Dvorakova, V. Korinek, and L. Prochazkova. 1961. Demonstration of the effect of the fish stock on the species composition of zooplankton and the intensity of metabolism of the whole plankton association. *Verh. Internat. Verein. Limnol.* 14:192–195.

Hutchinson, G. E. 1958. Concluding remarks. *Cold Spring Harbor Symp. Quant. Biol.* 22:415–427.

Hutchinson, G. E. 1961. The paradox of the plankton. *Amer. Natur.* 95:137–145.

Jackson, P. B. N. 1961. The impact of predation, especially by the tiger-fish (*Hydrocyon vittatus Cast.*) on African freshwater fishes. *Proc. Zool. Soc. London* 136:603–622.

Jaeger, R. G. 1970. Potential extinction through competition between two species of terrestrial salamanders. *Evolution* 24:632–642.

Jaeger, R. G. 1971. Competitive exclusion as a factor influencing the distributions of two species of terrestrial salamanders. *Ecology* 52:632–637.

Jonasson, P. M. 1971. Population studies on *Chironomus anthracinus. In* P. J. den Boer and G. Gradwell, eds., *Dynamics of Populations* (Proc. Advan. Study Inst. Dynamics of Number in Populations

(Oosterbeek, 1970), pp. 220–231. Centre for Agricultural Publishing and Documentation, Wageningen, The Netherlands.

Jones, N. S. 1948. Observations and experiments on the biology of *Patella vulgata* at Port St. Mary, Isle of Man. *Proc. Trans. Liverpool Biol. Soc.* 56:60–77.

Jones, N. S., and J. M. Kain. 1967. Subtidal algal colonization following the removal of *Echinus. Helgol. Wiss. Meeresunt.* 15:460–466.

Jordan, P. A., D. C. Shelton, and D. L. Allen. 1967. Numbers, turnover, and social structure of the Isle Royale wolf population. *Amer. Zool.* 7:233–252.

Kain, J. M. 1963. Aspects of the biology of *Laminaria hyperborea*. II. Age, weight and length. *J. Mar. Biol. Assoc.* 43:129–152.

Kitching, J. A., and F. J. Ebling. 1967. Ecological studies at Lough Ine. *Adv. Ecol. Res.* 4:197–291.

Kitching, J. A., J. F. Sloane, and F. J. Ebling. 1959. The ecology of Lough Ine. VIII. Mussels and their predators. *J. Animal Ecol.* 28:331–341.

Knight, F. B. 1958. The effects of woodpeckers on populations of the Engelmann Spruce beetle. *J. Econ. Entomol.* 51:603–607.

Koplin, J. R., and R. S. Hoffman. 1968. Habitat overlap and competitive exclusion in voles (*Microtus*). *Amer. Midl. Natur.* 80:494–507.

Korstian, C. F., and T. S. Coile. 1938. Plant competition in forest stands. *Duke Univ. School Forestry Bull.* 3:1–125.

Krebs, C. J., B. Keller, and R. Tamarin. 1969. *Microtus* population biology: demographic changes in fluctuating populations of *M. ochrogaster* and *M. pennsylvanicus* in southern Indiana. *Ecology* 50:587–607.

Levins, R. 1968. *Evolution in Changing Environments.* Princeton University Press, Princeton.

Levins, R., M. L. Pressick, and H. Heatwole. 1973. Coexistence patterns in insular ants. *Amer. Scientist* 61:463–472.

Lewis, J. R. 1964. *The Ecology of Rocky Shores.* English University Press, London.

Luckens, P. A. 1970. Breeding, settlement and survival of barnacles at artificially modified shore levels at Leigh, New Zealand. *New Zealand J. Mar. Freshwater Res.* 4:497–514.

Lutz, H. J. 1945. Vegetation on a trenched plot twenty-one years after establishment. *Ecology* 26:200–202.

Macan, T. T. 1965. Predation as a factor in the ecology of water bugs. *J. Animal Ecol.* 34:691–698.

MacArthur, R. H. 1960. On the relative abundance of species. *Amer. Natur.* 94:25–36.

MacArthur, R. H. 1968. The theory of the niche. *In Population Biology and Evolution,* pp. 159–176. Syracuse University Press.

MacArthur, R. H. 1972. *Geographical Ecology.* Harper and Row, New York.

McPherson, J. K., and C. H. Muller. 1969. Allelopathic effects of *Adenostoma fasciculatum,* "Chamise," in the California chaparral. *Ecol. Monogr.* 39:177–198.

Mech, L. D. 1966. The wolves of Isle Royale. U.S. Nat'l. Park Serv., Fauna Series No. 7.

Menge, B. A. 1972. Competition for food between two intertidal starfish species and its effect on body size and feeding. *Ecology* 53:635–644.

Morris, R. F. (ed.). 1963. The dynamics of epidemic spruce budworm populations. *Mem. Entomol. Soc. Canada* No. 31. 332 pp.

Muller, C. H. 1966. The role of chemical inhibition (allelopathy) in vegetational composition. *Bull. Torrey Botanical Club* 93:332–351.

Muller, C. H., and R. del Moral. 1971. Role

of animals in suppression of herbs by shrubs. *Science* 173:462–463.

Murdoch, W. W. 1966. Aspects of the population dynamics of some marsh Carabidae. *J. Animal Ecol.* 35:127–156.

Murdoch, W. W. 1971. *Environment: Resources, Pollution and Society.* Sinauer Assoc., Stanford, Conn.

North, W. J., M. Neushul Jr., and K. A. Clendenning. 1964. Successive biological changes observed in a marine cove exposed to a large spillage of mineral oil. *In* Commission Internationale par l'Exploration Scietifique de la Mer Mediterranée: Pollutions marines par les microorganismes et les produits petroliers. Symp. de Monaco (April 1964) Paris 1965, 335–348.

Orians, G. H., and G. Collier. 1963. Competition and blackbird social systems. *Evolution* 17:449–459.

Orians, G. H., and M. F. Willson. 1964. Interspecific territories of birds. *Ecology* 45:736–745.

Paine, R. T. 1966. Food web complexity and species diversity. *Amer. Natur.* 100:65–75.

Paine, R. T. 1971. A short-term experimental investigation of resource partitioning in a New Zealand rocky intertidal habitat. *Ecology* 52:1096–1106.

Paine, R. T., and R. L. Vadas. 1969. The effects of grazing by sea urchins, *Strongylocentrotus* spp., on benthic algal populations. *Limnol. Oceanogr.* 14:710–719.

Pearson, O. P. 1966. The prey of carnivores during one cycle of mouse abundance. *J. Animal Ecol.* 35:217–233.

Peterkin, G. F., and C. R. Tubbs. 1965. Woodland regeneration in the New Forest, Hampshire, since 1650. *J. Appl. Ecol.* 2:159–170.

Pitelka, F. A. 1951. Ecological overlap and interspecific strife in breeding populations of Anna and Allen hummingbirds. *Ecology* 32:641–661.

Pontin, A. J. 1969. Experimental transplantation of nest-mounds of the ant *Lasius flavus* (F.) in a habitat containing also *L. niger* (L.) and *Myrmica scabrinodis* Nyl. *J. Animal Ecol.* 38:747–754.

Porter, K. G. 1973. Selective grazing and differential digestion of algae by zooplankton. *Nature* (London) 244:179–180.

Putwain, P. D., and J. L. Harper. 1970. Studies in the dynamics of plant populations. III. The influences of associated species on populations of *Rumex acetosa* and *R. acetosella* in grassland. *J. Ecol.* 58:251–264.

Randall, J. E. 1961. Overgrazing of algae by herbivorous marine fish in Hawaii. *Ecology* 42:812.

Readshaw, J. L., and Z. Mazanec. 1969. Use of growth rings to determine past phasmatid defoliations of alpine ash forests. *Austral. Forestry* 33:29–36.

Reif, L. B., and D. W. Tappa. 1966. Selective predation: smelt and cladocerans in Harveys Lake. *Limnol. Oceanogr.* 11:437–438.

Reynoldson, T. B., and L. F. Bellamy. 1971. The establishment of interspecific competition in field populations with an example of competition in action between *Polycelis nigra* (*Mull.*) and *P. tenuis* (*Ijima*) (Turbellaria, Tricladida). *In* P. J. den Boer and G. Gradwell, eds., *Dynamics of Populations* (*Proc. Advan. Study Inst. Dynamics of Number in Populations* Oosterbeek, 1970), pp. 282–297. Centre for Agricultural Publishing and Documentation, Wageningen, The Netherlands.

Ricklefs, R. E. 1969. An analysis of nesting mortality in birds. *Smithsonian Contrib. Zool.* 9:1–48.

Sagar, G. R., and J. L. Harper. 1961. Controlled interference with natural populations of *Plantago lanceolata, P. major* and *P. media. Weed Research* 1:163–176.

Sheppard, D. H. 1971. Competition between two chipmunk species (*Eutamias*). *Ecology* 52:320–329.

Shirley, H. L. 1945. Reproduction of upland conifers in the lake states as affected by root competition and light. *Amer. Midl. Natur.* 33:537–612.

Slatyer, R. O. 1975. Structure and function of arid shrublands. Proc. U.S.–Australian Workshop on Range Management. *J. Range Management.* In press.

Smith, J. E. (ed.) 1968. '*Torrey Canyon*' *Pollution and Marine Life.* Cambridge University Press, London.

Southward, A. J. 1953. The ecology of some rocky shores in the south of the Isle of Man. *Proc. Trans. Liverpool Biol. Soc.* 59:1–50.

Southward, A. J. 1964. Limpet grazing and the control of vegetation on rocky shores. *In* D. J. Crisp, ed. *Grazing in terrestrial and marine environments,* pp. 265–273. Blackwell, Oxford.

Sprules, W. G. 1972. Effects on size-selective predation and food competition on high-altitude zooplankton communities. *Ecology* 53:375–386.

Stephenson, T. A., and A. Stephenson. 1949. The universal features of zonation between tidemarks on rocky coasts. *J. Ecol.* 37:289–305.

Stephenson, W., and R. B. Searles. 1960. Experimental studies on the ecology of intertidal environments at Heron Island. I. Exclusion of fish from beach rock. *Austral. J. Mar. Freshw. Res.* 11:241–267.

Stimson, J. S. 1970. Territorial behavior in the owl limpet, *Lottia gigantea. Ecology* 51:113–118.

Stimson, J. S. 1973. The role of the territory in the ecology of the intertidal limpet *Lottia gigantea* (Gray). *Ecology* 54:1020–1030.

Summerhayes, V. S. 1941. The effect of voles (*Microtus agrestis*) on vegetation. *J. Ecol.* 29:14–48.

Sutherland, J. P. 1970. Dynamics of high and low populations of the limpet *Acmaea scabra* (*Gould*). *Ecol. Monogr.* 40:169–188.

Tansley, A. G., and R. S. Adamson. 1925. Studies of the vegetation of the English chalk. III. The chalk grasslands of the Hampshire-Sussex border. *J. Ecol.* 13:177–223.

Terborgh, J. 1971. Distribution on environmental gradients: theory and a preliminary interpretation of distributional patterns in the avifauna of the Cordillera Vilcabamba, Peru. *Ecology* 52:23–40.

Thomas, A. S. 1960. Changes in vegetation since the advent of myxomatosis. *J. Ecology* 48:287–306.

Threinen, C. W., and W. T. Helm. 1954. Experiments and observations designed to show carp destruction of aquatic vegetation. *J. Wildlife Mgt.* 18:247–250.

Toumey, J. W., and R. Kienholz. 1931. Trenched plots under forest canopies. *Yale Univ. School Forestry Bull.* No. 30.

Vandermeer, J. H. 1972. Niche theory. *Ann. Rev. Ecol. Syst.* 3:107–132.

Warshaw, S. J. 1972. Effects of alewives (*Alosa psuedoharengus*) on the zooplankton of Lake Wononskopomuc, Connecticut. *Limnol. Oceanogr.* 17:816–825.

Watt, A. S. 1957. The effects of excluding rabbits from a grassland B (Mesobrometum) in Breckland. *J. Ecol.* 45:861–878.

Watt, A. S. 1960. The effects of excluding rabbits from acidiphilous grassland in Breckland. *J. Ecol.* 48:601–604.

Way, M. J. 1953. The relationship between

certain ant species with particular reference to biological control of the Coreid, *Therapuis* sp. *Bull. Entomol. Res.* 44:669–691.

Wells, L. 1970. Effects of alewife predation on zooplankton populations in Lake Michigan. *Limnol. Oceangr.* 15:556–565.

White, H. C. 1939. Bird control to increase the Margaree River Salmon. *Bull. Fish. Res. Bd. Canada* No. 58.

Whittaker, R. H. 1965. Dominance and diversity in land plant communities. *Science* 147:250–260.

Whittaker, R. H. 1967. Gradient analysis of vegetation. *Biol. Rev.* 42:207–264.

Whittaker, R. H. 1970. *Communities and Ecosystems.* Macmillan, New York.

Wilbur, H. M. 1972. Competition, predation, and the structure of the *Abystoma-Rana sylvatica* community. *Ecology* 53:3–21.

IV Outlook

17 Variations on a Theme by Robert MacArthur

G. Evelyn Hutchinson

Scientists are perennially aware that it is best not to trust theory until it is confirmed by evidence. It is equally true, as Eddington pointed out, that it is best not to put too much faith in facts until they have been confirmed by theory. This is why scientists are reluctant to believe in ESP in spite of indisputable facts. This is also why group selection is in such dispute among evolutionists. Only when a reasonable theory can account for these facts will scientists believe them. Ecology is now in the position where the facts are confirmed by theory and the theories at least roughly confirmable by facts. But both the facts and the theories have serious inadequacies providing stumbling blocks to present progress. (MacArthur, 1972, p. 253–254.)

The present essay considers four areas of ecology illustrating in different ways a little of what Robert MacArthur was writing about in the last work that he completed as the sole author. In the first section of this chapter some aspects of competitive exclusion are considered historically, because here we have a paradigmatic case of theoretical and empirical studies progressing dialectically in a thoroughly satisfactory manner. In the second section, on cyclical changes in populations, the relation between the empirical and the theoretical has been less happy, and has probably hindered the development of the subject. In the third section,

on some aspects of polymorphism, the very elementary theory imported from population genetics has been most useful, but it seems possible that something deeper may be waiting in the wings. In the fourth section, two rather large and relatively unexplored themes are considered; they are characterized by their ubiquity in ecology, yet are exceptional in that virtually no attention, theoretical or observational and experimental, has been paid to them.

The approach throughout is largely historical, but some new biological results are introduced.

"Unless it Doesn't"

The idea that two similar species are not likely to live together seems to have been held by several nineteenth- and early twentieth-century naturalists. As Hardin (1960) has pointed out, the idea is implicit in Darwin's writings, though never explicitly stated. Dr. Martin Cody has kindly called my attention to an expression of the idea in a paper by Hansmann (1857) on the warblers of Sardinia, where it was noticed by Dr. Hartmut Walter. Steere's (1894) empirical conclusion, that among the land birds of the Philippine Islands congeneric species did not occur together, suggests the same sort of conclusion. Grinnell (1904; cf. Udvardy, 1959) wrote

492

that "two species of approximately the same food habits are not likely to remain long evenly balanced in numbers in the same region. One will crowd out the other." This is an adequate qualitative statement of the principle of competitive exclusion. Ortmann (1906) concluded likewise that two allied species do not occupy the same range under identical ecological conditions. Hiltzheimer (1909), in a discussion of fossil European species of *Bison,* used the idea that two closely related species are not likely to live together. Monard (1920; cf. Macan, 1963) concluded from his study of the benthos of lake Neuchâtel that "dans un milieu uniform, restreint dans le temps et l'espace, ne tend à subsister q'une espèce par genre," which is the same as the conclusion reached by Steere, a quarter of a century earlier. Cabrera (1932), evidently influenced by Grinnell and by Ortmann, gave an excellent account of what he called the ecological incompatibility of allied coexisting species. As more of the immense body of writings of the late nineteenth- and early twentieth-century biologists is examined with competitive exclusion in mind, more examples of this sort are likely to be discovered. It is evident, however, that such concepts failed to elicit much deep interest.

It was only when Haldane (1924)[1] and, in an even more important paper, Volterra (1926) gave a clear mathematical demonstration that under certain reasonably natural conditions competitive exclusion would be expected, and Gause (1934, 1935) showed how in the laboratory actual systems in which competing organisms did exclude one another could be constructed, that people began to see not merely that such a generalization might be true but that, more importantly, it might be interesting.

Schrödinger,[2] I believe, said somewhere that Newton's First Law of Motion may be recast as: "A body perseveres at rest or uniform motion in a right line unless it doesn't." The last three qualifying words conceal the idea of force, which becomes explicit in the second law. Naturalists in the first decade or so of the twentieth century doubtless believed that one of two species of the same general biological characteristics, brought into a common habitat, would exclude the other unless it didn't. Except to people who had been led by their experience with birds, bison, or benthos, to believe in something like competitive exclusion, if perhaps in a rather naive way, the existence of counterexamples underlying the qualifying "unless it didn't" would make the generalization both tautological and insignificant. Only when there is a strong theoretical reason for the idea's being correct and a strong empirical reason for thinking that

---

[1] Although I was present in the audience when Haldane gave the verbal presentation of this paper to the Cambridge Philosophical Society, the ecological significance of the result, obtained as an unimportant case in his genetic argument, struck neither me nor, as far as I know, anyone else present.

[2] Neither Professor Martin Kline, Professor Saunders MacLane, nor I can find the source of the quotation, which may belong to folklore. Professor Gerald Holton and some of his colleagues suggest it is best, if somewhat frivolously, attributed to Newton himself.

in some cases it is incorrect, does the problem suddenly become important.

At the famous symposium of the British Ecological Society on *The Ecology of Closely Allied Species* held on 21 March 1944 "a distinct cleavage of opinion revealed itself on the validity of Gause's concept" (Harvey, 1945). The Oxford group—Elton, Lack, and Varley—were obviously stating something of importance in their implied insistence that what Gause did in the laboratory had great significance for what they were studying in nature. That extraordinary naturalist Captain Cyril Diver, then Clerk to the Select Committee on National Expenditure of the House of Commons, "who gave examples of many congeneric species of both plants and animals apparently living and feeding together," was also right in emphasizing his store of apparent counter-examples. MacArthur's (1958) warblers, which have played so great a part in the later development of the cluster of concepts that we are considering, at the time would have seemed to support Diver, though now we know that they conform to the principle of competitive exclusion in ways that would hardly have appeared possible in 1944.

The work of the past decade suggests that the known exceptions, apparent or real, to the principle of competitive exclusion, may be of four kinds.

1. Some exceptions are only apparent and can be rationalized in terms of the theory if the appropriate niche dimensions are considered. By far the most important of these cases is that in which predation, which can be quantified as a niche pa-rameter, keeps the populations of prey species at a low level, in such a way that what would be the dominant and ultimately the only species in the community without the predator suffers a proportionally greater reduction by predation than do the other prey species. The most dramatic recently studied cases concern marine benthic communities (Paine, 1966; Paine and Vadas, 1969; Porter, 1972) or pasture plants grazed by ungulates (Harper, 1969). This type of exception is discussed in detail by Connell in Chapter 16, and a new theoretical basis for it is provided by Levins through loop analysis in Chapter 1 (Figure 6b).

There may be a few other types of case that can be subsumed under the classical Volterra theory by choosing the correct, if rather unconventional, niche parameters. This may be the case locally with two species of towhee (Marshall, 1960; also later discussions by Hutchinson, 1965, and Cody, 1974).

2. The various situations involving vag-ile or fugitive species and more or less rapid or even random environmental changes, causing continuous alterations of the direction of competition between rap-idly reproducing species, provide a second type of apparent exception to competitive exclusion. Such situations in stream com-munities are discussed elsewhere in this volume by Patrick (Chapter 15). In these cases there is ideally no persistent uni-directional movement towards a particu-lar equilibrium position, owing to con-tinual changes in the environment which locally or generally change the direction of competition. The situation described by

Ross (1957) for the leaf-hopper *Erythroneura* and probably extremely common among small herbivorous insects belongs here. So in all probability does that presented by the autotrophic phytoplankton, the most advanced study of which is by Grenney, Bella, and Curl (1973). They present a theoretic analysis of an assemblage of phytoplankton species that can store nitrogen in a nitrate-limited environment, and conclude that since at "one point in time the combination of environmental conditions may be most suitable for one particular species and at another for an entirely different species," any species can persist provided a suitable combination occurs with sufficient frequency and duration. The persistence of all species thus depends on the probability of occurrence of certain events, but in most fairly old habitats natural selection will obviously have limited the co-occurring species to those that require the intervention of highly probable circumstances. Istock (1973) has recently discussed what seems to be a comparable situation in the water bugs of the family Corixidae.

3. A new type of "unless it doesn't" has recently been suggested by Ayala (1969, 1971) as the result of his studies of two pairs of species of *Drosophila*. In these cases rarity apparently confers a selective advantage. This could lead to coexistence within a single niche, as Ayala indeed believes to be the case. The effect, whatever it is due to, cannot be due to coevolution of particular kinds of symbiosis or commensalism, as *Drosophila pseudobscura* and *D. serrata* had presumably not experienced each other before meeting in

Ayala's cultures. Ayala, Gilpin, and Ehrenfeld (1973) have examined 10 possible kinds of models (they give 11 but model 3 is a special case of model 7), including the original one of Volterra, which itself constitutes a set of special cases of 8 of the 10. This population explosion of models can be rationalized somewhat in the following way:

For the general case write

$$dN_i/dt = (N_i r_i/K_i)[K_i \\ - f_{ii}(N_i) - f_{ij}(N_j)] \quad (1)$$

where the function f_{ii} expresses the intraspecific inhibitory effect of each individual within species i and the function f_{ij} expressing the interspecific effect of each individual of species j on species i. In general, the possible functions f_{ii}, f_{ji}, and the corresponding f_{ij} and f_{jj} will all be different.

If we expand each function as a power series and neglect all terms higher than those of the second order, we obtain the approximation

$$dN_i/dt = (N_i r_i/K_i)(K_i - \alpha N_i \\ - \beta N_i^2 - \alpha' N_j - \beta' N_j^2) \quad (2)$$

We assume $\alpha \simeq 1$, $\alpha' \simeq 1$. It will be noted that K_i now becomes a virtual upper limit that cannot be realized, even if N_j is absent, unless $\beta = 0$.

If $\beta = 0$ but β' is positive, eq. 2 reduces to the form given by Hutchinson (1948), in which the *interspecific* effect of each individual in competition is augmented by the presence of other conspecific individuals and therefore increases as the population increases (direct density-dependent competition). One example given by

Hutchinson from experiments with *Tribolium* probably has a different explanation (Ryan, Park, and Mertz, 1970), but in the case of terns (sea birds, Laridae) in which social facilitation has been studied on tropical islands in the Indian Ocean, a social intensification of competitive ability is reasonable and is doubtless a valid explanation of what has been observed in the field.

If $\beta' = 0$ but β is positive, eq. 2 reduces to eq. 5 of Ayala et. al. (1973), which they found fitted their experimental data well, though not better than some alternatives. In this case the *intraspecific* inhibition by crowding, due to each individual, increases as the population increases, so that when the species is rare each of its members competes interspecifically better than when it is common (inverse density-dependent competition). Of course β' could be positive but less than β, giving qualitatively the same effect but on a reduced scale. In practice one would probably usually find that all the possible βs are positive but unequal. In this formulation all the four qualitative outcomes of the Volterra equations can be obtained by adjusting the coefficients of the second-order terms. Gilpin and Ayala (1973) in their latest contribution regard the type of equation adopted here as unsatisfactory. They prefer a model in which the sigmoid growth curve of a single species can be made asymmetrical. This idea is interesting and doubtless not without merit; it rescues population ecology, they believe, from a state similar to that of astronomy before Kepler.

4. The possibility that coexistence may be due to commensal or symbiotic relationships has been explored by Gause and Witt (1935) in a paper which, though well known, never seems to have had the impact that it deserves.

The existence of nonobvious types of symbiosis and commensalism in nature has been little explored. The work of McNaughton and Harper (1960) and Harper and McNaughton (1962) seems to indicate that, of the five species of small red poppies of the genus *Papaver* found in Britain, only *P. rhoeas* habitually grows by itself. In mixed colonies density-dependent selection comparable to that later reported by Ayala and Gilpin certainly occurs, but this does not explain why the less abundant species are never found alone in nature. *P. rhoeas* evidently does not need its congeners in the way that they need it, but in competition it is apparently unable to displace them completely.

Among closely allied species that are not really full competitors, there are the three 17- or 13-year cicadas of the genus *Magicicada,* each of which occupies a slightly different habitat; the synchrony of the broods in any locality certainly indicates a symbiotic relationship that involves predator swamping (Beamer, 1931; Alexander and Moore, 1962; Lloyd and Dybas, 1966a, 1966b). Extraordinary kinds of symbiosis involving nest parasitism in tropical cow-birds have been reported by Smith (1968). An interesting suggestion, which might apply to species that are otherwise really in the same niche, has been made by Halkka, Raatikainen, Vasarainen, and Heinonen (1967), who note that if a common species is accompanied by a rarer one of the same general size, shape and behavior but with a different color pattern, the rarer species might be passed over by any predator that had formed a search image for the common species, as many predators would do. If the rarer species were the less effective competitor without the predator, the latter, simply by more frequently forming a search image for the more successful competitor, might create a balance between the two species. In a sense this reduces to predator-determined coexistence.

Looking over the whole history of the concept of competitive exclusion, we may distinguish three phases. In the first, lasting until the late 1920s, many people had ideas more or less equivalent to competitive exclusion, though others must have had a rich store of apparent counter-examples. The idea, though used occasionally, excited little interest. In the second phase from the late 1920s to the 1940s, there were initially mathematical demonstrations that, given certain quite reasonable postulates, competitive exclusion must occur in the establishment of equilibrium communities. The idea began to appear both acceptable and interesting. This led to much discussion, and apparent counter-examples were produced. In the third phase, from the 1940s on, the idea underwent modification to accomodate counter-examples, and gave rise to a large body of theory, though it is obvious that in its original form the Volterra equations are approximations to reality and competitive exclusion may not always take place. Robert MacArthur (verbal communication, 1964) seems to have concluded that the actual extermination of the less adapted species was likely to be so slow a process that a weak form of the principle, stating that one species is likely to become extremely scarce, is more justified than the strong form usually given. In his 1972 book he seems to imply that the coup de grace to the vanquished species, if it does disappear, is due to random population fluctuations or external events and not to competition.

The kind of history that we have just outlined may be regarded as characteristic of a healthy type of development in science. Examples compete with counter-examples, so that the theory is continually improved dialectically, until it is firm enough today to be the basis of exciting new developments. This ideal mode of development unfortunately does not characterize the whole area of contemporary ecology, as will be seen in the discussion of the following cases.

The Theory and Practice
of Oscillating Populations

It is a well-known conclusion of all studies of time lags and of prey-predator relationships, that it is very easy to make a mathematical model of an oscillating population (Lotka, 1925; Volterra, 1926; Nicholson and Bailey, 1935; Wangersky and Cunningham, 1957; May, 1973; and Leigh, Chapter 2, provide convenient land marks). Less effort has been expended in finding out how many natural populations really do oscillate in the regular periodic way that emerges from the mathematics.

The reason, in part, is that in the 1920s and 1930s, when the Hudson Bay Company's annual fur-trapping records first became widely available and well-known, the spectacular cases of the lynx and the snowshoe hare, along with much non-biological argument about the effects of sunspots as well as an uncritical use of Fourier analysis in studying variations in all sorts of time series from lake levels to stock market prices, led to a general acceptance of the idea of true periodicities where in fact they did not always exist. Then Cole (1954), in a critical analysis that was necessary to clear the air, showed that the properties of an ordered set of unequal random numbers lead automatically to a statistical quasi-periodicity, based solely on the fact that one in three of any triplet of unequal numbers, chosen at random, defines a maximum. Thus, in a long series of random numbers the mean period between peaks converges on 3. Later opinion, then, has swung to the other extreme, so that an equally unthinking rejection of real periodicities has, until recently, prevailed. Actually the development of computer technique makes possible and easy the use of the two appropriate and interrelated criteria for genuine periodicity, namely autocorrelation with a series of time lags and power spectrum analysis. The autocorrelation function gives a more impressive visual picture, but the autospectrum has some theoretical advantages. Moran (1949, 1953a, 1953b, 1954) first applied autocorrelation to the study of varying populations, and concluded that in the famous case of the populations of *Lynx canadensis*, as measured

by the Hudson Bay Company and Canadian Government fur statistics, a truly periodic oscillation is present. Finerty (1971), whose work is available so far only on microfilm and is therefore little known, found that there may be a comparable oscillation implied by the pelt data for the fisher *Martes pennanti* from the same sources (Figure 1). As is well known, the Canadian population of *Lepus americanus* fluctuates in a manner comparable to that of the lynx, though there are difficulties in interpreting the interactions of the two species (Leigh, 1968; Gilpin, 1973). Presumably the cycle is generated in the interaction of the hare with its environment. Finerty studied such data as are available (Keith, 1963) for the populations of *L. americanus* living south of the Great Lakes. The only series long enough to analyze (Figure 1) is from the game records collected by the State of Pennsylvania. Though Keith, among others, had suspected a ten-year cycle in these data, neither method of analysis discloses any statistically significant periodicity, and there is also no convincing evidence of periodicity to be obtained from an analysis of comparable data for the ruffed grouse *Bonasa umbellus* in the same state. Moran (1952) found no true periodicity in the records for four Scottish game birds, though there were significant (positive) interspecific correlations suggesting climatic control.

Andersen (1957) has studied the hunting statistics relating to the brown hare *Lepus europaeus* in Denmark, which give no indication of truly periodic variation when these data are freed from a secular

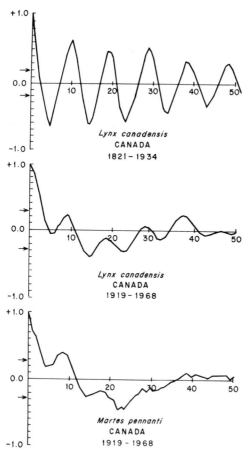
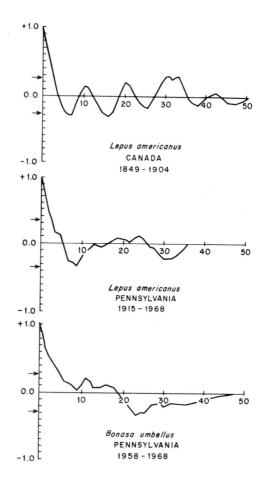

Figure 1 Autocorrelograms obtained by calculating the correlation coefficient (ordinate) of the logarithms of numbers of pelts reported for a series of years, with the same data displaced by 0, 1, 2, 3, . . . , n (abscissa) years. The lynx and the snowshoe hare in Canada are certainly, and the fisher probably, truly periodic. There is, however, no evidence of real periodicity for either the snowshoe hare or the ruffed grouse in Pennsylvania (after Finerty, 1971).

upward trend. Andersen found that just over half the variation that the data present can be explained in terms of three meteorological variables: x_1, the sum of

mean daily temperatures from March 1 to June 30: x_2, the number of days of frost from December 1 to March 1: and x_3, the rainfall from June 1 to July 31. These meteorological data are used in a regression equation

$$y_n = 1.3266 \, x_1 - 0.3003 \, x_2 + 0.3061 \, x_3 + 0.3428 \, y_{n-1}$$

in which y_{n-1} is an autoregressive term that takes care of survival and natality from the previous year. Andersen considers y_n to be essentially a random number generated by fluctuations in the three significant meteorological variables, so behaving as would be expected from Cole's hypothesis. The mean time between two successive peaks is 3.2 years, not far from the expected value of 3.0 years on Cole's hypothesis, whereas the incidence of high peaks gives a vague appearance of a nine-year cycle. The nature of the upward trend in the number of hares shot, if not in the total population, is unelucidated.

Finerty has examined the data for the annual take of arctic fox, *Alopex lagopus*, and red fox, *Vulpes fulva*, in Labrador (Elton, 1942), and the bounty records for foxes, including both *A. lagopus* and *V. vulpes*, in Norway (Johnsen, 1929). The populations of these predators are believed to vary with the sizes of the populations of lemmings on which the foxes largely feed, and to have a three- to four-year cycle. Such a cycle is initially suspicious because of the statistical basis for a quasi-periodicity demonstrated by Cole. No one working on lemmings in the field would doubt the reality of some sort of periodic rise and fall in the populations, but without long series of data the true periodicity of the rises and falls has remained doubtful. It is therefore satisfactory to find that Finerty has shown that

a real periodicity is implied by the Labrador data, and that in Norway this is true of the central third of the country (Nordland and Nord Trødelag) but not of the regions lying to the north or the south (Figure 2). There is no significant cross-correlation between the changes in the Labrador and Norwegian populations. As yet we have no reasonable explanation of the regularity of the cycles. In view of the evidence that the herd of moose, *Alces alces*, on Isle Royale in Lake Superior, is limited by sodium in its dietary intake (Botkin et al., 1973), much more work on possible biogeochemical determination of the cycles would be desirable.

Two further general points should be noticed. Firstly, the presence of a marked cycle in the central Norwegian populations of foxes but not in those to the north or south may be compared with a somewhat similar situation in the larch budworm *Zeiraphera diniana* in Switzerland. This moth shows fairly regular cyclical variations in population in optimal subalpine areas, but much less regular changes in lower areas at altitudes less than 1000 m above sea level (Baltensweiler, 1971).

Secondly, in view of the ease with which periodic variations can be generated in models by the introduction of time lags, which are inevitable in any real system, it is surprising that so few really good cases of clear periodic oscillations have been found in natural populations. It has been pointed out occasionally during the past quarter century (Hutchinson, 1948; Levins, 1957) that strong selection against long time lags might ordinarily be expected, and indeed from the rarity of ex-

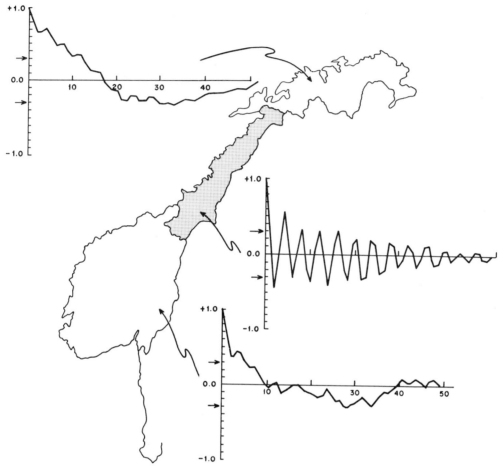

Figure 2 Autocorrelograms based on the returns for foxes taken in the northern, middle, and southern thirds of Norway, showing a significant truly periodic oscillation only in the middle (stippled) third, which is made up of the provinces of Nordland and Nord Trøndelag (modified after Finerty, 1971).

amples may be assumed to occur. This type of selection may be called τ-selection by analogy with r- and K-selection. It is conceivable, however, that in a few cases the oscillations can be used adaptively (e.g., by microtine rodents) and that τ-se-

lection might be reversed. Much more work is clearly needed, as it may prove that selection for or against time lags will prove to be of limited significance in predator-prey oscillations.

The most important aspect of definite

periodic variations in populations for the present discussion is that they call attention to an important aspect of the theoretical structure of ecology. It is possible to imagine very many different models of biological communities, each heavily dependent on the nature of the inevitably simplified assumptions on which it is based. If we have one model and one set of related observations, we can confirm or invalidate the theory. If the theory is false, we can modify it and try again. If we have several theories and many sets of observations, we can consider the frequency of confirmation of any model and can in fact build up a probability density or confidence map of a selected part of the subject. The features of this map would then provide material for second-order exploration. In the present case the observed probability of really regular oscillations of the sort predicted by models is in fact low. We must go on to ask ourselves why, but a general answer is not yet available. The methods of loop analysis introduced by Levins (Chapter 1) may help answer this problem. In summary, population fluctuations have to date been an area where experimental observations have been so much harder to obtain than theoretical predictions that the relations between the two have been unhappy.

Magpie Moths, Broom Beetles, and Spittlebugs

The idea that MacArthur and Wilson (1967) call *character release* and contrast with the well-known concept of character displacement can in a general way be traced back at least to the study by Ford

and Ford (1930) of the change in variability in the butterfly *Melitaea aurinia*. This change accompanied an increase in the butterfly's numbers after a period of decline; presumably the increase in population implies a decrease in the rigor of natural selection. Carson (1968) has stressed the importance of the phenomenon on general grounds. The detailed operation of character release or *niche variation,* as it is often though much less precisely called, has been studied recently in a number of papers that have given rise to some controversy (Van Valen, 1965; Willson, 1969; Van Valen and Grant, 1970; Soulé, 1970; Soulé and Stewart, 1970; Rothstein, 1973; see also Chapter 7 by Hespenheide). The phenomenon would seem to be a rather critical one that is easily obscured by other events. There is experimental evidence that the complementary effect of increased variability promotes the capacity of a species to invade the niche of a competitor (Shugart and Blaylock, 1973). So far the theory involved is of a completely common-sense kind.

In 1895 Alfred Russel Wallace published a long essay on *The Method of Organic Evolution* (Wallace, 1895), which is largely a review of William Bateson's *Materials for the Study of Variation* (Bateson, 1894), which had appeared during the previous year. Wallace (p. 436) wrote, "Referring to the case of two lady birds, the small *Coccinella decempunctata* being exceedingly variable, both in colour and spotting, the larger *C. septempunctata* very constant, he [i.e., Bateson] says 'To be asked to believe that the colour of *C. septempunctata* is constant because it matters to the species, and that the colour of *C. decempunctata* is variable because it does not matter, is to be asked to abrogate reason (p. 572).' I fear that I myself must be in this sad case for though I have not been asked to believe this unreasonable thing, yet I do believe it! Of course I may be wrong and Mr. Bateson right, but how is it that he is so absolutely sure that he is right?" Bateson (1895, p. 859), in a paper of which more hereafter, replied that "I do not find Mr. Wallace offering evidence and I am not aware that he has even hazarded a guess."

The ecologically-minded modern student of genetics (genetics being Bateson's word) would put the problem rather differently but would not be unsympathetic to half of Wallace's point of view. What Wallace was objecting to in Bateson was partly the latter's insistence on discontinuous or particulate variation, which we know to be formally correct, and partly to Bateson's implied scepticism of the adaptive significance of many of the characters exhibited by organisms, which to Wallace, the cofounder of the theory of natural selection, seemed almost axiomatically probable. At least we can all agree that if the pattern in one species was produced by natural selection, then selection has worked in a quite different way on the other species. It was inevitable that in passing from the pre-Mendelian to the post-Mendelian era, Wallace's views should have acquired a somewhat old-fashioned appearance, even though now his ideas, suitably rephrased, seem thoroughly up-to-date.

In the next few pages we shall consider three cases in which there is a possibility that the development of conspicuous adaptive coloration, usually aposematic, in insects that were originally cryptically colored has brought about a form of character release. Recall that some prey species have cryptic coloration and thereby avoid detection by predators; other potential prey species that are poisonous or distasteful are conspicuously colored (so-called aposematic coloration) and thereby warn predators that attacking and attempting to eat them is not worthwhile; and still other prey species maintain balanced color-pattern polymorphisms (so-called apostatic coloration) such that pattern variants, whose coloration is less cryptic than that of the commonest morph, derive a compensating advantage from the fact that predators learn to form a search image for the commonest morph and bypass the variants. Since the fine details of a warning pattern are less likely to be critical to the pattern's function than are the details of a concealing pattern, it will be suggested below that the evolution of aposematic or apostatic coloration may release a good deal of phenotypic variation. This release may occur because of reduced selection against homozygotes, the corresponding heterozygotes being slightly favored by some form of heterosis. In all three cases I shall discuss, there is more than a hint that some forms are adaptively conspicuous and other forms of the same species exhibit concealing patterns. This suggests that alternative adaptive strategies, self-advertisement or concealment, are available in existing genetic variants, whose proportions may then increase under selection in different environmental circumstances. In one of the three cases there is evidence of different dominance relationships in the two sexes, a condition that puts a brake on natural selection. This allows a small minority of individuals to continue for a long time to express phenotypically the gene against which selection is operating, so that there is nearly always something on which changed selection can operate whenever the environmental conditions change.

The first case is provided by the gooseberry, currant, or magpie moth of the Palaearctic, *Abraxas grossulariata.* This is a moderate sized insect with a wing span ordinarily of about 43 mm. It has creamy white wings with a pattern of black blotches and yellow fasciae, and a yellow body marked with black. The larva has comparable coloration, and the pupa is black and yellow. All stages are quite conspicuous and, unlike what occurs in many aposematic insects, this conspicuousness is probably maintained in twilight or moonlight, owing to the white ground color of the insect. All stages are unpalatable to insectivorous animals (Poulton, 1890); Fraser and Rothschild (1960) find the adults to be as distasteful to birds and mammals as the five-spot burnet *Zygaena lonicerae,* a very poisonous moth. *Abraxas* is usually placed in the subfamily Boarminae, most species of which, in common with the majority of the family Geometridae, are cryptically colored.

A. grossulariata is often regarded by moth collectors as the most variable of all the Macrolepidoptera of Western Europe. This, however, does not mean that most populations are conspicuously polymorphic, but rather that by patiently rearing large numbers of larvae, which used to be easy as they were pests of gooseberry and currant bushes, a great range of relatively rare color forms might ultimately be obtained. The range of variation in pattern may be appreciated from Figure 3, if it is borne in mind that the ground color can vary from creamy white to deep reddish orange, and that in one form the basal parts of the pale area of the fore wings are light blue. Probably 80 or 90 validly recognizable distinct phenotypes exist, based on at least 60 mutations. The genetics of the sex-linked aberration *dohrni* (Doncaster and Raynor, 1906) and of the variably dominant yellow aberration *lutea* (Onslow, 1919; Ford, 1940) have been the subject of classical studies, but few of the other genes are properly understood. There seem to be two (poorly known) dominants (Stovin, 1939, 1940), but most variant genes are doubtless autosomal recessives; as well as *dohrni*, a little known aberration *cockayni* (Hutchinson, 1974) may be sex-linked.

There is reason to believe that well over 10^5 but probably less than 10^7 individuals have by now been examined in nature in Europe (Hutchinson, 1969). We may therefore guess that when an aberration is known from a single wild-caught specimen, the frequency of the aberrant gene is usually somewhere around 0.001, or at least between 0.0003 and 0.003, and for all less rare autosomal recessives within the range 0.001–0.1. It is thus reasonable to suppose that at least in the cases of the more widespread, even if quite rare, aberrations, some degree of balanced polymorphism exists. The total range of diver-

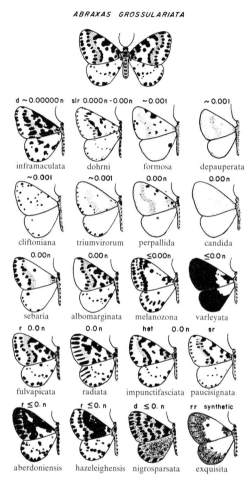

ABRAXAS GROSSULARIATA

d ~ 0.00000 n   slr 0.000 n - 0.00 n   ~ 0.001   ~ 0.001

inframaculata   dohrni   formosa   depauperata

~ 0.001   ~ 0.001   0.00 n   0.00 n

cliftoniana   triumvirorum   perpallida   candida

0.00 n   0.00 n   ≤ 0.00 n   ≤ 0.0 n

sebaria   albomarginata   melanozona   varleyata

r 0.0 n   0.0 n   het   0.0 n   sr

fulvapicata   radiata   impunctifasciata   paucisignata

r ≤ 0. n   r ≤ 0. n   d ≤ 0. n   rr synthetic

aberdoniensis   hazeleighensis   nigrosparsata   exquisita

Figure 3 *Abraxas grossulariata,* typical form *grossulariata* at top and a number of mutants, with rough estimates of the mean gene frequency in the populations of Europe. In the typical form the ground color is creamy white, with black blotches, some yellow at the base of the forewing, and a well-marked yellow fascia, stippled in the drawing. The two suspected dominants are marked *d,* the single sex-linked recessive *slr,* the only semirecessive *sr,* with its heterozygote *het;* all other mutants are assumed to be autosomal recessives, though there are data only for *varleyata, fulvapicata, hazeleighensis,* and *aberdoniensis,* the extreme expression of the last involving several genes. The double recessive *exquisita* marked *rr* is a synthetic combination of *dohrni* and *varleyata;* it is the only phenotype figured here not known to occur in nature. All figures about ⅓ natural size.

sity of patterns in this species is probably greater than in all the other species of the subfamily Boarminae taken together. The evolutionary potential of the whole set of mutants, if the patterns have any selective value, must be very great.

The most reasonable explanation of what we see in *Abraxas grossulariata* is that most of the heterozygotes have some slight selective advantage and that the homozygous recessives are at less of a disadvantage than would have been the case if the typical coloration had been cryptic, as in most of the allied members of the Geometridae. Rothschild (1963b) pointed out a correlation between distastefulness and conspicuous variability, and believed that in some cases this would promote the development of aposematic (warning) coloration. The idea put forward here is the complement of Rothschild's, namely that the development of distastefulness and of concomitant aposematic coloration provides a release mechanism that enhances the variability of the species. Quite significant variations in pattern or ground color would have little effect on the aposematic function of the pattern. As the distastefulness and aposematic coloration developed, the selection against rare homozygous recessives, maintained in the population by selection in favor of the heterozygotes, would diminish, and both heterozygotes and homozygous recessives would become somewhat commoner. A pool of very striking mutants might thus be revealed, some of which, as Rothschild suggests, might enhance the aposematic patterns.

We may also note that *Abraxas gross-*

ulariata appears, during the era of industrial melanism, to have made an abortive attempt to become cryptic. This involved the widespread occurrence of a form called *nigrosparsata*, in which the white areas of the wings are peppered with minute black dots. It seems to have appeared in South Wales at the end of the nineteenth century and to have been first noticed in Yorkshire around 1905. At Huddersfield in that county it became increasingly common until, in 1917, nearly 10% of the population was *nigrosparsata*, though to a very varying degree. The species then became extremely scarce for a few years, and when it reappeared, *nigrosparsata* occurred only casually (Porritt, 1921, 1926). Comparable, less well-documented changes occurred around the city of York (Hutchinson, 1974). Although this looks like a change from selection for aposematic or warning coloration to selection for cryptic or concealing coloration as the environment of industrial melanism developed, the case presents unresolved difficulties on account of the low heritability of *nigrosparsata*. It has been supposed (Stovin, 1940) to be a dominant, as are most other industrial melanic mutants, but penetrance of the gene must be low. It is conceivable that *nigrosparsata* is not exclusively determined by genetic constitution but that something else, such as a virus, is also involved. It is interesting that Rothschild (1963a) has noted the appearance of a dark form of the small magpie moth *Eurrhypara horticulata* near Oxford, along with a tendency towards darkening in several other moths, between 1957 and 1963, apparently a modified trend toward industrial melanism. *E. horticulata* belongs to a different family (Pyraustidae) from that of *Abraxas* but is distasteful, aposematic, and has a color pattern that is not unreminiscent of the larger *Abraxas*. Rothschild thinks that many insects "are both cryptic and aposematic at one and the same time."

The next two examples, *Philaenus* and *Phytodecta*, provide further cases of polymorphism in which one or more forms are primarily cryptically colored, while others are aposematic or apostatic. Consider first the homopteran *Philaenus spumarius*, which has been studied by many investigators, the most recent and important work being that by Halkka (1964) and his associates in Finland and elsewhere. The insect belongs to the Aphrophoridae and spends its nymphal life in a mass of froth on its food plant, which may be any one of a great number of dicotyledonous herbs, hence the popular name of spittlebug in North America and cuckoo-spit bug in Britain.

The nominotypical form *spumarius* is light brown, with two pairs of paler marks across the elytra. The pattern is more definite in males than in females. In *trilineatus* and related forms the pattern differs radically, the insect being straw-colored with three or one dark longitudinal stripes. There is a unicolorous yellow form *populi* which is clearly separated from *spumarius* in the males, much less clearly so in the females. Whittaker (1968) finds that in England the proportions of *trilineatus* and *populi* vary inversely. He also has experimental data suggesting that the transformation of one of these forms into the other may result from feeding on particular plants. According to Halkka, Heinonen, Raatikainen, and Vasaramen (1966) and Halkka, Halkka, Raatikainen, and Vasarainen (1968), the *trilineatus* gene is dominant in both sexes to all other genes involved in color patterns. Both *spumarius* and *trilineatus* are moderately cryptic.

There are also a number of striking dark forms of *Philaenus spumarius* (Figure 4). In at least the series *leucophthalmus, leucocephalus, lateralis,* and *marginellus* coloration is apparently controlled by alleles at a single locus (Halkka and co-workers, in Thompson and Halkka, 1973). In the female all these alleles are dominant to that for typical *spumarius* (Halkka *et al.,* 1966; Halkka *et al.,* 1968). Males of *marginellus* have never been found, but rare male specimens occur in the other dark forms. Halkka, Raatikainen, and Vilbaste (1967) suggest, almost by implication, that the *leucophthalmus, leucocephalus,* and *lateralis* alleles, dominant in the female, are recessive to typical *spumarius* in the male, and that the rare specimens of these forms in that sex are homozygotes; in *marginellus* the homozygote may be lethal.

There can be little doubt that the proportions of the various forms are near genetic and selective equilibrium in most though not quite all of the populations studied. Considering only the rarer female forms, Halkka, Raatikainen, and Vilbaste (1967) found the distributions given in Figure 4 for a locality in northeast Finland sampled in 1952 and 1964. Hutchinson (1963) gave data on rather small sam-

ples of both sexes taken together, from a locality near Cambridge, England, sampled in 1920 and again in 1962, which showed only a statistically insignificant ($P \simeq 0.1$) difference between the early sample and the one taken 42 years later. The difference that was observed probably occurred because the first sample was collected later in the season. There is quite good evidence, however, that *marginellus* has increased in Wisconsin between 1947 and 1970 (Thompson and Halkka, 1973).

There is a good deal of geographical variation in the proportions of the different forms. In the Baltic countries there is a tendency for lighter forms to be abundant in the southwest and darker forms to be commoner in the northeast. The form *trilineatus* is far commoner in England and in parts of North America than in any localities in the Baltic lands, except on occasional small islands where a founder effect is certainly involved (Halkka, Raatikainen, Halkka, and Lallukka, 1970; Halkka, Raatikainen, Halkka, and Lokki, 1971). All forms except *spumarius* (58.3%) and *populi* (34.8%) are rare on Cyprus (Lindberg, 1948).

Several suggestions as to the adaptive significance of the polymorphism have been made. Owen and Wiegert (1962) believed the fundamental advantage conferred on the species by the polymorphism to be apostatic, the variety of conspicuous color forms ensuring that any predator forming a search image would concentrate only on a minority of the species in its search for prey. Halkka accepts this suggestion but concludes that there are other types of selection at work, presumably

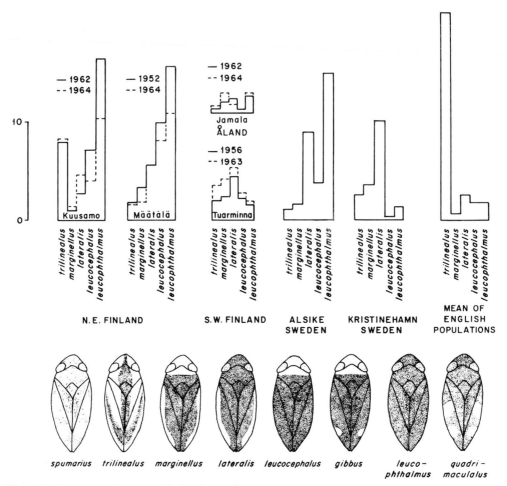

Figure 4 Percentage frequency distribution and appearance of some of the forms of *Philaenus spumarius,* ca. twice natural size. The Finnish and Swedish data are from the papers of Halkka and his associates, the English from Hutchinson (1964). The black forms with two (*albomaculatus*) or four (*quadrimaculatus*) white blotches, and the equivalent forms (*gibbus* and *flavicollis*) with pale heads, are included in *leucophthalmus* and *leucocephalus* respectively in Halkka's data; the form *vittata* is included with the form *trilineata* throughout.

involving climatic factors. The fact that in most English populations the combined frequency of the striking dark forms constitutes only a few percent of the total number of individuals suggests that the distribution of phenotypes can be far from optimal in providing apostatic adaptation. Thompson (1973), applying an idea of Lindroth (1971), thinks that the conspicuous color pattern of *marginellus* is an escape warning color. So long as the insect, which is a good jumper, was not too common, most predators would have nothing but disappointing encounters, as the potential prey would very often jump away. The predator, if it could easily recognize the potential prey, would learn that it was not worth the effort of abortive pursuit. If the conspicuous insect became very common, a small proportion, though possibly a significant absolute number, might be caught and the escape warning color selection would lead to balanced polymorphism. Thompson thinks that for the *P. spumarius* form *marginellus* in North America this happens when about 5% of all females belong to this form.

The last case is that of the chrysomelid beetle *Phytodecta* (*Spartoxena*) *variabilis* (Olivier) which lives in Spain on a species of broom (see Figure 5). Its populations were studied in premendelian days by Bateson (1895) in one of the first investigations of the frequency of various color forms in a polymorphic insect. Later Doncaster (1905) collected more field data. Breeding experiments that elucidated the rather peculiar genetics of the various forms were subsequently published in

Figure 5 *Phytodecta variabilis* about natural size. The form marked A has red elytra marked with black spots and is probably aposematic; the form marked B has greenish-grey elytra bearing dark stripes, and is probably cryptic (after Bateson).

an excellent but little known paper by de Zulueta (1925)[3].

There are five named forms of the insect which appear to be the expression of at least four alleles at the same locus, but a number of unelucidated modifying genes must also exist.

In f. *aegrota* Fabricius the elytra are greyish or yellowish green, marked with longitudinal stripes. In the rather misleadingly named f. *unipunctata* Olivier the ground color of the elytra is yellow, and each elytron has a large and a smaller discal spot. In f. *rubra* de Zulueta the pattern is similar but the ground color is red. In the nominotypical form *variabilis* a broadly striped and coalescent black pattern covers most of the elytra which in f. *koltzei* Weise are entirely black. According to de Zulueta, the pale striped *aegrota* is due to an allele recessive to all the others, whereas the allele causing the extension of the black pigment is dominant

[3]I am greatly indebted to my friend Professor Ramón Margalef of Barcelona for calling my attention to de Zulueta's paper. Before seeing it I had concluded that Bateson's results depended on reversal of dominance in the sexes. The food plant of the insect is given various names; de Zulueta uses *Retama sphaerocarpa*.

to the other alleles. The allele for the red background of *rubra* is dominant to that for the yellow of *unipunctata*. The locus involved is apparently on the sex chromosomes, but is carried by both X and Y, with little or no crossing-over in the male. With this arrangement, also known in some fishes, the phenotypic proportion of a dominant in the female part of the population can have any value, independent of the frequency in the male, while the frequency in the male can have any value equal to or greater than the gene frequency in the X-chromosome population. Selection for a dominant allele can therefore operate more or less independently in the two sexes, provided the dominant is advantageous in the male.

Near Madrid, de Zulueta found 58% of the females but only 0.5% of the males to be *aegrota*. Bateson in his well-studied populations from near Granada found 67 to 84% of the females and 19 to 38% of the males to be this form; Doncaster near Malaga found no great difference between the proportions in the two sexes. In general *rubra* is commoner than *unipunctata* and makes up most of the rest of the population, of which the very black forms constitute a few percent.

Bateson made a further observation, which is very relevant in the present context: "While the red-spotted forms are strikingly conspicuous objects, the striped greenish-grey forms resemble so nearly the colour of the twigs of the *Spartium* that it is impossible not to remark the likeness. If they were the only form known, the case might well be used as an illustration of a protective coloration. The red-spotted forms present some superficial likeness to the common Lady-bird (*C. bipunctata*[4]), a creature which exudes an acrid juice, and whose color has naturally been classed among 'warning colors'." He adds that *Phytodecta variabilis* "does not, as far as I know, possess any such irritant properties, but I have no information as to its enemies. As *Coccinella bipunctata* is not very common on the *Spartium*, probably no one will suggest that we have here an example of protective mimicry. I may mention, however, that *Coccinella septempunctata,* the larger scarlet species, occurs in vast quantities mixed with *Phytodecta variabilis.* Whether anyone would consider the resemblance to this species sufficiently close to constitute mimicry, I cannot say."

Doncaster's observations suggest that the males emerge before the females and that the cryptic coloration of the paler forms such as *aegrota* probably tends to become more effective as the season progresses. He believes, moreover, that a yellowish green, unmarked form may resemble the flowers of the broom. Since the males tend to emerge before the females, they would be subject to most intense selection when most conspicuous, whatever their color, and at this time pseudaposematic resemblance to a coccinellid might be advantageous. Later, as the bushes became thicker and with flowers, the selection for cryptic coloration on the now largely female population would operate. Considerable variability in these processes, dependent on local climatic

---

[4] Now known as *Adalia bipunctata* (Linn.)

conditions, is to be expected, so that the great regional differences in the frequency of the patterns are reasonable.

So far, although our investigation of these three cases has thrown some light on two previously known but very little appreciated phenomena, namely the association of variability with warning coloration, and the development of rather peculiar systems that involve sexual differences, nothing more than the simplest population genetics has been used. There is a possibility, however, that a very much deeper theoretical advance might be possible. The kind of character release that has been studied by Van Valen and others in insular birds is far less dramatic than what has just been described, and is recognized only by measurement and statistical study of the data so obtained. Yet it is probable that such small releases in the trophic structures of an organism would have more ultimate evolutionary potential than dramatic changes in coloration. Although certain groups of organisms may be characterized by the presence of particular pigments, taxonomists have traditionally tended to be suspicious of color characters, on the ground that such characters are usually not sufficiently constant to be taxonomically reliable. In a general way it is evident that it is much easier to alter a two-dimensional pattern without something's going wrong than it is to alter a three-dimensional structure. It is obviously much safer for a species to produce more than the range of color pattern present in the rest of its subfamily than it would be to produce the overall range of structure of the external genitalia in the subfamily, or even in the genus. It seems possible that the topological concepts of structural stability and catastrophe, developed and applied to other fields of biology by Thom (1970, 1972), could play a part in developing the theory of character release.

Hints for an Agenda: How Big is it and How Fast Does it Happen?

However well we may be able to treat certain situations in ecology—and our first set of examples shows that we can sometimes do a very good job—there remain large and challenging but initially rather indefinite problems, which appear rather like a landscape in a mist. The general form is apparent, though the scale may be distorted and most if not all the detail is obscured. Two such problems are briefly treated here. These particular problems were selected because they are always implicit in other kinds of work, and consideration of what has been done with them is instructive.

Alfred Russel Wallace (1858), in his contribution on natural selection communicated to the Linnaean Society, wrote: "The general proportion that must obtain between certain groups of animals is readily seen. Large animals cannot be so abundant as small ones; the Carnivora must be less numerous than the Herbivora; eagles and lions can never be so plentiful as pigeons and antelopes." These ideas may well have been entertained by other naturalists between 1858 and 1927, when they were organized by Elton (1927) as the pyramid of numbers and sizes. Elton's ideas became incorporated into the rapidly growing body of ecological knowledge and are now universally accepted. Lindeman's (1942) treatment of trophic dynamics considered the same pattern, though in terms not of individuals and species but of the set of trophic levels. Two questions may now be asked. The first question is, what determines the upper limit to the process? That is, how big and how scarce can the ultimate carnivore be? Obviously two antagonistic processes set the length of the food chain. There will always be an advantage in being at the top, as one is then free from Eltonian predation, though of course not from parasitism. At the same time the position is a very inefficient one relative to that of a herbivore or first-order carnivore, so that there is also a tendency for the top animal to usurp the role of the lower levels, as for instance in the evolutionary line that produced the whalebone whales. This situation raises the interesting question as to what governs the length of a food chain, as well as what controls the number and limits the size of the ultimate carnivore. Food chains of course tend to be longer in water than on land. This is clearly partly due to a buoyant, turbulent, liquid medium that allows the development of a rich phytoplankton of a kind that would be quite impossible in ordinary gases at atmospheric pressure. The largest animals feeding on phytoplankton, such as the anchovy *Engraulis ringens* of the Peru Current region, are obviously much smaller than the largest herbivores on land. The shortest food chains in the sea may involve homeotherms as their top members as on land, but there are cases in both fresh and marine water in which up to five or six levels exist, all poikilothermic and mostly fishes. On land the largest animals are, and probably always have been in the past, herbivores. They are subject to predation primarily when young (cf. Connell, Chapter 16). This is apparently as true of the Aldabra tortoise *Geochelone gigantea,* which is evidently eaten by land crabs when very young and under 20 cm long (Gaymer, 1968), as it is of baby African elephants, which are vulnerable to predators during the second half of their first year (Sikes, 1971, p. 271). Here the formal Eltonian system tends to break down, as it also does when there is control by parasites that cause diseases. These may be important in regulating the populations of very large animals, or moderately large but very intelligent animals such as *Pan* (Jane Goodall, personal communication), or *Homo* under primitive conditions.

Experience with large herbivores such as elephants does suggest that their unrestricted population growth can play havoc with the habitat, so that in a natural environment they would probably be resource-limited with a very long lag period in which the vegetation was rebuilt. Much movement from one area to another would have to be included in any model. Although it is quite likely that the existing elephants have experienced human disturbance, including predation, throughout their whole career as species, this would not have been true of their large ancestors and the even larger mammals, such as

Baluchitherium, of the middle Cenozoic. It is fashionable to look for external causes for the extinction of the spectacularly large animals of the past. However, the problem of the circumstances leading to inherent instability in the upper levels of a food web needs much further study, which should take into account the inevitable physiological differences and resulting different biological demands, between small and large, or between poikilothermal and homeothermal, members of the system under different climatic conditions. The techniques introduced by Levins in Chapter 1 may provide the needed springboard.

In contrast to discussion of the problems posed by large size, there is little to say about the opposite question of the lower limits of the dimensions of organisms, as the subject is hardly formulated. Morowitz (1967), developing an old speculative tradition going back to Clerk Maxwell (1875), Errera (1903), and D'Arcy Thompson (1917; omitted from second edition), has concluded from biochemical considerations that one would not expect complete organisms much less than 0.1 μ in diameter. In the greater part of the biosphere the total mass of organisms even of bacterial size must be quite small, enormous quantities being primarily associated only with interface regions such as soils or sediments. At slightly greater sizes we find two somewhat antagonistic processes, at least in aquatic environments. Small objects are more easily taken by microphagous feeders, as Munk and Riley (1952) have indicated, whereas large organisms are more easily eaten by

selective predators, as Brooks and Dodson (1965; Brooks, 1968, 1969) have emphasized. It is becoming more and more apparent that these relationships are of immense importance. From Brooks's work and from that of workers who have elaborated on his ideas (Zaret and Payne, 1973; Suffern, 1973), we may conclude that the whole structure of the food chains in any community is a major factor in determining the biomass and hence the productivity of every level including the primary producers.[5] The views from the pyramid erected by Elton, perhaps unwittingly in memory of Wallace, continue to show unfamiliar aspects of nature.

Looking at environmental biology even more generally, we might enquire what sets the rate at which evolution has happened. Darwin, in the first edition of the *Origin of Species* (Darwin, 1959) estimated (from probable assumptions about the rate of erosion in the Weald, the area between the North and South Downs in southern England) that the Cretaceous ended at least 300,000,000 years ago. In the second edition he tempered this estimate, giving a minimum of 100,000,000 years. The best contemporary estimate (Armstrong, in Flint, 1973) appears to be 63,000,000 years. Darwin's minimum estimate, therefore, is too great by less than a factor of two. In the third edition the matter was dropped. In 1862 Sir William Thomson (later Lord Kelvin) in a paper on the secular cooling of the earth, wrote (Thomson, 1862): "For eighteen years it

---

[5] This aspect of eutrophy was completely overlooked in my recent review (Hutchinson, 1973).

has pressed on my mind that essential principles of Thermodynamics have been overlooked by those geologists who uncompromisingly oppose all paroxysmal hypotheses." From a study of the known heat flux from the interior of the earth, he concluded that the crust must have consolidated from a molten state sometime between 20,000,000 and 400,000,000 years ago. Reverting to the subject in 1866 in a short paper entitled "The Doctrine of Uniformity in Geology briefly Refuted" (Thomson, 1866), he maintained categorically that "no hypothesis as to chemical action, internal fluidity, effects of pressure at great depth or possible character of substances in the interior of the earth, possessing the smallest vestige of probability, can justify the supposition that the earth's upper crust has remained nearly as it is," while large amounts of heat, which must have originally been stored in the planet, were dissipated. The argument of the second paper is less relevant to the problem of the age of the earth than is the first, but it is historically interesting as it includes the ex cathedra statement that there was not the smallest vestige of probability that the possible character of substances in the interior of the earth could falsify the author's reasoning. Marie Curie was born the following year.

Darwin reverted to the matter in the fifth edition of the *Origin of Species*. He suspected that the pre-Paleozoic might be as long as the period from the early Paleozoic to today, and he must have realized that Thompson's estimates were quite inconsistent with his own figure in the earlier editions. He looked Thompson's argument squarely in the face, and although it evidently made him uncomfortable, he passed on to other matters. He clearly felt that the shorter estimate would not give enough time for evolution. Actually he had no reason whatever to conclude this. The only fact mentioned is that little change has occurred since the beginning of the Pleistocene glaciation. But a time scale for the latter set of events has become available to us only in the past three decades. It is quite likely that the only firm data available to Darwin would have been the lack of change over two millenia in the characters of animals described by Aristotle. The difficulty was thus largely illusionary, as well as being based on a false physical conception of the heat balance of the earth. Not knowing these things, Darwin was however right not to become unduly troubled, because the evolutionary hypothesis explained so much that its rejection would have produced more difficulties than would the rejection of Kelvin's ideas as applied to the earth.

The problems raised about evolutionary rates and what determines them remain. At any given temperature, the velocities of ordinary physiological processes are presumably determined basically by chemical kinetics and diffusion. The rate of increase of a population, except initially, will ordinarily not be set by purely internal physiological factors determining the innate growth-rate constant, but by external factors involving direct environmental effects, interspecific and intraspecific competition, and other kinds of interaction. Moreover, though it is obvious that with a zero mutation rate no evolution would occur, most organisms seem to have a considerable reservoir of variability on which what Wallace (1968) calls soft selection can operate. The direction of this will vary with the environment. It would seem reasonable to suppose, and in fact the idea is usually implicitly taken for granted, that both evolutionary rates and directions are to a large extent determined ecologically.

It would seem that with much greater accuracy in dating the palaeontological record, a huge field for study should be opening up. Reasonably precise estimates of evolutionary rate should be possible, permitting a far greater knowledge of the effect of external ecological changes, body size, position in the food chain, and the like on the dynamics of evolution. Some very good work is being done, but this is an area in which evolutionary ecology can be directly studied over long periods of time, and it suffers from a considerable neglect.

Coda

We have now looked at four types of situations. In the first the relation of theory to observation and experiment has been very satisfactory and continues to develop fruitfully. In the second we come up against a situation in which good theory often does not work. This should lead us to conclude that the task of the theoretical worker is to investigate all possible models, whereas that of the observational and experimental ecologist is at least in part to find out how often the theoretician is right and what pattern underlies his hits or misses. Our third group of cases suggested that there may be quite unfamiliar concepts waiting to be imported into ecology, and our fourth group showed how needed ideas are sometimes still unborn.

MacArthur (1972) warned us not to become obsessed with our own methods and intellectual approaches to ecology, but to tolerate and encourage many different ways of research. He also felt that an initial insistence on too much generality might be misleading, giving as a possible example the idea that the communities of the two-dimensional habitats of the sea bottom might be controlled more by predators than those of the three-dimensional habitats of birds in forests, so that no general statement on the role of predation would be possible.

In the preceding pages I have considered a number of cases in which various models can be constructed or might be constructible to account for what is observed. The value of the model-making is that it provides us with a series of possibilities. We can then turn to nature and see which possibilities are realized and how often. Then we can ask ourselves more generally why some possibilities are highly probable and others very improbable. In this way a higher-order ecology might be constructed, free from the premature generality to which Robert MacArthur objected, having for its aim the understanding of the whole volume of animate nature. To do this, however, will require even greater detailed knowledge than ordinarily goes into ecological studies today, since such a program, in essence, is comparative. Many students get a fair idea of the contemporary status of ecological studies. Many ecologists of the present generation have great ability to handle the mathematical basis of the subject. Modern biological education, however, may let us down as ecologists if it does not insist, and it still shows too few signs of insistence, that a wide and quite deep understanding of organisms, past and present, is as basic a requirement as anything else in ecological education. It may be best self-taught, but how often is this difficult process made harder by a misplaced emphasis on a quite specious modernity. Robert MacArthur really knew his warblers.

Acknowledgments

I am greatly indebted to Dr. J. P. Finerty for permission to reproduce the autocorrelograms of Figures 1 and 2, and to Mrs. Virginia Simon and Miss Susan Klein for the care they have taken with the illustrations. Mrs. Mary Poulson has been her customary helpful self.

References

Alexander, R. D., and T. E. Moore. 1962. The evolutionary relationships of 17-year and 13-year cicadas, and three new species (Homoptera, Cicadidae, *Magicicada*). *Misc. Publ. Mus. Zool. Univ. Michigan* 1921:1–59.

Andersen, J. 1957. Studies in Danish hare-populations. I. Population fluctuations. *Danish Rev. Game Biol.* 3:8–131.

Ayala, F. J. 1969. Experimental invalidation of the principle of competitive exclusion. *Nature* (London) 224:1076–1079.

Ayala, F. J. 1971. Competition between species: Frequency dependence. *Science* 171:820–824.

Ayala, F. J., M. E. Gilpin, and J. G. Ehrenfeld. 1973. Competition between species: theoretical models and experimental tests. *Theor. Popul. Biol.* 4:331–356.

Baltensweiler, W. 1971. The relevance of changes in the composition of larch bud moth populations for the dynamics of its numbers. *In* P. J. den Boer and G. Gradwell, eds., *Dynamics of Populations* (Proc. Advan. Study Inst. Dynamics of Number in Populations. Oosterbeek. 1970), pp. 208–219. Centre for Agricultural Publishing and Documentation, Wageningen, The Netherlands.

Bateson, W. 1894. *Materials for the Study of Variation Treated with Especial Regard to Discontinuity in the Origin of Species.* Macmillan, London and New York.

Bateson, W. 1895. On the colour-variation of a beetle of the family *Chrysomelidae*, statistically examined. *Proc. Zool. Soc. London,* 1895:851–860.

Beamer, R. H. 1931. Notes on the 17-year cicada in Kansas. *J. Kansas Entomol. Soc.* 4:53–58.

Botkin, D. B., P. A. Jordan, A. S. Dominski, H. S. Lowendorf, and G. E. Hutchinson. 1973. Sodium dynamics in a northern ecosystem. *Proc. Nat. Acad. Sci. U.S.A.* 70:2745–2748.

Brooks, J. L. 1968. The effect of prey size selection by lake planktivores. *Syst. Zool.* 17:272–291.

Brooks, J. L. 1969. Eutrophication and changes in the composition of the zooplankton. *In Eutrophication: Causes, Consequences, Correctives*, pp. 236–255. National Academy of Sciences, Washington, D.C.

Brooks, J. L., and S. J. Dodson. 1965. Predation, body size and composition of plankton. *Science* 150:28–35.

Cabrera, A. 1932. La incompatibilidad ecológica: Una ley biológica interesante. *Anal. Soc. Cient. Argentina* 114:243–260.

Carson, H. L. 1968. The population flush and the genetic consequences. *In* R. C. Lewontin, ed., *Population Biology and Evolution*, pp. 123–127. Syracuse University Press, Syracuse.

Cody, M. L. 1974. *Competition and the Structure of Bird Communities*. Princeton University Press, Princeton.

Cole, L. C. 1954. Some features of random population cycles. *J. Wildlife Mgt.* 18:1–24.

Darwin, C. R. 1959. *The Origin of Species by Charles Darwin. A Variorum Text*, ed. M. Peckham. University of Pennsylvania Press, Philadelphia.

de Zulueta, A. 1925. La herencia ligada al sexo en el coleóptero *Phytodecta variabilis* (Ol.). Revist. Espan. Entom. 1:203–231.

Doncaster, L. 1905. On the colour-variation of the beetle *Gonioctena variabilis*. *Proc. Zool. Soc. London* 1905(2):528–536.

Doncaster, L., and Raynor, G. H. 1906. Breeding experiments with Lepidoptera. *Proc. Zool. Soc. London* 1906(1):125–133.

Elton, C. S. 1927. *Animal Ecology*. Sidgwick and Jackson, London; Macmillan, New York.

Elton, C. S. 1942. *Voles, Mice and Lemmings*. Clarendon Press, Oxford.

Errera, L. 1903. Sur la limite de petitesse des organismes. *Bull. Soc. Roy. Sci. Méd. et Nat. Bruxelles* 61:13–22.

Finerty, J. P. 1971. *Cyclic fluctuations in biological systems: A revaluation*. Ph.D. thesis, Yale University.

Flint, R. F. 1973. *The Earth and its History*. W. W. Norton, New York.

Ford, E. B. 1940. Genetic research in the Lepidoptera. *Ann. Eugen.* 10:227–252.

Ford, H. B., and E. B. Ford. 1930. Fluctuations in numbers and its influence on variation in *Melitaea aurinia*. *Trans. Roy Entomol. Soc. London* 78:345–351.

Frazer, J. F. D., and M. Rothschild. 1960. Defense mechanisms in warningly-colored moths and other insects. *XI. Int. Kongr. f. Entomol. Wien 1960. Verh. B.* III:249–256.

Gause, G. F. 1934. *The Struggle for Existence*. Williams and Wilkins, Baltimore.

Gause, G. F. 1935. Vérifications experimentales de la théorie mathématique de la lutte pour la vie. *Actual. Scient. Indust.* No. 277.

Gause, G. F., and A. A. Witt. 1935. Behavior of mixed populations and the problem of natural selection. *Amer. Natur.* 69:596–609.

Gaymer, R. 1968. The Indian Ocean giant tortoise *Testudo gigantea* on Aldabra. *J. Zool.* 154:341–363.

Gilpin, M. E. 1973. Do hares eat lynxes? *Amer. Natur.* 107:727–730.

Gilpin, M. E., and F. J. Ayala. 1973. Global models of growth and competition. *Proc. Nat. Acad. Sci. U.S.A.* 70:3590–3593.

Grenney, W. J., D. A. Bella, and H. C. Curl. 1973. A theoretical approach to interspecific competition in phytoplankton communities. *Amer. Natur.* 107:405–425.

Grinnell, J. 1904. The origin and distribution of the chestnut-backed chickadee. *Auk* 21:264–382.

Haldane, J. B. S. 1924. A mathematical theory of natural and artificial selection. Part 1. *Trans. Cambr. Philos. Soc.* 23:19–41.

Halkka, O. 1964. Geographical, spatial and temporal variability in the balanced polymorphism of *Philaenus spumarius*. *Hereditas* 19:383–401.

Halkka, O., L. Halkka, M. Raatikainen, and

A. Vasarainen. 1968. Transmission of genes for colour polymorphism in *Philaenus*. *Hereditas* 60:262–264.

Halkka, O., L. Heinonen, M. Raatikainen, and A. Vasarainen. 1966. Crossing experiments with *Philaenus spumarius* (Homoptera). *Hereditas* 56:306–312.

Halkka, O., M. Raatikainen, L. Halkka, and R. Lalluka. 1970. The founder principle, genetic drift and selection in isolated populations of *Philaenus spumarius* (L.) (Homoptera). *Ann. Zool. Fenn.* 7:221–238.

Halkka, O., M. Raatikainen, L. Halkka, and J. Lokki. 1971. Factors determining the size and composition of island populations of *Philaenus spumarius* (L.) (Hom). *Acta Entom. Fenn.* 28:83–100.

Halkka, O., M. Raatikainen, A. Vasarainen, and L. Heinonen. 1967. Ecology and ecological genetics of *Philaenus spumarius* (L.) (Homoptera). *Ann. zool. Fenn.* 4:1–18.

Halkka, O., M. Raatikainen, and J. Vilbaste. 1967. Modes of balance in the polymorphism of *Philaenus spumarius* (L.) (Homoptera). *Ann. Acad. Scient. Fenn.* Ser. A IV, No. 107.

Hansmann, A. 1857. Die Sylvien der Insel Sardinien. *Naumannia.* 7:404–429.

Hardin, G. 1960. The competitive exclusion principle. *Science* 131:1292–1297.

Harper, J. L. 1969. The role of predation in vegetational diversity. *In* G. M. Woodwell and H. H. Smith, eds., *Diversity and Stability in Ecological Systems,* Brookhaven Symposium in Biology No. 22, pp. 48–61. U.S. Department of Commerce, Springfield, Va.

Harper, J. L., and J. H. McNaughton. 1962. The comparative biology of closely related species living in the same area. VII. Interference between individuals in pure and mixed populations of *Papaver* species. *New Phytol.* 61:175–188.

Harvey, L. A. 1945. Symposium on 'The ecology of closely allied species.' *J. Ecol.* 33:115–116.

Hiltzheimer, M. 1909. Wisent und Ur in K. Naturalienkabinett zu Stuttgart. *Jahreshf. f. Ver. Vaterländische Naturkunde in Württemberg* 65:241–259.

Hutchinson, G. E. 1947. A note on the theory of competition between two social species. *Ecology* 28:319–321.

Hutchinson, G. E. 1948. Circular causal mechanisms in ecology. *Ann. N.Y. Acad. Sci.* 50:221–246.

Hutchinson, G. E. 1963. A note on the polymorphism of *Philaenus spumarius* (L.) (Homoptera: Ceropidae) in Britain. *Entomol. Month. Mag.* 99:175–178.

Hutchinson, G. E. 1965. *The Ecological Theater and the Evolutionary Play.* Yale University Press, New Haven.

Hutchinson, G. E. 1969. Some continental European aberrations of *Abraxas grossulariata* Linn. (Lepidoptera) with a note on the theoretical significance of the variation observed in the species. *Trans. Conn. Acad. Arts Sci.* 43:1–24.

Hutchinson, G. E. 1973. Eutrophication. *Amer. Scientist* 61:269–276.

Hutchinson, G. E. 1974. New and inadequately described aberrations of *Abraxas grossulariata* (Linn.) (Lepidoptera, Geometridae). *Entomol. Record* 86:199–206.

Istock, C. 1973. Population characteristics of a species ensemble of waterboatmen (Corixidae). *Ecology* 54:535–544.

Johnsen, S. 1929. Rovdyr og rovfuglstatistikken i Norge. *Bergens Mus. Årbok. Naturvitensk. Rekke,* No. 2.

Keith, L. B. 1963. *Wildlife's Ten-year Cycle.* University of Wisconsin Press, Madison.

Leigh, E. 1968. The ecological role of Volterra's equations. *In* M. Gerstenhaber, ed., *Some Mathematical Problems in Biol-*

ogy, pp. 1–61. Amer. Math. Soc., Providence.

Levins, R. 1957. *In* Discussion of P. J. Wangersky and W. J. Cunningham: Time lag in population models. *Cold Spring Harbor Symp. Quant. Biol.* 22:338.

Lindberg, H. 1948. On the insect fauna of Cyprus. Results of the expedition of 1939 by Harold Håkan and P. H. Lindberg. II. Heteroptera and Cicadina der Insel Zypern. *Commentationes Biologicae Helsingfors.* 7:23–175.

Lindeman, R. 1942. The trophic-dynamic aspect of ecology. *Ecology* 23:399–418.

Lindroth, C. H. 1971. Disappearance as a protective factor. *Entomol. Scand.* 2:41–48.

Lotka, A. J. 1925. *Elements of Physical Biology.* Williams and Wilkins, Baltimore (reissued as *Elements of Mathematical Biology,* Dover, 1956).

Lloyd, M., and H. S. Dybas. 1966a. The periodical cicada problem. I. Population ecology. *Evolution* 20:133–149.

Lloyd, M., and H. S. Dybas. 1966b. The periodical cicada problem. II. Evolution. *Evolution* 20:466–505.

Macan, T. T. 1963. *Freshwater Ecology.* Longmans Green and Co., London; Wiley, New York.

MacArthur, R. H. 1958. Population ecology of some warblers of northeastern coniferous forests. *Ecology* 39:599–619.

MacArthur, R. H. 1972. Coexistence of species. *In* J. A. Behnke, ed., *Challenging Biological Problems,* pp. 253–259. A.I.B.S., Oxford University Press, New York.

MacArthur, R. H. and E. O. Wilson. 1967. *The Theory of Island Biogeography.* Princeton University Press, Princeton.

McNaughton, I. H., and J. L. Harper. 1960. The comparative biology of closely related species living in the same area. I.

External breeding barriers between *Papaver* species. *New Phytol.* 59:15–20.

Marshall, J. T. 1960. Interrelation of Abert and brown towhees. *Condor* 62:49–64.

Maxwell, J. Clerk. 1875. Atom. *In Encyclopedia Britannica,* 9th ed., 3:36–49. A. and C. Black, Edinburgh.

May, R. M. 1973. *Stability and Complexity in Model Ecosystems.* Princeton University Press, Princeton.

Monard, A. 1920. La faune profonde du Lac de Neuchâtel. *Bull. Soc. Neuchâteloise Sci. Nat.* 44:65–236.

Moran, P. A. P. 1949. The statistical analysis of the sunspot and lynx cycles. *J. Animal Ecol.* 18:115–116.

Moran, P. A. P. 1952. The statistical analysis of game-bird records. *J. Animal Ecol.* 21:154–158.

Moran, P. A. P. 1953a. The statistical analysis of the Canadian lynx cycle. I. Structure and prediction. *Austral. J. Zool.* 1:163–173.

Moran, P. A. P. 1953b. The statistical analysis of the Canadian lynx cycle. II. Synchronization and meteorology. *Austral. J. Zool.* 1:291–298.

Moran, P. A. P. 1954. The logic of the mathematical theory of animal populations. *J. Wildlife Mgt.* 18:60–66.

Morowitz, H. J. 1967. Biological self-replicating systems. *Progr. Theoret. Biol.* 1:35–58.

Munk, W., and G. A. Riley. 1952. Absorption of nutrients by aquatic plants. *J. Mar. Res.* 11:215–240.

Nicholson, J. A., and V. A. Bailey. 1935. The balance of animal populations. *Proc. Zool. Soc. London.* 1935:551–598.

Onslow, H. 1919. The inheritance of wing color in Lepidoptera. I. *Abraxas grossulariata* var. *lutea* (Cockerell). *J. Genetics* 8:209–258.

Ortmann, A. E. 1906. Facts and theories in evolution. *Science* 23:947–952.

Owen, D. F., and R. G. Wiegert. 1962. Balanced polymorphism in the meadow spittlebug *Philaenus spumarius. Amer. Natur.* 96:353–359.

Paine, R. T. 1966. Food web complexity and species diversity. *Amer. Natur.* 100:65–75.

Paine, R. T., and R. Z. Vadas. 1969. The effects of grazing by sea urchins, *Strongylocentrotus* spp. on benthic algal populations. *Limonol. Oceanogr.* 14:710–719.

Porritt, G. T. 1921. The Huddersfield varieties of *Abraxas grossulariata,* with description of a new variety. *Entomol. Month. Mag.* 58:128–135.

Porritt, G. T. 1926. The induction of melanism in the Lepidoptera and its subsequent inheritance. *Entomol. Month. Mag.* 62:107–111.

Porter, J. W. 1972. Predation by *Acanthaster* and the effect on coral species diversity. *Amer. Natur.* 106:487–492.

Poulton, E. B. 1890. *The Colours of Animals, Their Meaning and Use.* Kegan, Paul, Trench, Trübner and Co., London.

Ross, H. 1957. Principles of natural coexistence indicated by leaf-hopper populations. *Evolution* 11:113–129.

Rothschild, M. 1963a. An aposematic moth, the small magpie (*Eurrhypara horticulata* (L.)) (Lep. Pyraustidae): A new example of industrial melanism? *Entomol. Month. Mag.* 98:203–204.

Rothschild, M. 1963b. Is the buff ermine (*Spilosoma lutea* (Hut.)) a mimic of the white ermine (*Spilosoma lubricipeda* (L.))? *Proc. Roy. Entomol. Soc. London,* ser. A., 38:159–164.

Rothstein, S. I. 1973. The niche-variation model—is it valid? *Amer. Natur.* 107:598–621.

Ryan, M. F., T. Parker, and D. B. Mertz. 1970. Flour beetles: responses to extracts of their own pupae. *Science* 170:178–179.

Shugart, H. H., and B. C. Blaylock. 1973. The niche-variation hypothesis: an experimental study with *Drosophila* populations. *Amer. Natur.* 107:575.

Sikes, S. 1971. *The Natural History of the African Elephant.* Weidenfeld and Nicolson, London.

Smith, N. G. 1968. The advantage of being parasitized. *Nature* 219:690–694.

Soulé, M. 1970. A comment on the letter by Van Valen and Grant. *Amer. Natur.* 104:590–591.

Soulé, M., and B. R. Stewart. 1970. The "niche-variation" hypothesis: a test and alternatives. *Amer. Natur.* 103:531–542.

Steere, J. B. 1894. On the distribution of genera and species of non-migratory landbirds in the Philippines. *Ibis.* 1894:411–420.

Stovin, G. H. T. 1939. *Abraxas grossulariata* ab. *inframaculata.* ab. nov. *Entomologist* 72:155–156.

Stovin, G. H. T. 1940. Some breeding experiments with *Abraxas grossulariata* (Lep. Geometridae). *Entomologist* 73:265–267.

Suffern, J. S. 1973. *Experimental Analysis of Predation in a Freshwater System.* Yale University, Ph.D. thesis.

Thom, R. 1970. Topological models in biology. *Towards a theoretical biology* 3. *Drafts,* pp. 86–116. Edinburgh University Press, Edinburgh.

Thom, R. 1972. *Stabilité Structuelle et Morphogénèse.* W. A. Benjamin Inc. Advanced Book Program, Reading, Mass.

Thompson, D' A. W. 1917. *On Growth and Form,* 1st. ed. Cambridge University Press, Cambridge, England.

Thompson, V. 1973. Spittlebug polymorphic for warning coloration. *Nature* 242:126–128.

Thompson, V., and O. Halkka. 1973. Color polymorphism in some North American

Philaenus spumarius (Homoptera: Aphro-phoridae) populations. *Amer. Midl. Natur.* 89:348–359.

Thomson, W. 1862. On the secular cooling of the earth. *Trans. Roy. Soc. Edin.* 23:157–169.

Thomson, W. 1866. The doctrine of uniformity in geology briefly refuted. *Proc. Roy. Soc. Edin.* 5:512–513.

Udvardy, M. F. D. 1959. Notes on the ecological concepts of habitat, biotope and niche. *Ecology* 40:725–728.

Van Valen, L. 1965. Morphological variation and width of ecological niche. *Amer. Natur.* 99:377–390.

Van Valen, L., and Grant P. R. 1970. Variation and niche width reexamined. *Amer. Natur.* 104:589–590.

Volterra, V. 1926. Variazioni e fluttuazioni del numero d'individui in specie animali conviventi. *Mem. Roy. Acad. Naz. dei Lincei* (ser. 6) 2:31–113.

Wallace, A. R. 1858. On the tendency of varieties to depart indefinitely from the original type. *In: On the tendency of species to form varieties; and on the perpetuation of varieties and species by natural means of selection,* by Charles Darwin . . . and Alfred Wallace . . . *J. Proc. Linn. Soc., London.* 3:53–62.

Wallace, A. R. 1895. The method of organic evolution. *Fortnightly Review* (N.S.) 57:211–224, 435–445.

Wallace, B. 1968. Polymorphism, population size, and genetic load. *In* R. C. Lewontin, ed., *Population Biology and Evolution,* pp. 87–108. Syracuse University Press, Syracuse.

Wangersky, P. J., and W. J. Cunningham. 1957. Time lag in population models. *Cold Spring Harbor Symp. Quant. Biol.* 22:329–338.

Willson, M. F. 1969. Avian niche size and morphological variation. *Amer. Natur.* 102:531–555.

Whittaker, J. B. 1968. Polymorphism of *Philaenus spumarius* (L.) (Homoptera, Cercopidae) in England. *J. Animal Ecol.* 37:97–111.

Zaret, T. M. and R. T. Paine. 1973. Species introduction in a tropical lake. *Science* 182:449–455.

18 Applied Biogeography

*Edward O. Wilson
and Edwin O. Willis*

The Diversity Ethic

In a world of shrinking faith and uncertain trumpets, very few precepts are any longer accepted as absolute. We can nevertheless hope that one of them will be the ethic of organic diversity—that for an indefinite period of time man must add as little as possible to the rate of worldwide species extinction and where possible he should lower it. This precept, which is based wholly on rational considerations, can also be the guiding principle of applied biogeography. It emerges from a recognition that man is the self-appointed but still profoundly ignorant steward of the world's natural resources, that the living part of the environment is still mostly unknown to him, and that he has therefore scarcely begun to conceive of the possible benefits that other organisms will bring in economic welfare, health, and esthetic pleasure. To sense the depth of that ignorance, consider that biologists do not even know to the nearest order of magnitude how many species exist. Ten years ago the popularly accepted figure for animals was C. B. Williams' estimate of three million, based on extrapolations of species abundance curves. Now some authors use the figure ten million, an order-of-magnitude conjecture advanced in the manner of physics. The reason for the upward revision is twofold: the dis-

covery that whole faunas, such as the marine annelids, abyssal benthos, and many insect taxa, are still in the earliest stages of Linnaean exploration; and the growing realization that large complexes of poorly defined sibling species are common even in the better-known animal and plant groups.

All this lack of information must be balanced by an equal amount of caution. Our best strategy is a holding operation, by which diversity is preserved through any reasonable means until systematics, ecology, and evolutionary theory work their way up from the stone age toward some degree of mastery of the essential subject matter. As an example of the worst thing that biologists might let slip by them, consider the possibility that the Atlantic and Pacific biotas could be mingled by migration through the new Panamanian sea-level canal proposed for construction in the 1980s. Three to five million years ago the emergence of the Panama Isthmus cut the straits that connected the Pacific Ocean and the Caribbean Sea, isolating the marine populations on either side. The existing ecological differences between the inshore habitats are substantial. The Atlantic coast has moderate tides, sandy beaches, mangrove swamps, and rich coral reefs. The Pacific side is characterized by strong tides, more silty water, periodic upwellings of cold nutrient-rich water, rocky shores created by extensive lava flows, and limited, depauperate coral reefs. Accelerated no doubt by such differences in the physical environment, evolution has proceeded mostly to the species level and beyond. Of the roughly 20,000 species of marine animals and plants that occur on both sides of the Panama Isthmus, perhaps no more than ten percent are held in common (Newman, 1972). In the extreme case of the fishes and mollusks, fewer than one percent are held in common. What would happen if free exchange of these faunas were permitted through a sea-level canal? On this point biologists have fallen into total disagreement. The following diversity of opinions has been expressed in various articles, seminars, and government hearings during the past ten years:

1. There would be only limited exchange of species, mostly from the Pacific to the Atlantic. The ecosystems would not be seriously disturbed (Topp, 1969; Voss, 1972).

2. The Atlantic marine biota is richer in species and hence possesses superior competitive ability. If allowed to invade through a sea-level canal, it would cause widespread extinction in the Pacific biota. The combined extinction rates of the Pacific and Atlantic elements might reach 5000 species (Briggs, 1969).

3. The Briggs argument (just cited) is based on the postulate that the greater the number of species, the greater their individual competitive ability. An alternative hypothesis that cannot be excluded with existing knowledge is that the greater fluctuation of the Pacific inshore environment induces the evolution of a higher proportion of opportunistic species, capable of wedging their way into existing biotas, especially within areas disturbed to some extent by man's activities. If this model is correct, and Briggs' conjecture wrong, the biotic flow would be predominantly from the Pacific to the Atlantic. In either case, the total impact on the two ecosystems cannot be predicted.

4. An exchange of biotas would be generally unpredictable and dangerous. Species could be removed not only by competitive replacement but also by overwhelming degrees of hybridization with imperfectly isolated geminate forms on the other side of the Isthmus (Rubinoff, 1965).

In fact, biogeography has neither the theory nor the previous experience to predict the outcome of an unimpeded exchange of faunas across the sea-level canal. This incapacity has become increasingly clear to concerned scientists who have tried to evaluate the evidence dispassionately, including Aron and Smith (1971). Therefore, a strongly cautious approach seems mandatory. It is necessitated not just by the very real possibility of widespread species extinction. The introduction of only one wrong species, such as the yellow-bellied sea snake from the eastern Pacific into the Atlantic Ocean (see Graham, Rubinoff, and Hecht, 1971), could inflict enough direct economic or ecological damage to justify the attempt to prevent any migration at all. Furthermore, changes in just a few species in a tightly integrated community could have widespread indirect effects by destabilizing the community (Levins, Chapter 1). Previous experience with the careless mixing of aquatic biotas, for example in the Great Lakes via the Erie and Welland canals, indicates that to permit the mixing of the rich tropical Pacific and Atlantic biotas would be playing ecological roulette with all cylinders loaded. Moreover, a unique biogeographic experiment of global proportions would thereby have been performed, without adequate preparation and in the wrong century. The natural setting for the experiment took millions of years to develop and cannot be repeated. Biology should be fully prepared before allowing it to proceed even piecemeal. For these reasons two groups of biologists, the University of Miami team supported by the Battelle Memorial Institute and the Committee of Ecological Research for the Interoceanic Canal (CERIC) of the National Academy of Sciences, have independently recommended that some kind of biological barrier be constructed across the canal before it is opened (Voss, 1972; Newman, 1972). The barrier can take any one or a combination of several forms: bubble curtains, ultrasonic screens, intrusions of heated or fresh water, and others. The details will be a straightforward exercise in engineering, infinitely simpler than the one biologists and the rest of humanity would face if the mixing is allowed to proceed.

The Design of Nature Preserves

Biogeographers cannot predict the outcome of mixing the Pacific and Atlantic biotas, except to say that it is dangerous, for the reason that this is one of the most complex problems they can ever conceivably face. Similarly, molecular biologists do not understand how metazoan tissues develop, and behavioral biologists cannot explain conscious thought, because these problems are also the Mount Everests of their respective disciplines. Like the rest of biology, however, biogeography is far from helpless when dealing with smaller, better-circumscribed units. The quantitative theory of island biogeography in which Robert MacArthur was so involved can be brought to bear on several kinds of problems of diversity maintenance. Preston (1962), Willis (1971), Wallace (1972), Diamond (1975), and Terborgh (1974a) have pointed out that the most straightforward application is in the design of natural preserves. Natural habitats have always been fragmented into island-like enclaves. With certain exceptions, such as the forests of New England, man has intensified this process, reducing the fragments in size and increasing their degree of isolation. The number of species belonging to a single taxon such as birds, ants, or flowering plants, equilibrates on a given island at a level that is a function of the area and the degree of isolation of the island (MacArthur and Wilson, 1967). Similar effects are seen on "habitat islands" within continents (Cody, Chapter 10). When the distance to the principal source area is held constant, whether that area is a continent, a set of islands, or just a similar habitat nearby, the number of species S increases approximately as a simple power function of the area, as follows: $\log S = a + z \log A$, where A is the area and *a* and *z* are fitted constants. When the independent parameter of isolation is increased, *z* rises at a rate characteristic of each taxon and the part of the world in which the relation is observed. In most cases *z* falls somewhere between 0.2 and 0.4 (cf. May, Chapter 4, Table 5; Diamond, Chapter 14, Figures 2 and 3). A very rough rule of thumb is that a tenfold increase in area results in a doubling of the number of species at equilibrium.

When a nature preserve is set aside, it is destined to become an island in a sea of habitats modified by man. The species number will shift from its original equilibrium because of the area and distance effects just cited. As years pass the diversity will decline, eventually reaching a new, lower steady state. An estimate of the loss can be made by comparing the reserve with the area-species curve of older systems, providing that appropriate systems exist under comparable conditions of isolation. Diamond (1972, 1973, and Chapter 14) has developed an elegant technique to estimate the relaxation rate and secondary equilibrium values in the case of island birds. He made use of land-bridge islands that were disconnected from New Guinea at known times in the recent geologic past. His results have been confirmed and extended by Terborgh (1974b) in parallel studies in the West Indies and Central America. Both Diamond and Terborgh discovered that significant drops in the number of species in newly disconnected islands take place over a period of decades in the smallest islands, which are comparable in area to small natural reserves on continents, and during centuries in islands comparable in size to our largest national parks.

Barro Colorado Island in Panama provides both a test of the theory and an alarming example of the high potential decrement rate on small islands. Barro Colorado actually consists of a hilltop of 15.7 km^2 of lowland tropical forest, which was isolated from surrounding forests about 1914 when Gatún Lake rose around it as part of the formation of the central part of the Panama Canal. Since 1923 the island has been a protected biological reserve, and its forests have been growing to maturity. Inserting the area of the island and its known period of isolation into an extinction model based on the West Indian studies, Terborgh (1974a) estimated that the number of resident bird species should have declined by nearly 10 percent. This is in close agreement with the decline actually observed. Let us examine the history of extinctions in some detail.

The birds of Barro Colorado, fortunately, have been well studied over the years. Chapman (1938) and others worked there during the 1920s and 1930s, Eisenmann (1952) and others visited from 1947 to date, and Willis studied there two to eleven months per year from 1960 to 1970. Of 208 species of birds breeding on the island in Chapman's time, 45 had disappeared by 1970. Several other species were down to one or a few individuals. A grebe and a gallinule have colonized the lake, and three species of the forest edge (two tanagers and a wren) are currently attempting colonization. Other species of the forest edge have attempted colonization but failed. No forest species has reached the island, although points on its edge are only about half a kilometer from sites on the mainland where the species occur.

For two reasons the record of extirpations must be interpreted with care. Early workers probably missed several species, including several tiny flycatchers and an elusive forest dove now present. Several sight records by early workers have to be doubted as possible vagrants or misidentifications. If some of these forest species, notably difficult to detect, were actually breeding earlier, there may have been a higher original avifauna and more extirpations than we think. On the other hand, if other species such as the tiny flycatchers have been successful colonists, the original avifauna may have been lower than we think. We regard a higher figure for the original avifauna to be the more likely, but neither possibility materially affects conclusions of this chapter.

The second reason for careful interpretation presents more of a problem. Barro Colorado was not just a tract of mature forest that became an island of mature forest. Rather, it was a mixture of mature forest and patches of second-growth forest, the latter now growing to maturity. Some 32 of the lost species of birds, or perhaps a few less, are birds of second-growth or forest edge (Willis, 1974). These species lost their habitats as the forest grew. They are "weed" species, abundant on the mainland and easily capable of colonizing patches of secondary forest there. Such opportunistic forms are in little danger of extinction even where cutting of the forests has reduced forests

to fragments. Thirteen lost species (Table 1), or perhaps a few more, regularly occur in extensive tracts of tall forest elsewhere, sometimes at lower density than in less mature forests. They can be expected to disappear as the forests are cut.

The particular, idiosyncratic causes of extinction are nearly impossible to pinpoint and are probably varied (Willis, 1974). Several of the lost species nest or feed on the ground. Perhaps leaf litter is reduced as monolayer trees take over the forest (see Horn, 1971), and very likely the sparser ground cover in mature forests provides relatively poor protection from predators. However, ground-living birds are also the ones that are least able to emigrate to the island. High densities of certain mammals, in part due to losses of large predatory mammals, could lead to destructive levels of predation on the nests and of some birds. Finally, because the range of habitats is limited, refugia do not exist during exceptional wet or dry years (cf. discussion of "hot spots" by Diamond, Chapter 14).

Two lost species belong to a group studied most intensively, the birds that follow army ants to feed on the arthropods flushed by these insect predators. Of the original seven species in this guild, the largest (*Neomorphus geoffroyi,* a ground cuckoo) was gone before Willis arrived in 1960. The second largest (*Dendrocolaptes certhia,* a woodcreeper) was down to two pairs in 1960 and disappeared by 1970. The third largest (*Phaenostictus mcleannani,* an antbird) decreased from 15 pairs to one female and a few males by 1970. The fifth species (*Gymnopithys bi-*

Table 1.   Forest birds extirpated from Barro Colorado Island

Species	Large for guild	Ground nester	Ground forager	Low density in tall forests	Immigration
Harpy eagle (*Harpia harpyja*)	+ +			*a*	*f*
Barred forest-falcon (*Micrastur ruficollis*)*	+			*b*	
Red-throated caracara (*Daptrius americanus*)*	+			*c*	*f*
Great curassow (*Crax rubra*)	+ +		+	*a*	*e*
Marbled wood-quail (*Odontophorus gujanensis*)	+	+	+	*b*	*e*
Rufous-vented ground-cuckoo (*Neomorphus geoffroyi*)	+ +		+	*a*	*e*
Barred woodcreeper (*Dendrocolaptes certhia*)*	+		+	*b*	*e*
Buff-throated Automolus (*Automolus ochrolaemus*)	+	+		*b*	*e*
Black-faced antthrush (*Formicarius analis*)	+		+	*b*	*e*
Sulphur-rumped flycatcher (*Myiobius sulphureipygeus*)				*d*	*e*
White-breasted wood-wren (*Henicorhina leucosticta*)		+	+	*b*	*e*
Nightingale wren (*Microcerculus marginatus*)			+	*b*	*e*
Song wren (*Leucolepis phaeocephalus*)*			+	*b*	*e*

Species marked with an asterisk disappeared during the 1964–1973 decade. Probably other species disappeared during 1971–1973. *a*, low density because large for the ecological guild. *b*, higher densities reached in less mature or dry forests. *c*, wasp-eating, wanders widely. *d*, nests over streams, which are uncommon on Barro Colorado. Several other flycatchers that nest over or near streams in second-growth have also been extirpated. *e*, immigration to Barro Colorado from the mainland unlikely. *f*, birds not now present on mainland, but could immigrate.

color, an antbird) decreased from some 50 to 20 pairs. Only the fourth, sixth, and seventh species, medium to small birds that often forage apart from the ants as well as near them, maintained substantial populations. Even over a ten-year period, reasons for declines were not evident; it can only be said that high predation on nests kept replacements below fairly high losses of adults despite frequent renestings.

Several other losses fit a pattern of early loss of large or specialized species, a pattern to be discussed later as "ecological truncation." The harpy eagle was the largest local raptor of Barro Colorado, the great curassow the largest local frugivore, the barred forest-falcon the largest insect-eater, the caracara the only wasp-eater, and the black-faced antthrush the largest litter insectivore. Some small and generalized birds are doing better than in similar mainland forests nearby, an example of the usual island pattern of "density compensation" (MacArthur, Diamond, and Karr, 1972) by which a large number of individuals of a few small species replace missing species (cf. Cody, Chapter 10; Brown, Chapter 13). The most abundant species on the island has been studied closely (Oniki, in preparation) and has high nest and adult losses but a wide range of foraging behaviors. Maturation of the forest has led to success for several large fruit-eaters and arthropod-eaters; average bird weights are higher in the half of the island covered by tall forest than in the remaining medium forest. Since losses of large species and density compensation go against this trend, small size of the island rather than forest growth is likely to be the main reason for losses. However, losses of small wrens may have followed from reduction of the isolated populations due to maturation of the forest.

Loss rates on Barro Colorado have remained steady or increased slightly, at about ten species per decade. Of these, about three have been forest inhabitants. The latter estimate is close to the one that Diamond (1971) made for losses on Karkar Island off New Guinea and Santa Cruz Island off California. Coiba, an island connected to Panama by a Pleistocene land bridge about 10,000 years ago, is much larger than Barro Colorado but has retained many fewer of the mainland species near it than has Barro Colorado (Wetmore, 1957). One can therefore predict that many more species will ultimately be lost from Barro Colorado.

All of this new information from island biogeography shows that planners and managers of national parks and other natural preserves will be prudent to take the spontaneous extinction rate into account and to choose appropriate measures to minimize it. The following basic procedures should be included:

1. Individual preserves must be made as large as possible. Since the areas of preserves will always be fixed by political compromise, estimates should be made of the extinction rates, as a function of time and area, of the most vulnerable taxa such as the birds and mammals. Then the minimal areas demanded should be the ones at which the initial and consequently highest extinction rates will be reasonably low. The projected rates should be such that only large increments of reserved land will lower them significantly further. In other words, land acquisition must reach the point of diminishing returns with respect to the most extinction-prone groups.

2. Unique habitats and biotas are best contained in multiple preserves, and these isolates should be located as closely together as possible. The reason is that extinction has a strong random component. Species seldom become extinct in every part of their range simultaneously. They tend to persist because ecologically suitable localities that lose them can be recolonized from other localities that are still occupied. Reciprocal intercolonization of preserves can proceed indefinitely through time and, if aided by deliberate transplantations, might extend the life of species well beyond what it would be under natural circumstances.

3. Because of the peninsula effect discovered by biogeographers, preserves of a fixed area should be as round in shape and as continuous as possible. (This principle and those embodied in the first two recommendations are illustrated in Figure 1.)

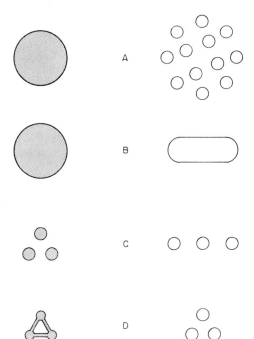

Figure 1 The geometrical rules of design of natural preserves, based on current biogeographic theory. The design on the left results in each case in a lower spontaneous extinction rate than the complementary one on the right. Both the left and the right figures have the same total area and represent preserves in a homogeneous environment. A: a continuous preserve is better than a fragmented one, because of the distance and area effects. B: a round design is best, because of the peninsula effect (cf. MacArthur and Wilson, 1967, pp. 115–116 and Figure 37). C: clumped fragments are better than those arranged linearly, because of the distance effect. D: if the preserve must be divided, extinction will be lower when the fragments can be connected by corridors of natural habitat, no matter how thin the corridors (Willis, 1974). Another principle, not incorporated in this figure, is that whatever the design of a given preserve, its extinction rate can be greatly lowered if similar preserves are located nearby.

4. Extinction models of the kind invented by Diamond and Terborgh should not be restricted to the most conspicuous or vulnerable organisms but should eventually be developed for all taxa. Those displaying the highest degrees of endemicity and vulnerability (the two phenomena are generally correlated) deserve first attention. No group, not even the humblest and most obscure among invertebrates and microorganisms, should be ignored. Thus, in addition to the full taxonomic surveys that are just getting underway in natural preserves, biologists should begin studies of species dynamics and species equilibria.

The Special Problem of Ecological Truncation

Truncation of ecological guilds, the well-known but seldom-emphasized early loss of specialists and of large species, is probably mainly due to the fact that most such species occur at very low densities, require large areas for sustenance, or both (see discussion of so-called incidence functions, the dependence of an island's species composition on its species number or area, in Chapter 14 by Diamond). "Density compensation" for truncation may lead to increased numbers of individuals of generalized small species (MacArthur, Diamond, and Karr, 1972). However, species diversities and biomasses will still be lower.

The ant-following birds of Barro Colorado show ecological truncation clearly, as shown earlier. Diamond (Chapter 14) found a large eagle (*Harpyopsis novaguineae*) and a distinctive kingfisher (*Clytoceyx rex*) gone from all the formerly connected islands off New Guinea. An even better example of truncation is presented by the disastrous extinctions of Hawaiian birds during the past century. These losses were evidently due at least in part to human cultivation of the lowlands and to the introduction of disease-vector mosquitoes (Warner, 1968). The first species to go were such large or specialized species as the big Kioea (*Chaetoptila angustipluma*) and both long-beaked Mamos (*Drepanis* spp.). Many of the species now approaching the end are characterized by intermediate size, distinctive adaptations, or both—the half-beaked *Hemignathus*, for example, the parrotbill *Pseudonestor*, and others. Few members of the endemic Hawaiian family Drepanididae other than small warbler-like species such as the Amakihi (*Loxops virens*) are doing well.

Truncation creates a particular difficulty for the planning of natural reserves, because multiple refuges of a given habitat tend to lose the same specialized species. Diamond (1972) found that the largest (3000 square miles) of the formerly connected islands off New Guinea has retained only 45 out of the 134 land-bridge bird species. Even if parks in the New Guinea of the future should be huge like these offshore islands, many species would still be lost.

The island effect and truncation, taken to their extremes, result in the domination of small parks by rats, cockroaches, sparrows, and similar invaders from nearby human areas. Practically no birds other than pigeons and starlings winter in wooded Tappan Square in Oberlin, Ohio (Margaret F. Smith, personal communication). A thousand such Squares would be far less valuable for the maintenance of biotas than one refuge of a thousand Square units.

Planned Biotic Enrichment

It is within the power of science not merely to hold down the rate of species extinction but to reverse it. Among the principal topics of community ecology now under intensive study is the species-packing problem (MacArthur and Wilson, 1967; MacArthur and Levins, 1967; MacArthur, 1972; Schoener, 1970; May, 1973). One of the more sophisticated recent developments is the specification of "assembly rules" by Diamond (Chapter 14). A central goal of this research is the identification of those traits that allow certain sets of species, but not others, to be fitted together in the same ecosystem without markedly increasing the species extinction rate. During colonization by undisturbed biotas, such congenial sets are gradually assembled by chance alone, raising the steady-state species number to what has been called the assortative equilibrium (Wilson, 1969). Theoretically, assortative equilibria can be planned that exceed any occurring in nature. Species might even be drawn from different parts of the world— not willy-nilly, as in the careless importations of the past, but after careful ecological analysis has identified them as candidates for insertion into new faunas. Some of the first and most important introductions would surely be "orphan species," those on the brink of extinction in their native range but capable of being fitted into certain alien communities elsewhere. We do not suggest that the state of the art is advanced enough for us to proceed with planned biotic mixing, only that species packing is one of the techniques of applied biogeography that seems likely to become practicable within the next several decades, on the basis of current and projected research.

Optimism is further justified by the favorable outcome of a few biotic mixtures that have already occurred haphazardly, indicating a degree of flexibility on the part of species that will provide biogeographers with some margin for error. The Kaingaroa Forest of New Zealand, for example, contains 250,000 acres of exotic conifers, including *Pinus radiata, P. ponderosa, P. contorta,* and *Pseudotsuga taxifolia* from North America, and *Pinus nigra* from southern Europe. Introduced birds mingle with endemic New Zealand species in this synthetic environment. Ecological differentiation is well marked; no two species have the same feeding habit, and the insectivorous birds exploit all of the major feeding niches except that of woodpeckers. The really surprising fact, however, is that some of the native species are now as abundant in the Kaingaroa Forest as in almost any native forest, and some are more abundant than in most of the remainder of their range (Gibb, 1961; personal communication, 1973). Furthermore, the invertebrate fauna of the forest consists mostly of native species (Rawlings, 1961). Two circumstances are special

in the case of the birds. First, the number of species is still small, largely because the New Zealand fauna was depauperate to start with, and the mixed community has probably not yet encountered many of the difficulties in packing that would be routine in large continental faunas. As discussed by Cody (Chapter 10), species of depauperate faunas are better able to colonize exotic habitats than are species of rich faunas. Second, forest birds are differentiated to a large degree by foliage height and profile rather than by the species of trees in which they live. Monophagous and oligophagous insects, particularly those specializing on hardwoods, would in most instances find it impossible to penetrate the Kaingaroa conifers. Yet the lesson is clear: what works in part by accident can be brought closer to perfection through design.

Ultimately, design might also include the artificial selection of strains, or even the creation of new species, for the purposes of biotic enrichment. If theory and experiment indicate that an orphaned species cannot be fitted into any existing communities, strains might be selected within captive populations of the species that could eventually be inserted into one or more communities. We do not seriously suggest that such a procedure will be followed in the foreseeable future for any but a very few of the organisms most valued by man. Furthermore, the genetic molding of communities is a technology that cannot be seriously contemplated until the inchoate discipline of population ecology has moved closer to a full solution of the species-packing problem.

The Creation of
New Communities

Many of the earth's major habitats are biological deserts: the open sea, the ice caps, some of the trace-element barrens, and the real deserts, the extremes of which are virtually abiotic. Quite by coincidence, technology is at this moment striving toward two major goals that could transform these areas: an unlimited or at least vastly greater source of energy, and, as one of the principal benefits of the first, the cheap desalinization of sea water. With the achievement of these goals, men will move increasingly onto the land deserts, carrying communities of organisms with them. We will not be satisfied, it is hoped, with limiting ourselves to a baggage of domestic animals, houseplants, pests, and commensals. It lies easily within our power to create wholly new parks and reserves where nothing existed before in historical times. But what will go into these de novo communities? Thought about this subject sharpens one's vision of the future of applied biogeography.

In fact, the deliberate creation of new biological communities has already begun. Large areas of desert-like barrens in Australia have been transformed into agricultural land by the simple addition of zinc, copper, or molybdenum, "trace" elements required for life that were previously present in abnormally low quantities (Anderson and Underwood, 1959). Marine biologists have discovered that artificial reefs, with rich complements of reef organisms, can be created just by dumping concrete rubble, abandoned automobiles, used automobile tires, and similar inert refuse onto the mud or sand floors of shallow marine waters. Successful experiments of this nature have been conducted off the shores of Florida and California (see Turner, Ebert, and Given, 1969). What these efforts engender are in effect habitat islands, the biotas of which grow and equilibrate according to the same laws of biogeography governing wholly natural islands. The communities are not likely to be as diverse as those that have evolved for millions of years in the natural islands, yet the process of enrichment can be speeded by the deliberate importation of compatible species to reach new and higher assortative equilibria. This is another aspect of biogeographic technology that ongoing basic research might render practicable during the next few decades.

Ecosystem Manipulation:
The Ultimate Game

The greatest misfortune that awaits the human intellect is to be no longer faced with something commensurate with its capacity for wonder. If the golden age of science really ends, and research shrinks to a few remote and arcane frontiers accessible only to specialists, the wonder will indeed be gone. By that time even the prescientific myths that sustained our ancestors, and intrigue us still, would have largely evaporated—having been accounted for in full, perhaps by the right kind of neurophysiological analysis of the limbic system and hypothalamus. But this possibility will not materialize during the lifetime of anyone now living. The ultimate complexity, offering an unexplored terrain of virtually infinite extent, lies in biology. Even after the cell has been torn down and put together again, and the labyrinthine mysteries of metazoan development followed to their ends, there lie ahead much more extensive challenges of ecology and biogeography. The full exploration of organic diversity is a prospect that suits the biocentric human brain, especially those emotive centers that evolved to make us superior hunters and agriculturists. The same instincts that motivate the birdwatcher, the butterfly collector, and the backyard gardener can indefinitely sustain the scientifically curious segment of a more sophisticated human population in the pursuits of ecology and biogeography. The very size of the world's biota, comprising millions of species, is itself a challenge that only generations more of study will encompass. The possibilities for ecosystems manipulation, outlined in this essay on applied biogeography, offer creative work that is orders of magnitude even more extensive.

References

Anderson, A. J., and E. J. Underwood. 1959. Trace-element deserts. *Sci. American,* January 1959, pp. 97–106.

Aron, W. I., and S. H. Smith. 1971. Ship canals and aquatic ecosystems. *Science* 174:13–20.

Briggs, J. C. 1969. The sea-level Panama Canal: potential biological catastrophe. *BioScience* 19:44–47.

Chapman, F. M. 1938. *Life in an Air Castle; Nature Studies in the Tropics.* D. Appleton-Century, New York.

Diamond, J. M. 1971. Comparison of faunal equilibrium turnover rates on a tropical island and a temperate island. *Proc. Nat. Acad. Sci. U.S.A.* 68:2742–2745.

Diamond, J. M. 1972. Biogeographic kinetics: estimation of relaxation times for avifaunas of southwest Pacific islands. *Proc. Nat. Acad. Sci. U.S.A.* 69:3199–3203.

Diamond, J. M. 1973. Distributional ecology of New Guinea birds. *Science* 179:759–769.

Diamond, J. M. 1975. The island dilemma: lessons of modern biogeographic studies for the design of natural reserves. *Biological Conservation,* in press.

Eisenmann, E. 1952. Annotated list of birds of Barro Colorado Island, Panama Canal Zone. *Smithsonian Miscellaneous Collections* 117(5):1–62.

Gibb, J. A. 1961. Ecology of the birds of Kaingaroa Forest. *Proc. New Zealand Ecol. Soc.* 8:29–38.

Graham, J. B., I. Rubinoff, and M. K. Hecht. 1971. Temperature physiology of the sea snake *Pelamis platurus:* an index of its colonization potential in the Atlantic Ocean. *Proc. Nat. Acad. Sci. U.S.A.* 68:1360–1363.

Horn, H. S. 1971. *The Adaptive Geometry of Trees.* Princeton University Press, Princeton.

MacArthur, R. H. 1972. *Geographical Ecology. Patterns in the Distribution of Species.* Harper and Row, New York.

MacArthur, R. H., J. M. Diamond, and J. R. Karr. 1972. Density compensation in island faunas. *Ecology* 53:330–342.

MacArthur, R. H., and R. Levins. 1967. The limiting similarity, convergence, and divergence of coexisting species. *Amer. Natur.* 101:377–385.

MacArthur, R. H., and E. O. Wilson. 1967. *The Theory of Island Biogeography.* Princeton University Press, Princeton.

May, R. M. 1973. *Stability and Complexity in Model Ecosystems.* Princeton University Press, Princeton.

Newman, W. A. 1972. The National Academy of Science Committee on the ecology of the interoceanic canal. *Bull. Biol. Soc. Washington* 2:247–259.

Preston, F. W. 1962. The canonical distribution of commonness and rarity: Part II. *Ecology* 43:410–432.

Rawlings, G. B. 1961. Entomological and other factors in the ecology of a *Pinus radiata* plantation. *Proc. New Zealand Ecol. Soc.* 8:47–51.

Rubinoff, I. 1965. Mixing oceans and species. *Natural History* 74:69–72.

Schoener, T. W. 1970. Nonsynchronous spatial overlap of lizards in patchy habitats. *Ecology* 51:408–418.

Terborgh, J. 1974a. Faunal equilibria and the design of wildlife preserves. *In* F. Golley and E. Medina, eds., *Tropical Ecological Systems: Trends in Terrestrial and Aquatic Research.* Springer Verlag, New York.

Terborgh, J. W. 1974b. Preservation of natural diversity: the problem of extinction prone species. *BioScience* 24:715–722.

Topp, R. W. 1969. Interoceanic sea-level canal: effects on the fish faunas. *Science* 165:1324–1327.

Turner, C. H., E. E. Ebert, and R. R. Given. 1969. Man-made reef ecology. *Fish Bull., Dept. Fish and Game, State of Calif.* No. 146, pp. 1–221.

Voss, G. L. 1972. Biological results of the University of Miami deep-sea expeditions. No. 93. Comments concerning the University of Miami's marine biological survey related to the Panamanian sea-level canal. *Bull. Biol. Soc. Washington* 2:49–58.

Wallace, B. 1972. *Essays in Social Biology. Volume 1. People, Their Needs, Environment, Ecology.* Prentice-Hall, Englewood Cliffs, N.J.

Warner, R. E. 1968. The role of introduced diseases in the extinction of the endemic Hawaiian avifauna. *Condor* 70:101–120.

Wetmore, A. 1957. The birds of Isla Coiba, Panamá. *Smithsonian Miscellaneous Collections* 134(9):1–105.

Willis, E. O. 1971. The loss of birds from a tropical forest reserve. Paper presented at Dauphin Island meeting, Wilson Ornithological Society.

Willis, E. O. 1974. Populations and local extinctions of birds on Barro Colorado Island, Panamá. *Ecol. Monogr.* 44:153–169.

Wilson, E. O. 1969. The species equilibrium. *Brookhaven Symp. Biol.* 22:38–47.

Index
Contributors

Index

Contributors

James H. Brown, Department of Biology, University of Utah, Salt Lake City. Present address: Department of Ecology and Evolutionary Biology, University of Arizona, Tucson

Martin L. Cody, Department of Biology, University of California, Los Angeles

Joseph H. Connell, Department of Biological Sciences, University of California, Santa Barbara

Jared M. Diamond, Physiology Department, Medical Center, University of California, Los Angeles

Madhav D. Gadgil, Centre for Theoretical Studies, Indian Institute of Science, Bangalore, India

Henry A. Hespenheide, Department of Biology, University of California, Los Angeles

Henry S. Horn, Biology Department, Princeton University, Princeton

G. Evelyn Hutchinson, Department of Biology, Yale University, New Haven

Frances C. James, University of Arkansas Museum, Fayetteville. Present address: Ecology Program, National Science Foundation, Washington

James R. Karr, Department of Biological Sciences, Purdue University, West Lafayette. Present address: Department of Ecology, Ethology, and Evolution, University of Illinois, Champaign

Egbert G. Leigh, Jr., Smithsonian Tropical Research Institute, Balboa, Canal Zone

Richard Levins, Harvard School of Public Health, Boston

John W. MacArthur, Marlboro College, Marlboro

Robert M. May, Biology Department, Princeton University, Princeton

Ruth Patrick, Academy of Natural Sciences of Philadelphia, Philadelphia

Eric R. Pianka, Department of Zoology, University of Texas, Austin

Michael L. Rosenzweig, Department of Biology, University of New Mexico, Albuquerque. Present address: Department of Ecology and Evolutionary Biology, University of Arizona, Tucson

William M. Schaffer, Department of Biology, University of Utah, Salt Lake City. Present address: Department of Ecology and Evolutionary Biology, University of Arizona, Tucson

Arthur M. Shapiro, Department of Zoology, University of California, Davis

Edwin O. Willis, Biology Department, Princeton University, Princeton

Edward O. Wilson, Museum of Comparative Zoology, Harvard University, Cambridge